MEDICAL RADIOLOGY

Diagnostic Imaging and Radiation Oncology

Springer
Berlin
Heidelberg
New York
Barcelona
Budapest
Hong Kong
London
Milan
Paris
Santa Clara
Singapore
Tokyo

Magnetic Resonance Angiography

With Contributions by

I.P. Arlart · G.M. Bongartz · H. Bosmans · C. Catalano · J.F. Debatin
R.R. Edelman · L. Guhl · M. Hauser · R. Hausmann · G.P. Krestin
A. Laghi · G. Laub · J.S. Lewin · W.J. Manning · G. Marchal · P. Pavone
B. Siewert · P. Van Hecke · R. Vosshenrich · P.A. Wielopolski · G. Wilms

Edited by

Ingolf P. Arlart, Georg M. Bongartz, and Guy Marchal

Foreword by

Albert L. Baert, Friedrich H.W. Heuck, and James E. Youker

Preface by Peter E. Peters

With 229 Figures in 491 Separate Illustrations

Springer

Professor Dr. INGOLF P. ARLART
Ärztlicher Direktor
des Radiologischen Instituts Stuttgart
Katharinenhospital
Kriegsbergstraße 60
70174 Stuttgart, FRG

Professor Dr. GUY MARCHAL
Dienst Radiologie
Katholieke Universiteit
Gasthuisberg
Herestraat 49
3000 Leuven, Belgium

Privat-Dozent Dr. GEORG M. BONGARTZ
Leitender Arzt für Röntgendiagnostik
der Universität Basel/Kantonsspital
Petersgraben 4
4031 Basel, Switzerland

MEDICAL RADIOLOGY · Diagnostic Imaging and Radiation Oncology

Continuation of
Handbuch der Medizinischen Radiologie
Encyclopedia of Medical Radiology

ISSN 0942-5373
ISBN 3-540-56896-4 Springer-Verlag Berlin Heidelberg New York

Library of Congress Cataloging-in-Publication Data. Magnetic resonance angiography/with contributions by I.P. Arlart . . . [et al.]: edited by Ingolf P. Arlart, Georg M. Bongartz, and Guy Marchal: foreword by Albert L. Baert, Friedrich H.W. Heuck, and James E. Youker: preface by P.E. Peters. p. cm. — (Medical radiology, ISSN 0942-5373) Includes bibliographical references and index. ISBN 3-540-56896-4 1. Blood-vessels—Magnetic resonance imaging. I. Arlart, I. II. Bongartz. Georg M., 1955- . III. Marchal, Guy, Prof. Dr. Med. IV. Series. [DNLM: 1. Magnetic Resonance Angiography. WG 500 M1959 1995] RC691.6.A53M334 1995 616.1′307548—dc20 DNLM/DLC for Library of Congress 95-31126

Cover design: Springer-Verlag, Design & Production

Typesetting: Best-set Typesetter Ltd., Hong Kong

SPIN: 10096249 21/3135/SPS - 5 4 3 2 1 0 - Printed on acid-free paper

Dedicated to

Gudrun, Oliver and Tobias	I.P. ARLART
Liese and Frank	G.M. BONGARTZ
Rita, Kathleen and Paul	G. MARCHAL

Foreword

On the basis of their wide knowledge in the field of radiological vascular diagnosis, including interventional radiology, the editors of this volume have achieved international as well as national recognition. Here they not only consider the technical features and clinical applications of magnetic resonance angiography in depth, but also document the results of comparative studies with other presently available imaging modalities.

Central themes in the scientific work of Ingolf Arlart are the problems of renovascular hypertension and vascular aneurysms, and the use of interventional radiology within the context of vascular disease.

Besides conducting intensive research and developmental work in various diagnostic areas of magnetic resonance imaging, Georg Bongartz has devoted particular attention to magnetic resonance angiography of the supra-aortic and brain-supplying arteries. The results of his clinical and radiological research found due recognition when he was presented with the Wilhelm Conrad Röntgen Prize in 1992 by the German Radiological Society, and he has also received various other awards. This background makes ideally suited to be among the editors of the current volume, which provides an up-to-date summary of progress in radiological vascular diagnosis.

Guy Marchal is well known for his innovative work concerning the use of contrast media in cross-sectional imaging modalities. He has published original contributions both on the use of iodinated contrast media in X-ray computed tomography and on the application of gadolinium, iron oxide, and other contrast media in magnetic resonance imaging of visceral organs and vessels. In 1993 he received the "Kernspintomographie Preis" (Magnetic Resonance Imaging Prize) in Garmisch Partenkirchen.

The tremendous rapidity in the development of medical radiology in all areas of diagnosis makes it essential that critical appraisals are compiled whenever concrete results of research have been achieved. The editors of, and contributors to, this volume have devoted themselves to this task with respect to magnetic resonance angiography, and have accomplished it most successfully. We hope that this important volume receives the audience it deserves in order that the results can be applied clinically as soon as possible, to the benefit of ill patients.

Leuven/Stuttgart/Milwaukee
June 1995

A.L. Baert
F.H.W. Heuck
J.E. Youker

Preface

The first visualization of vascular structures in humans using Röntgen's "new kind of rays" was performed in mid-January 1896 in Vienna by Maschek and Lindenthal. This took place about 2 weeks after Röntgen's preliminary report and prior to his first and only public presentation of X-rays before the Würzburg Physico-medical Society on January 23, 1896. Thus, we will soon commemorate the centenary of angiography – and it is very likely that the coming 100 years will see a dramatic change in the technique.

Maschek and Lindenthal needed an obscure type of contrast medium (Teichmann's mixture) and 57 minute exposure time with an unpredictably high radiation dose for their angiogram of an amputated hand. One hundred years later, magnetic resonance angiography (MRA) is ready for clinical application. The technique does not utilize ionizing radiation, nor selective catheterization of vessels to be imaged, and it can be carried out with or without specific contrast media. Unlike ultrasound techniques, MRA is not limited by acoustic windows and, thus, has a much wider range of clinical applications. MRA is, however, not a set technique but is constantly evolving. The significant improvement in image quality that has been achieved over the past few years is best demonstrated if one compares early MR angiograms with current state-of-the-art images.

The editors of this volume in the series *Medical Radiology* have been pioneers in MRA for many years. Their work, complemented by that of equally experienced contributing authors, guarantees a high scientific standard as well as sound clinical experience.

It is important to recognize that reading MR angiograms requires different skills than those employed in conventional studies. A profound knowledge of vascular anatomy is the only common prerequisite for both types of study. Sound interpretation of MR angiograms must be based on a thorough understanding of the underlying principles of MR physics. The same holds true for the selection of that MRA technique best suited to solving a given clinical problem.

The authors have succeeded in concisely compiling the current knowledge in the rapidly evolving field of MRA. Even perfectionists among angiographers must concede that the new noninvasive imaging has come fairly close to the type of quality to which one is accustomed – and we should never forget our patients: it is in their best interests that we are improving noninvasive angiography.

P.E. Peters, M.D.
Dept. of Radiology
Professor and Chairman
University of Münster

Introduction and Definition of Magnetic Resonance Angiography

I.P. Arlart and G. Bongartz

Diagnostic imaging of blood vessels plays an important role in the depiction of stenotic, occlusive, aneurysmal, and inflammatory vascular disease or vascular malformations. Malperfusion or ischemia caused by the vascular lesion may give rise to life-threatening conditions in the cerebrum and in other parenchymal structures. In addition, local complications such as vessel rupture or inflammation can develop at the site of the lesion. The general expectation of diagnostic vascular procedures is that they will permit accurate identification of the lesion and clarification of the status of adjacent vascular areas, with an image quality acceptable to the physician responsible for the patient's management.

Traditionally, conventional x-ray angiography is used to depict vascular lesions with high accuracy via intravascular administration of contrast material. A high-resolution technique ensures diagnostic angiograms even when the lesion is small and located very peripherally. Using this technique, exact anatomical information is obtained, as well as hemodynamic information by way of analysis of sequential images. Indirect methods such as intravenous administration of contrast agents to visualize arterial structures, as in intravenous digital subtraction angiography, have been widely abandoned due to poor image quality and the low signal-to-noise ratio. X-ray angiography equipment may be dedicated to intra-arterial angiographic studies as well as to interventional angiography that can be done under suitable circumstances immediately after the detection of a lesion. However, paravascular disorders that may influence intraluminal blood flow most frequently cannot be detected by conventional angiography, leading to interest in other imaging modalities which are able to demonstrate both the peri- and the intravascular space, such as B-scan ultrasonography (US) and contrast-enhanced computed tomography (CT). These techniques have become extremely important in routine clinical practice, and in many cases conventional angiographic examinations have to be complemented by these additional cross-sectional techniques. In particular, the recently developed helical CT technology enables reconstructed angiogram-like images to be obtained following intravenous administration of contrast media and has been shown to be particularly useful in diagnosing complex vascular disease of the aorta, the carotids, the intracranial vessels, the renal arteries, and the pelvic vessels. Compared to conventional catheter angiography, CT angiography represents a safer, faster, and more cost-effective diagnostic test for a wide variety of vascular problems. However, the quality of CT angiograms is susceptible to low cardiac output, and another drawback is the need for large volumes of contrast material. In contrast, US is completely noninvasive, and has the major advantage of being able to determine flow direction and to provide functional quantitative information on blood flow by means of the color Doppler technique. Areas in which US can be used successfully are the neck, the abdomen, and the extremities. However, in other regions of the body the value of US has been found to be limited by anatomical factors such as bone, calcifications, gas, deep location of vascular structures, and obesity of the patient.

With the invention of another completely noninvasive technique, nuclear magnetic resonance (NMR), and its implementation in medical imaging, new insights into morphology and pathology have been achieved. The ability of magnetic resonance imaging (MRI) to visualize reliably soft tissue alterations is based on the improved signal-to-noise ratio and the large variability in contrast enhancement in structures that are hardly visible with other imaging modalities. After the introduction of MRI into routine clinical practice, it was initially used to obtain cross-sectional images in all directions and in any region of the body. Neglecting its ability to provide further information on functional or pathological aspects, MRI was regarded as a potential substitute for CT that necessitated ionizing radiation. The advantages related to the entirely different signal and contrast mechanisms of MRI found only later acceptance, and although the feasibility of acquiring flow information by NMR had already been recognized during the 1950s, the technical prerequisites permitting sequences suitable for flow imaging were not met until during the past decade. In a way, imaging of vascular structures during the early phase was an unwanted side-effect including a variety of artifacts that were produced by vascular motion and pulsation. Through the development of modern MR technologies as well as further theoretical and practical user experience it became possible to compensate and suppress these artifacts and to visualize vascular structures. Intravascular blood motion was imaged with high contrast compared to the adjacent tissue either without intraluminal signal or with high signal intensity. Nevertheless, due to the relatively long acquisition time, all kinds of physiological motion during the measurement have the potential to create artifacts and distortion of the results. Two solutions to these problems have been validated: (a) improving the speed of MRI acquisition, thus avoiding the problems due to physiological motion, and (b) compensating for the motion-induced errors in signal registration by adjusting the measurement parameters to the motion. Both approaches are being followed; the latter offers the chance of visualizing vascular structures directly in a tomographic way on single slices. Additional application of reconstruction algorithms including maximum or minimum intensity projections can produce angiogram-like images of a certain area by suppressing the signal of stationary tissue. These images became known as "magnetic resonance angiograms" (MRA). However, the term refers to a broad variety of MR sequences by which intra-/extraluminal vessel contrast can be obtained. All MRA sequences are based on specific properties and characteristics of blood and blood flow. Minimum requirements for MRA are optimized contrast between vessel and background, a continuous delineation of defined vascular areas, and image information that is based on motion, i.e., flow.

Flow effects can be observed in any MR image. For instance, low signal intensity is commonly seen from flowing blood on spin-echo (SE) images. On the other hand, flowing blood usually appears bright on gradient-echo (GRE) images. These phenomena led to the adoption of the terms "black blood" and "bright blood" to describe techniques in MRA. In SE imaging, inflow of unsaturated protons generally results in high flow signal due to flow-related enhancement. However, this phenomenon is counteracted by the washout effect, which again reduces flow signal. The washout effect occurs when protons move out of the section without being affected by the 90° and 180° pulses. Hence, no signal can be produced. Flow void in SE sequences is more pronounced when thin slices or long echo times (TE) are used. Besides the washout effect, flow-related dephasing also may cause reduction of flow signal in SE sequences. In GRE imaging, flow-related enhancement is much more significant than in SE imaging because of the lack of washout effect due to the application of one radiofrequency pulse only. Flow-related dephasing that still causes a void in vascular signal can be minimized by the use of additionally applied magnetic field gradients (flow compensation or gradient motion rephasing). On the basis of GRE sequences two phenomena can be utilized for vascular imaging: the time-of-flight effect and phase effects.

Time-of-flight imaging is based upon differences in longitudinal magnetization between stationary and (in-)flowing spins. Blood entering the imaging section appears bright compared to stationary tissue that becomes saturated due to the repeatedly applied radiofrequency pulses. Flowing protons that enter the section are still unsaturated and produce a strong signal by flow-related enhancement.

In phase-contrast imaging, the MR signal is affected by alterations of the transverse magnetization. Normally, field gradients have no net effect on the phase of stationary protons because their phase effect is balanced at the time of the echo. However, when blood protons move along a gradient, they accumulate phase shifts which may vary with flow velocity and gradient amplitude. Signal intensity depends on accumulation or difference of the phase of the transverse magnetization.

Bearing in mind the phenomena briefly described above, MRA can be defined as a vascular MRI method based on physiological motion characteristics and influenced by the biochemical and biophysical properties of blood and blood flow. It represents a method for the functional imaging of blood vessels and has the ability to quantify blood flow, though precise anatomical depiction of vascular structures frequently cannot be guaranteed due to different flow patterns and velocities. To this extent, MRA resembles color duplex US rather than conventional angiography and potentially provides both morphological and physiological data from the arterial and the venous circulation in the assessment of a suspected vascular disease. On static SE images, blood flow appears dark, allowing depiction of both intra- and extravascular anatomy. Turbulent flow may generate a signal that cannot be differentiated from thrombus. Therefore, dynamic vascular studies using GRE techniques and phase velocity mapping techniques may help in assessing anatomical and flow abnormalities. MRA can yield information on flow direction, flow origin, and the presence or absence of collaterals. Veins and arteries can be imaged selectively due to their usually opposite flow direction by the use of presaturation techniques.

The major impetus for the development of MRA as a potential replacement for conventional angiography in the diagnostic imaging of vascular diseases is the morbidity and mortality associated with invasive catheter procedures. In particular when cerebral catheter angiography is performed, the occurrence of neurological complications cannot be excluded in spite of improved equipment, contrast agents, and techniques. Another motivation for developing MRA is cost reduction in public health care. Since MRA can be performed on an outpatient basis, the factor of hospitalization costs can be neglected. Furthermore, the combination of MRI and MRA within a single examination procedure will frequently avoid the need for multiple diagnostic imaging methods, such as catheter angiography, US, and CT. For instance, in a patient with head pain or transitory ischemic attacks, a routine brain study can easily be followed by an additional sequence for MRA.

Although MRA offers the advantage of providing a less invasive look at the vasculature, its clinical efficacy for many applications has not been proven to the satisfaction of many practitioners, and this is especially true for head and brain imaging. As a result, neurosurgeons often order an invasive catheter procedure after examining an MR angiogram. Informal comparisons of x-ray and MR-based angiography may eventually lead to a broader acceptance of MRA. Despite the lack of consensus on its real value, MRA is still widely used and its utility has been documented in a variety of instances. The major strengths of MRA until now lie in the depiction of the intracranial circulation and extracranial carotid vasculature, as well as in venography. MR sequences for these applications are well established, whereas a number of sequences developed for other parts of the body do not provide images of consistently high quality due to a variety of artifacts and a spatial resolution that is still significantly inferior to its conventional counterpart.

A prerequisite for the successful introduction of MRA into the routine diagnostic armamentarium is optimal MRA interpretation by well-educated and experienced medical personnel. Thus, two demands have to be fulfilled: (a) familiarity of the investigator with the principles of hemodynamics and MR physics, and (b) substantial experience of the investigator in the interpretation of catheter angiograms, including vascular anatomy and techniques. Moreover, experience in vascular US and vascular CT may be of value in order to evaluate MR angiograms more critically. It has to be recognized that MR angiograms are even more difficult to interpret than conventional catheter angiograms because anatomy and flow-related artifacts may alter the signal intensity and the resultant vascular image. Furthermore, some problems that limit the method remain to be solved:

1. Up to now no real quality control exists for MRA. Therefore, site-to-site and reader-to-reader variations occur.
2. Techniques have not yet been standardized and optimized.
3. It should be recognized that the technique of MRA is still evolving. This is true of the hardware, gradient field strength, data acquisition, software, application of contrast agents, and postprocessing techniques.

Because MRA has developed rapidly into a clinically useful diagnostic technique and is constantly evolving, it can be assumed that the aforementioned problems may be solved in the next few years when sufficient clinical experience becomes available on the basis of controlled scientific studies. However, until this time, indications for MRA cannot be clearly defined, useful selection of particular sequence design cannot be made, and the diagnostic accuracy of MRA cannot be evaluated reliably with regard to different vascular lesions in different vascular areas of the body. The well-known advantages of MRA over other imaging methods, such as the avoidance of ionizing radiation, the lack of need for contrast agents, and the wide field of view, may influence the spectrum of indications such that the focus is particularly on patients in whom these risk factors should be avoided.

This volume has been planned in order to present the actual state of development in MRA. The idea of this book was to give a general overview on MRA in a comprehensive format including basic principles of physics in MR, general descriptions of flow phenomena, imaging techniques, postprocessing techniques, hardware configurations and recent hardware developments, quantification of blood flow, and administration of contrast agents in MRA. Moreover, particular techniques of MRA are described and a practical "how to do" approach is adopted. Recommendations for typical examination protocols for different vascular areas and lesions are offered to facilitate the correct application of MRA. One major feature of the book is the critical discussion of the advantages and drawbacks of MRA, resulting in a catalogue of indications which are generally accepted today. Each chapter is systematically divided into a general introduction that includes alternative imaging modalities, a consideration of MRA techniques, a clinical part that refers to both anatomy and pathology, and a validation of the technique. A short summary at the end of each chapter characterizes the clinical and practical essentials for the area under consideration. Representative MR angiograms are provided to illustrate normal and pathological anatomy of the intra- and extracranial cerebral vasculature, the arterial systems of the chest, the abdomen, and the extremities, and the venous system of the body. In this respect, the book is primarily addressed to radiologists, angiologists, vascular surgeons, and neurosurgeons as well as to other physicians who are involved in vascular imaging and have only limited experience in vascular MRI.

Contents

Technical Considerations

Clinical Applications of Magnetic Resonance Angiography

Technical Considerations

1 Basic Principles of Nuclear Magnetic Resonance Imaging

P. VAN HECKE and G.M. BONGARTZ

CONTENTS

1.1 Introduction

Almost 50 years ago, in 1946, the phenomenon of nuclear magnetic resonance (NMR) was discovered by F. Bloch and E.M. Purcell independently. The characteristic resonance signals of different structures were used as a physicochemical analytical tool. It took another 30 years to introduce the technique into medical imaging. P. Lauterbur was the first to apply additional linear magnetic fields to the system in order to achieve spatial encoding of the signals.

NMR relies on the fact that some nuclei possess a spin, i.e., an intrinsic angular momentum due to their rotation about an internal axis. This angular momentum gives rise to an intrinsic magnetic moment which provides the nucleus with magnetic properties similar to those of a compass needle. The hydrogen atoms of the water and fat molecules which are very abundant in the human body have such a nuclear magnetic spin moment. When placed in a magnetic field, these hydrogen nuclei (or protons) align along the direction of the field about which they perform a periodic precession movement with frequency f_0.

The precession can be perturbed by external (electro-)magnetic energy at the resonance frequency f_0. This pulsed excitation energy delivered by a transmitter of electromagnetic energy is transferred to the spins of the volume to be imaged by a coil or antenna (electromagnetic induction). After the perturbation, the spins return to their equilibrium state and release the excitation energy as electromagnetic radiation at the resonance frequency, which is detected by the same antenna to deliver the NMR signal, which is amplified and processed.

This return towards equilibrium, also called *nuclear spin relaxation*, is characterized by a relaxation time which is specific for each biological tissue. The difference between relaxation times of various tissues is the source of contrast in NMR imaging.

The global excitation of all nuclear spins of the water and fat molecules of a volume of the human body does not provide any "spatial" information. In order to obtain an image of the distribution of the hydrogen spins, the electromagnetic excitation is combined with the application of a small additional magnetic field whose amplitude changes as a function of position (magnetic field gradient). This gradient modulates (encodes) the resonance frequency of the nuclear spins as a function of their position in

P. VAN HECKE, PhD, Dienst Radiologie, Katholieke Universiteit, Gasthuisberg, Herestraat 49, 3000 Leuven, Belgium
G.M. BONGARTZ, PD Dr., Dept. Medizinische Radiologic Universitätskliniken/Kantonsspital, Petersgraben 4, 4031 Basel, Switzerland

space. A spectral analysis of the nuclear spin signal after the excitation provides, through the frequency-space encoding, the spatial distribution of the nuclear spins, i.e., the NMR image, the intensity of which depends on the density and relaxation of the hydrogen spins in the particular tissue.

The basic instrumentation for the acquisition of NMR images follows from the above description:

1. A magnet, in which the subject is positioned; its magnetic field causes the hydrogen spins to precess about its direction.
2. A set of magnetic gradient coils to encode the precession frequency of the spins as a function of the position in the magnet.
3. A transmitter of electromagnetic energy to excite the spin system at its resonance frequency.
4. An antenna to transfer this energy to the spin system of the subject and to pick up the energy released by the spin system during relaxation.
5. A receiver to amplify this nuclear spin signal induced in the antenna.
6. A computer to digitize the signal and to calculate the image.

1.2 Nuclear Magnetism and Nuclear Magnetic Resonance

Magnetism is generated by motion of electric charges; therefore the intrinsic rotation (spin) of a nucleus (with an odd number of protons and/or neutrons, like hydrogen) also generates a local magnetic field. In medical imaging, the hydrogen nucleus (^1H), or proton, is optimally suited to the NMR experiment due to its bioavailability and its large magnetic moment.

In a strong external magnetic field ($\mathbf{B}_0$), the proton spin aligns either parallel or antiparallel to the direction of the external field. Although the parallel orientation is energetically more favorable than the antiparallel one, there is only a 0.0001% preponderance (at 1.5 T). The NMR experiment is entirely based on this small difference.

When protons are exposed to a well-defined amount of external electromagnetic energy, the low-energy state protons can be excited to the higher energy state. This transition is reversible and generates exactly the same amount of energy during its restitution. Energy transmission and energy absorption are achieved with electromagnetic coils which are introduced into the magnet and positioned close to the investigated volume.

1.2.1 Nuclear Spin Magnetic Moment

Atoms whose nuclei have an odd number of protons and/or neutrons are characterized by a spin angular momentum **I**. Since the nuclear spin is a physical quantity on the atomic scale, the formalism of quantum mechanics applies and states that the spin quantum number I is an integer for nuclei with an even mass number (^2H, ^{14}N), a half-integer for nuclei with an odd mass number (^1H, ^{13}C, ^{31}P, ^{23}Na), and zero for nuclei with even atomic and mass number (^{12}C, ^{16}O). To the spin angular momentum **I** of a nucleus corresponds a spin magnetic moment μ with the same direction as **I**:

$$\mu = \gamma \hbar \mathbf{I},$$

where γ is the gyromagnetic ratio of the nucleus and h is the Planck constant (in this chapter, divided by 2π).

For the hydrogen atom, $\gamma = 2.67 \cdot 10^8$ rad $\cdot$ s^{-1} $\cdot$ T^{-1} or 42.57 MHz/T. Nuclei characterized by a spin thus have magnetic properties. Clinical NMR imaging relies on the fact that the hydrogen nucleus ^1H, which is very abundant in the human body (water, fat), possesses a nuclear spin magnetic moment. In this introduction we will limit the discussion to the hydrogen nucleus, or proton, with spin $\mathbf{I} = {}^1/_2$.

1.2.2 Effect of a Magnetic Field on the Nuclear Magnetic Moment

In the absence of a magnetic field, the nuclear magnetic moments are oriented at random. When placed in a magnetic field, there is an interaction between the nuclear magnetic moments and the magnetic field which orients the spins in the direction of the field. The details of this interaction can be described by the quantum mechanical formalism or by the classical approach. The results of the quantum mechanical formalism useful for this introduction are briefly described prior to the classical description of NMR and its application to imaging.

1.2.3 Quantum Mechanical Description

1.2.3.1 The Isolated Spin

When placed in an external magnetic field $\mathbf{B}_0$ the nuclear spin magnetic moments orient along the direction of this magnetic field. In view of the quantization, only two orientations are allowed for the magnetic spin moment μ of a spin ${}^1/_2$, i.e., parallel or antiparallel to the direction of the magnetic field (z direction), with corresponding projections μ_z along the direction of the magnetic field given by:

$$\mu_{z+1/2} = +\tfrac{1}{2}\gamma h \text{ (parallel)}$$
$$\text{and } \mu_{z-1/2} = -\tfrac{1}{2}\gamma h. \text{ (antiparallel)} \tag{1.1}$$

The angle between the magnetic moment and the magnetic field is $\theta = \arccos(m\sqrt{3/4})$, with $m = \pm^1/_2$. To each of the two orientations corresponds a magnetic energy of:

$$E_{+1/2} = -\tfrac{1}{2}\gamma h B_0 \text{ (parallel)}$$
$$E_{-1/2} = +\tfrac{1}{2}\gamma h B_0 \text{ (antiparallel).} \tag{1.2}$$

The energy difference between these two orientations is:

$$\Delta E = \gamma h B_0. \tag{1.3}$$

"Excitation" of a spin from the lower energy level (parallel to the magnetic field) to the higher energy level (antiparallel to the magnetic field) thus requires an amount ΔE of electromagnetic energy. The frequency f_0 of this electromagnetic wave (also called resonance frequency) has to satisfy the relation:

$$\Delta E = 2\pi h f_0, \tag{1.4}$$

or with Eq. 1.3:

$$f_0 = \gamma B_0 / 2\pi. \tag{1.5}$$

This is the basic equation of NMR, which expresses that the resonance frequency f_0 of the spin, i.e., the frequency at which the spin absorbs or emits energy to change orientation state, is proportional to the amplitude of the magnetic field $\mathbf{B}_0$. In a field of 1.0 T, a typical value for a clinical NMR imager, the proton resonance frequency is 42.57 MHz (see Table 1.1).

The excitation energy ΔE required for NMR imaging is about 10^{12} times smaller than the energy involved in conventional x-ray imaging. This extremely low energy is one of the main assets of the NMR imaging technique.

Table 1.1. Resonance frequency f_0 of the hydrogen nucleus (proton) for typical values of the magnetic field $\mathbf{B}_0$ in clinical NMR imagers

$\mathbf{B}_0$ (T)	f_0 (MHz)
0.2	8.5
0.5	21.3
1.0	42.6
1.5	63.9
2.0	85.1

1.2.3.2 A System of Spins: the Macroscopic Nuclear Magnetic Moment

In reality one deals with a large quantity of spins, placed in a magnetic field, such as those of the subject to be imaged. At equilibrium, i.e., in the absence of electromagnetic excitation, the magnetic spin moments occupy the lowest energy state (parallel to $\mathbf{B}_0$). However, at room temperature, the thermal energy which activates the molecules that carry the proton spins is several orders of magnitude larger than the magnetic energy difference ΔE between the two orientations of the spins, so that the preferential orientation imposed by the magnetic field is scrambled. The exact distribution of the spins over the two energy levels is described by the Boltzmann law, which provides an expression for the ratio of the spin populations of the two levels:

$$n_{+1/2}/n_{-1/2} = e^{\Delta E/kT} \tag{1.6}$$

with $k = 1.38 \times 10^{-23}$ J/K, the Boltzmann constant, and T = the absolute temperature.

Since $\Delta E << kT$, the populations are about equal in the two levels, with a slight advantage for the lowest level. At the physiological temperature ($T = 37°$C or T = 310 K), $kT = 4.28 \times 10^{-21}$J. For $B_0 = 1.0$ T, one finds $\Delta E = 2.83 \times 10^{-26}$ J, which yields $n_{+1/2}/n_{-1/2} = 1.000007$. The relative population difference between the two levels amounts to 7 spins in 1 000 000! This (small) population difference leads to a net macroscopic magnetic moment M_0 along the direction of the magnetic field, equal to the difference of the z components of the $n_{+1/2}$ magnetic moments parallel to the field and the $n_{-1/2}$ magnetic moments antiparallel to the field (Eq. 1.1):

$$M_0 = (n_{+1/2} - n_{-1/2})\mu_z = (n_{+1/2} - n_{-1/2})\gamma h/2.$$

Appropriate techniques, described below, will lead this macroscopic magnetic moment to induce a signal in the antenna that will be used to generate an image of the distribution of the magnetic moments in the volume to be studied.

1.2.4 The Classical Description of the Motion of the Nuclear Magnetic Moment

1.2.4.1 The Isolated Spin in a Magnetic Field

In a magnetic field, the nuclear spin magnetic moment μ experiences a torque which results in a *precession* movement around the direction of the magnetic field (Fig. 1.1), described by the Larmor equation:

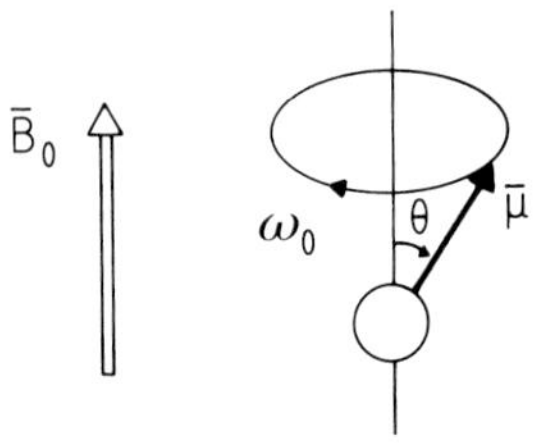

Fig. 1.1. Classical description of the motion of the single magnetic spin moment in a magnetic field $\mathbf{B}_0$. The magnetic spin moment $\boldsymbol{\mu}$ describes a precession motion around the direction of the magnetic field $\mathbf{B}_0$ with an angular frequency $\omega_0 = -\gamma\mathbf{B}_0$ (Larmor frequency)

$$d\boldsymbol{\mu}/dt = \boldsymbol{\mu} \times \gamma\mathbf{B}_0. \quad (1.7)$$

This precessional motion is similar to the rotational motion of a spinning top around its own axis, resulting from the interaction with the gravitational force of the earth. The angular precession frequency of the magnetic moment is given by the Larmor frequency:

$$2\pi f_0 = \gamma B_0. \quad (1.8)$$

This is the frequency which was already derived in Eq. 1.5 as the resonance frequency required to excite the spins to a higher energy state. The spin describes a motion on a cone with an aperture angle θ (see Sect. 1.2.3.1).

1.2.4.2 A System of Spins in a Magnetic Field

The system of N spins of the volume to be imaged is split into $n_{+1/2}$ spins, which precess on a cone parallel to the magnetic field, and $n_{-1/2}$ spins, which precess on a cone antiparallel to the magnetic field. At equilibrium, the precession frequency f_0 of the magnetic spin moments is the same and their phases, i.e., their position on the cone, are equally distributed (random phases). The sum of the components of the magnetic moments in the xy plane perpendicular to the direction z of the magnetic field $\mathbf{B}_0$ thus vanish. The sum of the components along the z axis parallel to the magnetic field $\mathbf{B}_0$ is the net macroscopic magnetic moment M_0 defined earlier (Fig. 1.2). The equilibrium state is thus characterized by a spin magnetic moment $\mathbf{M}$ at rest, aligned along the magnetic field $\mathbf{B}_0$, with components:

$$M_{x,y} = 0 \text{ (transversal component)}$$
$$\text{and } M_z = M_0 \text{ (longitudinal component).} \quad (1.9)$$

For a system of N spins $I = 1/2$, M_0 is given by:

$$M_0 = M_z = N\gamma^2 h^2 B_0/4kT.$$

The macroscopic magnetic moment is proportional to the number N of spins in the volume and to the strength of the magnetic field $\mathbf{B}_0$.

1.2.5 Excitation of the Spin System

Energy can only be exchanged if the external transmitter is at resonance with the system. The spin motion is defined by the external magnetic field strength and the so-called gyromagnetic ratio (γ). The frequency of this specific rotation is the Larmor frequency (f_0). Only if the transmitted electromagnetic wave is at the same frequency, will the resonance condition be fulfilled.

Radiofrequency (RF) pulses at the resonance frequency add energy to the system, thereby increasing the population of the higher energy level. When the external energy supply is switched off, the NMR signal can be received by the receiving coil as an electromagnetic wave with a frequency corresponding to the Larmor frequency of the nucleus. The exponential decay of the signal as a function of time is called the free induction decay (FID). The FID is the direct response of the system to a single external RF pulse. The FID is detectable in the xy plane of the NMR coordinate system, in which the z axis is oriented along the direction of the magnetic field $\mathbf{B}_0$.

The exploitation of the NMR phenomenon is based on the detection of the electromagnetic energy radiated by the spin system after a short excitation

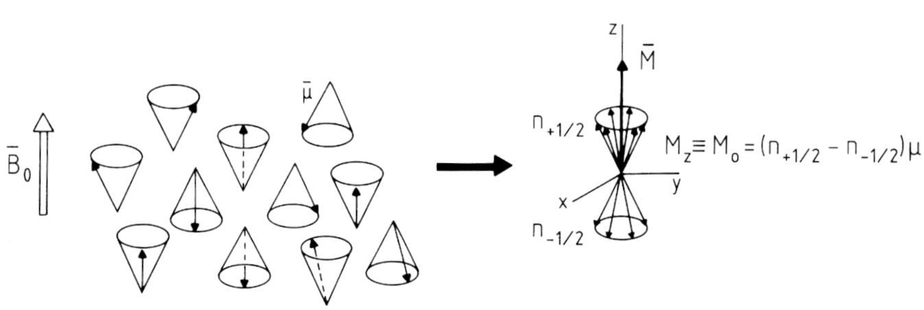

Fig. 1.2. System of N spins in a magnetic field $\mathbf{B}_0$, and equilibrium macroscopic magnetic moment M_0. The spins precess around the magnetic field $\mathbf{B}_0$ with a common angular frequency ω_0 and a random phase. The number of spins parallel to the field $\mathbf{B}_0$ exceeds the number of spins antiparallel to $\mathbf{B}_0$ and gives rise to a net macroscopic magnetic moment M_0 along $\mathbf{B}_0$ (z axis) since its component M_{xy} in the plane perpendicular to $\mathbf{B}_0$ is zero (random phases)

with an external source of electromagnetic energy at the resonance frequency f_0. The basic principles of excitation and subsequent observation of the NMR signal are described in classical terms below.

The excitation is maximal if the magnetic vector of the electromagnetic excitation wave, with magnitude B_1, is perpendicular to the magnetic field $\mathbf{B}_0$. The vector $\mathbf{B}_1$ thus rotates in the xy plane at a frequency ω_0. This is achieved by exciting a coil or loop with axis in the xy plane, with a sinusoidal current at the frequency $f_0 = \omega_0/2\pi$. Since the value of f_0 is in the radiofrequency range, the $\mathbf{B}_1$ field is called the *RF excitation field.*

The description of the motion of the macroscopic magnetic moment is simplified in a coordinate frame $(Ox'y'z')$ rotating at the frequency ω_0 around $\mathbf{B}_0$, the z direction of the fixed coordinate frame $(Oxyz)$. The two coordinate frames share the same origin O and z axis (Fig. 1.3). In this rotating frame, the RF excitation field $\mathbf{B}_1$ is thus stationary, its position being chosen along the x' axis. The magnitude of the external field $\mathbf{B}_0$ experienced by the magnetic moment in the frame rotating at the frequency f_0 is zero.

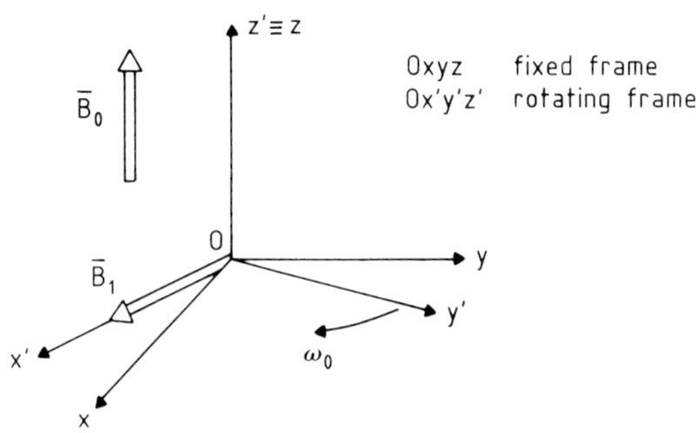

Fig. 1.3. Fixed and rotating coordinate frames. The coordinate frame $(Ox'y'z')$ rotates around the fixed frame $(Oxyz)$ with an angular frequency ω_0. The z and z' axes are common. In the rotating frame $\mathbf{B}_0$ vanishes and the vector $\mathbf{B}_1$ is stationary since it rotates around $\mathbf{B}_0$ (z axis) with the same frequency

In the absence of the RF excitation field $\mathbf{B}_1$, the equilibrium macroscopic magnetic moment $\mathbf{M}$ is oriented along the z axis in both coordinate frames. In the presence of $\mathbf{B}_1$, the magnetic moment $\mathbf{M}$ rotates (nutates) away from the z axis, around the direction of $\mathbf{B}_1$ (x'- axis) in the rotating frame, since $\mathbf{B}_1$ is the only acting magnetic field in the rotating frame. As soon as the excitation field $\mathbf{B}_1$ is removed, the nutation motion of $\mathbf{M}$ around $\mathbf{B}_1$ is discontinued. This classical description demonstrates that the macroscopic magnetic moment is tipped away from its equilibrium position along the z axis by the RF excitation pulse to generate a transversal component M_{xy}.

In NMR imaging the RF field $\mathbf{B}_1$ is applied as a short excitation pulse. A 90° *pulse* is an excitation of duration such that the equilibrium magnetic moment $\mathbf{M}$ is tilted from the z axis to the x' axis, i.e., over an angle of 90° (Fig. 1.4a). A 180° *pulse* is an excitation pulse of duration such that the magnetic moment is rotated over an angle of 180° around the direction of $\mathbf{B}_1$, e.g., from the $+z$ axis to the $-z$ axis (Fig. 1.4b).

At the end of the 90° pulse, the z component of the magnetic moment is zero. In terms of the population distribution over the energy levels, $M_z = 0$ implies that $n_{+1/2} = n_{-1/2}$. The 90° pulse thus equalizes the populations of the two energy levels and achieves the *saturation* of the spin system.

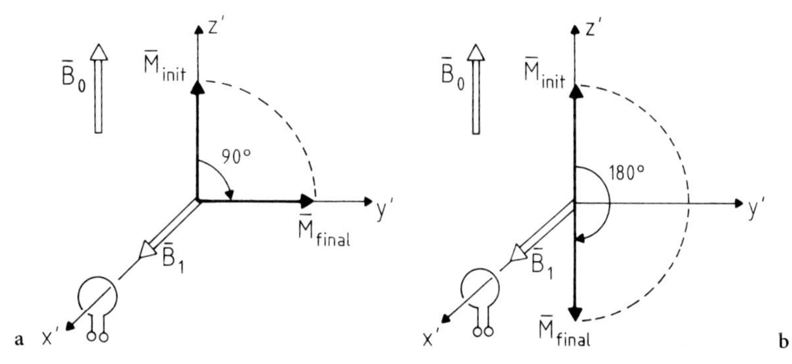

Fig. 1.4. **a** *90° pulse*: The initial position of the magnetic moment **M** is along the z axis. The duration of the RF excitation field $\mathbf{B}_1$ is such that **M** rotates (nutates) over 90° around $\mathbf{B}_1$ (x' axis) in the rotating frame and is tilted along the y' axis. In the fixed frame, **M** describes a spiral movement starting from the z axis towards the xy plane with a final position perpendicular to $\mathbf{B}_1$. **b** *180° pulse*: The initial position of the magnetic moment **M** is along the z axis. The duration of the RF excitation field $\mathbf{B}_1$ is such that in the rotating frame **M** rotates (nutates) over 180° around $\mathbf{B}_1$ (x' axis) towards its final position along the $-z$ axis. In the fixed frame, **M** describes a spiral movement from the z axis towards the $-z$ axis

1.2.6 The Free Precession Signal

The 90° excitation pulse tilts the magnetic moment in the *xy* plane, where it remains stationary in the rotating frame. In the fixed frame, however, the magnetic moment **M** precesses around the magnetic field $\mathbf{B}_0$ at the Larmor frequency f_0. This time-dependent magnetic moment induces a periodic voltage at the Larmor frequency f_0 in the coil used for the excitation pulse (Fig. 1.5).

The effect of the resonant 90° excitation pulse is to bring all spins to precess in phase, yielding a macroscopic magnetic moment **M** in the *xy* plane, with magnitude M_0. In reality, the magnetic field in the volume to be imaged is not homogeneous and the resonance frequency of the corresponding spins is not uniform. Spins located in a local magnetic field which is slightly higher than the average field have a slightly higher precession frequency than the average frequency, while the spins located in a lower magnetic field have a lower precession frequency.

Immediately after the excitation pulse, the individual spins which contributed to **M** start to precess around $\mathbf{B}_0$, in the *xy* plane, with slightly different frequencies leading to an increased dephasing or loss of coherence of the individual spins. Their vector sum, i.e., M_{xy}, decreases with time to ultimately vanish. The periodic signal $S(t)$ induced in the coil after a 90° excitation pulse is an exponentially damped sinusoidal signal with frequency f_0 and characteristic decay time T_2*:

$$S(t) = S_0 \sin(2\pi f_0 t)\exp(-t/T_2^*),$$

where S_0 is the initial amplitude of the signal at the end of the 90° pulse. The signal $S(t)$ is called the *free precession signal* and its exponential decay as a function of time is referred to as *free induction decay* (FID), as already mentioned.

1.2.7 The Spin Echo

Following the initial excitation which deflects the magnetization into the *xy* plane, the dephasing process starts immediately. Dephasing must be understood as the spread of the individual spin moments which form the net macroscopic vector in the *xy* plane. It is related to local inhomogeneities of the field. The compensation of the dephasing is called rephasing and can be achieved by the reversal of the phase of the spin moments by a 180° pulse (at the time TE/2). The effect of the inverting pulse is to reverse the spread direction of the spin moment vectors and to refocus them at the time TE, creating a maximal transverse magnetization, the so-called spin echo.

The dephasing of the spins due to the inhomogeneity of the magnetic field is reversible. A rephasing can be achieved by a 180° pulse or, as will be explained later, by a field gradient (see Sect. 1.4.2.1). The effect of a 180° pulse is easily visualized in the rotating frame (Fig. 1.6). After a time t = TE/2

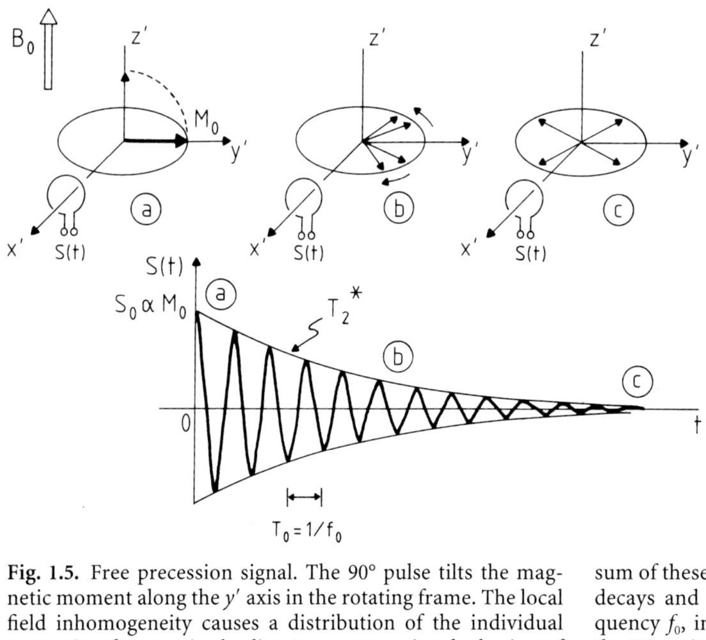

Fig. 1.5. Free precession signal. The 90° pulse tilts the magnetic moment along the *y'* axis in the rotating frame. The local field inhomogeneity causes a distribution of the individual precession frequencies leading to a progressive dephasing of the individual magnetic moments in the *xy* plane. The vector sum of these moments (the macroscopic transversal moment) decays and so does the amplitude *S* of the signal with frequency f_0, induced in the receiver coil. This signal following the RF excitation pulse is called the free precession signal, the exponential decay of which is characterized by a time T_2*

following the 90° pulse, the spins have acquired a dephasing angle, proportional to their departure from the nominal frequency f_0. The spins in a higher local field lead, and the spins in a lower local field lag the spins precessing at the resonance frequency (along the y' axis) in the nominal local field. The 180° pulse at time $t =$ TE/2 achieves a rotation of the individual spins around the direction of the excitation field $\mathbf{B}_1$ (y' axis) so that the spins with the largest lead have the smallest lead after the 180° pulse and the spins with the largest lag have the smallest lag after the 180° pulse. After the 180° pulse each spin continues to precess in the fixed frame with its own frequency in its local magnetic field since each spin occupies the same position before and after the 180° pulse. The spins with the smallest lead will thus catch the spins with the largest lead and those with the smallest lag will catch those with the largest lag; both groups, fast and slow spins, rephase exactly at a time TE/2 after the 180° pulse, leading again to the formation of the magnetic moment **M** and to the corresponding formation of a signal with maximum at time t = TE, after the 90° pulse, and subsequent decrease as the spins dephase again. This signal formed by the rephasing of the spins with a 180° pulse is called the *spin-echo* signal. This echo can be rephased again by another 180° pulse to form a second spin echo and so on.

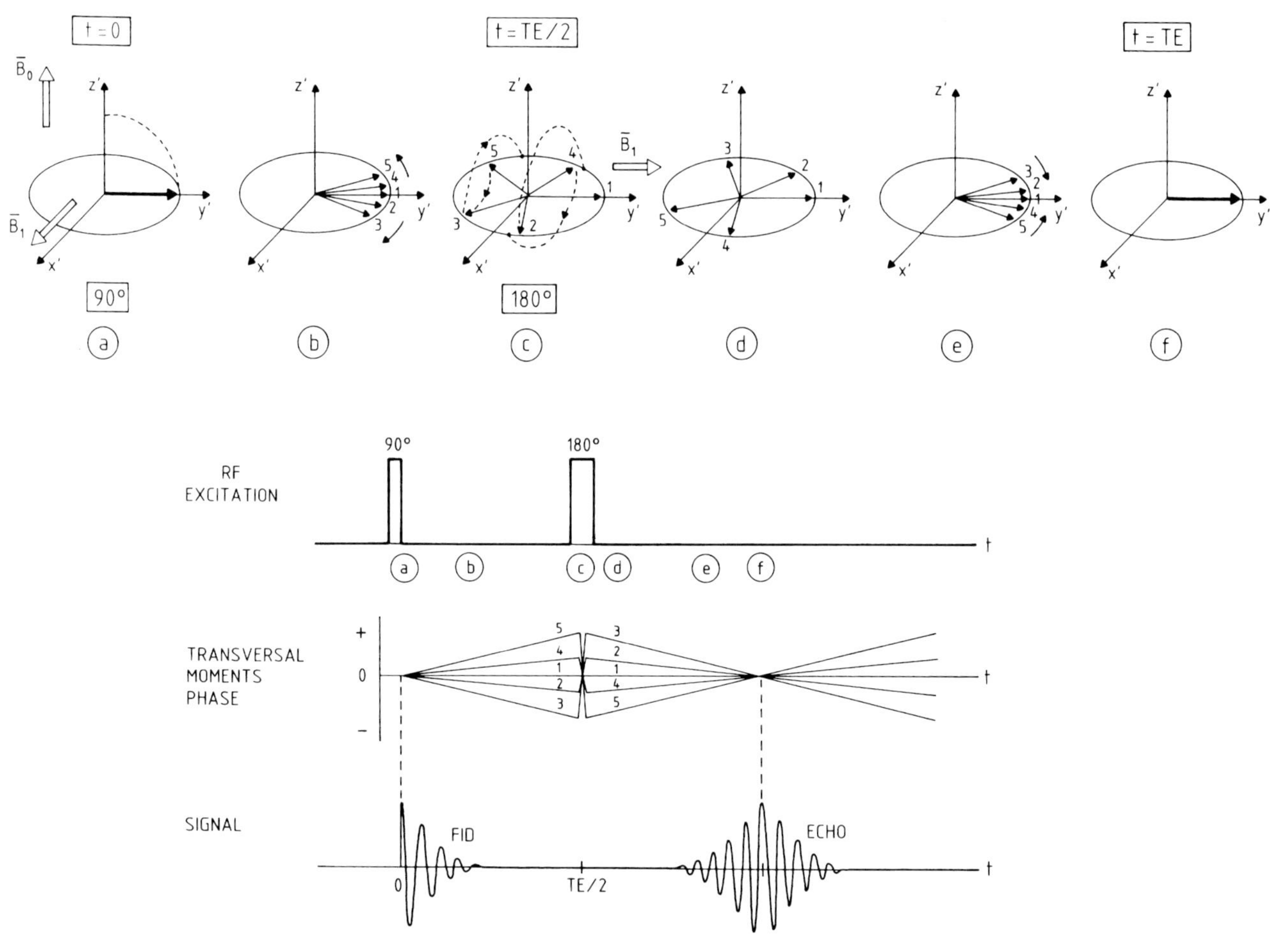

Fig. 1.6a–f. The spin-echo sequence {90°-TE/2-180°-TE/2 echo}. **a** The 90° pulse at $t = 0$ tilts the magnetic moment on the y' axis. **b** The local field inhomogeneity induces a progressive dephasing of the individual magnetic spin moments. **c** The spin moments are rotated about the y' axis by the 180° pulse at $t =$ TE/2. **d,e** The spins remain in their own local field and keep their individual frequencies. The spins with the largest lag (*5*) catch those with the smallest lag (*4*) and the spins with the largest lead (*3*) catch those with the smallest lead (*2*). After a delay TE/2 all individual magnetic moments are in phase to form the total magnetic moment **M** and the corresponding spin-echo signal (**f**). The evolution of the phase of the individual magnetic spin moments is also shown

1.3 Relaxation

The FID reflects the decay of the transverse magnetization (M_{xy}). In repetitive excitations, the signal amplitude of the FID will be determined by the recovery of the longitudinal magnetization (M_z) towards its equilibrium value. Both processes, the decay of the transverse component and the recovery of the longitudinal component, are called relaxation.

During the T_1 relaxation the magnetization vector M_z regains its amplitude by releasing energy. T_1 relaxation processes are due to energy exchange between the proton and the surrounding structure, the lattice. Finally, a stable equilibrium is reached. The rate of the T_1 relaxation of a specific proton depends directly on the molecular structure to which the proton is bound. Small molecules (like water) behave differently than larger molecules. The energy transfer of the T_1 process relies upon the Brownian molecular motion. In molecules like fat this motion has a frequency comparable to the Larmor frequency and therefore the energy exchange is accelerated as compared to water protons. The T_1 time of fat is shorter than the T_1 time of water. In standard imaging, structures with a short T_1 are displayed as a bright signal whereas those with a long T_1 are displayed as a dark signal (T_1-weighted images).

Differences in T_2 relaxation are due to interactions between neighboring spins in the magnetic field, resulting in minor differences in the precession frequency. These differences destroy the initial coherence of the spin phases in the *xy* plane. The sum vector of the transversal magnetization (M_{xy}) decreases due to the phase dispersion of its different components. Local inhomogeneities of $\mathbf{B}_0$ accelerate the T_2 decay. This effect is called T_2* decay, which is generally faster than T_2 decay. T_2* combines both tissue-dependent factors and $\mathbf{B}_0$-dependent factors. The $\mathbf{B}_0$ component can be compensated by the inversion of the spin system with a 180° pulse. This compensation for the field inhomogeneities results in purely T_2-dependent signals. In contrast to T_1-weighted imaging, structures with long T_2 times appear bright in T_2-weighted sequences while short T_2 times appear dark.

Both T_1 and T_2 relaxation times are tissue-dependent parameters which are used for the characterization of a certain structure. The close correlation of T_1 relaxation processes with the Larmor frequency explains why the T_1 time of a tissue is affected by the magnetic field strength, while this is hardly the case for T_2 processes.

After the 90° excitation pulse, the macroscopic magnetic moment **M** which was tilted in the *xy* plane has to return to its equilibrium position along the *z* axis with $M_z = M_0$. The transversal component M_{xy} of the magnetic moment hereby decreases from its initial amplitude M_0, immediately after the pulse, towards zero, and the longitudinal component M_z increases from the value zero immediately after the pulse to the equilibrium value M_0. This process whereby the electromagnetic excitation energy is dissipated by the spin system after the pulse is called *spin relaxation* and its duration is characterized by the *relaxation time*. Two distinct mechanisms contribute to this relaxation process: the relaxation of the transversal component M_{xy} and the relaxation of the longitudinal component M_z.

1.3.1 Relaxation of the Transversal Component M_{xy}

1.3.1.1 Spin-Spin Relaxation

The decay of the amplitude of the magnetic moment in the *xy* plane due to the inhomogeneity of the magnetic field (characteristic time T_2*) is a reversible effect as shown by the formation of the spin echo after a 180° pulse. The mechanism of *spin-spin relaxation*, however, causes the transversal component of the magnetic moment to decay irreversibly towards its zero equilibrium value. The amplitude of the rephased magnetic moment at the maximum of the echo is thus smaller than the original amplitude of the magnetic moment after the 90° pulse. This irreversible decay of the amplitude of the transversal magnetic moment (as measured at the maximum of the echo at t = TE) is described by an exponential function with characteristic time T_2, the spin-spin relaxation time (Fig. 1.7):

$$M_{xy}(\mathrm{TE}) = M_0 \exp(-\mathrm{TE}/T_2). \tag{1.10}$$

The relaxation time T_2 is an intrinsic parameter of the tissues (typical value 100 ms), while T_2* is determined by the inhomogeneities of the magnetic field (due to tissue susceptibility and/or external field homogeneity). The NMR signal induced in the coil is proportional to M_{xy}. The observed decay of the amplitude S_e(TE) of the top of the echo signal is thus characterized by the same exponential function and relaxation time T_2 as M_{xy} (Eq 1.10):

$$S_e(\mathrm{TE}) = S_0 \exp(-\mathrm{TE}/T_2), \tag{1.11}$$

with S_0 the echo signal amplitude for TE = 0. The decay of the amplitude of the successive echoes of a

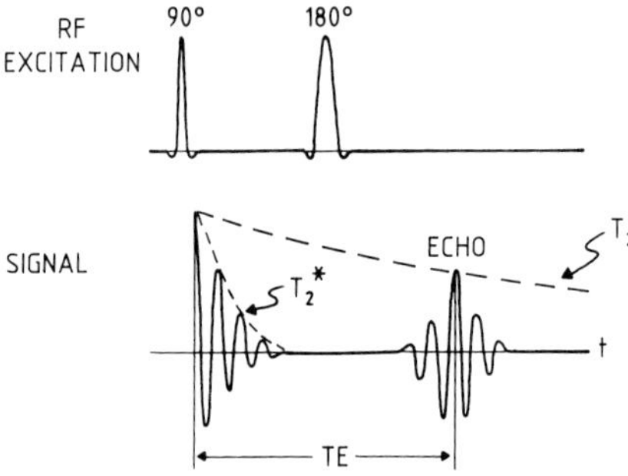

Fig. 1.7. Spin-spin relaxation. Despite the rephasing achieved by the 180° pulse, the amplitude of the echo exponentially decreases due to the spin-spin relaxation of the transverse component of the magnetic moment towards its zero equilibrium value, in a characteristic time T_2, called the spin-spin relaxation time

multiple echo sequence is also described by this function.

The physical mechanism of the spin-spin relaxation is related to the exchange of magnetic energy between the spin magnetic moments. The individual spins are subjected to a a weak magnetic field induced by the magnetic moments of their neighbors. The 90° excitation pulse brings all spins in phase to form a transversal macroscopic magnetic moment. Due to the fluctuating local field induced by the neighboring spins (some neighbors have a spin moment up and some neighbors a spin moment down), the precession frequency of the spins is altered in a random way, leading to an irreversible scrambling of their phase in a time T_2.

1.3.1.2 Relation Between T_2 and T_2*

The decay of the transversal magnetic moment M_{xy} and of the free precession signal induced in the coil after the excitation pulse is described by a characteristic time T_2*, which is a function of the intrinsic spin-spin relaxation time T_2 (interaction between the spins) and of a characteristic extrinsic time T_{2extr}, which describes the dephasing of the spins due to the inhomogeneity of the external magnetic field (interaction between the spins and the magnetic field). This relation can be described with the relaxation rates (the inverse of the relaxation time) as:

$$1/T_2^* = 1/T_2 + 1/T_{2extr}, \tag{1.12}$$

with $1/T_{2extr} = \gamma\Delta\mathbf{B}_0$, where $\Delta\mathbf{B}_0$ = the mean deviation of the local magnetic field due to variations of the external magnetic field and differences in tissue magnetic susceptibilities. For biological tissues and for the inhomogeneity of the clinical magnets, $T_{2extr} << T_2$, so that $T_2^* \sim T_{2extr}$. The decay of the free precession signal (T_2*) is thus largely determined by the magnetic field inhomogeneity and the susceptibility differences. The decay of the spin echo, however, is determined by T_2 since the spin-echo sequence has the property to refocus the phase differences due to magnetic field inhomogeneities.

1.3.2 Relaxation of the Longitudinal Component M_z

1.3.2.1 Spin-Lattice Relaxation

The spin-spin relaxation achieves a distribution of the magnetic excitation energy in the spin system leading to a loss of phase coherence and a decay of the transverse magnetic moment. The return to the thermal equilibrium state requires the dissipation of the magnetic energy into mechanical (thermal) energy, whereby the longitudinal component M_z of the magnetic moment recovers towards the equilibrium value M_0. This process is called the *longitudinal relaxation* or *spin-lattice relaxation.*

In NMR imaging sequences the 90° excitation pulse tilts the magnetic moment in the *xy* plane so that the longitudinal component M_z is zero. The spin system is saturated since the populations of the two energy levels are equal. The growth of M_z from the value zero to the equilibrium value M_0 follows an exponential recovery characterized by the spin-lattice relaxation time T_1:

$$M_z = M_0[(1-\exp(-t/T_1)]. \tag{1.13}$$

The recovery of M_z following a 90° pulse is called the *saturation recovery*. If M_z is inverted to the $-z$ axis by a 180° pulse, the subsequent exponential recovery of M_z towards M_0 is called the *inversion recovery.*

The conversion of the magnetic excitation energy of the spin system into mechanical energy is accomplished by the interaction of the spins with their surrounding (or "lattice"), i.e., by the rotational and translational motion of the atoms and molecules to which the spins are attached. In biological systems, the values of T_1 range from a few hundred milliseconds to a few seconds and are specific for a tissue or fluid.

1.3.2.2 Observation of the Longitudinal Component M_z

The longitudinal component M_z of the magnetic moment cannot be observed since it is stationary and perpendicular to the axis of the RF coil. In order to monitor its recovery during the relaxation process, M_z has to be tilted in the *xy* plane with a second 90° pulse. The amplitude of the signal induced in the coil is then proportional to the amplitude of M_z just before the second 90° pulse. NMR imaging sequences exploit the recovery of the longitudinal component M_z during the T_1 relaxation.

In the saturation-recovery sequence (Fig. 1.8), the recovery of M_z is observed at a time t = TR after the 90° pulse, with a second 90° pulse which tilts M_z(TR) in the *xy* plane. In NMR imaging, the observation of M_z(TR) is achieved with the spin-echo sequence. During the acquisition of an image, the spin system is subjected to a periodic (TR) excitation with the spin-

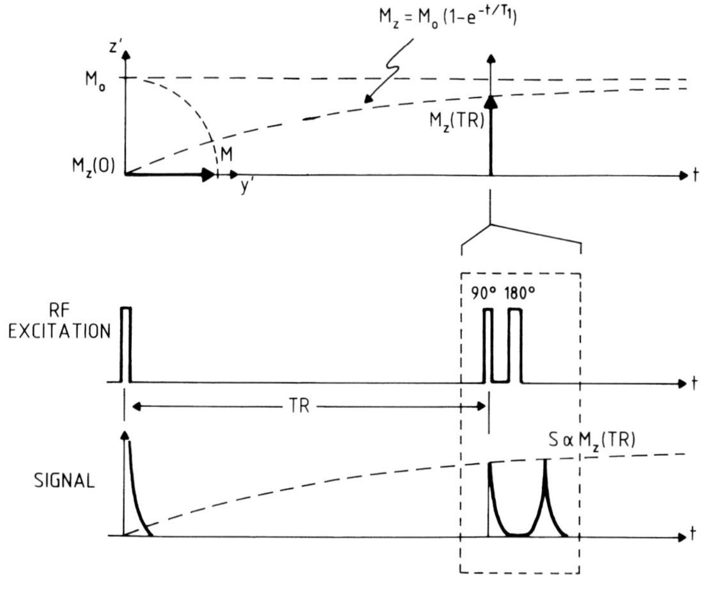

Fig. 1.8. Spin-lattice relaxation and saturation-recovery sequence. The 90° pulse tilts the magnetic moment in the $x'y'$ plane. The longitudinal component M_z recovers exponentially from zero towards its equilibrium value M_0 with a characteristic time T_1, the spin-lattice relaxation time. The observation of M_z during the relaxation process requires the application of a second 90° pulse (or of a spin-echo sequence) to tilt M_z in the $x'y'$ plane for observation

echo sequence. After a few excitations, M_z reaches a stationary value determined by the excitation period TR and the spin-lattice relaxation time T_1. The amplitude of the echo in this excitation train is proportional to the value of M_z just before each excitation pulse. After each excitation, M_z recovers from its zero value. The complete saturation-recovery sequence used in NMR imaging is denoted {TR-90°-TE/2-180°-TE/2-echo-}$_N$.

The amplitude of the echo signal S(TR) is proportional to M_z(TR) (Eq. 1.13) and depends on the echo time TE (Eq. 1.11):

$$S(\mathrm{TR}) = S_0[1 - \exp(-\mathrm{TR}/T_1)] \exp(-\mathrm{TE}/T_2), \quad (1.14)$$

where S_0 is the echo amplitude corresponding to M_0. If the echo time is short (TE $\ll T_2$), there is no T_2 effect and the amplitude of the echo signal is proportional to M_z(TR):

$$S(\mathrm{TR}) = S_0[1 - \exp(-\mathrm{TR}/T_1)]. \quad (1.15)$$

1.4 Sequences

The NMR (echo) signal can be altered by appropriate timing of RF and gradient pulses in order to achieve foreseeable contrasts (T_1-weighted, T_2-weighted, and proton density-weighted images). Moreover, physical effects like flow or magnetic susceptibiliy can also alter the signal. An MRI sequence is the combination of different RF pulses and gradient pulses, corresponding to a particular timing. The resonating spin system is perturbed and altered by voluntary manipulation of the magnetization vector. The timing of the different pulses allow T_1 and T_2 processes to contribute to the final signal. The predictable behavior of the spin system differs according to the type of excitation.

1.4.1 SE Sequences

The sequence of a 90° and a 180° pulse which gives rise to a spin echo is called the *spin-echo sequence* and is denoted {90°-TE/2-180°-TE/2-echo}. It is the basic sequence for NMR imaging. The observation of the NMR signal makes use of the spin echo rather than the FID after a 90° pulse, because the FID signal cannot be acquired immediately after the pulse due to the presence of magnetic field gradients required for imaging and to the minimum delay of the receiver electronics needed before the acquisition of the signal. In SE sequences the signal and the achieved contrast are affected by both T_1 and T_2 relaxation processes simultaneously.

To generate image contrast based on differences in the T_1 relaxation of the tissues, TR must be kept shorter than the expected T_1 to avoid complete T_1 relaxation taking place before the signal is read out (T_1-weighted imaging). On the other hand, contrast based on T_2 differences can only be achieved if all T_1 relaxation processes are completed (long TR) and the differences in T_2 decay are brought into evidence by the prolongation of TE (T_2-weighted imaging). The combination of long TR and short TE avoids relaxation effects and is called proton density-weighted imaging. The signal is maximized with such a parameter combination while the contrast is rather low.

1.4.2 GE Sequences

The reduction of the imaging time remains a major goal of NMR imaging research. The penalty is a lower signal-to-noise ratio (S/N) and often a more difficult interpretation of the image contrast as the image intensity is a more complicated function of the relaxation times, repetition period, and pulse angle. The most important type of imaging sequences with short TR implemented on all conventional NMR imagers belong to the family of *gradient-echo (GE) sequences.*

The measuring time is defined by the factors TR, matrix (number of phase encoding steps), and the number of excitations. Reduction of any one results in faster imaging capabilities. In SE methods, the reduction of TR is limited by the recovery of the longitudinal magnetization according to the T_1 time of the specimen. Maximal shortening of TR would result in insufficient M_z and tremendously reduce the overall signal (low S/N). Two modifications in the sequence design enable a much faster acquisition and still preserve sufficient signal. First, in GE sequences, RF pulses with an angle smaller than 90° are applied to tilt M_z into the *xy* plane. Hence, only a part of the longitudinal magnetization is used for the excitation, while the rest is still available for the next pulse. The drawback, of course, is a smaller M_{xy} component and a lower signal induced in the coil. Secondly, the 180° inversion pulse of the SE sequence is replaced by a gradient pulse to generate the echo. GE excitations cannot compensate for field inhomogeneities. Therefore, GE sequences are affected by T_2* effects rather than by T_2 effects.

For a given acquisition time the overall SNR in GE sequences is superior to that in SE sequences. The contrast in GE imaging depends not only on TR and TE but also on the applied flip angle (FA).

1.4.2.1 Principle

Gradient-echo sequences do not use a refocusing 180° pulse. The echo after the 90° pulse is formed using a bipolar gradient in the read-out direction. After the excitation pulse, the spins are dephased in the transverse plane by the negative lobe of the read-out gradient and are subsequently rephased by the positive lobe of the read-out gradient. Maximal echo signal occurs at time TE, when the area of the positive lobe equals the area of the negative lobe of the read-out gradient (Fig. 1.9). The positive read-out gradient is maintained beyond this time to allow the (symmetrical) acquisition of the entire echo signal. The GE imaging sequence makes use of the same gradients for slice selection and phase encoding as the SE sequence.

The decay of the GE amplitude as a function of TE is characterized by T_2*, i.e., by the field inhomogeneity. Indeed, the read-out gradient is added to the local field and the inversion of the polarity of this gradient does not influence the dephasing accumulated by the spins due to the local field inhomogeneities. The GE sequence is thus very sensitive to field inhomogeneities, such as those due to magnetic susceptibility differences between adjacent tissues or structures (air/tissue, bone/tissue).

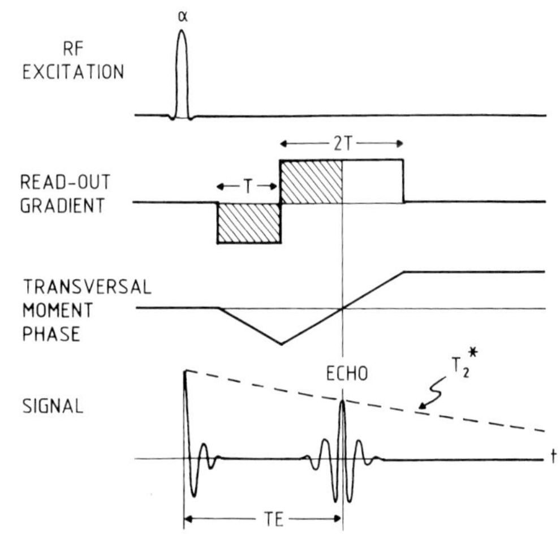

Fig. 1.9. Formation of the gradient-echo. The spin system is excited with an RF pulse with angle α ($\alpha < 90°$). The echo is formed at time TE after the excitation pulse, by the reversal of the polarity of the gradient in the read-out direction. The echo maximum occurs when the area of both gradient lobes is equal, corresponding to the rephasing of the spins. The amplitude of the gradient echo at t = TE decays with time T_2*

The ability of the GE sequence to achieve fast imaging relies on the use of a single RF excitation pulse, the angle of which can be adjusted. Fast imaging implies repetition times TR much shorter than T_1, the relaxation time needed for the magnetic moment to recover to its equilibrium value along the *z* axis. For a 90° pulse, the full longitudinal component is tilted in the *xy* plane, so that by the next excitation, only a small fraction of the full M_z component is available since TR << T_1. This small component induces an equally small signal in the coil. An excitation pulse with angle $\alpha < 90°$, however, leaves a component $M_0\cos\alpha$ along the *z* axis, which is available for the next excitation (Fig. 1.10). The NMR signal is then proportional to the transversal component $M_0\sin\alpha$. After a few pulses the longitudinal component of the magnetic moment and the corresponding amplitude of the signal induced in the coil reach a steady state.

For a given TR and T_1, the signal is maximal for an excitation angle α, such that:

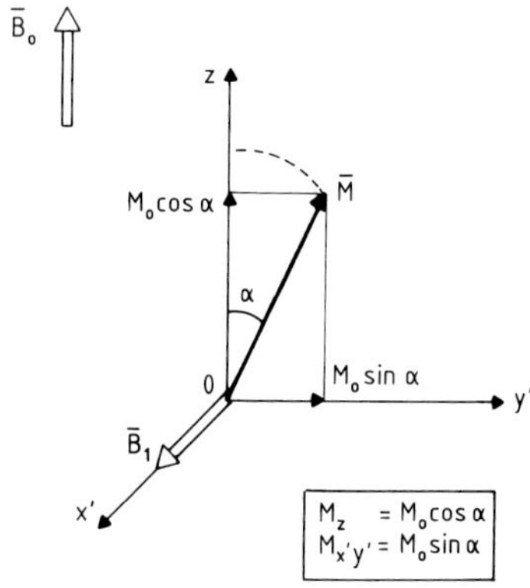

Fig. 1.10 Components of the magnetic moment after excitation with a pulse with angle $\alpha < 90°$. For $\alpha = 30°$, 87% of the z component of the magnetic moment is conserved and available for the next excitation; the transversal component conserves 50% of its maximum value

$$\cos\alpha = \exp(-\mathrm{TR}/T_1). \text{ (Ernst angle)} \tag{1.16}$$

In fast imaging the ratio TR/T_1 is small, which requires a small angle α for maximal signal. Typical values for fast imaging GE sequences are TR from 20 to 100 ms, TE from 5 to 20 ms, α between 10° and 60°, and the acquisition time between 5 and 25 s.

Since TR can be as short as, or shorter than T_2^*, the transversal component of the magnetic moment has not completely decayed before the next excitation. The many variants of the fast GE sequences differ in the way this residual transversal component is dealt with. A few typical GE sequences routinely used in clinical imaging are briefly discussed below.

1.4.2.2 GE Sequence with Dephasing Gradient

With this type of sequence the residual transversal magnetic moment is destroyed after the acquisition of the echo signal, using a strong dephasing (*spoiler*) gradient. An example is the FLASH (fast low angle shot) sequence (Fig. 1.11). The use of the spoiler gradient ensures that the amplitude of the steady-state magnetic moment along the z axis at the end of each cycle is independent of T_2.

Another option is to randomly alter the phase of the excitation pulse in each cycle, so that the phase of the transversal component is also random from cycle to cycle, leading to the cancellation of this component.

1.4.2.3 GE Sequence with Rephasing Gradients

This type of sequence incorporates the residual transversal component of the magnetic moment at the end of the cycle into the next cycle. Both the longitudinal and the transversal components thus reach a steady state and contribute to the echo signal. The dephasing of the transversal component caused by the various gradients of the imaging sequence is compensated by additional refocusing gradients in each direction, after the echo acquisition. The use of gradients symmetrical with respect to the center of the sequence (at the echo time TE) achieves the rephasing of the transversal component at the end of the sequence and maintains the coherence of the transversal component from cycle to cycle,

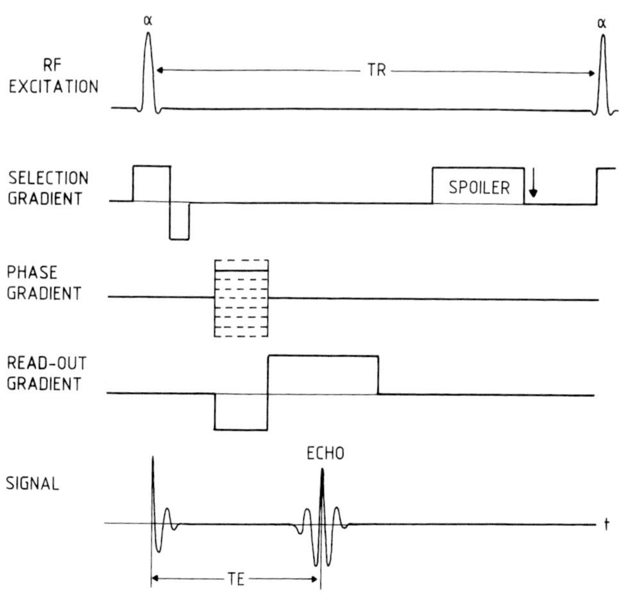

Fig. 1.11. FLASH sequence. In addition to the slice selective, phase-encoding and read-out gradients, a spoiler gradient is added at the end of the sequence to destroy the residual transversal magnetic moment

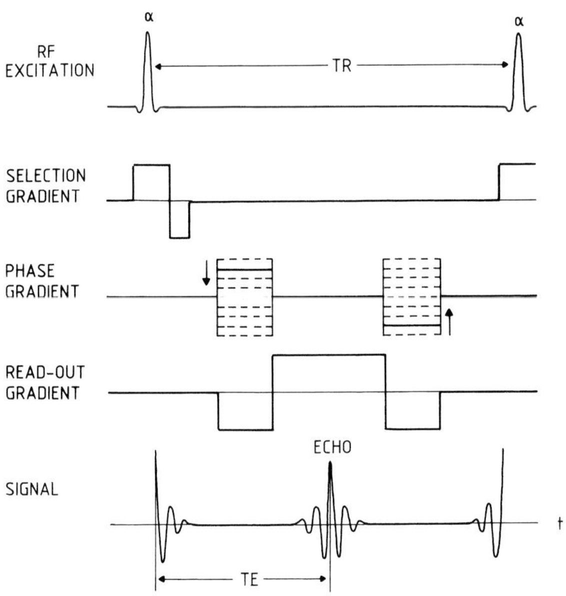

Fig. 1.12. FISP sequence. Additional rephasing gradients are introduced in each direction in order to compensate at the end of the cycle for the dephasing introduced by the slice selective, phase-encoding, and read-out gradients (t = TE)

such as in the FISP (fast imaging with steady-state precession) sequence and the GRASS (gradient recalled acquisition in the steady state) sequence (Fig. 1.12).

The amplitude of the echo signal for these sequences is a more complicated function of the parameters T_1, T_2*, TR, TE, and α. When TR << T_2*, the residual transversal component is important and makes a major contribution to the signal, a characteristic feature of such sequences. One shows that the echo amplitude is independent of TR and that the sequence can be used for very short TR, which makes it very suitable for 3D acquisition. For TR > T_2^*, the sequence behaves like the previous one in view of the absence of a transversal component, which makes this method less interesting for multislice applications. Also, when the angle is reduced, the transversal component is reduced and the sequence behaves like the previous one.

1.5 Spatial Encoding and Fourier Analysis

The signal in an NMR experiment derives from the multitude of different spins or magnetization components in a particular volume. In MRI, the frequency components must also be analyzed according to their spatial distribution within that volume.

To obtain spatial information, a small magnetic field that increases linearly with distance (the so-called magnetic field gradient) is superimposed on the main magnetic field ($\mathbf{B}_0$). The Larmor frequency of the spins within the specimen will then increase accordingly along this gradient axis. This enables the spatial identification of a spin along this axis by the analysis of its precession frequency.

The second axis of the spatial localization derives from the spatial dependence of the phases of the spins. This spatial variation of the spin phase is achieved by the application of very short gradient pulses between the different RF pulses (in SE imaging) or between RF pulse and gradient inversion (in GE imaging). Since this phase-encoding gradient pulse defines only one single phase step of the imaging matrix, it requires the repetition of the sequence for each phase step according to the resolution in phase direction.

The third spatial axis is called slice select direction. The position of the slice is achieved by the spatial shift of the initial RF pulse, which is a combination of the frequency of the pulse and the amplitude of the slice selection gradient along the third spatial axis. Depending on the frequency shift of the pulse, only a certain slice will resonate since the adjacent structures are affected by RF pulses which do not meet the Larmor frequency.

The different phases and frequencies of the detected signal in an MRI experiment are analyzed by means of a Fourier analysis. The Fourier transformation characterizes the various waves which form the resulting wave of the echo. It resolves both the phases and frequencies as a function of the spatial coordinates and thus assigns the signals to their spatial locations.

1.5.1 Principle

The principle of NMR imaging is to make the external magnetic field $\mathbf{B}_0$ spatially dependent over the volume to be imaged. The resonance frequency f_0, which is proportional to the magnetic field (Larmor equation), is thus likewise made spatially dependent (spatial encoding of f_0). After the excitation of all spins the NMR signal is composed of the signals of the individual spins, each with their own frequency determined by their location in the (position-dependent) magnetic field. A spectral analysis of the NMR signal provides the individual frequencies and corresponding amplitudes of the spins which lead to the reconstruction of the position (frequency) and the intensity (amplitude) of the elements (pixels) of the image. This reconstruction is based on the frequency-space relation derived from the Larmor equation and from the field-space variation implemented in the imager magnet by the manufacturer.

1.5.2 Magnetic Field Gradient

The variation of the amplitude of the magnetic field as a function of spatial position is called the *magnetic field gradient*. In NMR imagers a small linear gradient is added to the magnetic field $\mathbf{B}_0$ along each of the three orthogonal directions x, y, and z to achieve a three-dimensional encoding of the resonance frequency. These gradients are obtained using small currents flowing through appropriate gradient coils mounted in the magnet bore.

1.5.3 Spatial Encoding

1.5.3.1 Slice Selection

The selection of an anatomical slice, say perpendicular to the z direction (axial slice), is achieved with a gradient $G_z = dB_0/dz$ applied in the z direction. The magnetic field thus increases in the z direction (Fig. 1.13). Its value $B(z)$ at a position z is given by:

$$B(z) = B_0 + G_z \cdot z$$

and the corresponding resonance frequency at location z is given by:

$$f_0(z) = \gamma(B_0 + Gz \cdot z)/2\pi. \quad (1.17)$$

For each value of z, the gradient defines a plane perpendicular to the z axis, in which all spins have the same resonance frequency given by Eq. 1.17. In general, the direction of the magnetic field gradient (x, y, z, or arbitrary) determines a plane (sagittal, coronal, axial, or a combination) in which all spins resonate at the same frequency. Excitation of the spin system at the corresponding frequency (Eq. 1.17) yields an NMR signal which originates solely from that particular slice. The ability to generate an image in a plane with arbitrary orientation is a major advantage of the NMR imaging technique.

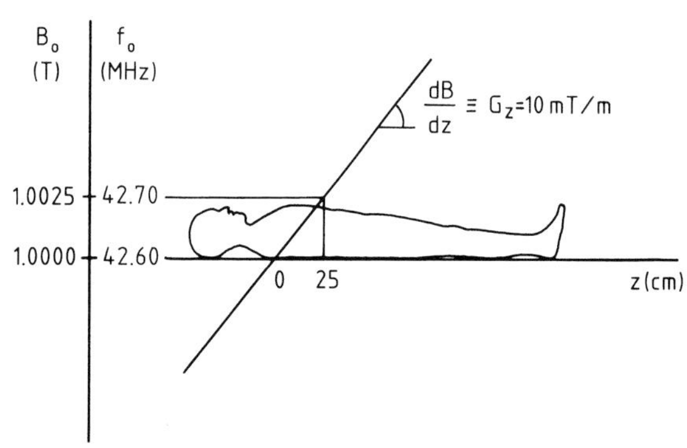

Fig. 1.13. Magnetic field gradient in the z direction and 2D axial slice selection in the z direction. The magnetic field gradient G_z achieves a small linear variation of the magnetic field in the z direction around the center of the magnet. Here, at the center of the magnetic ($z = 0$), $B_0 = 1$ T or $f_0 = 42.60$ MHz. For $G_z = 10$ mT/m, the excitation of the spins of an axial slice at distance $z = 25$ cm requires an excitation frequency $f_0 = 42.70$ MHz (Eq. 1.17)

1.5.3.2 Thickness of the Slice

The thickness Δz of the slice is determined by the value of the gradient G and by the bandwidth Δf_0 of the excitation pulse through an expression similar to Eq. 1.17:

$$\Delta f_0 = \gamma G \Delta z / 2\pi. \quad (1.18)$$

The thickness can be adjusted by the choice of the proper pulse bandwidth in a gradient G, or by the proper gradient for a pulse bandwidth Δf_0.

1.5.3.3 Slice Selective Pulse

The ideal slice selective pulse, in combination with the corresponding gradient, should excite only those spins within the slice to provide a rectangular spatial slice profile. From Fourier analysis it is shown that such profile is obtained for an excitation pulse with frequency f_0 and a time-dependent shape function $\sin(t)/t$, abbreviated sinc(t) (Fig. 1.14). The 90° and 180° excitation pulses of the spin-echo sequence are slice selective and both have a sinc(t) shape.

1.5.3.4 Rephasing Gradient

A slice with thickness Δz can be visualized as comprising a stack of thin parallel slices in which the resonance frequency is the same, though different from that in the neighboring slice. At the end of the excitation pulse the spins of the various thin slices are not in phase since they precess at different frequencies. Their dephasing continues as long as the slice selection gradient is applied. This dephasing can be compensated by inverting the direction of the gradient after the 90° pulse (Fig. 1.15) for a the duration needed to refocus the transverse magnetic moment (the area of the rephasing gradient equals the area of the slice selective gradient calculated from the center of the 90° pulse). For a 180° slice selective pulse, the selection gradient should be equal before and after the pulse. The basic spin-echo sequence adapted for slice selection is shown in Fig. 1.15.

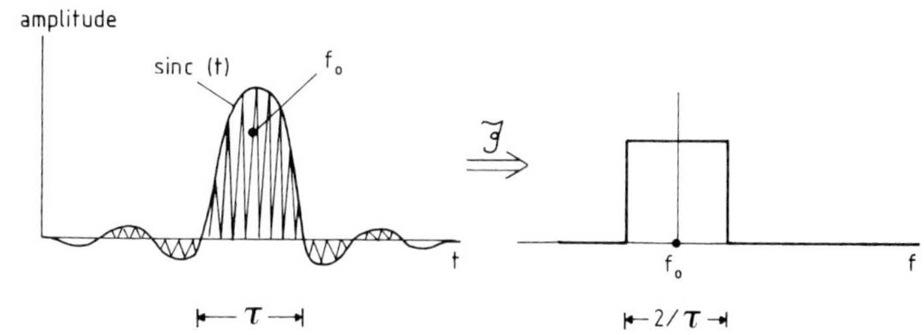

Fig. 1.14. Selective excitation pulse. Excitation spectrum of a pulse of frequency f_0 and shape [sinc(t)]. The pulse is selective since the resulting frequency spectrum around f_0 is rectangular; amplitudes of frequencies outside this window are zero

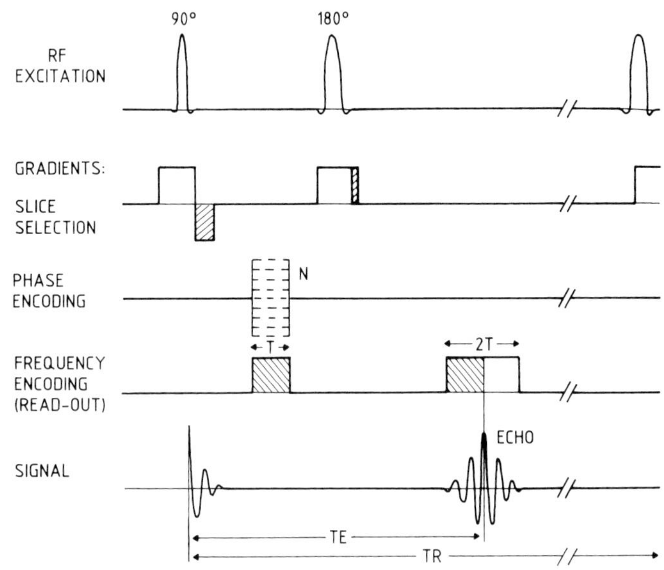

Fig. 1.15. 2D spin-echo imaging sequence. The selective 90° and 180° pulses applied jointly with the slice selection gradient form a spin-echo whose signal originates exclusively from the spins located in the selected slice. Dephasing during the 90° pulse is compensated with the rephasing gradient (*hatched*). Imperfections of the slice selection gradient during the 180° pulse are compensated with the additional gradient (*hatched*). The echo signal is acquired in the presence of the frequency-encoding or read-out gradient, which provides the spatial information along the x axis after a first Fourier transform. Dephasing due the read-out gradient is compensated by the read-out rephasing gradient (*hatched*) applied between the 90° and the 180° pulses. The sequence is repeated N times for N increments of the phase-encoding gradient which modulate the echo signal of the spins as a function of their position along the y axis. A second Fourier transform of the N signals yields the amplitude of the $N \times N$ pixels of the image

1.5.3.5 2D Image Acquisition of the Slice

1.5.3.5.1 Frequency and Phase Encoding. The slice selective sequence excites all the spins in the slice. The reconstruction of the 2D distribution of the spins within the slice makes use of the same principle of spatial frequency encoding with magnetic field gradients, in this case along the x and y directions of the plane, perpendicular to the z axis. It is a double encoding of the *frequency* of the echo signal (x direction) and of the *phase* of the signal (y direction). The 2D information cannot be obtained with one excitation/acquisition but requires N repetitions of the basic excitation sequence, corresponding to N different values of the phase-encoding gradient (for an $N \times N$ image matrix). N discrete samples of each echo signal are acquired in the presence of the frequency-encoding gradient in the x direction (Fig. 1.15). The frequency of the spins in the xy plane thus increases linearly in the x direction and the echo signal is a superposition of signals whose frequency depends on the spatial position along the x axis. The amplitude of the frequency component corresponding to location x along the x axis is proportional to the number of spins along the line perpendicular to the x axis at location x.

Acquisition of spatial information in the y direction, for each x value, requires the application of the phase-encoding gradient in the y direction. This gradient is applied before the read-out gradient, for N different values leading to N excitations of the spin system with the imaging sequence (Fig. 1.15). This gradient imposes a dephasing on the spins, proportional to their position on the y axis. The dephasing is incremented with each of the N increments of the phase-encoding gradient. Spins far away from the origin will be rephased (multiples of 360°) very quickly, while spins close to the origin will be rephased much more slowly. This phase periodicity is thus a function of the position of the spin on the y axis. The phase increment imposed on the spins during each phase-encoding gradient pulse is memorized by the transversal component. The signal corresponding to the individual locations on the x axis is thus phase-modulated by the N increments of the y gradient (pseudo-time). Fourier transformation of these N signals then yields the spin distribution along the y axis and thus the 2D distribution of the spin density over the $N \times N$ pixels.

1.5.3.5.2 Rephasing of the Read-out Gradient. The frequency-encoding gradient applied during the read-out (or sampling) of the echo signal entails a rapid dephasing of the transversal magnetic moment and thus an important signal loss. This dephasing can be compensated by application of a gradient of the same amplitude and reversed polarity before the read-out gradient (Fig. 1.15). If the duration of this gradient is half the duration of the read-out gradient, the spins dephased by this first gradient will be rephased by the read-out gradient at the time TE of the echo maximum. In the basic spin-echo imaging sequence, this first gradient is placed between the 90° and 180° pulses. Since the 180° pulse already inverts the phase of the spins, the dephasing and the read-out gradients have the same polarity. The role of all gradients in the imaging sequence can, of course, be interchanged.

1.5.3.6 Multislice Acquisition

The acquisition of a 2D image composed of $N \times N$ pixels requires the consecutive application of N spin-echo imaging sequences, with a period TR, of the order of the relaxation time T_1, to allow sufficient recovery of the magnetic moment prior to each new excitation. The imaging time is thus $N \times$ TR, or for $N = 256$ and $T_1 = 1$ s, about 4 min. Since the RF and gradient pulses and the signal acquisition do not take more than a few tens of milliseconds, the remainder of the period TR can be filled with the excitation and acquisition of successive parallel slices, in the order:

1st TR period: slice 1 ($N = 1$), slice 2 ($N = 1$), . . . , slice n ($N = 1$),
2nd TR period: slice 1 ($N = 2$), slice 2 ($N = 2$), . . ., slice n ($N = 2$),
. . .
Nth TR period: slice 1 (N), slice 2 (N), . . . , slice n (N).

Depending on the value of TR, between 10 and 20 slices of the volume to be imaged can be acquired within the single-slice imaging time $N \times$ TR.

1.5.4 Three-Dimensional Imaging

Usually, gradients are switched separately and consecutively during the imaging sequence so that the variations in frequency and phase of the signal occur independently from each other. The lines of the matrix (k-space) are excited one by one, each with a different phase step. The frequency encoding of the rows is performed at once by the fixed read-out gradient for each phase line. Therefore, the phase encoding determines the duration of a measurement, whereas frequency encoding does not affect imaging time.

The third dimension, i.e., the slice thickness in 2D imaging, suffers from the nonideal definition of the RF pulse profile. An ideal rectangular spatial profile cannot be achieved and the real shape usually resembles a gaussian curve with shoulders on both sides. These shoulders must be taken into account in the definition of the actual slice thickness. The slice thickness in a 2D MRI experiment is an average value rather than an exact measure. The shoulders of the RF pulse profile create additional excitation in adjacent slices (cross-talk), which can affect their signal behavior. These drawbacks are avoided in 3D imaging.

In a 3D acquisition the in-plane resolution follows the rules of the 2D technique. Whereas in 2D imaging slice thickness is defined by the shape of the RF pulse and the strength of the slice selection gradient, in 3D methods an additional phase-encoding gradient is used to acquire multiple contiguous thin slices. In practice, a broad RF pulse excites a rather thick slab which is partitioned by a second phase-encoding gradient (in the third direction) into thin parallel slices of exact thickness and with hardly any cross-talk. The benefits are the sharp slice definition and the possibility of obtaining slices as thin as 0.6–1 mm with a good S/N. The 3D acquisition enables secondary reconstruction of the images along any orientation within the excited volume.

The additional phase-encoding step along the slice select orientation also affects the measuring time. The imaging time increases linearly with the number of phase steps in either phase direction.

The benefits of 3D techniques with respect to resolution and postprocessing are useful for MRA applications.

The 2D imaging method provides the image of a single slice or of a set of parallel slices acquired consecutively during the TR period. The acquisition of the 2D slices of a volume requires the excitation of contiguous slices (no gap), which presents technical problems. Moreover the acquisition of thin slices is not possible due to the poor S/N. 3D acquisition makes it possible to acquire at once the (3D) signal of all spins in the volume and to reconstruct the stack of contiguous 2D slices (partitions). In the 3D imaging sequence, the slice selection gradient applied during the excitation pulse is replaced with a phase-encoding gradient in the z direction, after the excitation pulse, similar to the phase-encoding gradient in the y direction. The RF pulse thus excites the whole volume rather than a single slice. The read-out (x direction) and phase-encoding gradients (y direction) are identical to those in the 2D sequence (Fig 1.16).

The 3D spatial information is acquired after N increments of the phase-encoding gradient in the y direction for N' increments of the phase-encoding gradient in the z direction. The 3D image acquisition time is thus $N \times N' \times \text{TR}$. For TR = 1 s and $N = N' = 256$, the image acquisition would amount to 1.092 min, which is, of course, prohibitive. 3D acquisition makes use of fast gradient-echo sequences to keep the acquisition time within acceptable clinical examination limits. To reduce the imaging time, the number of phase-encoding steps N' in the z direction, i.e., the number of partitions, is reduced. If the z dimension is reduced accordingly to keep the size of the volume elements (voxels) constant (*isotropic*), the volume imaged is a parallelepipedum rather than a cube. If, however, the z dimension of the volume is kept constant, the z dimension of the voxel is larger than the x and y dimensions and the voxel is *anisotropic*. For TR = 50 ms, the acquisition time for a volume with 32 partitions is $256 \times 256 \times 32 \times 50$ ms = 410 s or 6.8 min.

The 3D spatial distribution of the spins in the volume is reconstructed with a 3D Fourier transformation and can be displayed as a stack of 2D images corresponding to the slicing of the volume in N' partitions, corresponding to the N' increment of the phase-encoding gradient in the z direction. With the available 3D information the volume can also be resliced along any other direction, a major advantage of this technique. Most applications, such as MRA, require isotropic voxel acquisition for the isotropic voxel reconstruction along any direction.

Another major asset of the 3D technique is the availability of contiguous very thin slices (thickness equal to the resolution in the z direction). For equal slice thickness, the S/N of the 3D slice is improved with respect to the S/N of the 2D slice, with a factor equal to the square root of the number of phase-encoding steps N'.

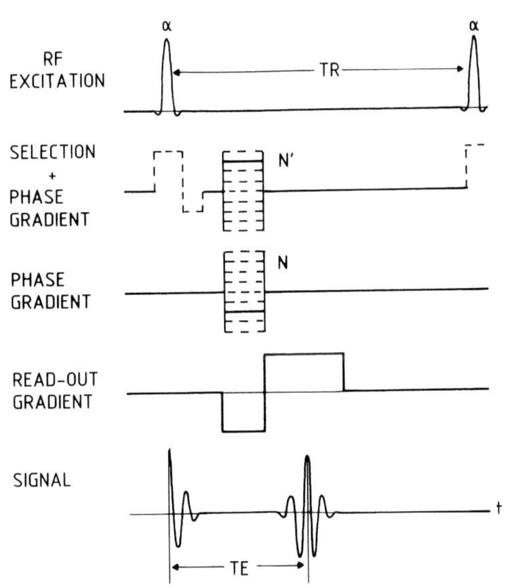

Fig. 1.16. 3D gradient-echo sequence. In the 2D sequence, the slice selective gradient is replaced with a second phase-encoding gradient in the z direction, applied after the excitation pulse. In practice, the RF pulse is selective and excites, jointly with a slice selection gradient, a slab with thickness equal to the z dimension of the volume to avoid a folding in the images of the structures located outside the volume

1.6 NMR Image Acquisition Parameters

1.6.1 Sampling Frequency and Bandwidth

The echo signal contains all the information for the reconstruction of the image. It is acquired as a series of N discrete values, sampled at an interval ΔT (Fig. 1.17), and digitized for processing in the NMR imager computer. The faithful reproduction of all frequency components contained in the echo signal requires that the sampling rate $f_s = 1/\Delta T$ be equal or larger than twice the largest frequency f_{max} which is contained in the NMR echo signal (Nyquist criterium) and which corresponds to the largest spatial dimension to be imaged

$$f_s = 1/\Delta T = 2f_{max}. \tag{1.19}$$

If this criterion is not satisfied, signals corresponding to frequencies larger than $f_s/2$ will appear at the wrong position in the image and give rise to *aliasing* or *folding* artifacts. The receiver chain which amplifies the echo signal should have a bandwidth Δf at least equal to the largest frequency f_{max}.

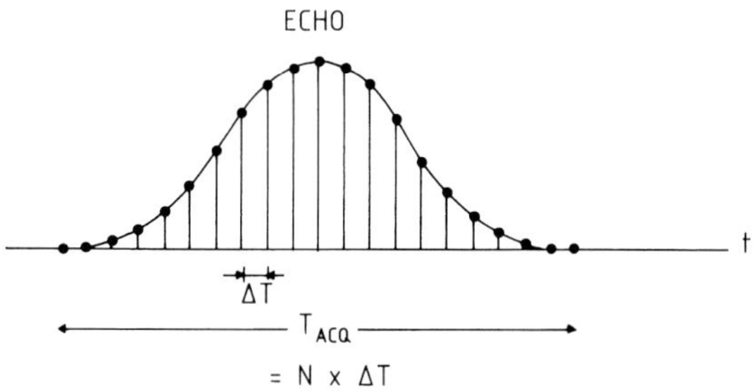

Fig. 1.17. Sampling of the echo signal. The sampling of the analogue echo signal at intervals ΔT provides N discrete values which are digitized in the computer. The sampling frequency $f_s = 1/\Delta T$ must satisfy the Nyquist criterion $f_s=2f_{\max}$, with $f_{\max}$, the largest frequency in the spectrum, corresponding to the largest spatial dimension to be imaged. The echo signal acquisition time is $T_{\mathrm{acq}} = N \cdot \Delta T$

1.6.2 Frequency-Encoding Gradient and Field-of-view

The relation between the frequency-encoding gradient G_x, the largest frequency in the NMR signal, and the corresponding spatial dimension $L_x/2$ can be derived from Eq. 1.17:

$$f_{\max} = \gamma G_x L_x / 4\pi. \tag{1.20}$$

L_x is the dimension of the image or the *field-of-view* (FOV) in the direction of the corresponding gradient. At a fixed sampling rate, the amplitude of the gradient has to be calculated for the desired FOV.

The acquisition of N samples of the echo signal provides, after discrete Fourier transformation, N frequencies or N elements of the image (pixels). The duration of the acquisition of the echo signal is then $T_{\mathrm{acq}} = N \cdot \Delta T$ (Fig. 1.17). Making use of Eq. 1.19, Eq. 1.20 is then written:

$$2\pi = \gamma G_x L_x T_{\mathrm{acq}} / N. \tag{1.21}$$

1.6.3 Phase-Encoding Gradient and Field-of-view

Incrementation of the phase of the echo signal in each of the N cycles is achieved by the N step incrementation of the amplitude of the phase-encoding gradient G_y, in the basic sequence from $-G_y$ to $+G_y$. The discrete phase variation and the second Fourier transform required to convert the phase periodicity into spatial location in the y direction impose a sampling criterion on the phase increment, equivalent to that for the frequency sampling. i.e., at least two samples per period. The gradient increment G_y/N should lead to a maximal phase increment of 180° at the largest spatial dimension L_y, yielding an expression similar to that for the frequency encoding (Eq. 1.21):

$$2\pi = \gamma (G_y/N) L_y T, \tag{1.22}$$

where T, is the duration of the G_y gradient. T is fixed and G_y is calculated to provide the FOV in the y direction.

1.6.4 Spatial Resolution of the Image

The size of a pixel in the x direction is defined as $\Delta x = L_x/N_x$, with L_x the FOV and N_x the number of points in the x direction. The size of the pixel in the y direction is $\Delta_y = L_y/N_y$, with L_y the FOV and N_y the number of points in the y direction.

1.7 Image Quality

1.7.1 The Signal-to-Noise Ratio of the NMR Images

The basic criterion for the quality of the image is the signal-to-noise ratio (S/N), i.e., the ratio of the useful spin signal S to the unwanted random noise signal N which originates from the sample and the electronics. The amplitude S of the signal of a pixel is proportional to the macroscopic magnetic moment of the spins, i.e., to the number of spins in the corresponding volume element (voxel) $V_{\mathrm{vox}} = \Delta x \Delta y \Delta z$ (Fig. 1.18):

$$S \propto \Delta x \Delta y \Delta z, \tag{1.23}$$

with Δz the thickness of the slice and $\Delta x \Delta y$ the size of the pixel in the 2D plane, or (see Sect. 1.6.4):

$$S \propto (L_y/N_y)(L_x/N_x)\Delta z. \tag{1.24}$$

The NMR signal and the noise signal superimposed on that signal have different properties which can be exploited to enhance the S/N. The addition of a number NA of identical (coherent) signals yields a total signal proportional to NA. From signal theory it is known, however, that the addition of NA (incoherent) noise signals yields a total signal proportional to $\sqrt{NA}$. The S/N of the sum of NA noisy NMR signals is then:

$$\mathrm{S/N} \propto NA\sqrt{NA} = \sqrt{NA}. \tag{1.25}$$

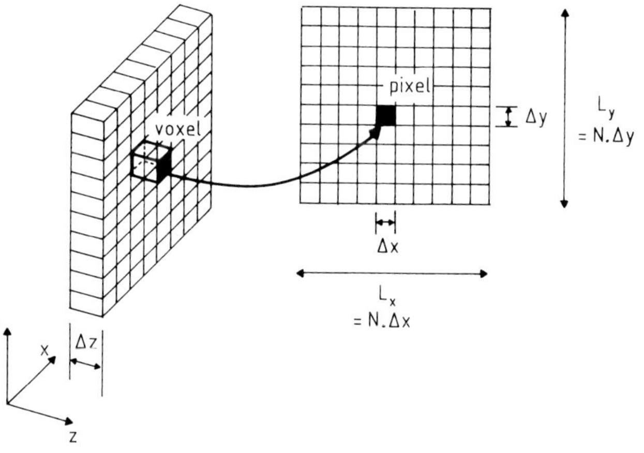

Fig. 1.18. 2D image of slice. The imaging sequence (N excitations) provides a 2D image ($N \times N$ matrix) of the slice with thickness Δz. The dimensions of the field of view are $L_x \cdot L_y$. The dimensions of the image element (pixel) of the corresponding spatial volume element (voxel) of the slice are $\Delta x = L_x/N$ and $\Delta y = L_y/N$

The S/N increases as the square root of the number of accumulations of the same signal, i.e., as the square root of the total imaging time. A twofold improvement in S/N requires a fourfold increase in number of acquisitions, i.e., in imaging time.

Since each additional pixel in the phase-encoding direction of the image matrix requires an additional (phase-encoding) excitation, it implies a longer imaging time accompanied by an increase in S/N. The S/N is thus proportional to the square root of the number of pixels (i.e., the number of excitations) in the phase-encoding direction:

$$S/N \propto \sqrt{N_y}. \qquad (1.26)$$

Finally, the largest frequency of the NMR signal, f_{max}, determines the bandwidth of the receiver chain. The frequency spectrum of the noise is uniform over the bandwidth of the signal; from signal theory it is known that the noise signal N is proportional to the square root of the bandwidth of the signal spectrum, i.e., of the bandwidth of the amplifier:

$$N \propto \sqrt{f}. \qquad (1.27)$$

The expression for the S/N can thus be written as:

$$S/N \propto L_x L_y \Delta z \sqrt{NA}/\sqrt{N_y N_x}. \qquad (1.28)$$

The effects of the choice of the image parameters on the S/N of (the individual pixels of) the NMR image can be calculated from this expression (Table 1.2) and are discussed in the following section.

1.7.2 Influence of the Imaging Parameters on the S/N of the Image

1. *FOV.* If the dimensions L_x and L_y of the FOV are reduced by a factor of 2, for N constant, the pixel area $\Delta x \Delta y$ is reduced by 4. The spatial resolution of the image is improved by a factor of 2 in both directions but the S/N is reduced by 4, which requires a 16-fold acquisition time to restore the original S/N.

2. *Slice thickness.* A reduction by a factor 2 in the thickness Δz of the slice is a twofold reduction of the voxel of the corresponding pixel and a twofold reduction of the S/N. The advantage of a thinner slice is better resolution in the z-direction and the reduction of *partial volume effects* due to the presence of several tissues or structures in the thicker slice.

3. *Number of acquisitions.* The addition of a number NA of identical NMR signals improves the S/N but the method is time consuming. It is also a

Table 1.2. S/N vs matrix size (Eq. 1.24)

$N_y \times N_x$	S/N	$\Delta y \Delta x$ mm² (L = 256 mm)	Relative resolution		Relative acquisition time
			Phase (y)	Frequency (x)	
128 × 128	2.83	2.0 × 2.0	50%	50%	50%
128 × 256	1.41	2.0 × 1.0	50%	100%	50%
192 × 256	1.15	1.3 × 1.0	75%	100%	75%
256 × 256	1.00	1.0 × 1.0	100%	100%	100%
256 × 512	0.50	1.0 × 0.5	100%	200%	100%
512 × 512	0.35	0.5 × 0.5	200%	200%	200%

	T1 SIGNAL	T1 IMAGE	T2 SIGNAL	T2 IMAGE
HYPERACUTE (first hours) (OXYHEMOGLOBIN)	HYPOINTENSE		HYPERINTENSE	
SUBACUTE (hours to 1 week) (DEOXYHEMOGLOBIN)	ISO - or SLIGHTLY HYPOINTENSE		center: areas of HYPOINTENSITY periphery HYPERINTENSE	
SUBACUTE (2nd week) (INTRACELLULAR METHEMOGLOBIN)	center HYPOINTENSE periphery HYPERINTENSE		HYPOINTENSE	
SUBACUTE (3rd week) (EXTRACELLULAR METHEMOGLOBIN)	HYPERINTENSE		HYPERINTENSE	
EARLY CHRONIC (macrophages, neoangiogenesis, hemosiderin)	HYPERINTENSE		center HYPERINTENSE periphery HYPOINTENSE	
LATE CHRONIC (hemosiderin)	HYPOINTENSE		HYPOINTENSE	

Fig. 2.1. Phases of evolution of a hematoma

In the *acute phase* (up to 1 week after hemorrhage), blood demonstrates low to medium signal intensity on T1-weighted images and high signal intensity on T2-weighted images. Sedimentation may occur within an acute hematoma, causing fluid-fluid levels. At this stage, MRI findings are nonspecific, because both abscesses and tumors may have similar appearances.

In the *subacute phase* (more than 1 week after hemorrhage), blood demonstrates high signal intensity on both T1- and T2-weighted images and has similar characteristics for a variable period thereafter. On T1-weighted images hematomas sometimes demonstrate the "concentric ring sign," a central area of medium signal intensity with a periphery composed of two rings. The inner ring demonstrates high signal, which has been ascribed to the presence of the hemoglobin degradation product methemoglobin. The outer ring remains of very low signal intensity on all pulse sequences and is thought to represent hemosiderin within macrophages around the edge of the hematoma. This characteristic appearance may develop any time after 3 weeks following the onset of hemorrhage. Gradually, the central portion of the hematoma assumes uniform high signal intensity on T1-weighted images and becomes indistinguishable from the inner ring.

In the *chronic phase*, as a result of the ingestion of the blood products, hemosiderin is found within the lysosomes of the macrophages. Hemosiderin creates a susceptibility effect that makes the area of hemosiderin black on T2-weighted images (Gomori and Grossman 1986).

2.3 Signal Intensity of Flowing Blood in Spin-Echo and Gradient-Echo MR Images

2.3.1 Nature of Flow

In order to understand flow appearance in MR images, it is important to understand some basic concepts of flow hemodynamics (McDonald 1974; Milnor 1982). Blood is a non-Newtonian fluid, that is, viscosity is not independent of velocity. In the case of static blood, the viscosity tends to increase, whereas in the case of high-velocity flow a reduction in the viscosity occurs. At higher velocities, the flow becomes turbulent where vortices appear and the adjacent layers of fluid tend to mix with each other. The flow profile of a turbulent flow is typically rectangular and it is referred to as a plug flow profile.

At flow velocities of a few centimeters per second a laminar flow results, where adjacent layers of fluid glide past each other without mixing. Laminar flow is typically seen in small arteries and veins. However, due to vascular pulsation and to the size and shape of major arteries, a plug flow profile, rather than a laminar one, is produced. In plug flow, velocities are uniform across the vessel diameter, in contrast to the parabolic profile typical of laminar flow (Fig. 2.2).

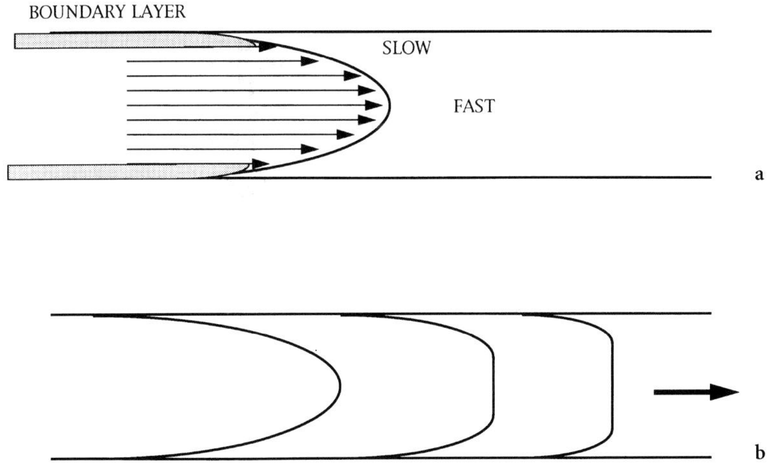

Fig. 2.2. **a** Flow of constant velocity in a vessel produces a parabolic flow profile, with fastest flow toward the middle of the vessel. **b** A rapid plug flow produces a flat flow profile, where all the fluid elements move at the same velocity

The transition between laminar and turbulent flow is determined by several factors, lumped together into a unitless number called the Reynolds number (Re):

$$\mathrm{Re} = (\rho \times v_{av} \times R/\eta),$$

where ρ = density, v_{av} = mean flow velocity, R = radius of the vessel, and η = viscosity of the fluid. The Reynolds number determines the threshold at which laminar flow ceases and is replaced by turbulence. At $\mathrm{Re} \geqslant 2000$, the flow is turbulent. Turbulent flow is very important because it causes a signal loss. It is typical of the vessel tract distal to a stenosis, where a signal loss artifact can be produced (HARRISON and MARSHALL 1983).

2.3.2 Inflow Phenomena

Saturation or time-of-flight effects are due to changes in longitudinal magnetization, which are a function of applied radiofrequency pulses [spin-echo (SE) and gradient-echo (GE)] and T1 relaxation time. These phenomena are the basis of the well-known effects of flow on signal intensity observed in conventional MR images, i.e., flow-related enhancement or paradoxical enhancement (MILLS et al. 1984; BRADLEY and WALUCH 1984; KAUFMAN et al. 1983) and flow void or high-velocity signal loss or wash-out effect (BRADLEY and WALUCH 1984). The occurrence of each of these phenomena depends on the course of a vessel within the imaging plane (they are more evident if the vessel lies perpendicular to the plane) and on the blood flow velocity. If blood moves so fast as to be outside of the imaging plane in the time between the 90° and the 180° pulses, high-velocity signal loss will occur; if the blood flow is slow enough that a part of the blood is replaced by fully magnetized protons coming from the outside of the imaged volume, flow-related enhancement will result (Fig. 2.3).

2.3.3 Spin-Phase Phenomena

Spin-phase phenomena were first studied by Carr and Purcell in 1954. In 1961 Hahn used flow-induced phase shifts to measure motion of seawater. Phase shift techniques are based on the accrued phase shift of spins moving along a magnetic field gradient (SINGER 1978; VON SCHVLTHESS and HIGGINS 1985; WEHRLI et al. 1984; VALK et al. 1985).

The 90° RF pulse gives rise to a well-known phenomenon: it localizes all the spins in the same phase or, put more precisely, in phase coherence. After a while the phase coherence starts to be lost because of spin-spin interactions and magnetic field inhomogeneities.

Any kind of magnetic field inhomogeneity produces phase dispersion and therefore loss of signal. The linear field gradients used for imaging are an obvious source of inhomogeneities and are responsible for the phase effects related to flow. Upon adding a magnetic field linear gradient, a local increase in resonance frequency takes place. Each spin accumulates a phase shift according to the following equation:

$$\phi = \gamma G x T,$$

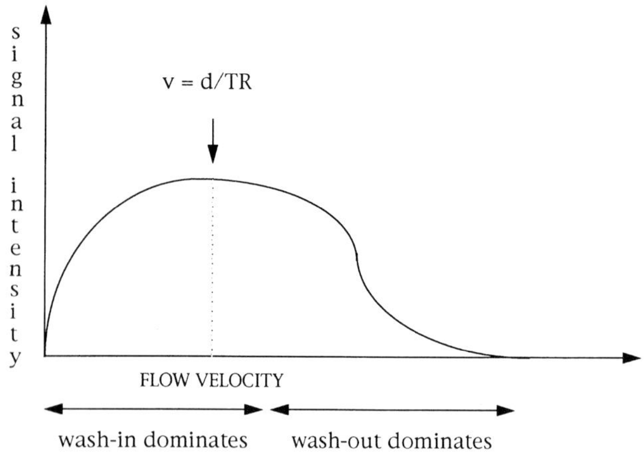

Fig. 2.3. Signal modulation in SE images depends on flow velocity

where: γ = gyromagnetic ratio, G = gradient amplitude, x = position of the spin along the axis of the gradient, and T = gradient duration.

When the gradient is turned off, protons start to spin again at the usual frequency, though maintaining their phase shift (Fig. 2.4). After a while a second identical gradient, but with opposite direction, is applied, causing an opposite phase shift of the protons. As a consequence, stationary spins are submitted to a bipolar gradient with the same strength and duration but with opposite direction. Therefore, the net phase shift at the end of the application of the bipolar gradients will be zero. Something different happens with flowing blood spins. In fact, in the interval between the initial excitation with the RF pulse and the following gradient, flowing spins move linearly along the gradient. Therefore, when the second gradient is turned on they are located in a place where, because of the presence of the gradient, there is a relatively higher magnetic field. For this reason, moving spins start precessing at a frequency higher than stationary tissues (Fig. 2.5). The phase shift of moving protons can be calculated according to the following equation:

$$\phi = \gamma G v T \tau,$$

where: γ = gyromagnetic ratio, G = gradient amplitude, v = velocity of moving spin, and T = gradient duration. This phase difference is maintained by turning the gradient off: we can consequently talk of movement-related phase shift (Fig. 2.6).

The phase shift is measured in degrees in respect to zero, which corresponds to the protons' phase shift after the original RF pulse. As we have seen before, the phase shift can be related and is proportional to the intensity of the gradient and to the increased protons' position in the interval between the RF pulse and the onset of the gradient.

The phase shift phenomenon due to different velocities and acceleration of spins inside the voxel produces a signal loss in MR images because of the

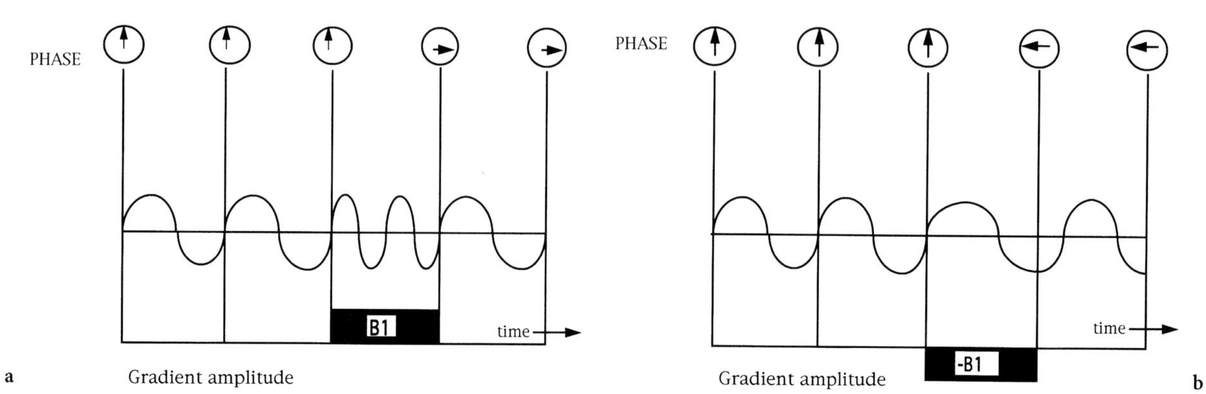

Fig. 2.4. **a** Influence of a positive gradient on phase. When the gradient is applied, a higher (B_1) magnetic field is present. The resonance frequency increases for the time the gradient is on. When the gradient is turned off, the result is a more anterior position of the spins (a positive phase shift). **b** Influence of a negative gradient on phase. The effect is the same but opposite, and a negative phase shift results

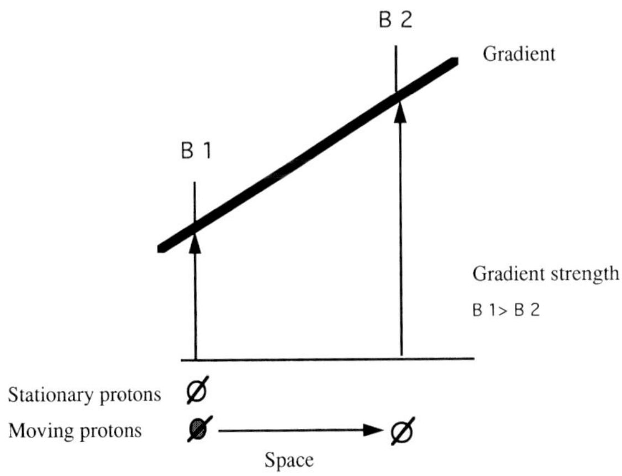

Fig. 2.5. Magnetic field gradient with higher intensity for moving spins

destructive interference of the magnetization vectors of precessing spins. This phenomenon, called intravoxel phase dispersion, will be discussed below.

2.3.4 *Signal Intensity of Flowing Blood*

In conventional MR images blood signal intensity can be either high or low, according to different phenomena. In this section these phenomena are discussed in turn.

2.3.4.1 Black Blood

2.3.4.1.1 Flow Void due to Wash-out Effect. A wash-out effect typically produces a signal void from the vessels. It occurs during the acquisition of an SE sequence (Axel 1984; von Schulthess 1992). During an SE sequence, if flow is fast, spins submitted to the 90° pulse will have already left the slice at the moment of the 180° pulse, by which they are not influenced. Protons that in the meanwhile have reached the slice will not have been submitted to the 90° pulse and therefore the 180° pulse has no effect on them: the vessel will present no signal (signal void). Such images are called "black blood images" and are specifically designed to make flowing blood appear dark (Fig. 2.7).

Flow void occurs if blood flow velocity exceeds a critical value according to the following equation:

$$v > s/T,$$

where: v = velocity, s = slice thickness, and T = time at which the rephasing pulse 180° occurs, which equals TE/2 for the first echo. Knowing the TE employed in the sequence and the slice thickness, the maximum velocity of the blood spins to be submitted to both of the pulses can be calculated.

The relationship between signal intensity, slice thickness, and flow velocity can be expressed as follows:

$$\mathrm{SI} = 1 - (\mathrm{TE}/2 \times v \times 1/d),$$

where: v = velocity of blood flow perpendicular to the imaging plane and d = slice thickness.

Of course, in SE sequence there is always a balance between inflow and wash-out effects. The occurrence of either paradoxical enhancement or flow void depends on flow velocity, the sequence acquired, the slice thickness, the position of the slice in the stack, and the order of the acquisition in multislice sequences. Wash-out effects are reduced and vascular signal is enhanced by increasing the slice thickness, shortening the TE, or using diastolic gating to minimize flow velocity. On the other hand, the combination of flow presaturation and thin slices maximizes flow void.

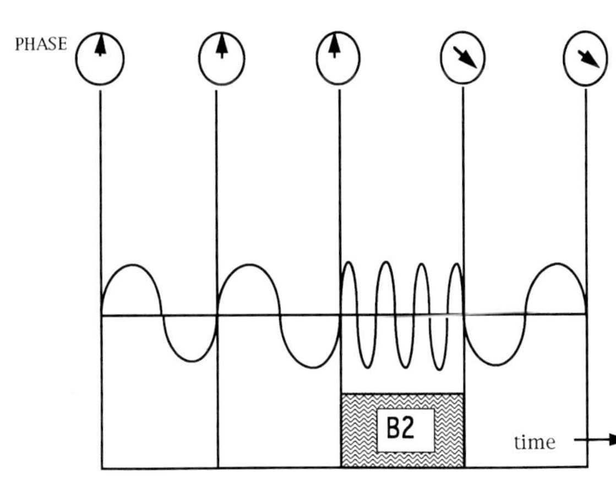

Fig. 2.6. Due to movement in the space, flowing spins are subjected to a higher magnetic field (B2 > B1). Therefore, the increase in phase is higher, with a resulting phase shift versus the stationary spins

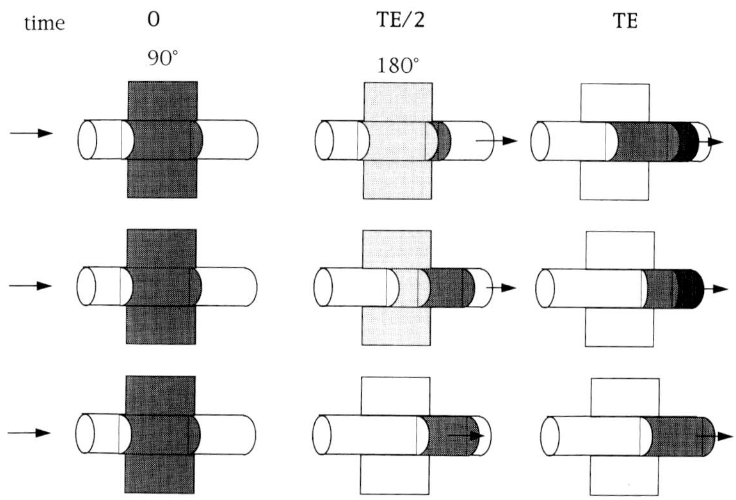

Fig. 2.7. Flow void phenomenon in SE images

2.3.4.1.2 Presaturation Technique. Presaturation is a method available with commercial MRI equipment which produces a low vascular signal (Felmlee and Ehman 1987; Edelman et al. 1988). Increased signal secondary to flow-related enhancement (FRE) can hinder interpretation of MR images. FRE is caused by the inflow of fresh spins of unsaturated protons. The saturation of these protons can be used to suppress the high signal intensity. Several techniques are available to saturate protons. All are based on additional RF pulses, usually 90°, though application of larger or smaller flip angles before the beginning of either an SE or a GE sequence can be highly effective. The presaturation pulse is applied outside the imaging volume, so as not to affect spins generating MR images (Figs. 2.8, 2.9).

Although presaturation eliminates FRE and ghosting artifacts, if blood flow is very slow and blood remains inside the imaging volume for a long time, remagnetization can occur and a signal can be produced in the inner slices. This problem is especially relevant for vessels with slow flow velocity, such as veins, but it can also occur in arteries due to slow flow during diastole.

2.3.4.1.3 Phase Dispersion. Intravoxel phase dispersion is due to motion-induced phase shift phenomena secondary to frictional forces between blood spins and the vessel wall (von Schulthess 1992). A gradient of velocity is created inside the voxel, resulting in different phase accumulation for different spins. As a consequence signals coming from different spins tend to cancel each other and the result is a decreased signal intensity from the voxel (Fig. 2.10). The phenomenon is more pronounced along the vessel wall, producing a typical appearance in a vessel cross-section called the "donut sign." In order to obtain a greater coherence inside the voxel, it is necessary to have a smaller number of spins, with reduction of voxel inhomogeneities. Therefore, by using a smaller voxel, a lasting coherence can be obtained, giving rise to a greater relative signal than with a

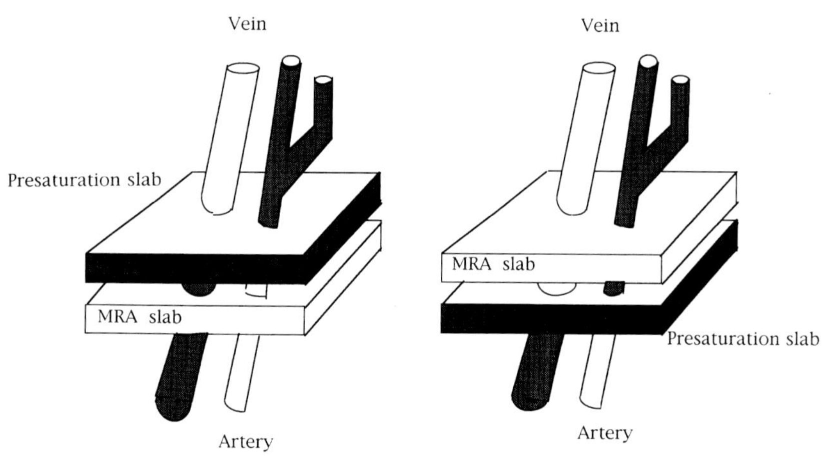

Fig. 2.8. Presaturation above the imaging volume (*left*) eliminates venous signal. Presaturation below the imaging volume (*right*) eliminates arterial signal

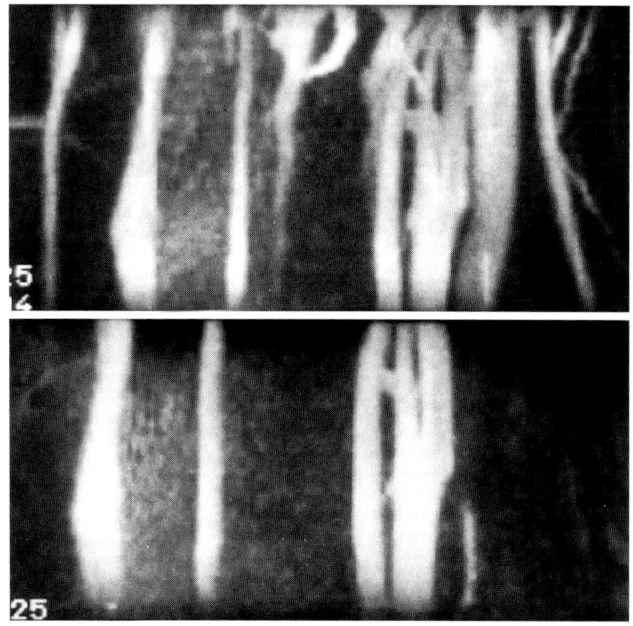

Fig. 2.9. MRA images acquired without (**a**) and with (**b**) presaturation pulses. The carotid and vertebral arteries are well defined only after cancellation of the venous structures

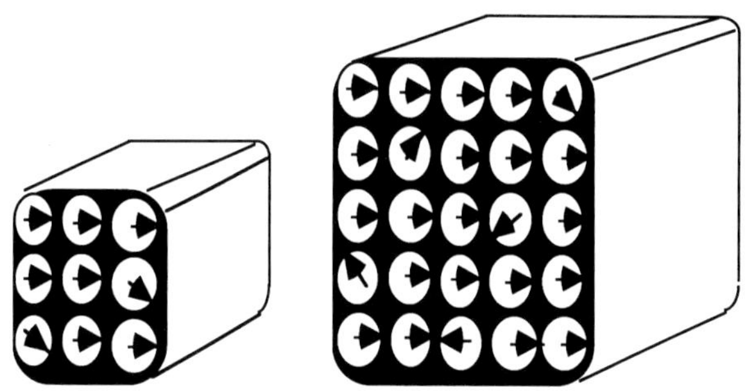

Fig. 2.10. Intravoxel phase dispersion

bigger voxel, where dephasing is easier and signal loss is more evident. Therefore, it is necessary to apply the smallest possible slice thickness and maximal matrices.

2.3.4.2 Bright Blood

2.3.4.2.1 Flow-Related Enhancement (FRE). FRE or paradoxical enhancement is a common phenomenon during an MR study (Mills et al. 1984; Bradley and Waluch 1984; Kaufman et al. 1983). Basically it is due to the difference in saturation between protons from stationary tissues and flowing blood, that is moving protons. During the image acquisition stationary spins are submitted to repeated RF pulses. If the time interval between two pulses (TR) is long, stationary tissues will have enough time to recover longitudinal magnetization and therefore will produce a high signal. If TR is very short, as during pulse sequences used for MRA, the protons of the stationary tissues will recover less and less longitudinal magnetization. Therefore signal will be progressively lost and the tissues will be saturated: no signal derives from protons of the stationary tissues (Fig. 2.11). In the case of blood flow, when spins come into the imaging volume, they will not have been submitted to previous excitations and they can be considered as fresh spins. Therefore they produce a large signal. Since these spins are continuously replaced by new fresh protons flowing into the imaging volume, they always have a high signal compared with stationary tissues, whose saturation is variable, depending on their specific relaxation time and the TR used (Fig. 2.12). This effect is known as flow-related enhancement, since flow seems to produce a vascular enhancement similar to that seen

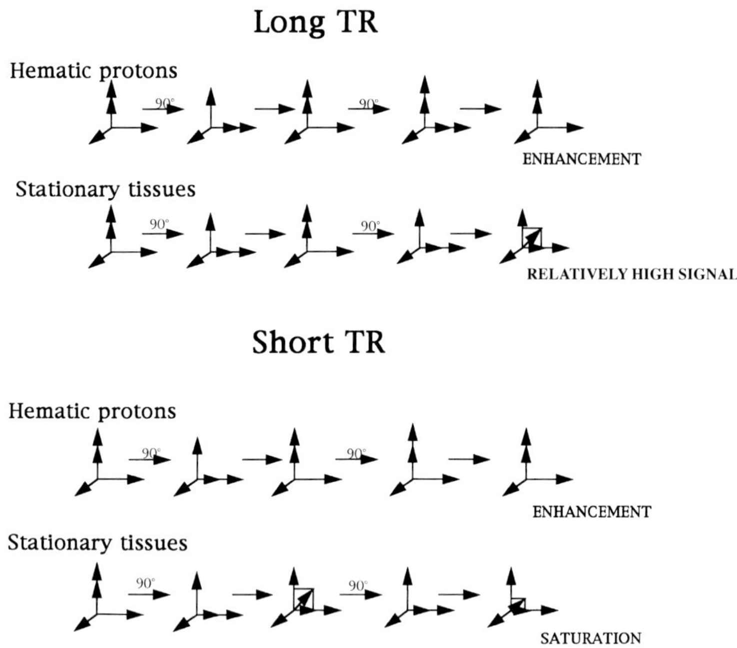

Fig. 2.11. Different recovery of longitudinal magnetization of tissues according to different TR. With a long TR stationary tissues can recover almost completely the longitudinal magnetization, whereas with a short TR saturation occurs. On the other hand, hematic protons, that is moving protons, always display the maximum signal ("enhancement")

following the administration of an intravenous contrast agent.

Flow-related enhancement is much more significant during a single-slice imaging acquisition, since blood flows out of the imaging volume in a very short time, and the signal intensity will be constantly high. In multislice imaging acquisition, flow-related enhancement will be evident in several slices, but a signal loss will occur in the last slices due to saturation of the flowing spins. Therefore, progressive saturation of blood protons can be observed in the case of vessels lying in the same plane of the imaging section, due to the longer period that blood remains within the imaging volume (Fig. 2.13).

Finally, the high signal intensity of blood during acquisition of a GE sequence is due to the entry slice effect (Fig. 2.14). This is because the short TR in GE does not permit the longitudinal magnetization to return to values close to its maximum; rather it is only a small fraction thereof when the next pulse is applied. If a substantial fraction of blood is replaced during the TR of the sequence, entry slice effects can easily lead to a very strong signal increase, and because high-velocity signal losses do not occur (there is no 180° pulse), this signal increase is evident even at high velocities and is not counterbalanced; as a result intravascular signal will be very bright.

The high signal intensity on GE images (Mills et al. 1984) is also due to the fact that even if the spins, tilted by the dephasing gradient (≤90° flip angle), have left the slice, they are submitted to the inversion of the gradient (rephasing gradient); in fact the dephasing and rephasing gradients are applied to

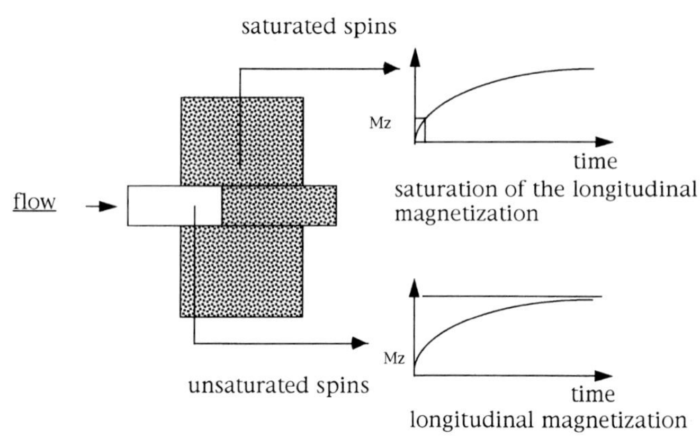

Fig. 2.12. Inflow phenomenon: fresh unsaturated blood spins present maximal signal intensity

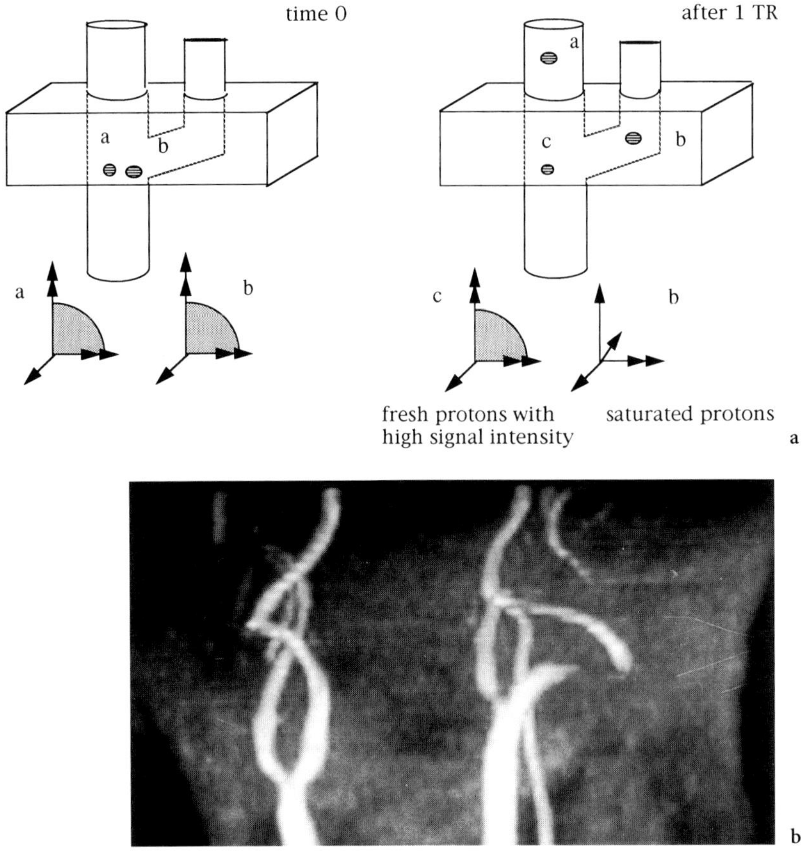

Fig. 2.13. a In-plane saturation; this can occur when the vessels have a course parallel to the slice. **b** Due to the horizontal course of the internal carotid artery, cancellation of the signal occurs

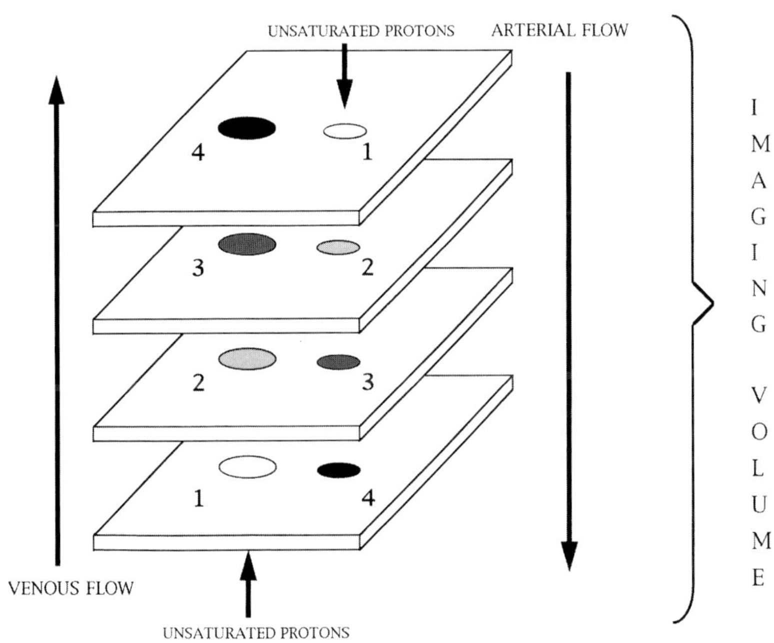

Fig. 2.14. Flowing protons entering the imaging volume are unsaturated; therefore a high signal intensity results. A progressive saturation in the volume occurs, with a decrease in signal intensity

the entire volume of examination and not only to the slice. Therefore signal is received even if the spins have left the slice under examination.

2.3.4.2.2 Gradient Refocusing. This technique is also called flow compensation (PATTANY et al. 1987; HAACKE and LENZ 1987; EHMAN and FELMLEE 1990). It is used in order to reduce phase shift phenomena caused by motion and to avoid intravascular phase dispersion. A series of repeated and inverted gradients are applied; the effect is a phase shift which is equal but of opposite sign to that produced by blood flow (Fig. 2.15). In this way signal from blood vessels is reobtained. This technique has been very successful in reducing flow artifacts in GE images, but it is less useful for eliminating flow artifacts in SE images. In fact in an SE sequence, mechanisms of wash-in and wash-out are responsible for flow artifacts, in addition to phase dispersion, and they are not addressed by gradient moment nulling. In some cases a paradoxical increase in artifacts can be produced by using gradient moment nulling during an SE acquisition, probably due to the increased peak vascular signal generated by the elimination of intravoxel phase dispersion.

2.3.4.2.3 Even-echo Rephasing. In special cases, such as constant-velocity motion of blood flow, so-called even-echo rephasing (EER) can occur for flow along the direction of a symmetric gradient field, such as the read-out gradient (BRADLEY and WALUCH 1984; ALVAREZ and HYMAN 1986). In a multiecho sequence, phase dispersion is reduced on the even (second, fourth, etc.) echoes as compared with the odd (first, third, etc.) echoes. For even echoes, the phase change is zero for both stationary and moving spins, resulting in an absolute increase in flow signal.

It is very uncommon to observe real EER during real pulse sequences. This is due to the arterial flow, which does not have a constant velocity, and phase dispersion generated by acceleration and turbulence is not canceled on even echoes. Signal loss is also due to the wash-out phenomenon, which is not altered by EER. Finally, EER occurs only for symmetric gradients, which are rarely used in real pulse sequences.

2.3.4.2.4 Diastolic Pseudogating. Diastolic pseudogating is a flow artifact that results in arterial high signal intensity under particular conditions. It is explained as follows (VON SCHULTHESS 1992): during a cardiac cycle, blood flow, due to its pulsatile nature, is alternatively rapid (during systole) and slow (during diastole). Since the TR is independent of the patient R-R interval, data are acquired randomly during the cardiac cycle. An occasional synchronization of MR images with diastolic phase can occur; during diastolic phase flow velocity is very slow and therefore a high signal intensity will be produced even if the vessel has a high peak systolic velocity.

2.3.4.2.5 Coincidental Phase Cancellation. Coincidental phase cancellation is an uncommon artifact caused by cancellation of the phase shifts due to frequency-encoding and slice-selection gradients (VON SCHULTHESS 1992). Sometimes with a particular orientation of blood vessels it can happen that one gradient will produce a certain positive phase shift while another will produce a negative one of the same degree. As a result, the signal intensity of the vessel will be quite similar to the background, and the blood vessel tends to disappear.

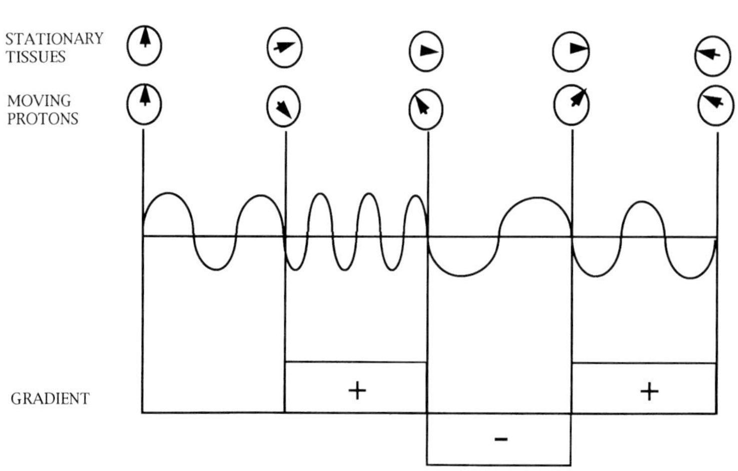

Fig. 2.15. Gradient refocusing. The phase shift between stationary tissues and flowing spins is cancelled when particular gradient-wave forms are employed

2.4 Flow Artifacts

2.4.1 Displacement

The position of a blood vessel inside the imaging volume is determined by the frequency- and phase-encoding gradients, as for any other structure. However, unlike other structures, blood flow can move in the time between excitation by the gradients and read-out. This movement causes a displacement of the vessel from the true position and is dependent on the flow velocity, direction, and pulse sequence. The displacement along the frequency-encoding axis (Δx) and phase-encoding axis (Δy) is expressed as:

$$\Delta x = v \times \mathrm{TE};\ \Delta y = v \times t,$$

where: t = time delay between excitation and phase encoding. Obviously, the higher the TE, the larger the displacement.

Another cause of displacement is the fact that phase-encoding and frequency-encoding gradients are not contemporary; rather there is a slight delay between them. As a result, after the phase-encode position has been determined, the spin moves along the frequency-encoding axis, leading to a relative displacement (Barth et al. 1985).

2.4.2 Ghost Artifacts

Slight variation in flow velocity over time causes superimposition on the magnetization, which varies sinusoidally because of the effects of phase encoding. The result is signal scattered along a line in the phase-encoding direction containing the voxel with

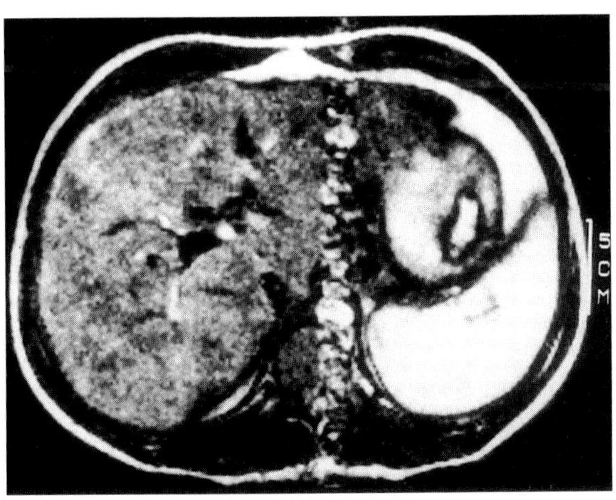

Fig. 2.16. Ghost artifact

the slightly varying magnetization. This phenomenon is known as ghosting (Wood and Henkelman 1985; Perman et al. 1986) (Fig. 2.16). The higher the intensity of the vascular signal and the flow velocity variation, the worse is the ghosting. This artifact can be successfully suppressed by using cardiac gating, flow compensation, or presaturation, the last-mentioned being most effective. Increasing the number of signal averages reduces ghost artifact according to the following formula:

$$\text{reduction of ghost artifact} \propto \sqrt{\mathrm{NEX}}.$$

References

Alvarez O, Hyman RA (1986) Even echo rephasing in the diagnosis of giant intracranial aneurysm. J Comput Assist Tomogr 10:699–701

Axel L (1984) Blood flow effects in magnetic resonance imaging AJR 143:1157–1166

Barth K, Deimling M, Friyschy P (1985) Visualization and measurement of flow with MRI. Biomed Tech 30:12–17

Bradley WG, Waluch V (1984) MR even echo rephasing in slow laminar flow. J Comput Assist Tomogr 8:594–598

Edelman RR, Atkinson DJ, Silver MS (1988) FRODO pulses: a new method for elimination of motion, flow, and wraparound artifact. Radiology 166:231–236

Ehman RL, Felmlee JP (1990) Flow artifacts reduction in MRI: a review of the roles of gradient moment nulling and spatial presaturation. Magn Reson Med 14:293–307

Felmlee JP, Ehman RL (1987) Spatial presaturation: a method for suppressing flow artifacts and improving depiction of vascular anatomy in MR imaging. Radiology 164:559

Friedman GB, Sandhu HS (1981) NMR in blood. Magn Res Rev 6:247–307

Gomori JM, Grossman RI (1986) Mechanisms responsible for the MR appearance and evolution of intracranial hemorrhage. Radiology 161(P):364

Gomori JM, Grossman RI, Goldberg HI et al. (1985) High field magnetic resonance imaging of intracranial hematomas. Radiology 157:87–93

Haacke EM, Lenz GW (1987) Improving MR image quality in the presence of motion by using rephasing gradients. AJR 148:1251

Harrison MJG, Marshall J (1983) Does the geometry of the carotid bifurcation affect its predisposition to atheroma? Stroke 14:117

Kaufman L, Crooks LE, Sheldon P, Hricak H, Herfkens R, Bank W (1983) The potential impact of nuclear magnetic resonance imaging on cardiovascular diagnosis. Circulation 62:251–257

McDonald DA (1974) Blood flow in arteries, 2nd edn. Williams and Wilkins, Baltimore

Mills CM, Brant-Zawadski M, Crooks LE et al. (1984) Nuclear magnetic resonance – principles of blood flow imaging. AJR 142:165–170

Milnor WM (1982) Hemodynamics. Williams and Wilkins, Baltimore

Pattany PM, Phillips JJ, Chiu LC, Lipcamon JD, Duerk JL, McNally JM, Mohapatra SN (1987) Motion artifact suppression technique (MAST) for MR imaging. J Comput Assist Tomogr 11:369–377

Perman WH, Moran PR, Moran RA, Bernstein MA (1986) Artifacts from pulsatile flow in MR imaging. J Comput Assist Tomogr 10:473–483

Singer JR (1978) NMR diffusion and flow measurements and an introduction to spin-phase graphing. J Physiol 11:281–291

Valk PE, Crooks LE, Kaufmann L, Roos MS, Orthendahl DA, Higgins CB (1985) Blood flow in magnetic resonance imaging: correlation of image appearances with spin echo phase shift and signal intensity. AJR 146:931–939

von Schulthess GK (1992) Blood flow. In: Higgins CB, Hricak H, Helms CA (eds) Magnetic resonance imaging of the body. Raven Press, New York, pp 313–337

von Schulthess GK, Higgins CB (1985) Blood flow imaging with MR: spin-phase phenomenon. Radiology 157:687–695

Wehrli FW (1988) Principles of magnetic resonance. In: Stark DD, Bradley WG (eds) Magnetic resonance imaging. Mosby, St. Louis, pp 3–23

Wehrli FW, MacFall JR, Axel L et al. (1984) Approaches to in-plane and out-of-plane flow imaging. Noninvas Med Imaging 1:127–136

Wood ML, Henkelman RM (1985) MR image artifacts from periodic motion. Med Phys 12:143–151

3 Imaging Techniques of Magnetic Resonance Angiography

R. HAUSMANN

CONTENTS

3.1 2D and 3D Time-of-flight

Visualization of flowing blood with MR is possible by means of two different methods, i.e., time-of-flight (TOF, also called inflow) and phase-contrast (PC, also called phase-sensitive) techniques. TOF in its two- or three-dimensional mode is very robust, fast, and easy to apply. This chapter will describe the basics of TOF together with the techniques employed in its realization. Shortcomings and possible artifacts will be discussed, and advice will be given on how to avoid misinterpretations by the proper choice of parameters and acquisition modes.

3.1.1 Principles

The movement of blood is the basis of MRA, the visualization of blood vessels with the aid of MR without use of contrast agents (MILLS et al. 1983). Within the last 4 years, this technique has developed towards a new application of MR which allows the generation of three-dimensional angiograms of arbitrary regions of the body within reasonable scan times (MASARYK et al. 1989).

Two basic effects can be used with MR to visualize flowing spins (i.e., blood or even CSF): *amplitude* and *phase* effects. The first effect is used for the TOF methods (AXEL 1984) whereas the second is used in PC methods (MORAN 1982). While most TOF techniques work with only one acquisition, PC techniques always require more than two (and up to six) acquisitions to sensitize the sequence to more than one flow direction and to suppress stationary background signal. This can lead to significant differences in scan time between these techniques.

All TOF techniques use the same idea of inflow of labeled spins into the imaging volume. This effect was first described by SINGER (1959). In the simplest form, the labeling is done passively by allowing unsaturated (or fully relaxed) spins to enter the excitation volume. In this way, full magnetization is available for the blood volume and gives rise to high signal intensity. The contrast between these labeled (or tagged) spins and the stationary background tissue is then increased further by heavy suppression of background using short repetition times. Other forms of labeling are inversion of longitudinal (parallel the main magnetic field) magnetization (NISHIMURA et al. 1988) and saturation of inflowing spins (DUMOULIN et al. 1989a), but these are not widely used.

The TOF method can be implemented in different modes, using 3D single-volume (LAUB and KAISER 1988; RUGGIERI et al. 1989), 2D sequential (GULLBERG et al. 1987; KELLER et al. 1989; EDELMAN et al. 1989a), and 3D multivolume techniques (PARKER et al. 1991). Each of these has its own benefits and drawbacks depending on the flow velocities and the size of the acquisition volume. However, for almost all clinical needs an optimized parameter setting exists for one of these three techniques that will allow reliable depiction of vessels.

3.1.2 Flow Sensitivity

3.1.2.1 Phase Effects

Although the TOF method is based on amplitude effects caused by the differences in longitudinal mag-

R. HAUSMANN, PhD, Siemens AG, Medizinische Technik, Henkestraße 127, 91052 Erlangen, FRG netization, it is important to understand the phase effects for this technique as well. Phase effects can cause severe artifacts in TOF MRA, if these are not properly avoided. Pulsation and accelerated blood flow are the principal reasons for artifacts. Phase effects are based on the fact that spins moving along magnetic field gradients acquire a phase shift in the transverse magnetization, which is directly related to the velocity of the flowing spins (CARR and PURCELL 1954).

The phase shift introduced in the transverse magnetization by additional gradient pulses is explained in Chap. 2. Basically it can be divided into a stationary phase term, which is usually rephased to form the echo at the echo time TE, a phase term originating from motion with constant velocity, and other higher order terms caused by accelerated motion, irregular flow, etc. In order to gain maximum signal for flowing blood, ideally all higher order terms proportional to velocity, acceleration, etc. should be minimized. Most importantly, the velocity term must be nulled at the echo time. This is achieved by so-called gradient moment rephasing (GMR) or flow compensation (HAACKE and LENZ 1987). Figure 3.1 shows a typical gradient rephasing scheme for the read-out gradient in comparison with a non-compensated pulse form. The requirements for gradient performance (especially rise times) increase with increasing motion compensation, since echo time should remain short for TOF MRA. These hardware issues will be discussed in Chap. 6.

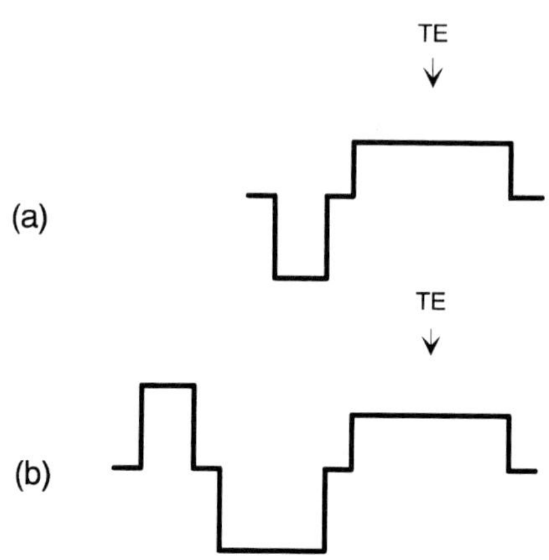

Fig. 3.1a,b. Gradient moment rephasing. **a** With two lobes stationary phase components can be nulled at the echo time **b** Upon adding an additional bipolar gradient wave form, velocity phase components (first-order flow compensation) are canceled It is obvious that nulling more and more phase components of higher order increases the minimal possible echo time TE. Therefore, in most of the TOF applications, only first-order flow compensation is used to achieve short echo times and minimize intravoxel phase dispersion

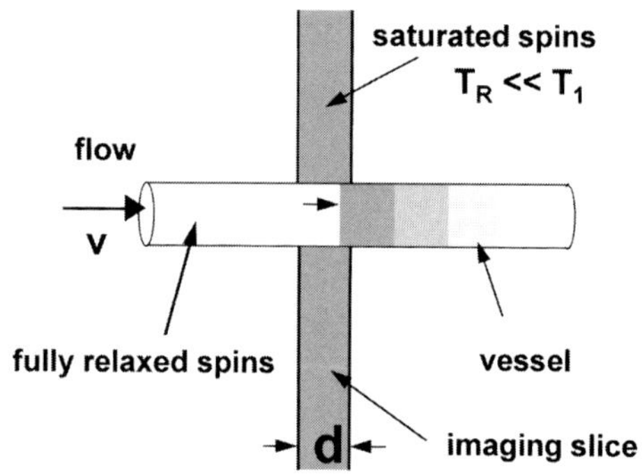

Fig. 3.2. TOF technique: Unsaturated spins with full longitudinal magnetization M_0 flow into the imaging slice and result in maximum signal. Stationary spins become saturated as long as T_R is small compared to T_1. Complete refreshment of spins is achieved when the flow velocity is at least $v = d/\mathrm{TR}$

All of these uncompensated phase effects lead to signal cancellations. Basically, if higher orders of motion are not rephased at the echo time, different phase behaviors will be present within one volume element (voxel) and can lead to diminished signal or even complete signal cancellation. Such effects can be present in poststenotic regions or at bifurcations of vessels. This has to be carefully taken into consideration when diagnosing stenosis with TOF MRA.

3.1.2.2 Amplitude Effects

3.1.2.2.1 Contrast Generation in TOF MRA. The dominant effect on which TOF MRA is based is an amplitude effect related to the longitudinal magnetization. The basic idea is straightforward: The contrast between blood vessels and background tissue is based on the contrast between high signal originating from unsaturated spins flowing into the excitation region (slice in 2D techniques or volume in 3D techniques) and the low signal (and contrast) in stationary tissue. Figure 3.2 illustrates the contrast mechanism. To ensure significant saturation effects within the stationary tissue, the repetition time TR must be small compared to the T1 relaxation times of stationary material. This can be achieved with gradient-echo pulse sequences. With short TR times of 40 ms, spins have insufficient time to recover between successive excitation pulses. This leads to low signal within the stationary tissue. The situation is different for spins flowing through the imaging slice: if the flow velocities are high enough and/or the slices are thin enough, the affected spins "feel" only a few excitation pulses. For a complete refreshment of spins inside the imaging slice, the velocity has to be at least $v = d/TR$, where d is the slice thickness. For a realistic TR of 40 ms and slice thickness d of 2 mm, the refreshment velocity is only 5 cm/s. Therefore, and due to the partial flip angle of the excitation pulse, flowing spins contribute with their full longitudinal magnetization to the signal. This mechanism leads to high contrast between flowing blood and stationary background signal and is often referred to as flow-related enhancement or even paradoxical enhancement (Fig. 3.3).

The model shown in Fig. 3.2 is oversimplified. In reality, an optimal compromise is required between TR, flip angle, slice thickness, flow velocities, relaxation time T1, and the necessary coverage.

3.1.2.2.2 Saturation of Flowing Blood. Similar to the saturation of stationary tissue, there is saturation of flowing blood if larger excitation regions (volumes) are used. The larger the volume, the more excitation pulses are induced in the flowing blood. The same effect occurs if blood flows inplane for a certain amount of time. Therefore, spin saturation inside vessels has to be considered as a potential source of signal loss. Each of the TOF techniques (2D or 3D) has its particular advantages and disadvantages and

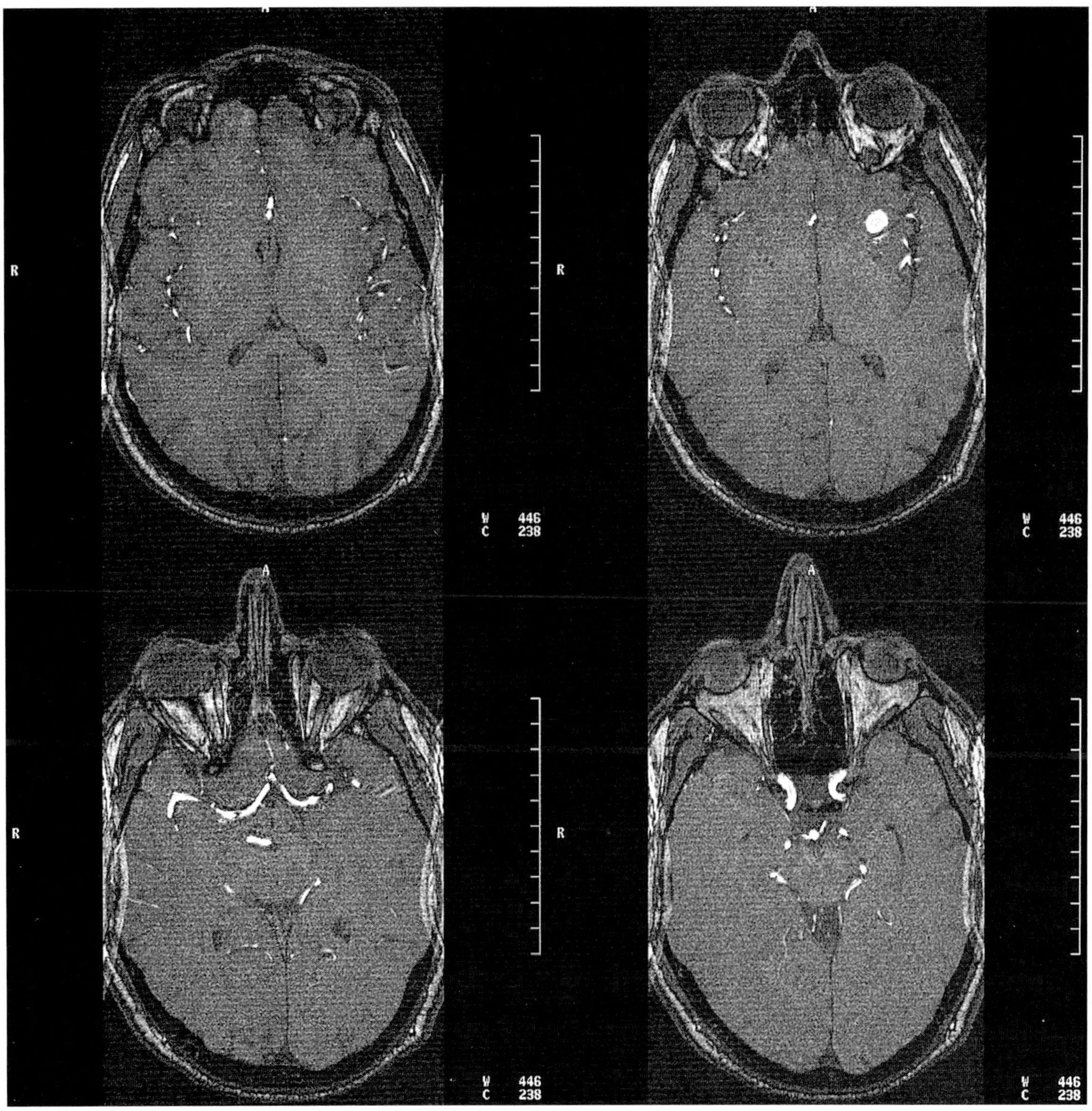

Fig. 3.3. Individual slice of a 3D TOF angiogram: high signal within blood vessels and low signal and contrast in stationary tissue

is therefore prone to special artifacts, and different optimal parameter sets exist.

3.1.2.2.3 Field Strength Dependency of Optimal Echo Time. Since water and fat spins have a slightly different resonance frequency, there are echo times where signal from fat and water is in opposed-phase condition and echo times for which both water and fat are rephased. If within one voxel there is water and fat content, signal can be eliminated if opposed-phase conditions are met. For TOF MRA it is advantageous to use opposed-phase echo times, since fat is always in the stationary tissue and this should be of low signal intensity. For a field strength of 1.5 T, the first opposed-phase echo time is 2.2 ms, which would be optimal for minimal dephasing but cannot be achieved with present hardware due to gradient limitations. The next one is 6.6 ms and this is the typical echo time for 1.5-T TOF MRA. Table 3.1 gives an overview of the different field strengths and the corresponding in-phase opposed-phase echo times. Note that for 0.5 T the first opposed-phase condition is 6.6 ms, identical to the second one at 1.5 T. For minimal dephasing, the shortest possible echo time is desirable, but for regions of the body where significant fat tissue is present (as in the neck, chest, or abdomen), somewhat longer echo times in opposed conditions are preferable.

3.1.2.2.4 High Signal for Tissue with Short T1. Since TOF sequences use short TR times, typically all stationary tissue can be saturated sufficiently to suppress stationary signal. However, for tissue with extremely short T1 values, as is present in hemorrhagic lesions after a certain time (methemoglobin), signal is high and can be confused with flowing material, e.g., in large aneurysms. This effect has to be seen as an artifact in TOF MRA or as additional information. The real differentiation between flow and T1 effect has to be made with phase-contrast MRA, since this technique is sensitive to flow only. Typically, a rather fast, thick-slice 2D phase-contrast scan in the region of interest can resolve the question.

Table 3.1. Echo times in ms for in-phase (IN) or opposed-phase (OP) conditions at different field strengths. For optimal signal and minimal dephasing, the shortest possible echo time should be used. If fat tissue is present, the opposed phase echo time is preferable since signal from fat is canceled more effectively. The echo times in italics are those most used at present

Field strength (T)	OP	IN	OP
1.5	2.2	4.4	*6.6*
1.0	3.3	6.6	*9.9*
0.5	*6.6*	13.2	–
0.2	*16.5*	33.0	–

3.1.2.2.5 Flow Separation Using Saturation Pulses. With all methods it is possible to suppress the venous or arterial flow using appropriate presaturation pulses (EDELMAN et al. 1989b). This is demonstrated in Fig. 3.4. For example, to suppress the jugular vein in studies of the carotid arteries a 90° presaturation pulse is used superior to the imaging slices. Venous spins reach the imaging volume in completely saturated form and cannot contribute to the signal. This method provides a powerful tool for separating veins from arteries in cases of opposite flow direction (neck and extremities).

3.1.3 Application of Time-of-flight Techniques

As already indicated, the most frequently used MRA technique is the single-volume TOF method (Fig. 3.5) owing to (a) the fact that it can be applied from the intracranial circulation through the thoracic arteries and down to the renal arteries, and (b) its ease of use. Sequential 2D techniques are applied mainly in slower flow situations and for breath-hold studies (Fig. 3.6). An optimal compromise between these two techniques is the sequential 3D multivolume method (Fig. 3.7). This method combines the benefits of 3D [high, isotropic resolution, high signal-to-noise ratio (S/N) and ideal slice profile] and 2D techniques (high contrast, low spin saturation for flowing spins), and will be described in more detail in Chap. 4 (LEWIN and LAUB 1992).

3.1.3.1 3D Volume Acquisition

In 3D volume techniques one large volume is excited and then imaged in contiguous slices by using an additional phase-encoding table in the slice direction. This leads to thin slices well below 1 mm. Together with high-resolution inplane (e.g., 384 × 512 and rectangular field of view), this leads to very small isotropic voxels. 3D sequences have an inherently higher S/N due to the larger number of excitations and allow shorter echo times because of the rather relaxed conditions for gradient design in the slice select direction compared to thin-slice 2D sequences. This leads to minimal intravoxel phase dispersion and optimal signal, especially in regions of moderate

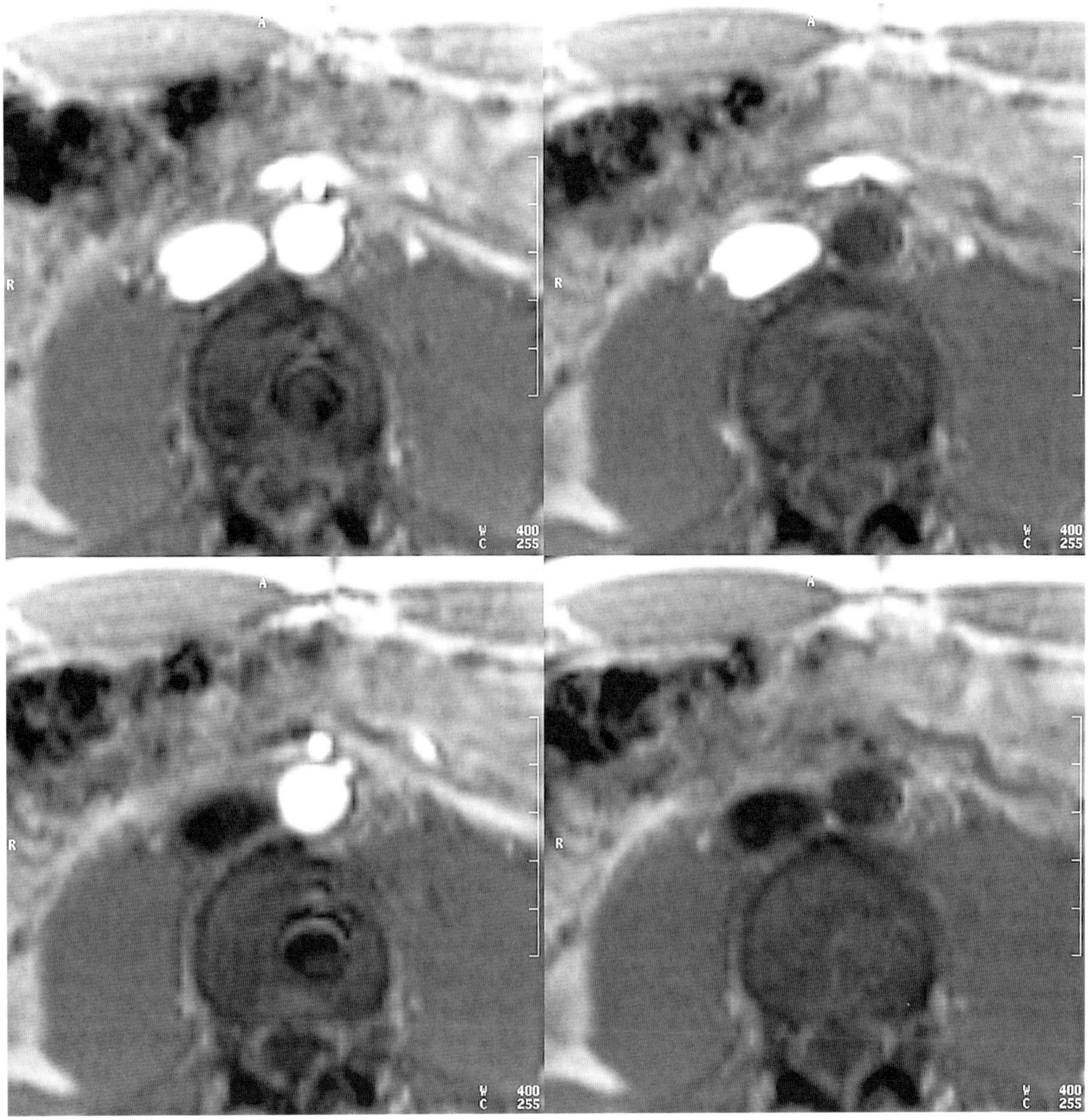

a b c d

Fig. 3.4a–d. Selective MRA: **a** Without presaturation pulses both the abdominal aorta and the vena cava are visualized. **b** Placing the presaturation above the imaging volume saturates selectively thoes spins which cross the presaturation volume before entering the imaging volume. **c** Placement at the other side saturates the complementary vessels. **d** Parallel saturation regions saturate both the aorta and the vena cava

and fast flow. However, when using the 3D technique care has to be taken that the main bloodstream is perpendicular to the volume orientation. This avoids significant spin saturation within the blood vessels, since spins are progressively saturated on their way through the volume.

The saturation of flowing spins depends on different sequences as well as physiological effects. On the one hand, larger flip angles and short TR times are optimal for good background suppression. On the other hand, exactly the opposite, namely small flip angles and longer TR times, are ideal for avoiding saturation of flowing blood. This dilemma can only be resolved by optimal parameter choice in particular applications. Acquisition times have to be considered as well. Gradient-echo sequences like FLASH (or spoiled GRASS) or FISP (GRASS) with small flip angles between 15° and 25° and short repetition times in the order of 40–50 ms are routinely used. New techniques like MTS, TONE, or multiple-thin-

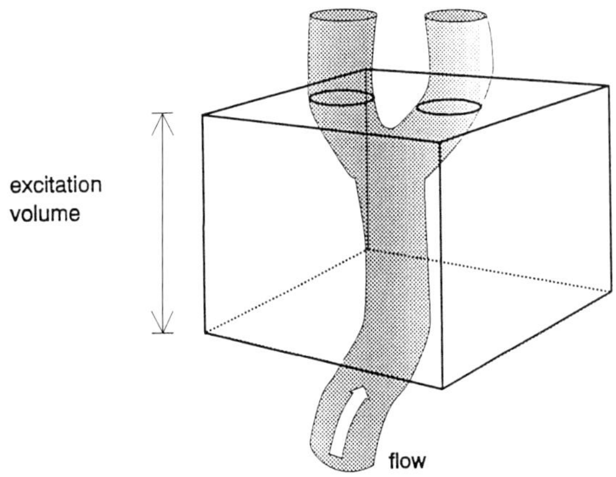

Fig. 3.5. 3D single-volume technique: Excitation of a single volume and partitioning of the whole volume by use of an additional phase-encoding gradient in the slice select direction. It is important that the main flow direction is perpendicular to the volume orientation

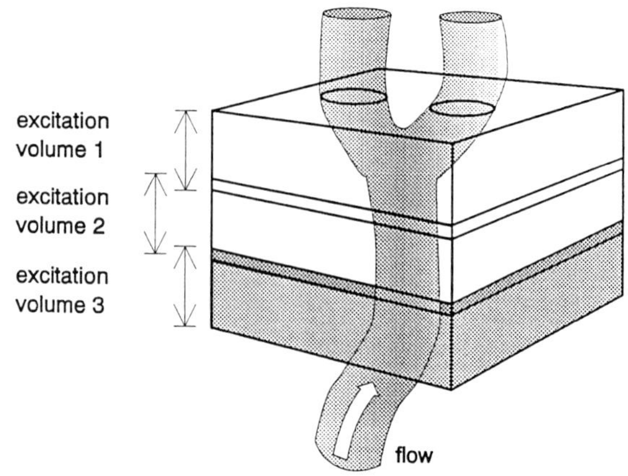

Fig. 3.7. 3D multivolume technique: Successive acquisition of thin overlapping 3D volumes. Within the overlap region the more central slices of the neighboring slabs are used to compensate the nonideal slab profile

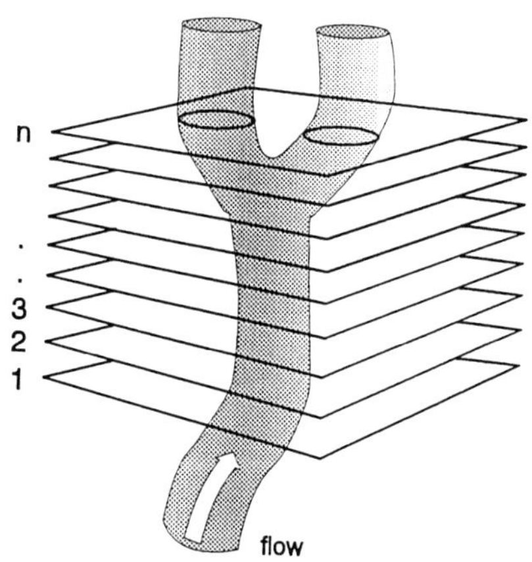

Fig. 3.6. 2D sequential slices: Sequential acquisition of successive overlapping slices. It is important that slices are acquired in ascending order and not interleaved. Otherwise. a zipper artifact may occur due to patient motion (e.g., in the neck region)

volume techniques help and will be described in Chap. 4.

Because of the high sensitivity for fast flowing blood, 3D volume acquisition is particularly well suited for imaging of the arterial vasculature (Haacke et al. 1990; Lewin et al. 1991). Typical applications are axial studies of the intracerebral arteries, as shown in Fig. 3.8.

3.1.3.2 2D Sequential Slice Acquisition

The progressive spin saturation can be avoided by acquiring multiple thin 2D slices in a sequential way. Overlapping slices allow a smooth depiction of vessels, avoiding the zebra-like artifact due to imperfect slice profiles. Although S/N is larger for 3D TOF, contrast-to-noise ratios are higher with 2D TOF techniques, since higher flip angles can be used. Even at medium flip angles of 40–50°, flowing spin saturation is not present when using thin slices. However, considerable spin saturation within vessels can occur, when vessels run inplane or reverse their direction.

Larger flip angles can also induce artificial reduction in vessel diameter, since at the edges of vessels significantly slower flow may be present, which saturates even within one slice. This effect has to be taken seriously, for example, in the depiction of the portal vein or other abdominal vessels.

The 2D technique is very well suited for breath-hold studies since only 5–10 s is necessary for one slice. Typically, 2D gradient-echo sequences (FLASH or spoiled GRASS) with slice thicknesses of 2–5 mm are used. Echo times are somewhat longer than in 3D TOF sequences. Since the slice thickness in this technique is usually larger than the in-plane resolution, there are restrictions on the maximum-intensity projection (MIP) postprocessing in terms of view direction. The nonisotropic resolution of 2D TOF is the biggest disadvantage of the technique. On the other hand, 2D TOF allows better delineation of vessels in highly stenotic areas and when there is very slow flow, such as in veins.

3.1.3.2.1 Tracking Presaturation Pulses. Tracking presaturation pulses are important for effective separation of arteries and veins. Instead of placing one stationary presaturation pulse at one side of the slice

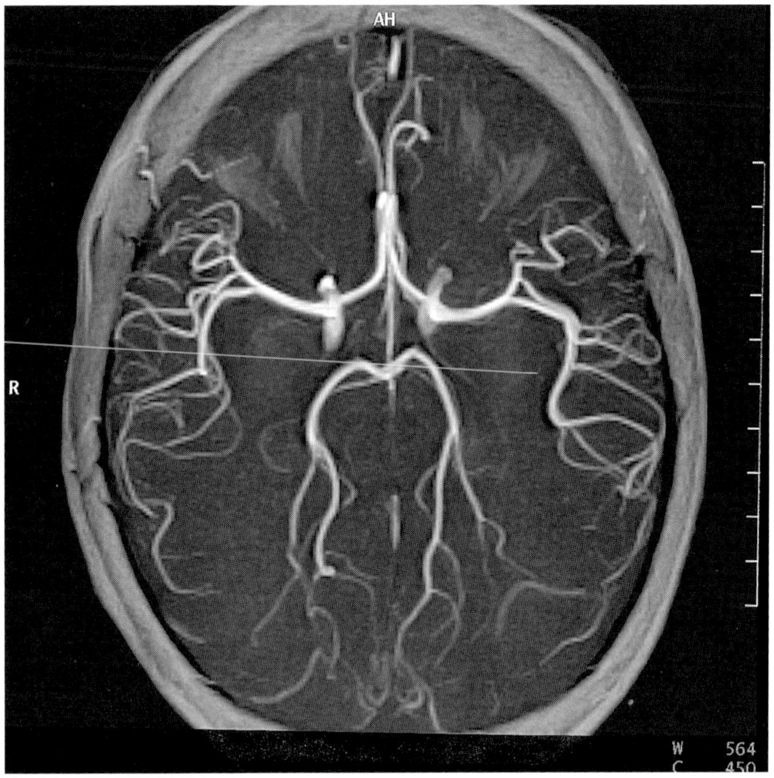

Fig. 3.8. Axial MIP of a 3D data set with 64 partitions (healthy volunteer): FISP 3D sequence, $T_R = 42$ ms, $T_E = 10$ ms, flip 20°, FOV = 180 mm × 180 mm, matrix = 192 × 256, volume thickness 72 mm, scan time 8.38 min, 1 T field strength

stack, the presaturation pulse has to keep a fixed distance from the individual slice. Otherwise, flowing material can recover magnetization, since the distance between presaturation pulse and imaging slice becomes too large. However, care has to be taken not to saturate the vessel of interest in cases of heavy pulsating flow. If the presaturation pulse is too close to the imaging slice, retrograde flow becomes saturated and signal drop out occurs. Variable distances between saturation and imaging slice help, as does triggering the sequence to the cardiac cycle.

In situations where heavy pulsation is present, as in the arteries of the leg, severe ghosting artifacts can occur. These artifacts increase with flip angle. Therefore, a medium flip angle of 30–50° has to be used.

3.1.3.2.2 Thin Overlapping Multivolume Acquisition. By combining the advantages of the 2D and 3D techniques described in the preceding sections, the multivolume sequential 3D TOF technique allows the coverage of larger volumes without saturation effects. The advantages of 3D techniques like isotropic resolution and high S/N remain, as do the advantages of 2D when using thin volumes of 10–20 mm. Since overlapping slabs have to be acquired to circumvent the nonideal volume profile, the acquisition time is larger by 30% compared to a single-volume scan with the same coverage. Problems can occur with this method if the patient moves between successive scans.

3.1.4 Summary

Time-of-flight angiography is certainly the technique mostly used in clinical practice. The reasons are manifold: short acquisition times, ease of use, feasibility of implementation with practically all MR scanners (i.e., from different manufacturers), and the wide clinical experience already gained in its use. 3D techniques are mostly used for arterial MRA, whereas 2D techniques are used in cases of problematic flow, as in highly diseased vessels or in veins. The advantage of short acquisition times for individual slices in 2D makes it an ideal technique for breath-hold studies. Unlike conventional angiography, TOF MRA does not always depict the full lumen of vessels. Different flow conditions and hardware limitations lead to signal variations which reflect the hemodynamic effects more than they do the anatomical reality. With this in mind, however, good diagnosis can be achieved with TOF MRA.

Optimization of sequences and methods and improvements in hardware (e.g., greater gradient strength) will allow MRA to achieve standards reached by conventional methods today (like 512 or higher resolution). The limited S/N due to smaller voxel sizes can be compensated by techniques like magnetization-transfer MRA or TONE techniques (LAUB and PURDY 1992) or by using gadolinium DTPA. The possibility of shortening the echo times will reduce the inherent artifacts associated with dephasing.

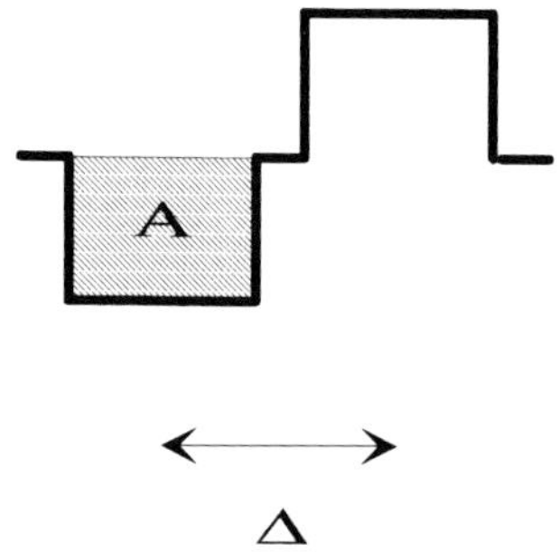

Fig. 3.9. Bipolar gradient pulse for flow encoding

3.2 2D and 3D Phase-Sensitive Techniques

This section describes the second major method of acquiring MR angiograms: phase-contrast techniques. The major advantage of phase-sensitive techniques is their sensitivity to flowing blood only. Unlike in TOF sequences, no other factor, such as T1 shortening, can lead to high signal. Additionally, full background suppression can be achieved and therefore very low flow velocities can be visualized. Rather long acquisition times and therefore limited resolution are the price that has to be paid. This section describes the principles, the major terminology, and the application of the two different phase-sensitive techniques in detail.

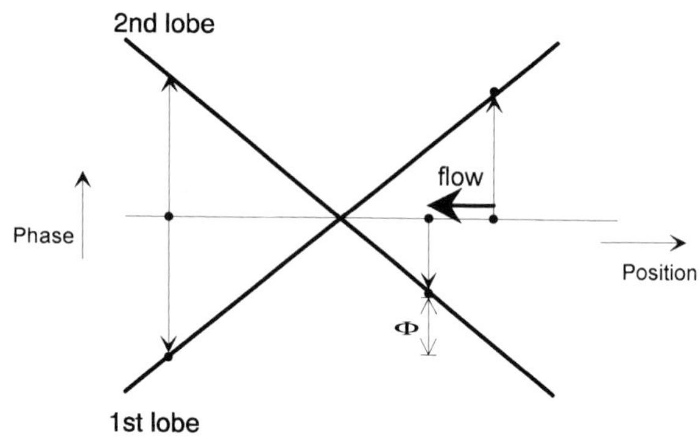

Fig. 3.10. The effect of a bipolar gradient pulse on the phase behavior of stationary and flowing spins

3.2.1 Principles

Whereas phase effects are unwanted in TOF MRA and therefore sequences are optimized to minimize resulting artifacts, phase sensitivity can be employed to generate MR angiograms. The idea is based on the fact that spins moving along magnetic field gradients acquire a phase shift in the transverse magnetization which is directly related to the velocity of the flowing spins (CARR and PURCELL 1954).

Each hydrogen nucleus has a precession frequency inside a magnetic field which is directly proportional to the local field at the position of the nucleus. The main magnetic field is to a good approximation constant, so that all hydrogen nuclei have the same precession frequency if no additional magnetic field gradients are applied.

If a bipolar gradient pulse (Fig. 3.9) is switched on and off between the excitation and detection of the MR signal, position-dependent phase shifts are induced. Figure 3.10 indicates the effect of the first lobe of the bipolar gradient on the phase behavior of a stationary spin as well as on that of a moving spin. In our terminology, the word "phase" means nothing more than the fact that due to a gradient pulse the spins at a particular position precess faster (or slower) that those in the pure main magnetic field. In the rotating frame of reference this results in a phase angle (Fig. 3.11). The second lobe of the bipolar gradient induces exactly the opposite phase shift, if the spins retain their position (i.e., in stationary material). If the spins have moved between the two gradient lobes, the second lobe cannot reverse the effect of the first and a residual flow phase is induced. This flow phase is directly related to the velocity of spins, the strength of the gradient pulses, and the time distance between the two lobes.

A bipolar gradient pulse induces no residual phase shift for stationary spins, but a defined velocity phase for constant flow v along the gradient direction. This is the basis for all phase-sensitive techniques, including flow quantification. In reality, the velocity v is rarely constant. Higher order terms like acceleration, change of direction, and pulsation can be present. Therefore, the model described here is oversimplified. Assuming small voxel sizes and short sensitization periods, the assumption of one velocity v within a voxel is nevertheless valid to a good approximation. With this technique, flow sensitivity is only achieved in the direction in which the bipolar gradient pulse is applied. Therefore, three different measurements are necessary to obtain flow sensitivity in all three directions.

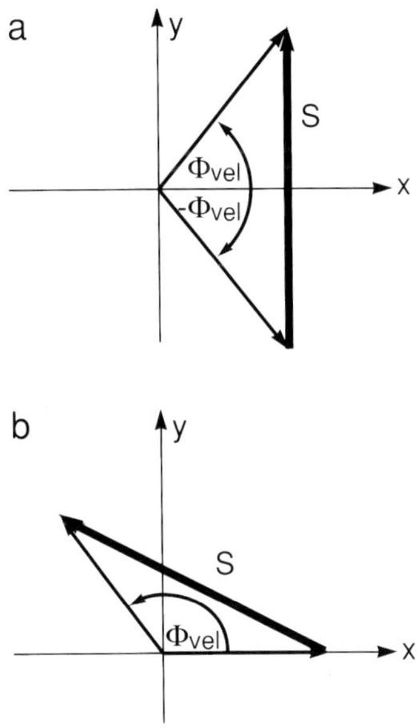

Fig. 3.11. Transverse magnetization at the echo time for bipolar pulses with opposite polarity (**a**) and for a combination of flow-compensated and flow-encoded sequences (**b**)

3.2.2 *Application of Phase-Sensitive Techniques*

Phase-sensitive techniques basically fall into two categories, and the underlying situations are shown in Fig. 3.12. The first, called phase contrast (DUMOULIN and HART 1986; DUMOULIN et al. 1989b), is based on the assumption that within one voxel there is a well-defined induced flow phase (velocity) and that this phase is transferred by subtraction techniques (complex or phase) to a certain gray-scale value (high signal for flowing blood). In this method, signal intensity corresponds to the phase within the volume element (hence the term phase contrast).

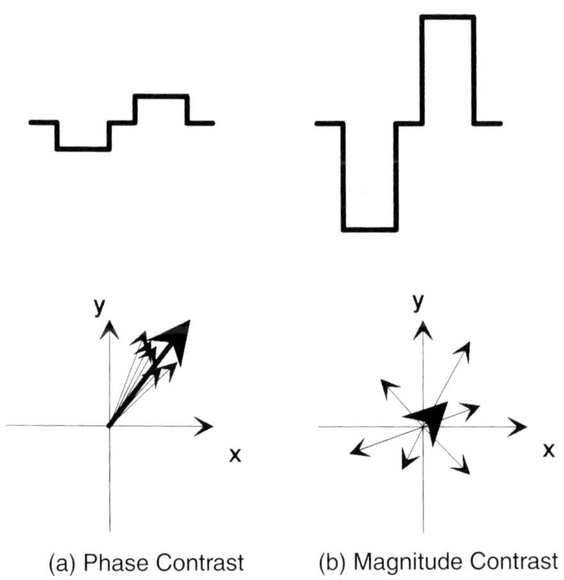

Fig. 3.12a,b. The different phase situations for magnitude- and phase-contrast MRA: Small voxel sizes and moderate encoding values allow alignment of flow phases within one voxel (**a**). Using larger voxel sizes and strong dephasing gradients, large phase dispersions are present and lead to signal cancellation (**b**)

The second technique, called magnitude contrast or the rephased/dephased method (AXEL and MORTON 1987), is based on the assumption that within one voxel there are several flow velocity components and that this induces a phase dispersion (magnetization vectors pointing in different directions). This incoherent effect leads to a signal cancellation in the presence of flow (which is a magnitude effect, hence the term magnitude contrast).

3.2.2.1 Phase-Contrast MRA

The standard method for suppressing the stationary background signal and phase shifts due to imperfections in the main magnetic field is to acquire two data sets with opposite bipolar gradient pulses for each flow direction. The two data sets are then subtracted either by calculating the phase angle between the two complex signals (phase difference method) or by a real complex subtraction (BERNSTEIN and IKEZAKI 1991). The difference signal is Fourier transformed to obtain the final flow image. With this method, a total of six measurements (six-point method) are required to produce an angiogram sensitive to all directions of flow. The three images sensitive to flow in the orthogonal directions are combined to provide an image sensitive to flow in all three directions.

This phase-contrast angiography (PCA) technique leads to significantly longer scan times compared to TOF MRA. Even assuming minimal TR times, which typically cannot be used with TOF techniques due to saturation, three times longer scan times are necessary for PCA. To a certain extent the phase-contrast method avoids the problem of progressive spin saturation that leads to the limitations of imaging volume thickness and poor slow-flow sensitivity when using 3D TOF MRA techniques. This is simply due to the fact that stationary background signal is completely suppressed and even partially saturated spins can be observed. These vessels would disappear using TOF techniques since the vessel signal is lower than the background signal.

3.2.2.1.1 Flow Sensitivity. The sensitivity of the sequence to a particular flow velocity range is an important feature of PCA. Assuming constant flow velocities within one voxel, which can be achieved using very small voxel sizes, the flow-encoding bipolar gradient pulses should be designed in such a way that the flow phase for the maximal velocity component is smaller than 180° [this velocity-encoding value is often referred to as the VENC (velocity encoding) value]. If this is not done, the signal from flowing spins can be sttenuated, leading to reduced vascular signal (aliasing effect). A range of velocities within the voxel can lead to signal attenuation as well. This velocity dispersion is employed in rephased/dephased or magnitude contrast techniques, but should be avoided in phase-contrast techniques.

3.2.2.1.2 Four-Point Methods. Information about all three flow directions can also be acquired using only three flow-sensitized sequences and one reference scan (HAUSMANN et al. 1991; PELC et al. 1991). In the simplest form, one flow-compensated data set is acquired and subtracted from all the other three flow-sensitized data sets. This is schematically depicted in Fig. 3.11b. The flow-compensated data set can be used as an amplitude (parenchymal) image. A more complex form of a four-point method involves use of a Hadamard multiplexing scheme (DUMOULIN et al. 1991; PELC et al. 1991). In this method, flow encoding is performed for each direction in each acquisition. The reconstruction of individual flow images sensitive in one particular direction then has to be a combination of all four data sets. Both methods differ in their sensitivity to pulsation as well as coherent noise accumulation. The simple four-point method has many flow-compensated parts of the sequence and is therefore fairly insensitive to pulsation. In all of these four-point methods the measurements are acquired in an interleaved mode to eliminate the possibility of patient motion between separate measurements.

3.2.2.1.3 2D and 3D Acquisitions. Different phase-contrast techniques can be employed using state-of-the-art scanners. The most widely used is 3D phase contrast (PERNICONE et al. 1990), which allows, for example, depiction of the total vasculature of the brain in 10–20 min, depending on resolution (Fig. 3.13). Good depiction of abdominal vessels is also possible using 3D phase contrast due to its inherent averaging effect (DUMOULIN et al. 1990).

Short acquisition times with the advantages of phase-sensitive techniques, such as sensitivity only

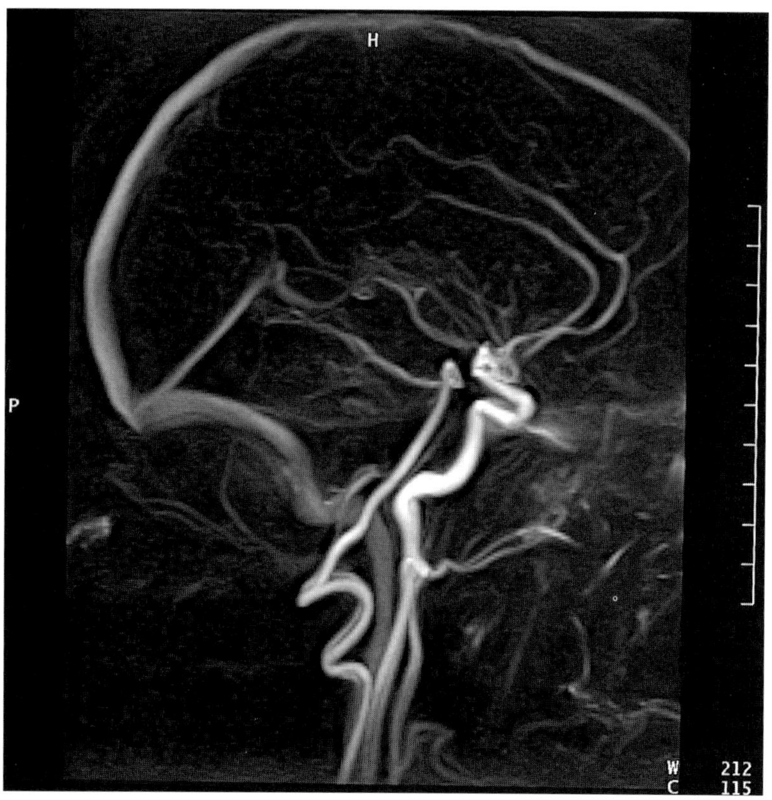

Fig. 3.13. Sagittal MIP of a 3D phase contrast angiogram of a central head section (healthy volunteer). $T_E = 8.6$ ms, VENC = 30 cm/s, flip 20°, FOV = 200 mm × 200 mm, matrix = 128 × 256, volume thickness 150 mm (64 partitions), scan time 10.41 min, 1 T field strength

to flow, can be achieved using the thick-slice 2D phase-contrat method. As a projective method this sequence allows 3 to 4-cm projections (typically with many acquisitions) within ca. 2 min and excellent depiction of even tiny vessels, as shown in the example of a hand angiogram in Fig. 3.14.

3.2.2.2 Magnitude-Contrast MRA

The second phase-sensitive method is based on the dephasing of spins, if no flow compensation is applied. Due to the different flow velocities present within one voxel, the signal is attenuated for vessels compared to stationary tissue. This effect is typically increased by using a bipolar dephasing gradient leading to complete signal drop-out within the vessels ("black blood imaging"). If these images are subtracted from a second flow-compensated data set the stationary background signal drops out and only the vessels remain. Note that in this method the flow sensitivity is achieved in only one spatial direction (readout direction). Due to the interleaved realization of the combined acquisition and the subsequent subtraction, artifacts from moving stationary tissue are suppressed. The flow sensitivity direction can be changed by swapping the readout and phase-encoding gradients. With this method a slightly larger voxel size results in higher contrast of vessels compared to background since dephasing is more pronounced in this case.

The applications of this technique are limited to regions of mainly unidirectional flow, e.g., in the legs (Fig. 3.15). It is obvious that sufficient signal loss (and therefore contrast between vessels and background) can only be achieved when using large bipolar (dephasing) gradients. Of course, this imposes some requirements on the gradient system.

3.2.3 Summary

Phase-contrast techniques have benefits and drawbacks compared to TOF. In clinical routine and for high resolution, 3D TOF is the most widely used technique. However, in special cases and to clasify flow direction and velocity, phase-contrast techniques can additionally be employed. The unique advantage of phase contrast is its sensitivity only to flow, as a result of which on T1 values or other tissue parameters influence the contrast between blood and stationary tissue (Fig. 3.16).

Phase-contrast methods are somewhat more sensitive to dephasing resulting in flow void due

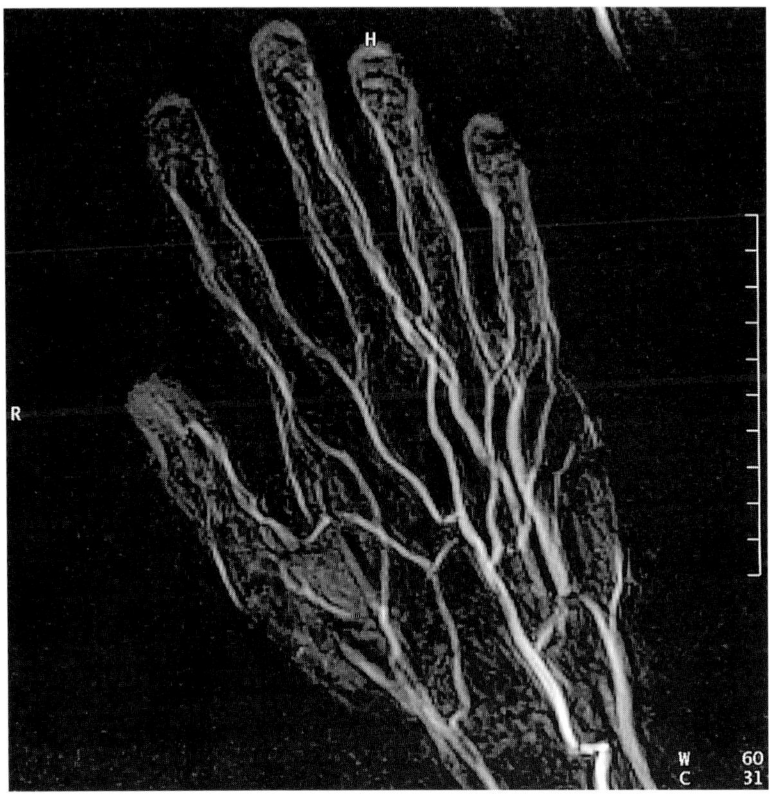

Fig. 3.14. Coronal 2D phase contrast angiogram of the hand. Using 16 acquisitions, sufficient signal to noise can be achieved. Note the excellent depiction of small digital vessels and ideal background suppression using a 4-cm-thick slice (projection). VENC = 10 cm/s, FOV = 220 mm × 220 mm, matrix = 192 × 256, scan time 2.17 min, 1 T field strength

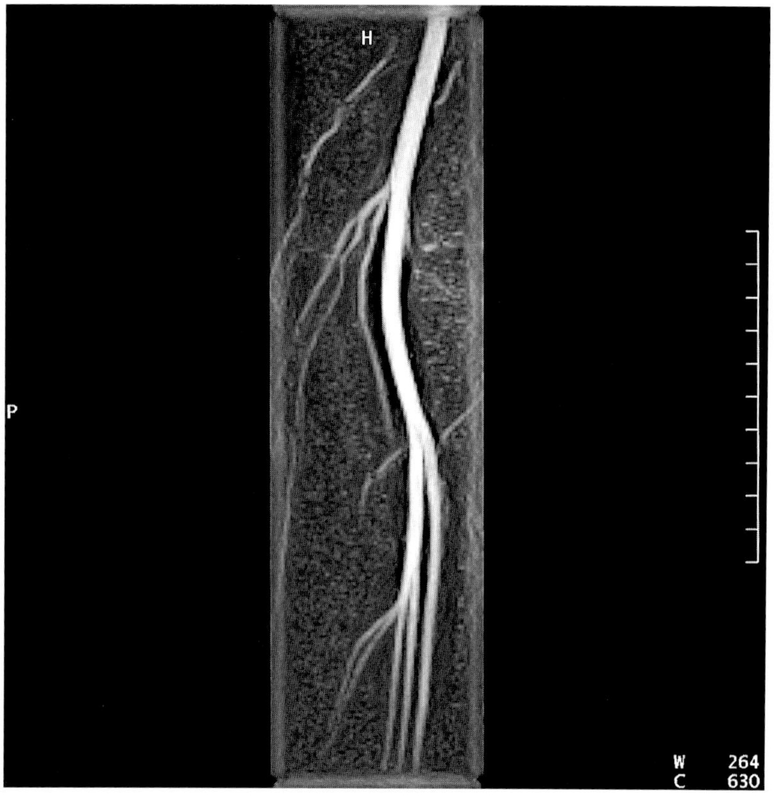

Fig. 3.15. Sagittal MIP of a 3D magnitude contrast angiogram of the popliteal trifurcation. 3D sequence, T_R = 25 ms, T_E = 14 ms, flip 15°, FOV = 240 mm × 240 mm, matrix = 192 × 256, scan time 10.18 min, 1 T field strength

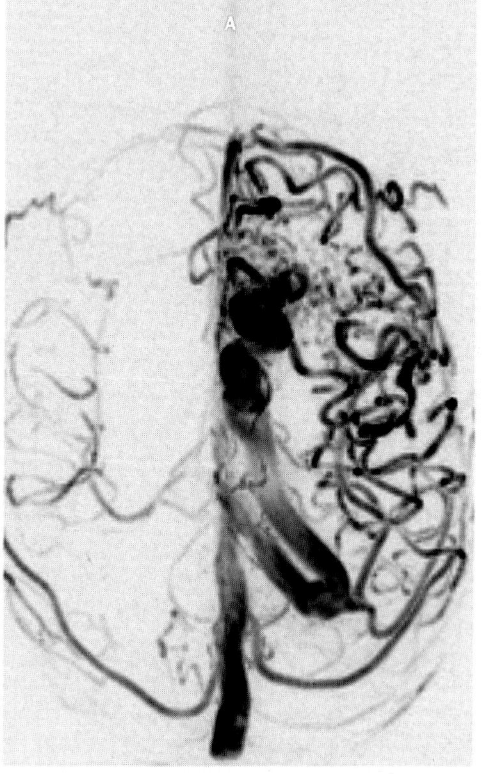

Fig. 3.16. 3D phase contrast angiogram of an arteriovenous malformation. Note the excellent depiction of the pathology and the draining vein into the sagittal sinus. VENC = 10 cm/s, T_E = 11 ms, flip 20°, FOV = 200 mm × 200 mm, matrix = 128 × 256, volume thickness 80 mm, scan time 12.11 min, 1 T field strength. (Courtesy of Dr. PETERSEN, University of Tübingen, Germany)

to nonuniform flow, since gradient motion refocusing cannot be applied in the flow sensitivity direction.

Imperfections of the system, like transient eddy current fields, can lead to phase errors in the individual measurements. When different gradient schemes are used for the individual acquisitions, as in the phase-contrast sequences, and complex data are subtracted, these phase errors result in suboptimal background suppression. Therefore a high-performance gradient system with optimal eddy current compensation is required for PCA.

Another important application of phase-sensitive methods is the quantification of flow velocities. Since the relationship between phase and velocity is to a certain extent linear, direct measurement of flow velocities is possible. The theoretical basis of these sequences is identical to that for PCA and will be described in Chap. 7 in more detail.

Acknowledgements. The author would like to thank Gerhard Laub and Uwe Böttcher from Siemens Medical Engineering Group for very fruitful discussions and Jonathan S. Lewin of Case Western Reserve University, Cleveland for the time spent together working on MRA of the abdomen.

References

Axel L (1984) Blood flow effects in magnetic resonance imaging. AJR 143:1157–1166

Axel L, Morton D (1987) A method for imaging blood vessels by phase compensated/uncompensated difference images. J Comput Assist Tomogr 11:31–34

Bernstein MA, Ikezaki Y (1991) Comparison of phase-difference and complex-difference processing in phase-contrast MR angiography. J Magn Reson Imaging 1:725–729

Carr HY, Purcell EM (1954) Effects of diffusion on free precession in NMR experiments. Phys Rev 94:630–638

Dumoulin CL, Hart HR (1986) Magnetic resonance angiography. Radiology 161:717–720

Dumoulin CL, Cline HE, Souza SP et al. (1989a) Three-dimensional time-of-flight magnetic resonance angiography using spin saturation. Magn Reson Med 11:35–46

Dumoulin CL, Souza SP, Walker MF, Wagle W (1989b) Three-Dimensional phase contrast angiography. Magn Reson Med 9:139–149

Dumoulin CL, Yucel EK, Vock P et al. (1990) Two- and three-dimensional phase contrast MR angiography of the Abdomen. J Comput Assist Tomogr 14:779–784

Dumoulin CL, Souza SP, Darrow RD et al. (1991) Simultaneous acqusition of phase-contrast angiograms and stationary-tissue images with Hadamard encoding of flow-induced phase shifts. J Magn Reson Imaging 1:399–404

Edelman RR, Wentz KU, Mattle HP (1989a) Projection arteriography and venography: initial clinical results using MR. Radiology 172:351–357

Edelman RR, Mattle HP, Kleefield J, Silver SM (1989b) Quantification of blood flow with dynamic MR imaging and presaturation bolus tracking. Radiology 171:551–556

Gullberg GT, Werli FW, Shimakawa A, Simons MA (1987) MR vascular imaging with a fast gradient refocusing pulse sequence and reformatting images from transaxial sections. Radiology 165:241–246

Haacke EM, Lenz GW (1987) Improving MR image quality in the presence of motion by using rephasing gradients. AJR 148:1251–1258

Haacke EM, Masaryk TJ, Wielopolski PA et al. (1990) Optimizing blood vessel contrast in fast three-dimensional MRI. Magn Reson Med 14:202–221

Hausmann R, Lewin JS, Laub G (1991) Phase-contrast MR angiography with reduced acquisition time: new concepts in sequence design. J Magn Reson Imaging 1:415–422

Keller PJ, Drayer BP, Fram EK et al. (1989) MR angiography with two-dimensional acquisition and three-dimensional display: work in progress. Radiology 173:527–532

Laub G, Kaiser WA (1988) MR angiography with gradient motion refocusing. J Comput Assist Tomogr 12:377–382

Laub G, Purdy DE (1992) Variable-tip-angle slab selection for improved three-dimensional MR angiography. J Mogn Reson Imaging 2:86c

Lewin JS, Laub G (1992) Intracranial MR angiography: a direct comparison of three time-of-flight techniques. AJNR 12:1133–1139

Lewin JS, Laub G, Hausmann R (1991) Three dimensional time-of-flight MR angiography: applications in the abdomen and thorax. Radiology 179:261–264

Masaryk TJ, Modic MT, Ross JS et al. (1989) Intracranial circulation: preliminary clinical results with three-dimensional (volume) MR angiography. Radiology 171:793–799

Mills CM, Brant-Zawadzky M, Crooks LE et al. (1983) NMR: principles of blood flow imaging. AJNR 4:1161–1166

Moran PR (1982) A flow velocity zeugmatographic interlace for NMR imaging in humans. Magn Reson Imaging 1:197–203

Nishimura DG, Macovski A, Jackson JI et al. (1988) Magnetic resonance angiography by selective inversion recovery using compact gradient echo sequence. Magn Reson Med 8:96–382

Parker DL, Yuan C, Blatter DD (1991) MR angiography by multiple thin slab 3D acquisition. Magn Reson Med 17:434–451

Pelc NJ, Bernstein MA, Shimakawa A et al. (1991) Encoding strategies for three-direction phase-contrast MR imaging. J Magn Reson Imaging 1:405–413

Pernicone JR, Siebert JE, Potchen EJ et al. (1990) Three-dimensional phase contrast MR angiography in the head and neck: preliminary report. AJR 155:167–176

Ruggieri PM, Laub GA, Masaryk TJ, Modic MT (1989) Intracranial circulation: pulse sequence considerations in three dimensional (volume) MR angiography. Radiology 171:785–791

Singer JR (1959) Blood flow rates by nuclear magnetic resonance measurements. Science 130:1652–1653

4 Contrast Improvement, Artifacts, and Artifact Reduction

G. LAUB

CONTENTS

4.1 Introduction

As discussed in preceding chapters, the basic rationale underlying time-of-flight (TOF) MRA is the utilization of inflow enhancement while compensating for flow-induced phase shifts. For high-resolution studies with isotropic voxels a three-dimensional, or volume acquisition technique is recommended, with the major flow along the slab select direction (MASARYK et al. 1989). While inflow enhancement and resulting vessel contrast is good enough on the side where the vessels enter the volume, the blood signal becomes progressively smaller towards the exit side of the volume. Therefore, the visibility of blood vessels decreases as blood penetrates the volume, as shown in Fig. 4.1. Ultimately, at some point there is no longer any contrast between blood and surrounding stationary tissue. The slab thickness and corresponding vessel coverage cannot be selected on an anatomical basis but rather are determined by contrast issues. The reason for the loss of blood signal is related to the decrease in the spins' magnetization as they enter the imaging volume and experience the rf pulses in the sequence. This effect is demonstrated in Fig. 4.2 in the form of a numerical simulation. The graph shows the transverse magnetization M_x which corresponds to the signal from the blood as a function of the number of rf pulses. For spins moving at a constant velocity v the number of rf pulses directly corresponds to the spins' position in the imaging volume when multiplying the velocity by the pulse repetition time TR and pulse number. Three different curves are shown for different values of the rf excitation pulse.

A large flip angle of 40° results in a high signal amplitude at the entrance plane which rapidly approaches a low steady-state value within the slab. On the other hand a low flip angle of 10° maintains a low signal from blood over a long distance. A reasonable compromise between absolute signal intensity, contrast, and depth penetration is obtained at a flip angle of 20°, which is usually used in many clinical imaging protocols. Still, at 20° progressive saturation is observed, limiting the vessel coverage in clinical applications.

Several approaches have been studied in the past to overcome or reduce the limitation of vessel coverage in three-dimensional TOF MRA protocols. In this paper the most practical techniques will be presented in more detail as they have been shown to be very beneficial in standard flow imaging protocols.

G. LAUB, PhD, Siemens AG, Medizinische Technik, Henkestraße 127, 91052 Erlangen, FRG

4.2 Ramped rf Pulses

A more effective use of the magnetization is possible when using variable flip angles across the slab, as shown in Fig. 4.3. At the entrance plane when unsaturated spins enter with full magnetization, a relatively small flip angle of, for example, 10° will still provide sufficient signal without reducing the longitudinal magnetization too much. Next, the flip angle is increased by a small amount to maintain or even increase the signal from blood spins on their way through the volume, and so forth. The technique using ramped rf pulses is referred to as "tilted optimized nonexcitation" (TONE) rf pulses (ATKINSON et al. 1994). In principle, it is possible to shape the flip angle distribution over the entire imaging volume according to the specific flow velocity and vessel coverage. A detailed analysis of variable flip angle excitation is shown in Fig. 4.4. The major improvement as compared to the standard rf pulses with a constant flip angle across the volume is

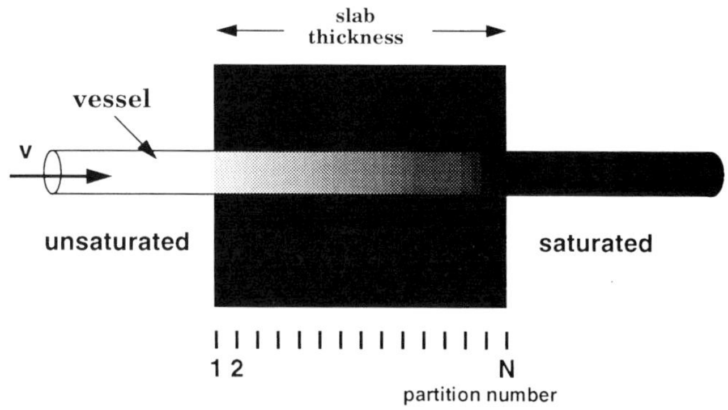

Fig. 4.1. Unsaturated spins are entering the imaging volume in a 3D TOF MRA study. The contrast between blood and surrounding stationary tissue decreases progressively owing to the saturation of blood when it is moving in the excitation volume of the sequence

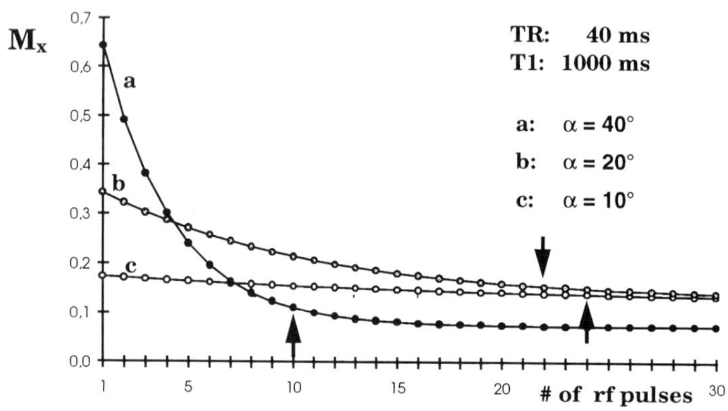

Fig. 4.2. Saturation of spins. The signal from spins entering the imaging volume of a 3D sequence decreases continuously to reach a steady-state value. With larger flip angles the signal starts from a high value and decreases rapidly to reach a low steady-state value. Lower flip angles maintain a small signal over a longer distance. *Arrows* indicate when blood signal is equal to the steady-state signal from surrounding stationary tissue, e.g., gray matter with a T1 of approximately 800 ms. From this simulation a flip angle of 20° is seen to be optimal regarding vessel contrast and depth penetration

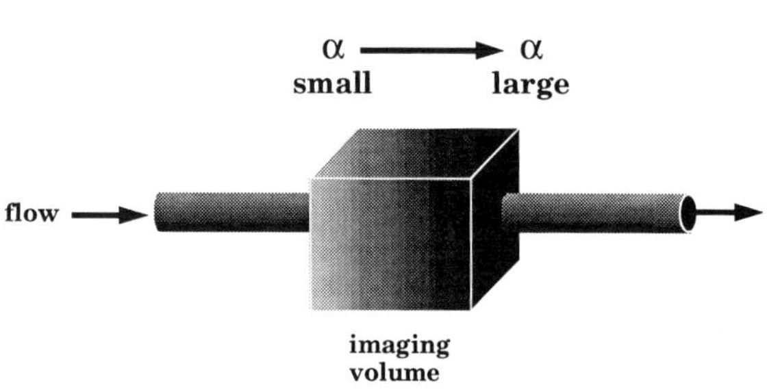

Fig. 4.3. Principle of TONE (tilted optimized nonexcitation) rf pulses. For unidirectional flow a modified rf pulse is used which produces small flip angles at the entrance side of the imaging volume to avoid saturation of the spins when entering the volume. The flip angle is gradually increased towards the exit plane of the volume to compensate for the partial saturation of the spins when moving through the volume

a more constant distribution of blood signal intensity over the imaging volume. In summary, the magnetization of blood is used more effectively, due to the larger flip angles towards the exit side of the imaging volume, which also cause greater saturation of the stationary tissues at this position. It is important to note, however, that TONE pulses are to be used only in those cases where the blood flow is mostly uniformly directed, such as in the carotid artery or aorta. A typical example of TONE pulses versus standard excitation pulses is shown in Fig. 4.5. With the standard rf pulses most of the magnetization of moving spins is lost in the aorta before entering the renal artery. Therefore, the contrast of the blood in the renal artery is very small. With TONE pulses, spin saturation in the aorta is much reduced, resulting in better visibility of the renal artery. An-

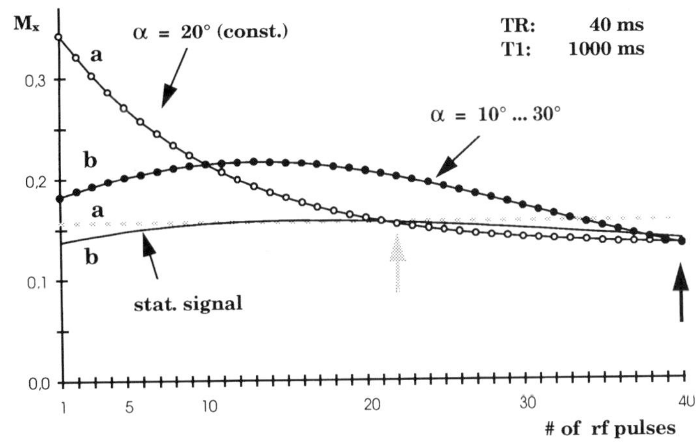

Fig. 4.4. Transverse magnetization of spins as a function of rf pulses. For constant flip angles throughout the imaging volume, e.g., 20°, there is a constant decrease until the magnetization has approached its steady-state value. In the case of a variable flip angle the signal reaches its maximum value after about 22 excitations (*gray arrow*). Consequently, the contrast of blood is maintained over a longer distance in the imaging volume (*black arrow*)

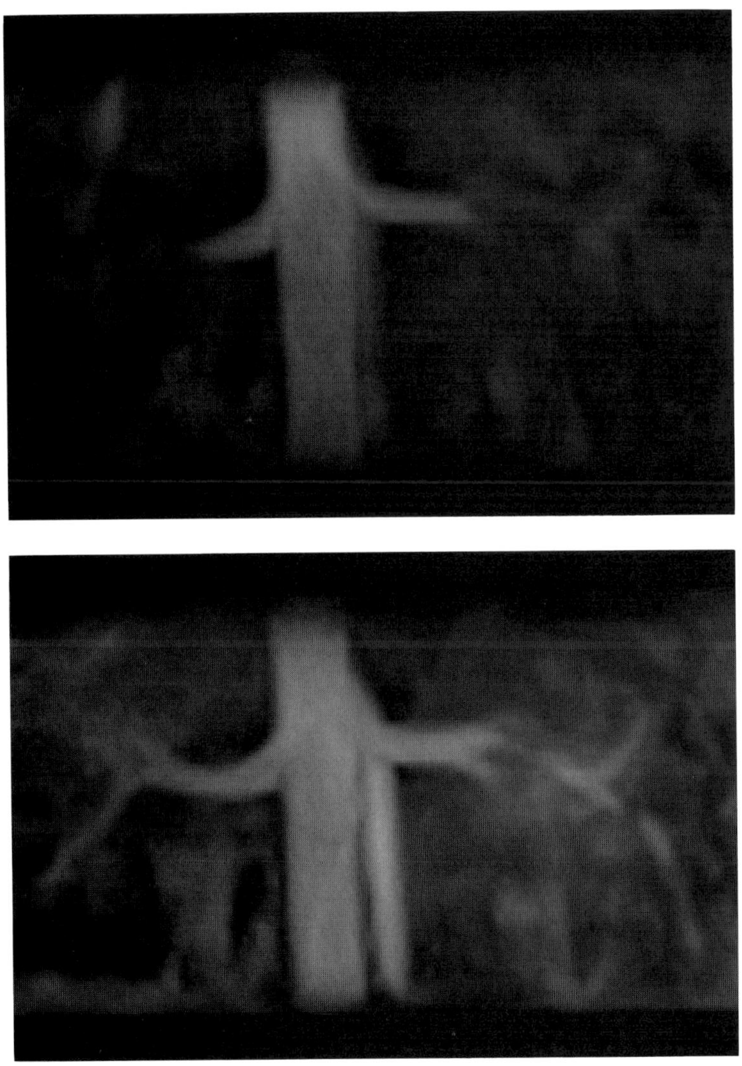

Fig. 4.5. The effect of TONE pulses on the signal intensity of blood in TOF MRA. *Top*: standard rf pulses with a constant flip angle of 20°. Most of the blood's magnetization is creating signal on the entrance side of the imaging volume. *Bottom*: TONE rf pulse with a linear increase of flip angle from 10° to 30°. In this case the blood signal is more uniformly distributed over the length of the aorta in the imaging volume, resulting in a better delineation of the renal artery. Other sequence parameters were kept the same in both cases

other area which benefits from the application of ramped rf pulses is the intracanial vasculature, including the circle of Willis. The incorporation of TONE pulses into the sequence reduces spin saturation when blood moves through the carotid siphon, resulting in a better contrast of blood in the more peripheral parts of the intracranial vasculature (see also Fig. 4.7).

4.3 Magnetization Transfer Suppression

A further improvement in vessel contrast is possible by the application of magnetization transfer pulses (Wolff and Balaban 1989; Lin et al. 1993). The idea is to use off-resonance rf pulses which do not directly affect the mobile protons that usually create the signal in MRI. Protons with restricted mobility, however, do become saturated, and because of cross-correlation or chemical exchange processes the magnetization will be transferred to some biological tissue such as gray or white matter, resulting in a partial saturation (Fig. 4.6). This mechanism is referred to as magnetization transfer contrast (MTC) and is typically used for a modification of tissue contrast. In MRA magnetization transfer pulses can be applied, too, as blood will not be affected by the MTC pulses, and as a result there will be more contrast between blood and stationary background signal. As signal from stationary tissue like gray and white matter in the brain is reduced by magnetization transfer pulses, this technique is referred to as magnetization transfer suppression (MTS). MTS is particularly useful in the brain, as the background signal from gray and white matter tissue is significantly reduced while the signal from blood flowing into the imaging volume remains the same as in standard MRA sequences. A trade-off in MTS MRA is the prolongation in TR which accompanies the application of MTS pulses. Typically, the minimum TR of an MRA sequence is increased by about 10 ms. However, MRA protocols do not operate at the shortest TR possible, but at a TR of 30–40 ms to maximize the contrast of inflowing blood. At these TR times, the incorporation of additional MTS pulses is possible without further compromises.

A clinical example of MTS MRA is shown in Fig. 4.7. The application of MTS pulses can be seen from the higher fat signal in the orbits. This is due to the fact that protons in lipid tissues are not affected by the MTS pulses. Therefore, orbital fat usually shows up as bright spots in the MIP image in MTS MRA.

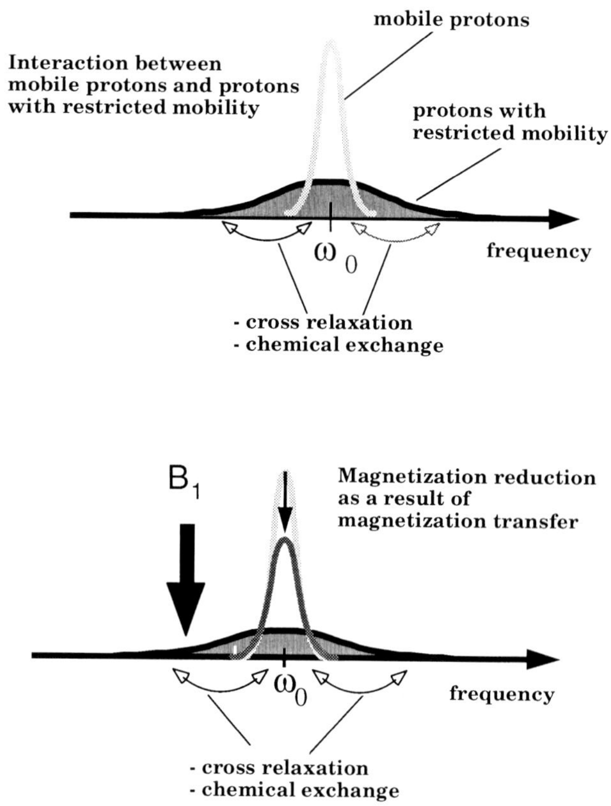

Fig. 4.6. Principle of MTS MRA. *Top*: Frequency spectra for biological tissue with a narrow peak for mobile protons and a broad peak for protons with restricted mobility (i.e., protons in macromolecules). These two pools interact with each other via cross-correlation and chemical exchange. *Bottom*: The pool of protons with restricted mobility is saturated with an off-resonance rf pulse B1. As a result of the magnetization transfer the magnetization of mobile protons decreases

4.4 Fat-Suppressed MRA

As discussed previously, magnetization transfer pulses can be used effectively to reduce signal from stationary background tissue such as gray or white matter in the brain. However, MTC does not affect the signal intensity of fat, as there is no interaction between the mobile protons of lipid tissues and immobile protons of solid components (e.g., proteins and hydration water molecules). As a result, fat shows up as bright spots on MR angiograms incorporating MTC pulses for contrast improvement. A possible way of removing the signal from fatty tissues which may interfere with the blood signal, depending on the projection direction, is to apply additional fat saturation pulses to remove any signal from fatty tissues within the volume of interest (Lin et al. 1993;

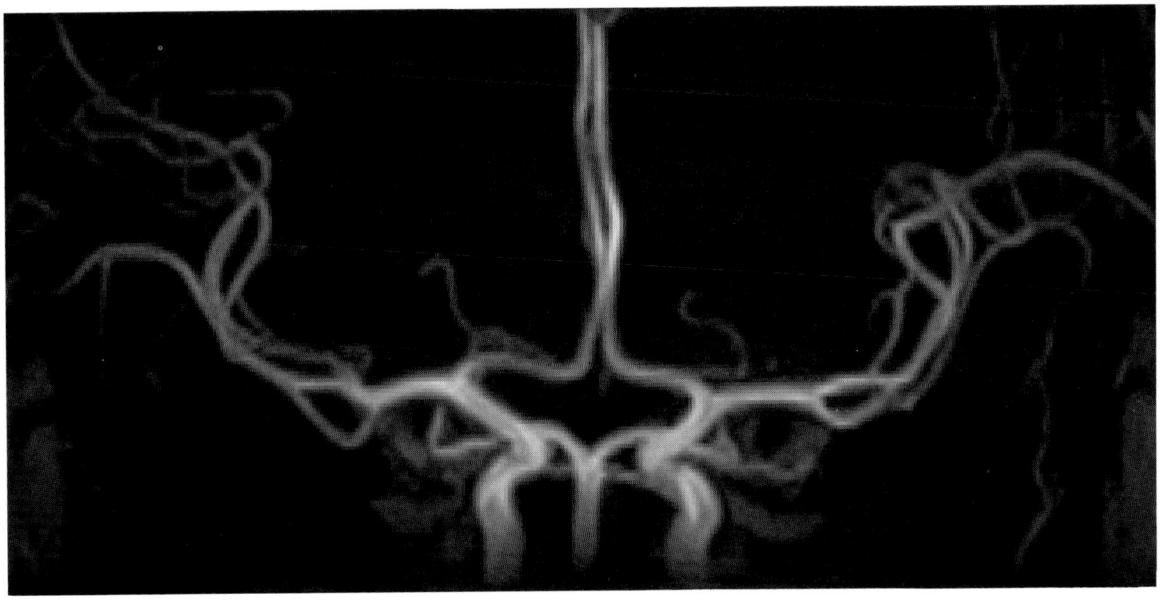

Fig. 4.7. Application of TONE and MTS for intracranial MRA

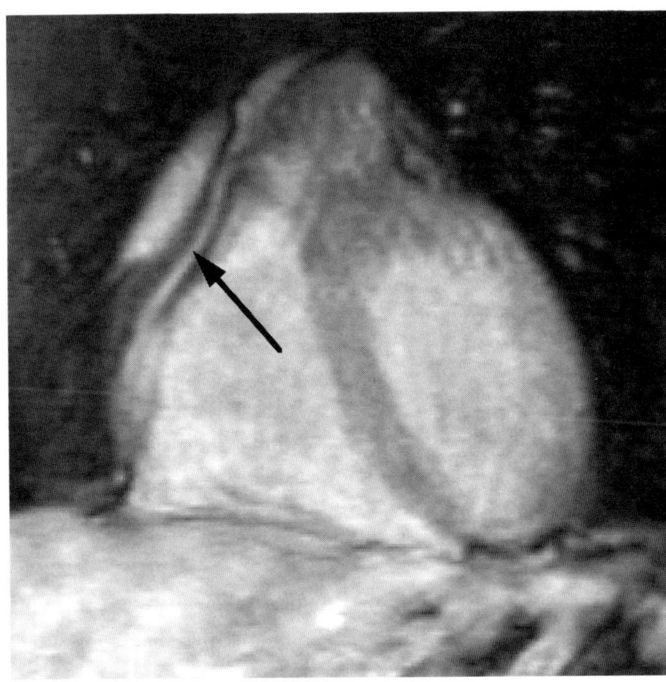

Fig. 4.8. Fat saturation pulses help to improve the visibility of coronary arteries

TKACH et al. 1993). While this technique appears to be straightforward and simple to implement, it still presents inherent difficulties mostly related to the magnetic field homogeneity. To work successfully the field homogeneity must be better than chemical shift, which is about 3.5 ppm. This amount of homogeneity must be maintained not only over the volume of interest, but also over the entire volume of inflow to avoid a presaturation of the inflowing blood which would result in a loss of blood signal intensity within the volume of interest. In practice it is difficult to achieve reliable and sufficient homogeneity in all patients. Therefore, fat saturation techniques are limited to some specific applications where fat is located close to the vessels to be displayed, such as the coronary arteries (EDEL MAN et al. 1991). An exam-

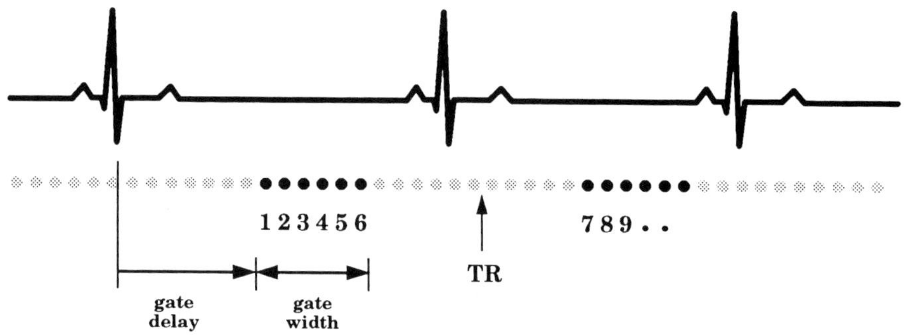

Fig. 4.9. Principle of gated MRA. A gate delay defines the start of data acquisition after the R wave. During the gate width several lines of k-space are acquired in the usual order at an interval of TR. The total scan time is increased as data are acquired only during a portion of the cardiac cycle

ple of this technique is shown in Fig. 4.8. Efficient fat saturation is necessary for this technique to create sufficient contrast between the vessels and surrounding lipid tissues.

4.5 Gated MRA

A major source of artifacts and signal losses of blood is related to the pulsatility of blood flow. Any variations in the blood flow velocity during the data acquisition will cause signal modulations in both amplitude and phase which will result in signal losses or flow voids, as well as ghost images. A common method of reducing these artifacts is to synchronize the data acquisition with the cardiac cycle, as shown in Fig. 4.9. Data are acquired only within a window, or gate, during the cardiac cycle (Selby et al. 1992). Within this window flow will be relatively constant, thus reducing most of the artifacts associated with an ungated study. As a side-effect, the scan time will be increased, depending on the ratio between gate open and gate closed. In an extreme case only one line of k-space is scanned per heart beat in order always to perform k-space scanning at the same point during the cardiac cycle. In this way it is possible to acquire data even in the heart, where one may expect most of the variations during the cardiac cycle. However, it also takes a significant amount of time to acquire one image, as only one line of k-space is acquired per heart beat. In practice, for MRA a reasonable compromise must be chosen between the length of the acquisition window, or number of lines per heart beat, and the resulting scan time. For peripheral MRA, as an example, 13 lines are measured every heart beat, resulting in an acquisition window of 200 ms. This is short enough to avoid blood flow variations throughout the data acquisition window. To acquire one slice with a 182 × 256 image matrix takes 14 heart beats. This procedure is then repeated at different slice positions until sufficient coverage is achieved. The resulting 3D data set is postprocessed with a maximum-intensity algorithm to produce a projection-type image of the vasculature as shown in Fig. 4.10.

4.6 Multislab TOF MRA

Three-dimensional TOF MRA is a technique which provides excellent spatial resolution with isotropic voxels and a high signal-to-noise ratio. On the other hand, vessel contrast decreases due to progressive spin saturation when spins are moving through the volume of interest. As a result, vessel coverage may be limited, even in combination with MTS and TONE. A simple solution to this problem is to repeat the same study at a different slab position (Lewin and Laub 1991). This idea is demonstrated in Fig. 4.11. In this particular example, four slabs are acquired in a sequential order to obtain the same coverage as is achieved with a single slab with the same coverage. With the single-slab technique, however, significant spin saturation is observed, which is less visible in the multislab technique owing to the smaller slab thickness. The excitation profile of each slab is not perfectly rectangular; instead, there is a drop-off of flip angle on both sides of the slab, resulting in a corresponding loss of signal intensity. Therefore, individual slabs cannot be positioned next to each other; rather, they need to overlap each other. The amount of overlap depends on the exact form of excitation profile which is used in the sequence. With optimized rf pulses, an overlap of typically 30% is

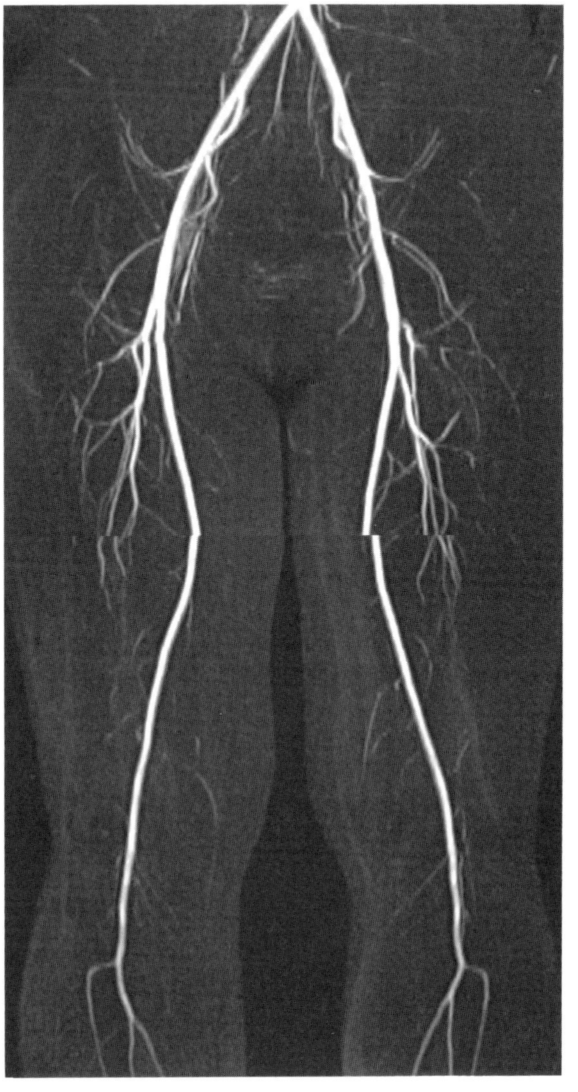

Fig. 4.10. Demonstration of gated MRA. A series of transverse slices are acquired with ECG gating to reduce the sensitivity to the pulsatility of the blood flow. Thirteen lines of k-space are acquired every heart beat. Each slice, therefore, takes 14 heart beats to measure a 182 × 256 imaging matrix. All slices are used to calculate projection-type images using MIP at any orientation. A travelling presaturation pulse is incorporated into the sequence to remove signal from the veins

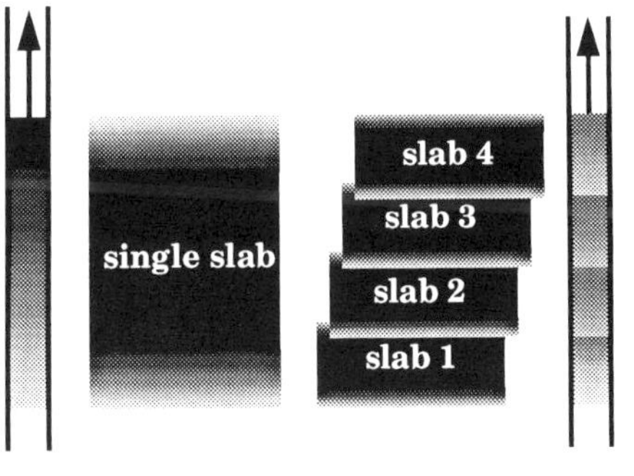

Fig. 4.11. Principle of multislab MRA. A single slab results in severe spin saturation when blood is penetrating the imaging volume. The same coverage can be achieved by multiple thin slabs. Slabs are acquired one after the other. Within each slab there is less saturation, resulting in a more homogeneous signal intensity distribution of blood. Due to the imperfect slice profile of each slab, with a decrease in flip angles on either side of the slab, individual slabs need to be overlapped

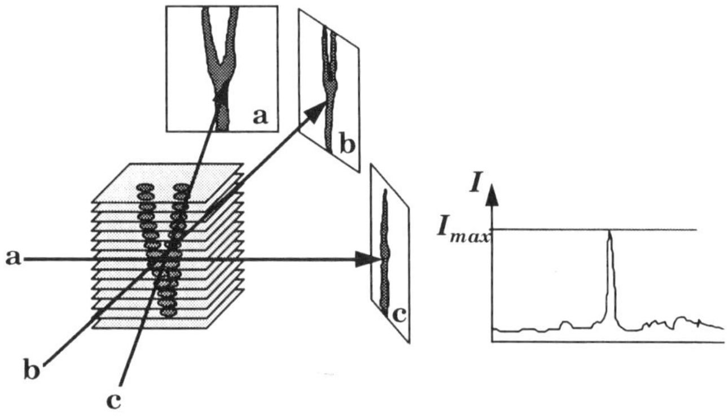

Fig. 5.1. Schematic representation of the maximum-intensity projection (MIP) method. A set of parallel rays is penetrating through the volume (only the center ray is shown for the sake of simplicity). Along each ray the signal intensity is evaluated, and the maximum-intensity value is used for the corresponding point in the projection image. Three different projection directions are shown in this figure

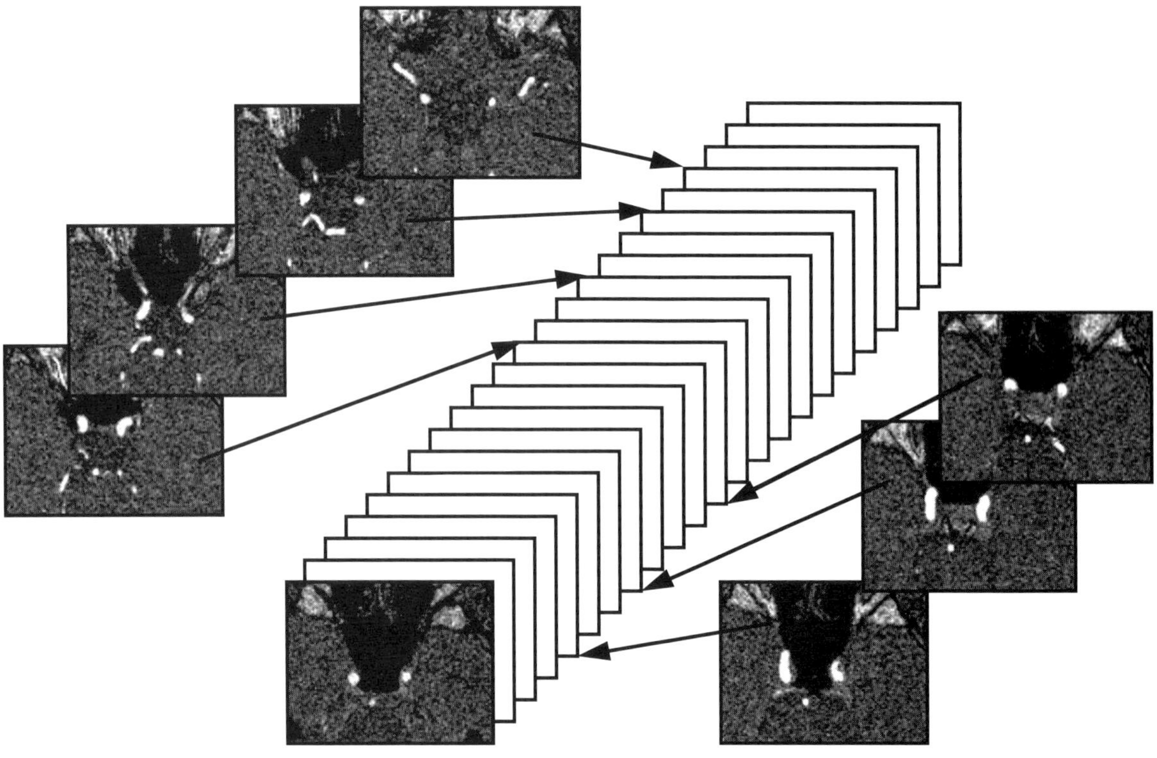

a

Fig. 5.2a–d. Calculated MIP from the three-dimensional data set of the carotid artery. sixty slices were acquired in sequential order and transverse orientation with a 160 × 256 matrix and 3/4 rectangular field-of-view. Total acquisition time was 4 min 48 sec for all 60 slices. **a** Source images. **b** Coronal projection at +20°. **c** Coronal projection. **d** Coronal projection at –20°

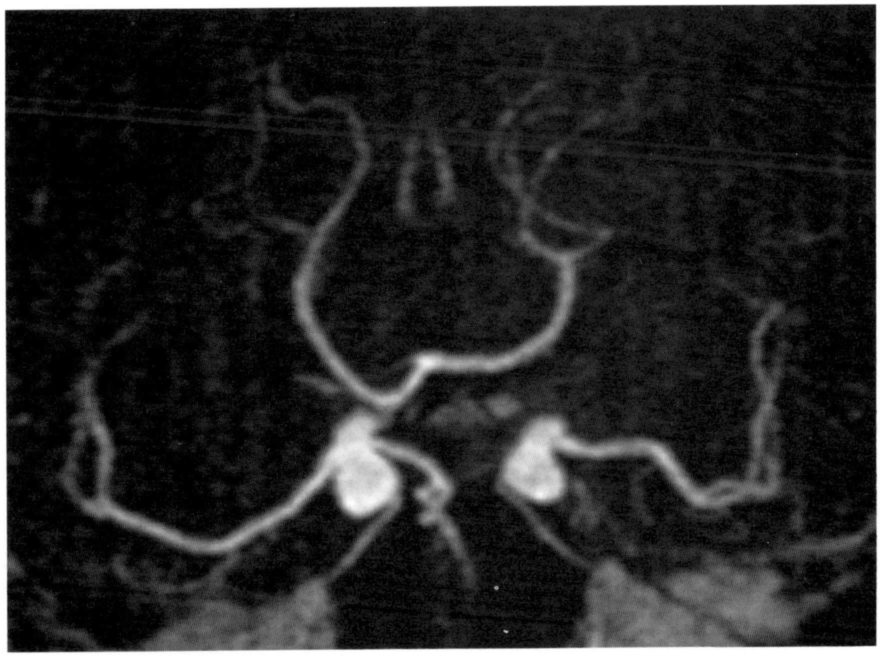

b

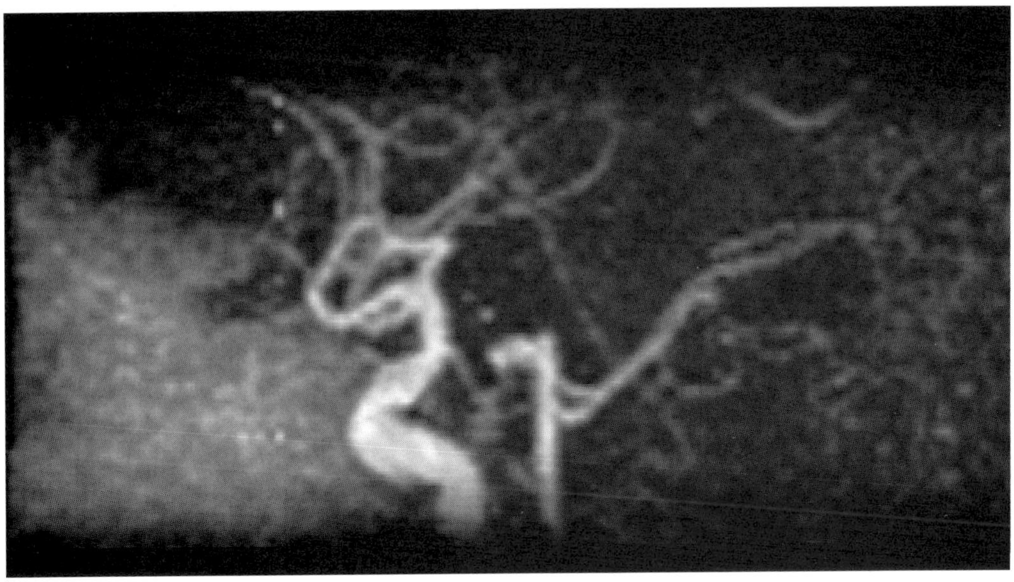

c

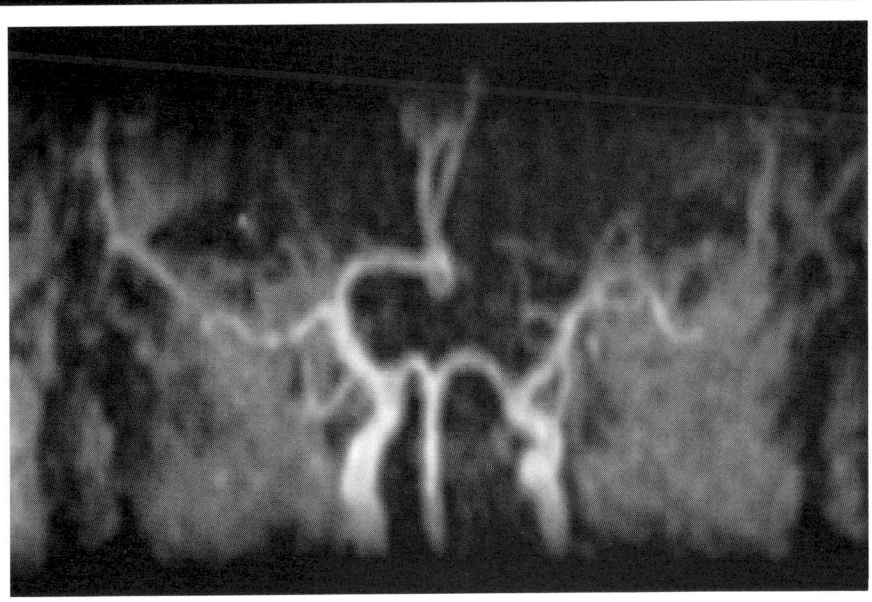

d

Fig. 5.2b–d

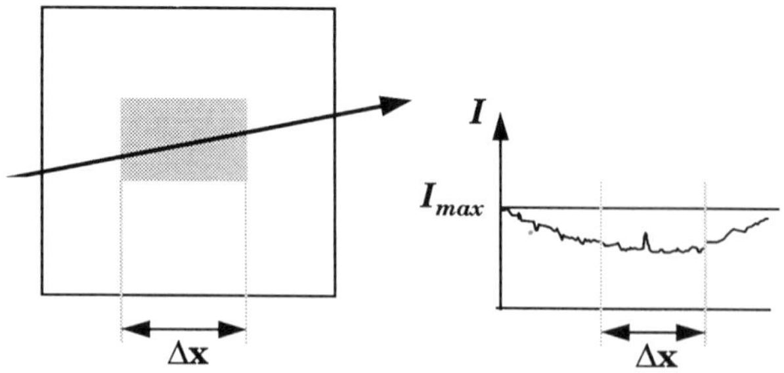

Fig. 5.3. Principle of targeted MIP. A subvolume, or region of interest, is selected. Only data points within this "target" are considered in the ray-tracing algorithm. A signal point with a local maximum in signal intensity is picked up by the targeted MIP of size Δx, but not by a full MIP, as the local maximum is smaller than the overall maximum signal intensity I_{max}

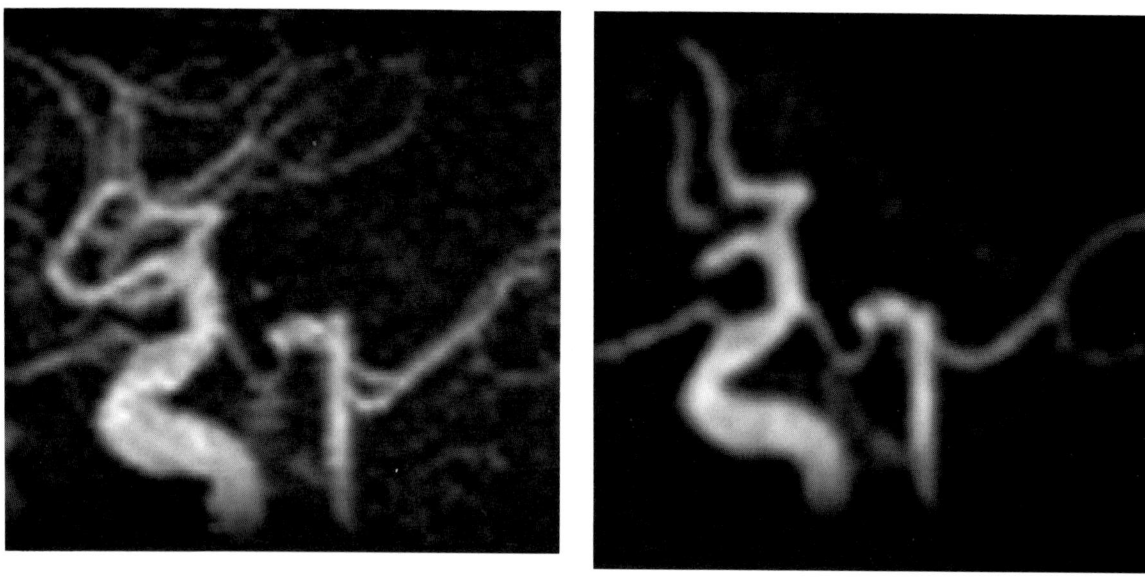

Fig. 5.4. Effect of targeted MIP on small-vessel details. The 3D data set was acquired with a FISP sequence including MTS and TONE (TR/TE/α = 40 ms/6.5 ms/10–30°). Total scan time was 8 min 20 s. *Left*: MIP of all pixels within the 3D data set. *Right*: Targeted MIP including a small region of interest covering the posterior communicating artery. Due to the smaller volume in the targeted MIP fewer background pixels contribute to the projection image, producing greater contrast for small vessels

a targeted MIP on a real data set is demonstrated in Fig. 5.4. With targeted MIP the contrast of smaller vessels is clearly improved as compared to the full MIP. Additionally, the targeted MIP offers the advantage of shorter calculation times, as fewer data points need to be considered in the MIP algorithm.

In all cases, however, the maximum-intensity algorithm works well only for date sets which provide sufficient contrast between blood vessels and stationary background. It is only the signal intensity of image pixels which can be used for the identification of blood vessels Therefore, all efforts need to be taken to increase the signal intensity of blood while reducing the signal intensity of stationary tissues.

Alternatively, other postprocessing techniques like vessel tracking (Li et al. 1992), data-adaptive reprojection techinques (DART; Korosec et al. 1992), or connected voxel algorithms (Saloner et al. 1991) may be applied to improve the overall vessel delineation. These techniques make use of more sophisticated assumptions regarding the size and texture of blood vessels, and operate similarly to image segmentation techniques. The goal is to have blood vessels separated from background tissue. After this step, standard projection techniques can be used to create projection-type images from the vasculature. Additionally, surface rendering techniques can be applied to create images which reveal the three-dimensional structure of the vessel anatomy (Bosmans et al. 1991).

References

Anderson CM, Saloner D, Tsuruda JS et al. (1990) Artifacts in maximum-intensity-projection display of MR angiograms. Am J Roentgenol 154:623–629

Bosmans M, Hoehne KH, Lanb G, Pommert A, Tiede U (1991) Improvement of 3D Acquisition and Visualization in MRI. Magn Reson Imaging 9:597–609

Cline HE, Dumoulin CL, Lorensen WE et al. (1991) Volume rendering and connectivity algorithms for MR angiography. Magn Reson Med 18:384–394

Korosec FR, Weber DM, Mistretta CA et al. (1992) A data adaptive reprojection technique for MR angiography. Magn Reson Med 24:262–274

Laub GA (1990) Display for MR angiography. Magn Reson Med 14:222–229

Li W, Haacke EM et al. (1992) Automated local maximum-intensity projection with three-dimensional vessel tracking. J Magn Reson Imaging 2:519–526

Saloner D, Hanson WA, Tsuruda JS et al. (1991) Application of a connected-voexel algorithm to MR angiographic data. J Magn Reson Imaging 1:423–430

6 Hardware Configurations

R. HAUSMANN

CONTENTS

6.1 Introduction

Like all MR techniques, flow imaging depends on the particular hardware configuration, including field strength, radiofrequency (RF) coils, gradient performance, and proper software implementations. In this chapter a short overview is given of those hardware components that affect the performance of MRA. Basically all of the various MRA techniques require optimal hardware conditions, and great benefits have derived from recent improvements, such as higher gradient performance, active shielding of gradients, and optimal coils. Different optimization strategies can compensate for particular effects like lower field strength; however, all of these strategies have their drawbacks, e.g., in terms of acquisition time.

6.2 Field Strength

6.2.1 Parameter Dependencies

6.2.1.1 Signal-to-Noise Ratio

In MRI the available signal is directly related to the strength of the main magnetic field. The higher the field strength, the higher the signal-to-noise ratio in MR images. This is, of course, also true for MRA and holds only for a given sequence and resolution. Because of this, low-field MRA has some drawbacks in terms of signal-to-noise ratio as compared to high-field applications. In fact, up to now most experience has been acquired with 1.5-T systems. In the last few years, however, more and more mid- and low-field MR systems have been installed and it has been shown that MRA is feasible at 0.5 or even 0.2 T as well (PAVONE et al. 1992; SADAMOTO et al. 1991).

6.2.1.2 T1 Effects

The longitudinal relaxation times T1 also depend on field strength: the higher the field strength, the longer is T1. Therefore, for a given repetition time, T1 relaxation is more pronounced at low field strength (short T1) than at high field strength. This leads to a higher background signal level for low-field MRA. Similar effects are introduced in flowing blood; therefore, the effect on contrast-to-noise is only limited compared to high-field MRA.

More important is the effect of field strength on the attenuation of fat signal at different echo times. This effect is described in Sect. 3.1.2.2 in more detail. Echo times have to be chosen such that the fat and water signal vectors point in the opposite direction and therefore signal is canceled within a voxel. This leads to a better signal attenuation in the background.

6.2.1.3 T2 Effects

The third effect of field strength on imaging parameters is on the transversal relaxation time T2. Although most MRA techniques are basically T1-weighted sequences, their nature as gradient-echo sequences makes them sensitive to changes in local susceptibility, resulting in signal attenuation in boundary regions between air and tissue. This effect basically is not present at low field strengths and increases with field strength.

R. HAUSMANN, PhD, Siemens AG, Medizinische Technik, Henkestraße 127, 91052 Erlangen, FRG

6.2.2 MRA at Different Field Strengths

There are basically three different levels of field strength which need to be discussed separately. As already mentioned, most experience has been acquired with high field strengths ranging from 1.0 to 1.5 T. The second level can be defined as the mid field strength, typically 0.5 T, while the third is 0.2 or 0.3 T. For the first two levels from 0.5 to 1.5 T the possibilities of MRA are technically very similar, meaning that with all of the systems 2D and 3D time-of-flight (TOF) and 2D and 3D phase contrast (PC) sequences are implemented. The differences in quality and resolution mainly depend on the signal-to-noise ratio and can be compensated by scan time to a high degree.

6.2.2.1 MRA at 0.2 T

At very low field strength, acquisition strategies have to be changed to compensate for the signal-to-noise factor. This leads to both longer scan times and less flexibility in terms of applications and techniques. Most experience at very low field strength has been acquired with 0.2-T systems from different manufactures and with different designs.

The inherently lower signal-to-noise ratio can be compensated by using proper sequences with longer read-out intervals. Typically, longer acquisition times have to be accepted due to the increased number of averages (typically two instead of one). The method of choice for 0.2-T MRA is 2D TOF. These sequences have the advantage of low saturation effects within blood vessels and therefore allow for large flip angles (up to 90°) and a higher signal-to-noise ratio. Pixel size is also somewhat larger than in 3D imaging and this helps to improve further the signal-to-noise ratio. The compensation of the lower signal-to-noise ratio is, however, incomplete, and high-field-strength MRA still provides better images, especially at higher resolution in 512 matrix applications. The main applications of low-field MRA lie in the imaging of great vessels such as the carotid arteries, the aorta, or major branches of the intracranial vasculature.

6.3 Gradient Performance

6.3.1 Gradient Strength and Rise Time

MRA sequences are typically flow-compensated (TOF) or flow-sensitized (PC) gradient-echo sequences. For optimal visualization of signal, including complicated flow patterns, short echo times are mandatory. Since phase effects scale with an at least quadratic power of the echo time, small reductions in echo time lead to large improvements, especially in pathological conditions. To decrease the echo time for a given resolution, short rise times of the gradient pulses are very important. Table 6.1 gives an overview of TE realizations for a given sequence structure (3D TOF with GMR; identical resolution, asymmetry, and slice thickness).

It is very important that the gradient motion refocusing schemes (the gradient pulse trains) are as short as possible and as close as possible to the actual center of the echo. Therefore, even if a longer echo time has to be chosen to ensure optimal fat-water cancellation, the short rise times make it possible to decrease phase effects.

Another important feature of a gradient system is the available peak power. Typical values are 10–15 mT/m. Recently, through the introduction of echo-planar systems with increased gradient performance, 23–25 mT/m is available on standard whole-body scanners. Doubling the peak gradient power increases resolution by a factor of 2. Therefore, 512 matrix imaging and even 1024 matrix imaging is feasible with these new gradient systems. This allows the acquisition of much more detail and the depiction of smaller vessels on a routine basis, and, even more importantly, it decreases the intravoxel phase dispersion (dephasing effects). In the long run, the benefits of extremely high resolution (similar to conventional angiography) will favor high-field systems due to their inherent advantages in terms of signal-to-noise ratio.

6.3.2 Fidelity of Gradient Wave Forms

Although the amplitude and rise time of a gradient pulse are important, it is much more essential that the pulse has exactly the predefined form. Figure 6.1

Table 6.1. Different gradient specifications and their effect on echo time TE of a 3D TOF sequence. All sequence parameters such as bandwidth, minimal FOV (180 mm), and echo asymmetry were kept constant

Gradient strength	Rise time	Echo time TE
10 mT/m	1 ms	6.8 ms
15 mT/m	1 ms	5.5 ms
25 mT/m	0.6 ms	3.8 ms
25 mT/m	0.3 ms	3.3 ms

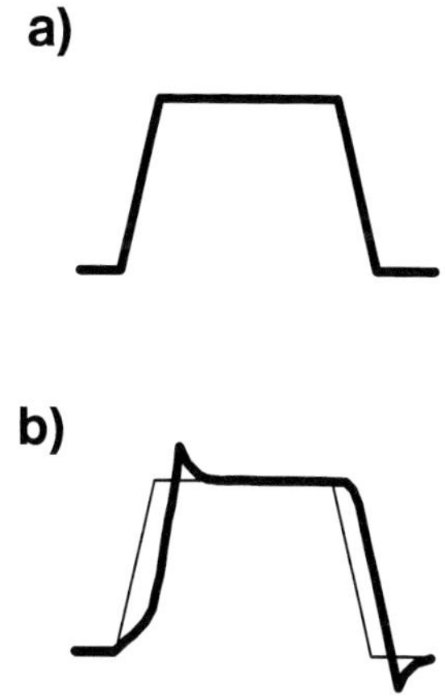

Fig. 6.1. Ideal (a) versus nonideal gradient wave form (b). Different effects like gradient delay (temporal shift, distortions due to eddy currents) can lead to substantial distortions in gradient wave forms and therefore to extreme phase effects, especially in flow-compensated sequences

shows an ideal pulse shape and different distortions which can arise from various imperfections and external effects on the gradient pulse.

Eddy currents are additional gradient fields induced when switching the actual imaging gradient pulses. These effects can be compensated by proper premodulation of the actual gradient current, so that the wave form becomes ideal again. This is only possible if the eddy currents are "well behaved", meaning that their time decay is well known. A better way of avoiding eddy currents is the use of actively shielded gradient coils. These coils compensate the influence of the gradient field on metallic structures outside the bore (where eddy currents are typically induced) by means of a secondary coil. All state of the art MR units presently on the market are equipped with such actively shielded gradient coils.

Other distortions such as time delay and imperfections due to the specific design of the gradient amplifier are possible. Proper adjustment of the hardware is mandatory for MRA. Especially, phase-sensitive MRA techniques, e.g., PCA, are extremely sensitive to any kind of distortion. Typically, even when using actively shielded gradient coils, additional phase corrections by software have to be done to avoid subtle phase distortion effects.

6.4 Radiofrequency Coils

The choice of radiofrequency coils influences the quality of MRA images in various ways. Firstly, local RF coils improve the signal-to-noise ratio and allow higher resolution imaging, as with normal parenchymal imaging. Array coils guarantee a high signal-to-noise ratio over larger fields of view and are better for imaging the abdominal vessels. With regard to signal-to-noise ratio, it is of no great importance whether receive-only coils (with RF transmission performed via the whole-body resonator) or transmit and receive coils are used.

Signal from blood vessels in TOF angiographic sequences is maximal when unsaturated blood can enter the imaging slice or volume. Therefore, orientation of the imaging volume is important. However, not only the orientation but also the method of selecting the appropriate imaging volume can affect the signal intensity. This is shown in Fig. 6.2. Using a transmit and receive head coil, optimal inflow conditions are secured even when a coronal or sagittal slab orientation is employed since the excitation volume is limited to the coil volume only. The situation is different when using a receive-only coil, as illustrated by Fig. 6.2b: Due to the large excitation volume it is likely that, when employing coronal or sagittal orientations, blood will already be completely satu-

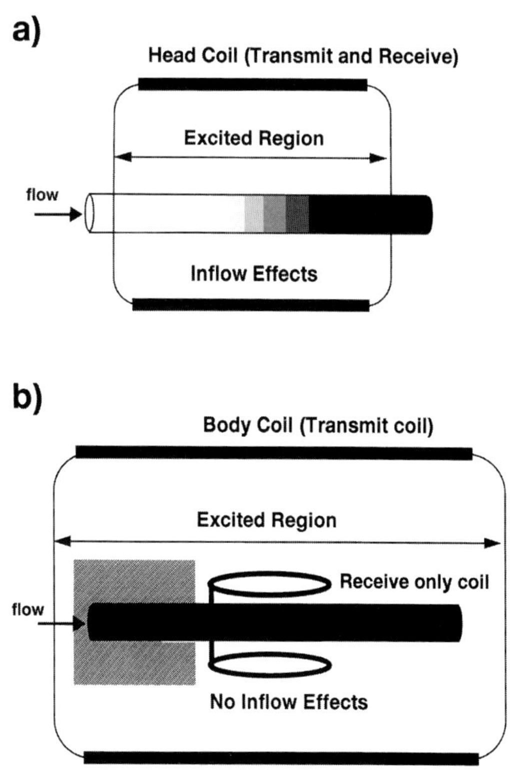

Fig. 6.2. Comparison of transmit and receive coil, e.g., head resonator (a) with receive-only coil, e.g., neck coil (b). If the excitation volume covers the shaded area, inflow is lost and signal is already saturated when entering the imaging volume

rated before entering the receive coil at all. Therefore, axial slice orientation is mandatory in this configuration! Thus the use of receive-only coils can limit the choice of slice orientations for MRA.

6.5 Physiological Synchronization

Since the flow velocities and pattern change within the cardiac cycle, it can be very helpful to gate or trigger MRA sequences. Especially in peripheral MRA, results are significantly improved by using triggered sequences. Triggering to the proper cardiac phase ensures optimal inflow conditions and therefore optimal signal and contrast.

Since ECG triggering always prolongs the scan, new acquisition modes like segmented k-space are important. These techniques will be described in Chaps. 7, 12, 13. In 3D imaging, gating to the optimal inflow window helps to improve the contrast. The increase in scan time can be tolerated.

All state of the art MR units allow ECG triggering and gating. It is essential that new techniques like segmented k-space or other advanced image acquisition and processing methods become available once clinical benefit is established at research sites. In these developments, software flexibility and proper hardware specifications represent the basis for good MRA results.

References

Pavone P, Marsili L, Catalano C et al. (1992) Low field MR angiography of the carotid arteries. Radiology 184:401–404

Sadamoto K, Miki H, Niidi A et al. (1991) Clinical applications of 3D MR angiography using 0.5 T MRI. Neuroradiology 33 (Suppl):520–521

7 Quantification of Blood Flow

J.F. Debatin and M. Hauser

CONTENTS

7.1 Introduction

Magnetic resonance techniques can go beyond the morphological assessment of the cardiovascular system by allowing direct quantitative characterization of flow dynamics. Reflecting the inherent motion sensitivity of MRI with regard to both amplitude and phase, two different MR techniques are available for flow measurements: one is based on the influence of motion on signal amplitude, the other on the influence of motion on signal phase. They can be combined with various gating schemes to achieve excellent temporal resolution throughout the cardiac cycle (Pelc et al. 1991a). Compared to Doppler ultrasound, both techniques are less operator dependent and do not require the presence of an appropriate acoustic window.

J.F. Debatin, MD, MRI Center, Department of Medical Radiology, Zürich University Hospital, Ramistraße 100, 8091 Zürich, Switzerland

M. Hauser, MD, MRI Center, Department of Medical Radiology, Zürich University Hospital, Ramistraße 100, 8091 Zürich, Switzerland

Pioneered by Wehrli et al. (1986), the amplitude technique exploits a time-of-flight effect. It is based on the application of a thin saturation band across the vessel of interest and tracking of the flow-induced displacement of this band. This technique, referred to as bolus tracking, requires in-plane visualization of the vessel. While it has been shown to accurately measure flow velocities in large thoracic vessels as well as the portal venous system (Edelman et al. 1989a,b), the technique is handicapped by saturation phenomena, volume averaging, and insensitivity to slow flow, particularly in small vessels.

The phase technique provides velocity maps that directly determine the spatial mean velocity within each voxel across the lumen of a vessel. By analyzing the vessel of interest in cross-section, it overcomes many of the limitations inherent to bolus tracking and Doppler ultrasound (Pelc et al. 1991a). This is particularly true with regard to flow volume quantitation. Doppler ultrasound and MRI bolus tracking sample flow parameters only in a single plane of a vessel's cross-section. Quantitation of blood flow volume therefore requires the additional determination of the vessel's diameter as well as assumptions with regard to the actual flow pattern within the vessel. Although the characterization and classification of the frequently complex flow dynamics seen throughout the vascular system have been the subject of some controversy, there is general agreement that physiological velocity profiles usually constitute a mixture of different classic flow profiles (Caro et al. 1978). Modeling of in vivo flow is hence rather difficult and can potentially introduce considerable measurement error. Due to the voxel-based flow analysis inherent to MRI phase mapping, the phase-contrast (PC) technique remains totally insulated from the ramifications of this discussion. Hence, accurate flow and velocity quantitation is possible throughout virtually all vascular structures with the phase-based flow-mapping technique (Pelc et al. 1991a).

Following technical refinements in gradient and sequence design, the phase technique has thus emerged as the method of choice for flow quantitation with MRI (Haacke et al. 1991). This chapter will focus on the physical principles and clinical applications of this phase-based flow quantitation technique, referred to as PC flow mapping.

7.2 Technical Considerations

As with any new sophisticated technique, achieving good results with PC flow mapping requires a thorough understanding of the physical principles underlying the flow measurements and careful attention to detail in the performance of the method.

7.2.1 Principle of Phase

Each MRI raw data set produces an image that is based upon the transverse magnetization within each voxel. This transverse magnetization is a vector quantity defined by both magnitude and direction (Fig. 7.1). The latter represents the phase of a particular spin. PC flow mapping exploits the fact that the precessing frequency of spins is dependent on the strength of the magnetic field. Spins moving through magnetic field gradients will therefore achieve a different precessing frequency and thus a different phase. In fact, in the presence of a magnetic field gradient, constant velocity motion produces a change in phase (ϕ) proportional to velocity (v):

$$\phi = v(\gamma M_1), \tag{7.1}$$

where M_1 is the first moment of the gradient wave form "$G(t)$" evaluated at the time of the center of the echo and γ is the constant gyromagnetic ratio.

Phase information is not merely affected by velocity, but also by many other processes such as magnetic field inhomogeneities, eddy currents, and pulse sequence tuning. Insensitivity to other sources of phase shift can be achieved by obtaining two measurements with different velocity-induced phase shifts, but the same phase effects due to other sources. The two data sets with different first moments along one direction are acquired in an interleaved fashion. If two complete data sets are acquired, the difference in phase ($\Delta\phi$) in each voxel is:

$$\Delta\phi = v(\gamma \Delta M_1), \tag{7.2}$$

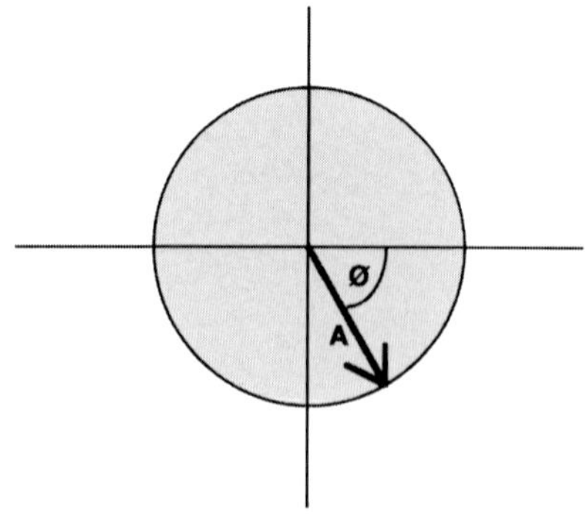

Fig. 7.1 Magnetization on the transverse plane: signal consists of an amplitude (A) and a direction ($\varnothing$ = phase)

where ΔM_1 is the change in first moment and v is the velocity component traveling in the direction of ΔM_1. The moment change causes spins moving in the encoded direction to acquire a phase shift proportional to velocity (Pelc et al. 1991a). Since static structures exhibit no phase change, the subtraction eliminates undesired phase effects. Hence the resultant "phase contrast" image merely reflects motion in the direction of ΔM_1. This concept can be generalized for the measurement of flow in any direction (Pelc et al. 1991b; Hausmann et al. 1991). For reasons of measurement accuracy it is crucial, however, that quantitative flow measurements are always performed in the slice select direction.

7.2.2 Flow Quantitation

The PC image depicts pixel values that are dependent only on motion in the slice select direction. By properly scaling the phase shift data, velocity images can be formed. Velocity data are expressed on a gray scale with negative values displayed dark, positive ones bright, and zero velocity as gray. By convention, flow from right to left, anterior to posterior, and superior to inferior is defined as "positive," while flow in the opposite direction is considered "negative." To suppress noise in regions of low signal, vascular images are frequently "magnitude masked" by multiplying the phase shift with the signal magnitude in each pixel (Pelc et al. 1991a).

Phase-contrast velocity measurements (cm/s) can be used directly, much like Doppler measurements. For flow that is oblique to the imaging plane, the velocity values must be corrected by $\cos(\theta)$, where θ is the angle between the flow direction and the slice select direction. For through-plane flow, the flow rate (ml/s) is defined by the product of measured flow velocity (cm/s) and pixel area (cm^2):

$$\text{velocity (cm/s)} \times \text{area (cm}^2\text{)} = \text{flow rate (ml/s)}. \quad (7.3)$$

The flow volume rate through any vessel can be obtained by summing the flow rates in all pixels within a region of interest that includes the entire vessel lumen.

7.2.3 Flow Sensitivity Adjustment

The strength of the flow-encoding gradient and hence the sensitivity of the PC technique to flow is controlled by ΔM_1. It can be characterized by the phase change per unit velocity and expressed as radians per cm/s. A more intuitive parameter is the "velocity encoding value" (VENC), defined as the velocity that produces a phase shift of π radians or 180°. Hence, VENC controls the dynamic range of velocities that can be measured.

Since phase is unique only over 360°, the phase shift computed from different phase reconstructions is forced to lie in the ±180° range. Any actual phase shift outside this range is erroneously depicted by a multiple of 360°. A velocity of 1.1 VENC will be erroneously depicted as –0.9 VENC. This phenomenon is referred to as velocity aliasing (Fig. 7.2), and must be considered analogous to spatial wrap-around artifacts in MRI. To avoid such aliasing, VENC should always be chosen to exceed the expected maximal velocity within the vessel of interest. However, VENC should not be chosen too high, since it is inversely proportional to the signal-to-noise ratio (SNR) in PC images (Pelc et al. 1991b). Thus Evans et al. (1993) suggest that VENC should be chosen to exceed the maximal expected velocity by not more than 25%.

7.2.4 Strategies for PC Synchronization

Phase-contrast velocity data are temporal averages of instantaneous velocities. The degree of temporal averaging is determined by the length of the data acquisition window. For nongated sequences, the latter is defined by the total imaging time. While this is sufficient for the quantitative characterization of constant flow found in much of the venous system, pulsatile arterial flow requires measurements at multiple points along the cardiac cycle. This can be achieved by combining the PC technique with various forms of prospective and retrospective gating.

Cine-PC imaging combines phase modulation with gradient-echo cine-imaging. Similar to conventional cine, cine-PC employs a short repetition time

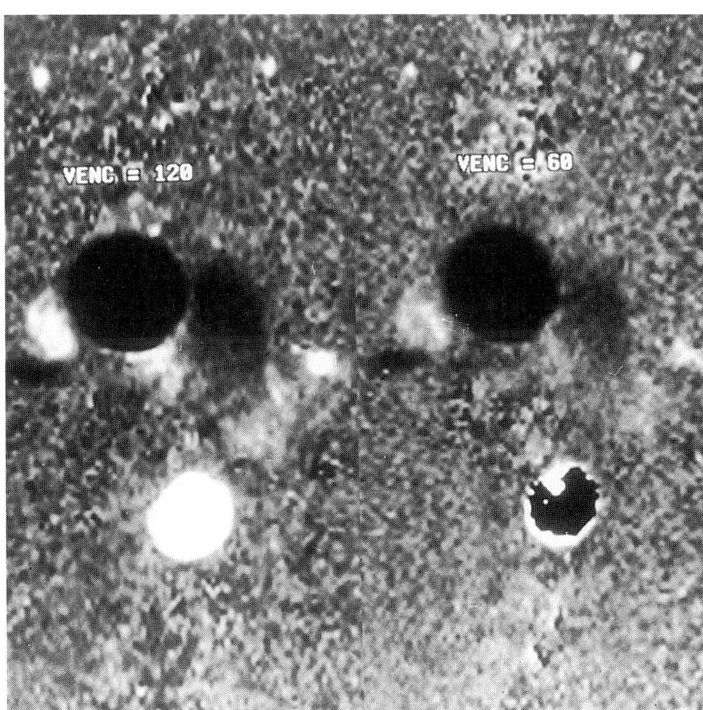

Fig. 7.2. The velocity encoding value (VENC) must be chosen appropriately to avoid aliasing. Defined as the velocity that produces a phase shift of π radians or 180°, VENC should always be chosen to exceed the expected maximal velocity within the vessel of interest

and gradient-refocused echoes. Pulse repetition is independent of the ECG signal, which is monitored and used to increment the spatial phase encoding at the beginning of each cardiac cycle retrospectively. For a single section and a single flow-encoding direction, two sequences with alternating first moments are interleaved during each cardiac cycle, yielding a two-dimensional velocity map in a vascular cross-section. The temporal resolution is defined by twice the length of the repetition time.

The number of cine-PC frames which can be obtained within a cardiac cycle is inversely proportional to the length of the repetition time and the subject's heart rate. However, the data acquired in any one cardiac cycle can be interpolated to the desired number of "n" frames (usually 16) regardless of the actual number of acquired phases. The total cine-PC scan time is determined by the product of desired phase encoding steps (128–256) and the length of the cardiac cycle.

To permit breath-held data acquisitions, the PC technique has recently been implemented in segmented k-space acquisitions (Debatin et al. 1994a; Fredrickson and Pelc 1994) as well as ultrafast echo planar data acquisition strategies (Debatin et al., to be published, a). With these techniques, between 6 and 20 quantitative data points can be obtained throughout a cardiac cycle within a comfortable breath-hold.

7.2.5 Flow Averaging Effects

Both temporal and spatial data averaging may adversely impact on PC measurement accuracy. The number of data points necessary to properly characterize a flow profile is dependent on the degree of pulsatility. Flow volume measurements without sufficient temporal resolution may contain errors. These errors are unpredictable and may over- or underestimate true flow. Temporal resolution should therefore always be maximized, providing at least 16–20 phases for flow quantification in highly pulsatile vessels such as the aorta or superior mesenteric artery. To maximize temporal resolution, flow quantification should be limited to through-plane flow, encoded in a single direction in a single imaging location.

Spatial averaging effects are somewhat less intuitive. Intravoxel partial voluming occurs along the periphery of a vessel within voxels containing both static and flowing spins. Due to flow enhancement, the signal intensity of moving spins exceeds the signal intensity of static spins. For moving spins, the total signal intensity fraction is thus far greater than the pixel volume fraction. This results in an overestimation of true flow. The degree of error is directly proportional to the number of such peripherally located voxels relative to the total number of voxels containing flowing spins. Hence the effect increases proportionally with increases in vessel angulation, slice thickness, or voxel size (Pelc et al. 1991a).

Another form of spatial averaging occurs in vessels subject to motion perpendicular to their flow direction (Debatin et al. 1994a). This applies particularly to the quantitative analysis of intra-abdominal vessels like the portal vein and renal arteries. Motion causes blurring of the target vessel, resulting in overestimation of the apparent vessel size (Fig. 7.3). Around the vessel's periphery a number of voxels contain flowing and static spins for varying degrees of time. This too causes overestimation of the flow rate, the extent of which is dependent on the degree of motion relative to the vessel size (Debatin et al. 1994a).

7.2.6 Optimal Quantitative Analysis

In order to maximize the accuracy of quantitative PC flow mapping analysis, a number of ground rules apply:

1. *Repetition time (TR)*: The shortest possible repetition time should be chosen (20–40 ms) to assure maximal temporal resolution.
2. *Echo time (TE)*: The shortest possible echo time should be chosen (5–10 ms) to minimize dephasing artifacts.
3. *Flip angle*: A flip angle of 30–45° renders good intravascular signal, while preserving sufficient signal in the stationary tissue does not exacerbate intravoxel partial volume effects. A smaller flip angle should be chosen for ultrafast turbo gradient PC sequences.
4. *Imaging plane*: To maximize measurement accuracy and precision, vessels should be imaged perpendicular to their course. To assure an optimal PC imaging plane perpendicular to the vessel of interest, double oblique localizing strategies are imperative.
5. *Sectional thickness*: To limit partial voluming, the sectional thickness should be chosen as thin as possible, while maintaining adequate SNR. A thickness of 3–6 mm is recommended.

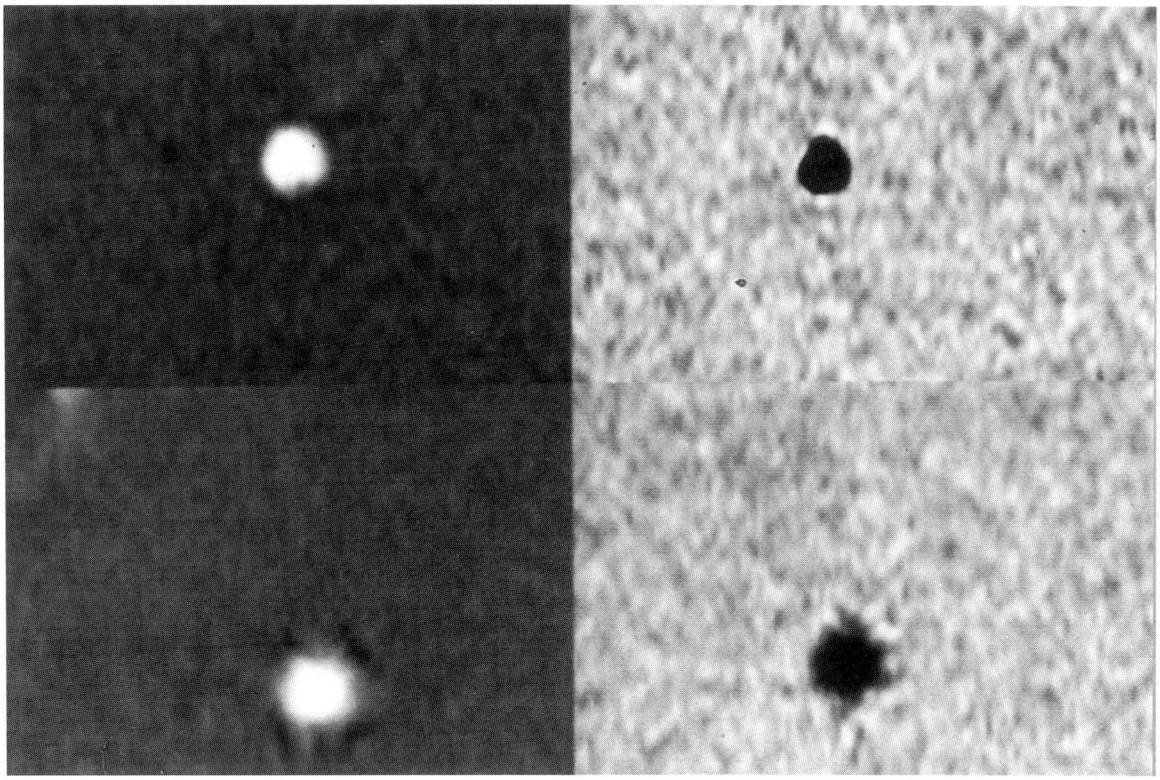

Fig. 7.3 Amplitude (*left*) and corresponding phase image (*right*) of a flow phantom acquired with (*top*) and without (*bottom*) simulated respiratory motion. Motion induces blurring of the vessel margin, which results in an artifactual increase in vessel size

6. *In-plane resolution*: Field-of-view and matrix should be chosen to assure a minimum of 12–16 pixels within the lumen of the vessel of interest.
7. *Temporal resolution*: Temporal resolution should be maximized by only mapping through-plane flow and encoding in a single direction.
8. *Velocity-encoding value (VENC)*: VENC should be chosen to exceed the maximum expected flow velocity by 25%. Values summarized in Table 7.1 should be viewed as guidelines.
9. *Frequency-encoding direction*: The frequency-encoding direction should be chosen to avoid signal contamination in the vessel of interest of pulsatility artifacts.
10. *Saturation bands*: Phase artifacts from surrounding pulsatile structures may be reduced by the placement of spatial presaturation bands. These bands should be placed carefully to avoid inadvertent saturation of inflowing spins in the vessel of interest.
11. *Data acquisition strategy*: While conventional data acquisition is sufficient for the quantitative analysis of most vessels, fast and ultrafast acquisition strategies should be chosen to quantitate flow in vessels subject to respiratory motion. To assure sufficient accuracy, these data must be acquired within a comfortable breath-hold.
12. *Flow data analysis*: To reduce noise, the measurement region of interest (ROI) should be defined carefully to include only the vessel volume. These definitions should be performed separately for each frame on the magnitude images, using a threshold approach based on the maximal intravascular signal intensity. While the correct threshold value depends on a variety of factors, including vessel size, velocity profile, and signal intensity of the surrounding tissues, it ranges between 30% and 50% of the maximal intraluminal signal intensity for most vascular structures.

7.3 Clinical Applications

The number of clinical applications for quantitative PC imaging has been growing rapidly. The ensu-

Table 7.1. Velocity encoding value (VENC) guidelines

Carotid arteries	100 cm/s
Cardiac valve plane/proximal ascending aorta	200 cm/s
Internal mammary artery/venous grafts	60 cm/s
Thoracic aorta	150 cm/s
Abdominal aorta	120 cm/s
SMA/splenic artery	120 cm/s
Renal arterles	100 cm/s
Iliac arteries	120 cm/s
Femoral arteries	100 cm/s
Popliteal arteries	80 cm/s
Tibial/peroneal arteries	40 cm/s
Pulmonary arteries	80 cm/s
Vena cava	40 cm/s
Portal vein	40 cm/s

ing discussion will be limited to representative examples and must by no means be considered comprehensive.

7.3.1 Cerebrovascular System

In the nondiseased state, the majority of the clinically important arteries of the head and neck are characterized by a laminar flow profile and are only mildly pulsatile. These favorable conditions make them ideally suited for quantitative PC flow mapping (Fig. 7.4), complementing the already well-established vascular MRI methods. In contrast to other quantitative techniques, such as stable xenon computed tomography, single-photon emission tomography, and positron emission tomography, which yield total or regional cerebral blood flow (Burt et al. 1992; Fazekes et al. 1988; Meyer et al. 1981; Shirahata et al. 1985), the PC technique allows for the assessment of individual vessels. Localized disease can thus be detected. Furthermore it allows for the comprehensive evaluation of patients with subclavian steal syndrome. In addition to quantifying the degree of subclavian steal, PC flow mapping can assess the efficacy of vascular reconstructive surgery (Bornstein et al. 1988).

Phase-contrast flow analysis has considerable potential in the evaluation of flow-restrictive extracranial lesions as well as intracranial lesions. Beyond distinguishing the two, PC flow analysis may help determine the degree of flow compromise and might be used to evaluate the progression or stability of disease.

With regard to vascular malformations, potential goals of PC flow analysis include the identification of typical hemodynamics and the assessment of therapeutic interventions. Marks et al. (1992) noted carotid arterial flow to be increased bilaterally in 16 patients with large arteriovenous malformations (AVMs). Following embolization of the AVMs, repeat flow analyses revealed considerable reductions in total cerebral blood flow (Marks et al. 1992).

Phase-contrast techniques have also been used to measure flow in the sagittal sinus (Jordan et al. 1992). By using a combination of MRA and PC flow analysis techniques, it may be possible to determine the degree of venous compromise and to evaluate the flow rate through the affected venous channel.

7.3.2 Cardiac Applications

7.3.2.1 Overall Cardiac Function

Phase-contrast MRI provides an ideal means for the quantitative assessment of cardiac function through flow volume measurements of the ascending aorta (left ventricular function) or the main pulmonary artery (right ventricular function) corresponding to cardiac output (Fig. 7.5). The imaging plane (perpendicular to the vessel) and velocity encoded value (150 cm/s in the aorta; 60 cm/s in the pulmonary artery) must be chosen carefully. PC-based velocity and flow volume determinations have been found to be highly accurate (Kondo et al. 1991; Evans et al. 1993; Maier et al. 1989). To avoid dephasing artifacts the measurement plane in the ascending aorta must be chosen somewhat distal to the valve plane. Coronary flow therefore cannot be accounted for, resulting in a slight underestimation of true aortic flow (Kondo et al. 1991; Evans et al. 1993).

7.3.2.2 Valvular Heart Disease

Valvular heart disease, including aortic regurgitation, has been studied extensively using the PC technique (Dulce et al. 1992; Mostbeck et al. 1992; Honda et al. 1993). The amount of antegrade and retrograde flow in the ascending aorta can be quantified with PC flow analysis (Fig. 7.5). Retrograde diastolic flow in the aorta represents a combination of normal regurgitant coronary flow, estimated at between 2% and 6% of cardiac output (Dulce et al. 1992; Bogren et al. 1989a), and blood escaping retrograde into the left ventricle through the aortic valve (Honda et al. 1993). Patients with aortic insufficiency will have increased retrograde flow during di-

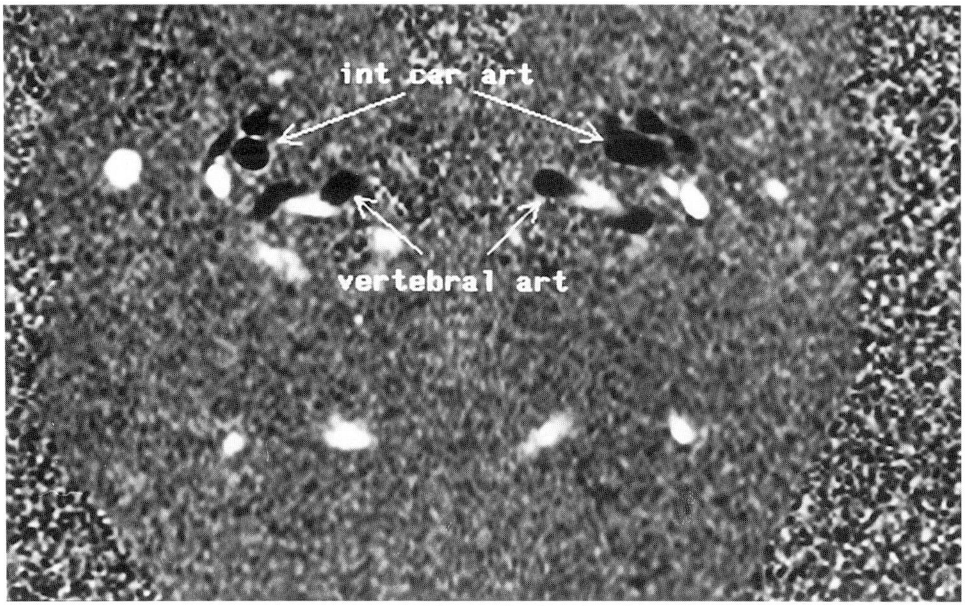

a

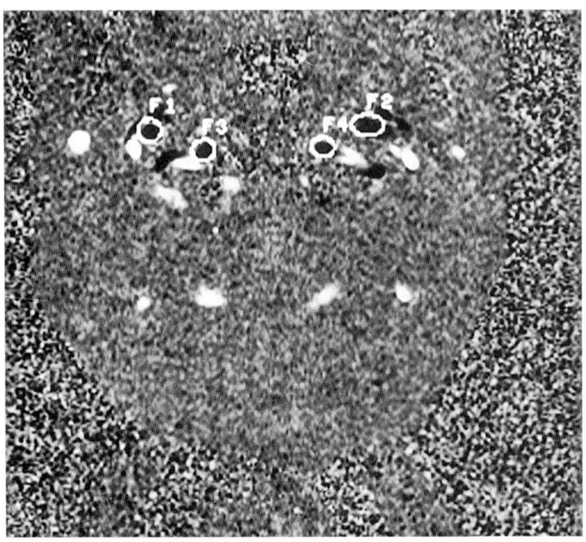

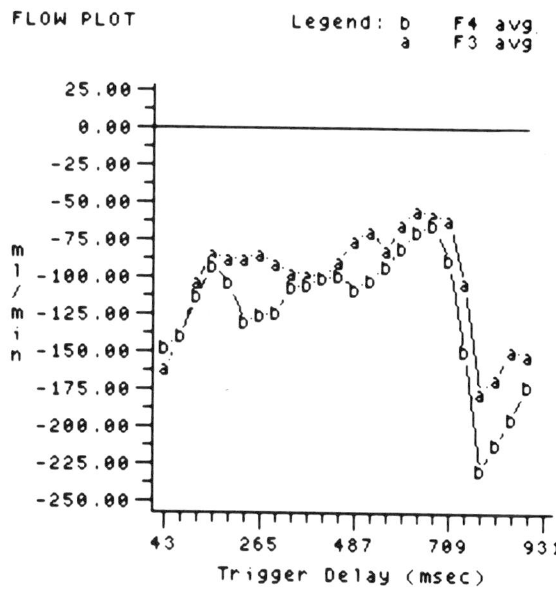

FLOW RATE / CYCLE (ml/min)

	F3	F4
Avg:	-101.4	-122.3

FLOW DATA (ml/min)

TRIG DLY (msec)	F3 AVG	F4 AVG
43	-160.6	-147.2
80	-142.4	-138.7
117	-102.7	-112.0
154	-84.2	-91.5
191	-87.7	-103.6
228	-86.9	-129.3
265	-84.8	-125.5
302	-90.9	-122.7
339	-96.3	-107.0
376	-97.8	-105.4
413	-99.4	-100.5
450	-89.3	-98.1
487	-74.6	-107.7
524	-68.7	-101.3
561	-81.5	-92.4
598	-64.4	-79.5

b

Fig. 7.4a–c. Internal carotid arteries (ICA) and vertebral arteries (VA)In a normal subject. SI encoded PC images shortly above the carotid bifurcation (**a**) demonstrating the ROIs for the ICA and VA flow measurement (**b**, *left*). The table in **b** shows average flow rates (RVA 100 ml/min, LVA 122 ml/min), and flow data are also illustrated. **c** Volume flow rates (*F1*, RICA; *F2*, LICA; *F3*, RVA; *F4*, LVA)

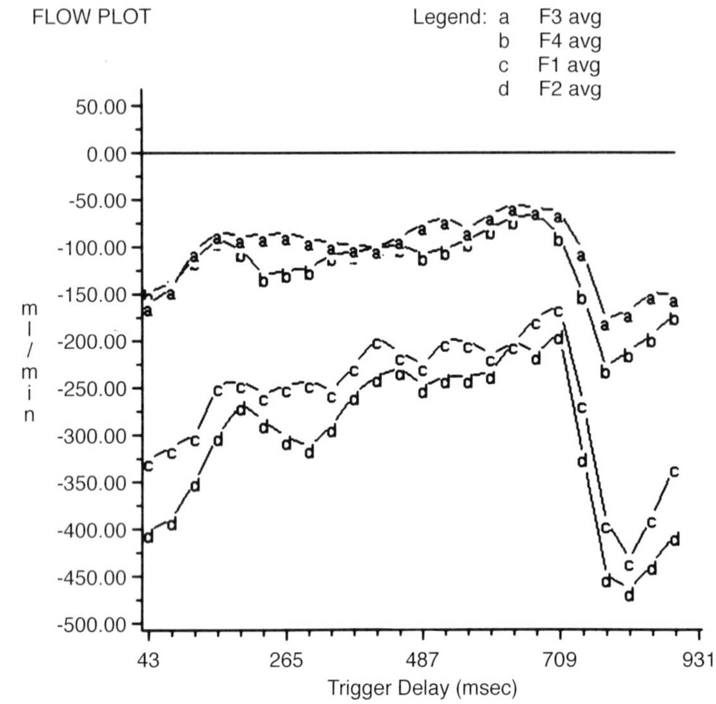

Fig. 7.4c

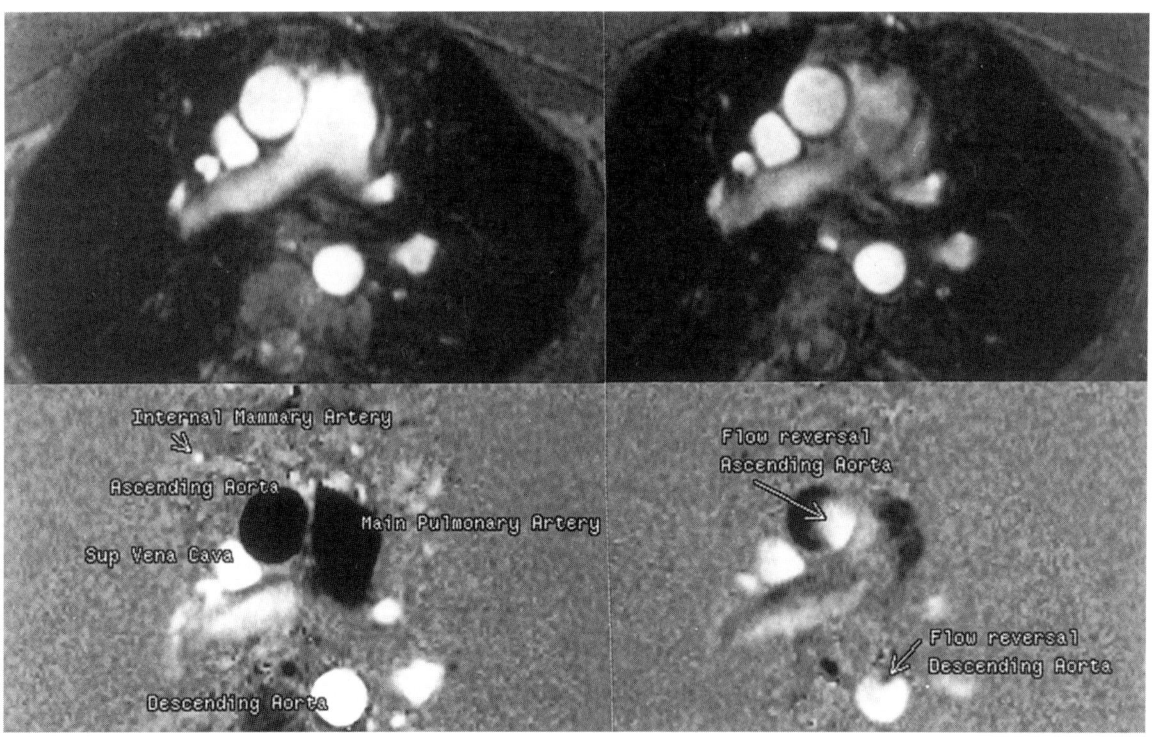

a

Fig. 7.5a,b. Ascending and descending aorta in a normal subject: axial cine-PC acquisition at the level of the pulmonary artery bifurcation, using ECG triggering and a VENC of 150 cm/s. Magnitude and SI encoded PC images of two frames acquired in early systole (**a**, *left*) and mid diastole (**a**, *right*) are displayed with the flow volume profiles (**b**). Maximal flow in the ascending aorta occurs during early systole, at which time it is homogeneously antegrade. Toward the end of systole and during diastole, a channel of retrograde flow develops in the ascending aorta along the left posterior wall (**a**, *right*), while there is continued antegrade flow along the anterior wall

Fig. 7.5b

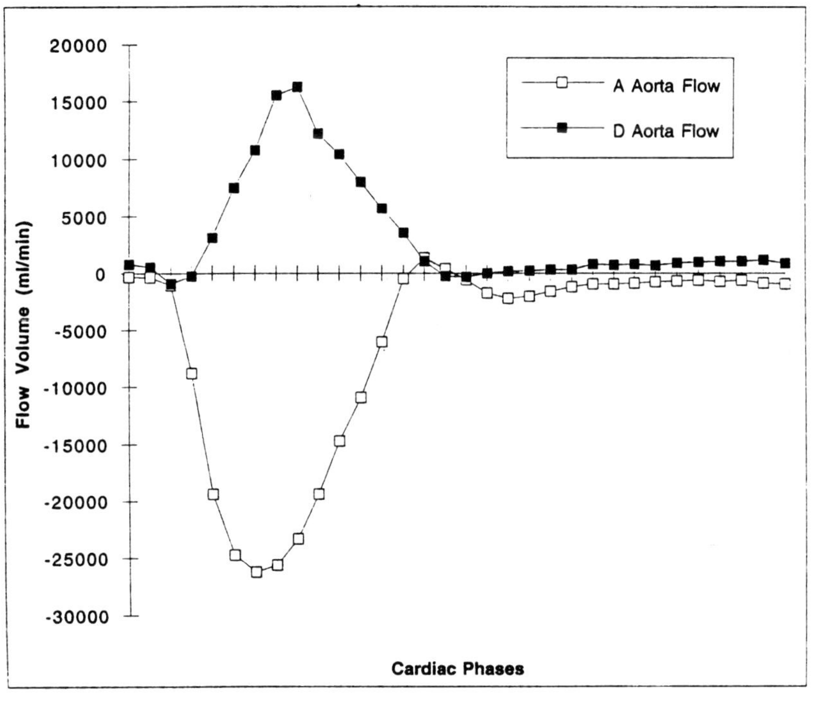

astole regardless of whether the aortic regurgitation is isolated or part of a complex multivalvular disease process.

In patients with severe mitral valve insufficiency (Mohiaddin et al. 1991a; Globits et al. 1992) the PC flow profile across the mitral valve is altered. Instead of the normal biphasic antegrade flow, these patients have reverse flow in the pulmonary veins during systole (Mohiaddin et al. 1991a; Galjee et al. 1991).

Flow patterns in patients with valvular stenosis contain higher-order-motion terms causing complex flow in and distal to the area of stenosis (Sondergaard et al. 1992). Peak systolic jet velocities up to 5.6 m/s and 2.4 m/s have been seen in patients with aortic stenosis and mitral stenosis, respectively (Eichenberger et al. 1993). By aligning a short-axis PC imaging plane perpendicular to the aortic jet within 1 cm of the aortic valve, one investigating group has reported encouraging results with regard to quantification of aortic flow (Eichenberger et al. 1993). The initial velocity encoding value was set at 500 cm/s and adjusted upwards if aliasing occurred. Based on PC-measured maximum instantaneous aortic jet velocities, pressure gradients can be calculated using the simplified Bernoulli equation. PC-based gradient determinations correlated closely with Doppler and cardiac catheterization measurements, both in normal subjects and in patients with aortic stenosis (Eichenberger et al. 1993). At this time, however, the clinical utility of the PC-based parameters remains ill defined.

7.3.2.3 Congenital Heart Disease

The noninvasive evaluation of postsurgical patients with congenital heart disease is clearly desirable. Shunt volumes can be easily assessed by determining right and left ventricular outputs (Pelc et al. 1991a; Rees et al. 1989) (Fig. 7.6).

Ideal evaluation of surgical shunts should go beyond the determination of patency. PC imaging provides an ideal means for the quantitative measurement of flow in surgical shunts. Shunt volume determinations over time as well as in differing physiological conditions provide most valuable clinical information (Kilner et al. 1991; Sieverding et al. 1992) (Fig. 7.7). Surgical grafts must be sufficiently well localized to permit a PC imaging plane perpendicular to the conduit clear of any metallic clips, which may induce dephasing artifacts.

7.3.3 Thoracic Aorta

The velocity-dependent contrast inherent to PC imaging can differentiate between fast flow, slow

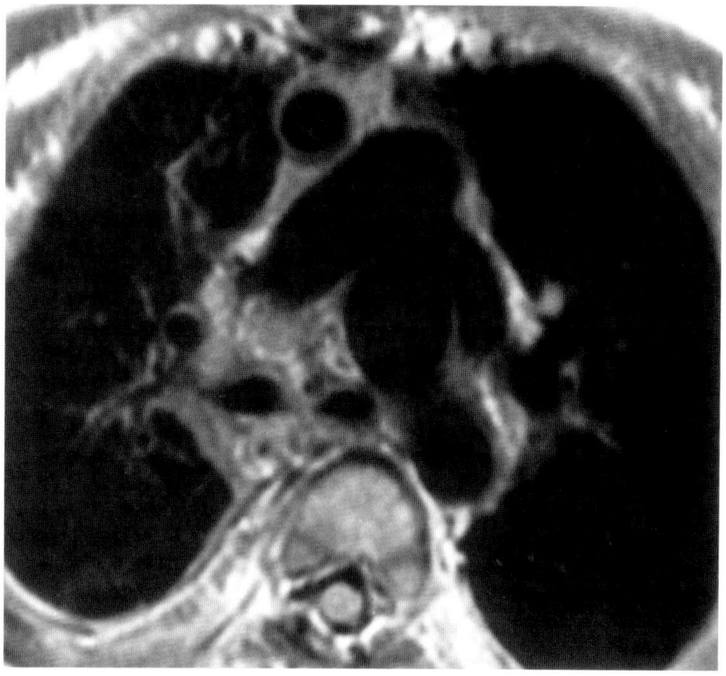

a

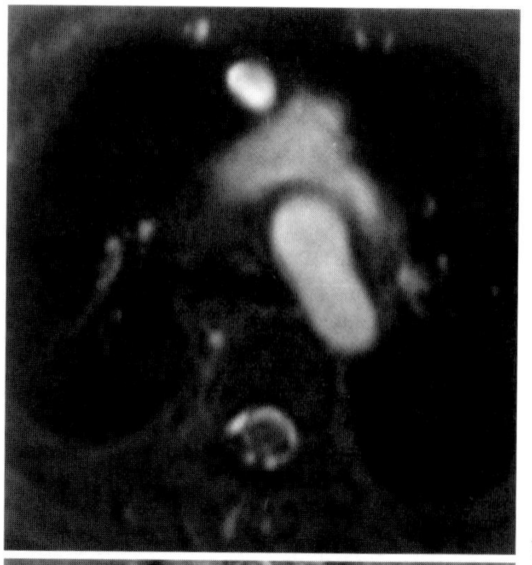

b

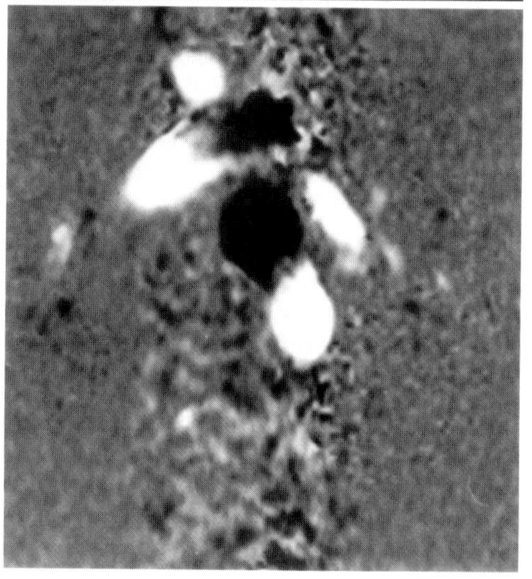

c

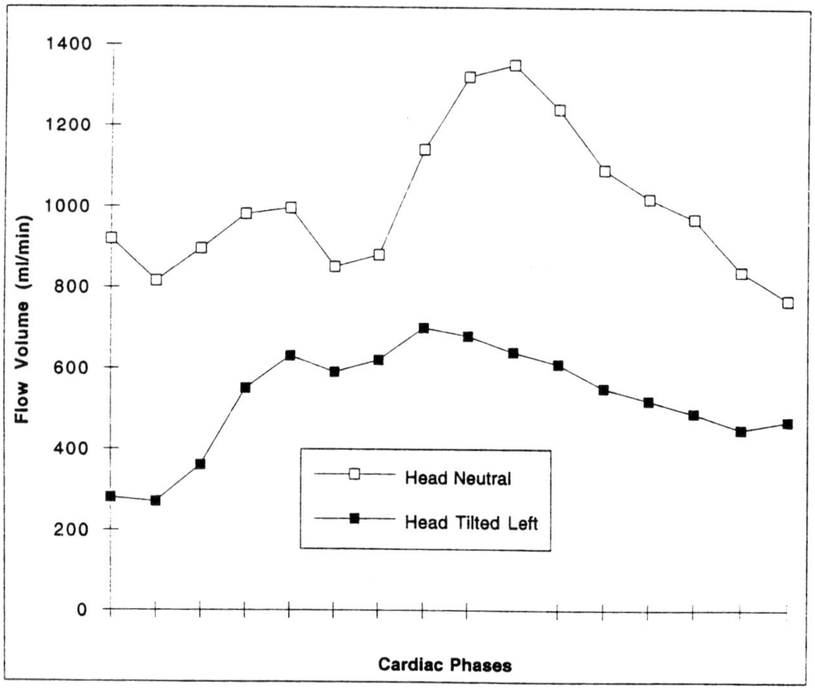

d

Fig. 7.7a–d. Fifteen-year-old patient with complex congenital heart disease: status post placement of a surgical jugulocaval shunt for obliteration of the superior vena cava. At the time of the examination the patient presented with positional transient ischemic attacks. **a** Axial spin-echo image at the level of the pulmonary artery bifurcation depicts the graft located posterior to the sternum. On the magnitude (**b**) and corresponding PC image (**c**) of an axial SI flow encoded cine-PC acquisition, obtained at acomparable level, the conduit is shown to contain flow in the SI direction. At this level, cine-PC acquisitions were performed with the patient's head neutral and tilted to the left. Compared to graft flow with the patient's head in a neutral position, flow in the left-side down position was markedly reduced (**d**). Positional kinking of the graft at the site of anastomosis was confirmed surgically

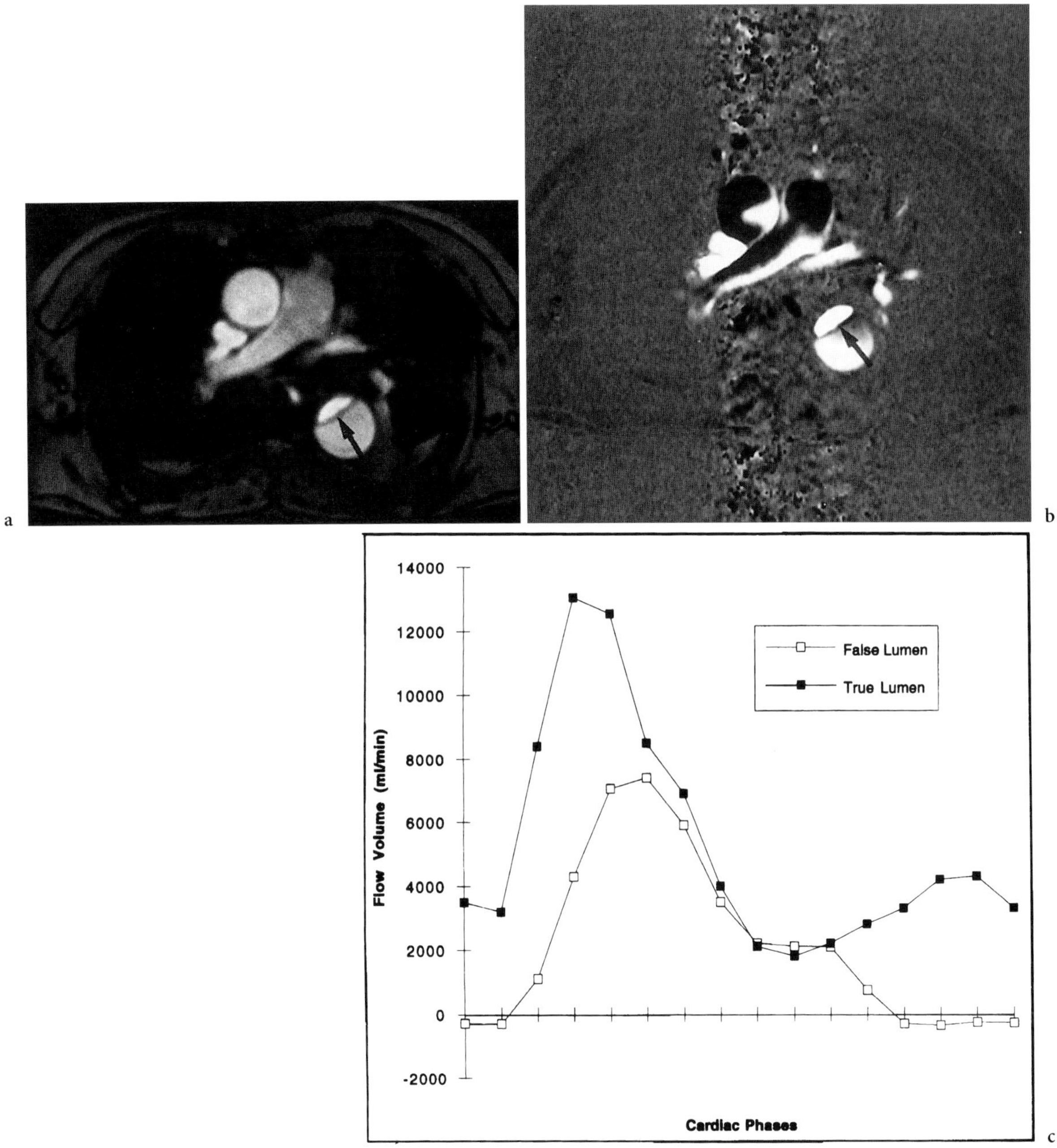

Fig. 7.8a–c. Axial cine-PC acquisition obtained at the level of the pulmonary artery bifurcation in a 41-year-old in a patient with a Stanford type B aortic dissection. **a** Magnitude and **b** SI encoded PC images of a single systolic frame demonstrate the descending aorta to be aneurysmally dilated. **c** Flow volume in the true and false lumen plotted for the 16 acquired frames. The intimal flap is visible on both images (**a**,**b**), separating the larger false lumen posterolaterally (*arrow*) from the compressed true lumen. On the PC image, signal in the false lumen is more heterogeneous, with actual reversal of flow seen as dark signal. Overall the flow in the false lumen is lower than in the true lumen, however

perfusion data, PC flow mapping is capable of calculating blood flow to each lung in addition to reliably determining the cross-sectional area of the central pulmonary vessels. In order to help determine which lung should be transplanted, the accurate measurement of flow to each lung is particularly useful in the planning stages of single lung transplantation (LEVINE et al. 1990). Such measurements may also be useful for postoperative follow-up as well as in the assessment of a wide variety of pulmonary disease states (CAPUTO et al. 1991).

Flow profiles in patients with pulmonary arterial hypertension differ from those in normal subjects (Fig. 7.9). In a recent study, both pulmonary peak systolic velocities and blood flow were significantly lower in patients with pulmonary arterial hypertension (30 cm/s vs 59 cm/s and 48 ml vs 63 ml) despite similar cardiac outputs. Furthermore, in patients with pulmonary hypertension, blood flow at peak systole was significantly more heterogeneous, with wide differences reported between maximum central and spatial mean velocities (KONDO et al. 1992). Finally, the percentage of retrograde flow in patients with pulmonary arterial hypertension is significantly greater than that of healthy subjects (17.3% vs 3.3%). The retrograde channel occurs along the right posterior wall of the main pulmonary artery in mid-late systole or early diastole (KONDO et al. 1992). While these observations appear promising, the exact sensitivity and specificity of these abnormal flow charac-

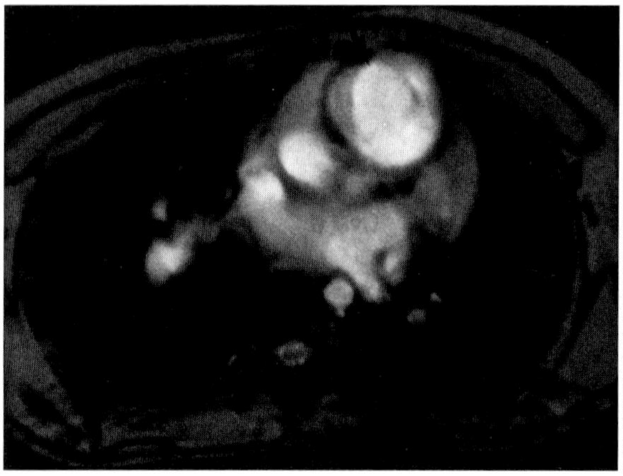
a

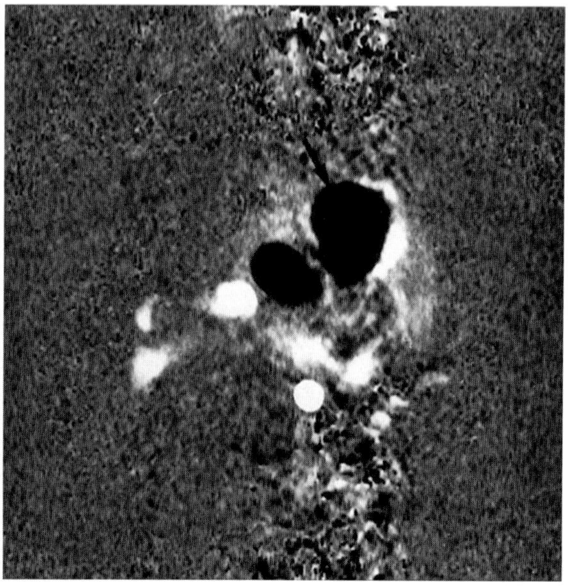
b

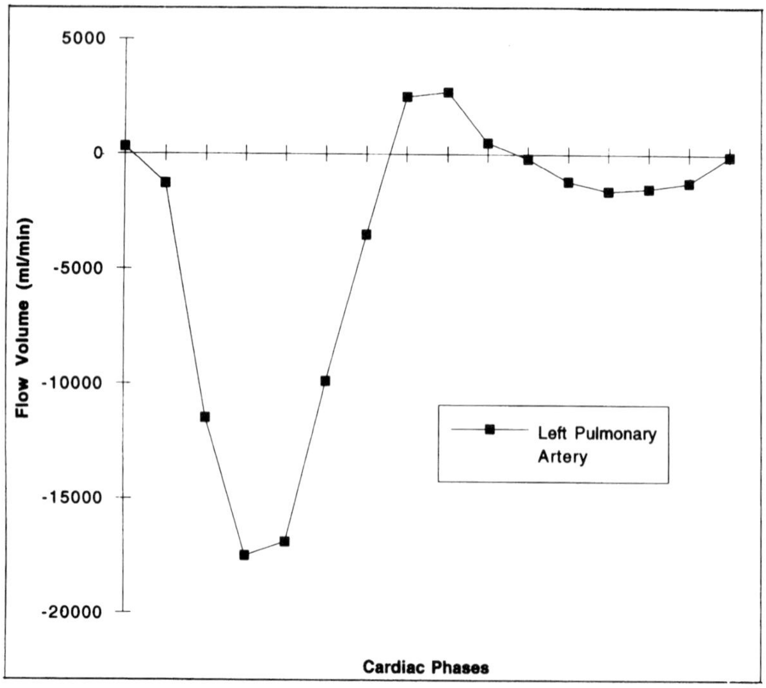

c

Fig. 7.9a–c. Forty-year-old female with primary pulmonary arterial hypertension: axial cine-PC acquisition obtained at the level of the pulmonary outflow tract. **a** Magnitude and **b** SI encoded PC images of a single systolic frame demonstrate a significantly dilated pulmonary outflow tract (*arrow*). The flow pattern of the pulmonary outflow tract and the left pulmonary artery is abnormally pulsatile, raflecting the increased pulmonary pressures (**c**). There is merely minimal flow during diastole. Flow reversal during early diastole indicates a small component of pulmonary valvular insufficiency teristics with regard to the identification and classification of pulmonary arterial hypertension still remain to be established.

7.3.5 Native and Grafted Internal Mammary Arteries

Normal internal mammary artery (IMA) flow is antegrade throughout the cardiac cycle with highly variable peak systolic velocities, ranging from 37 to 101 cm/s in normal subjects. Flow volume is similar from side to side (DEBATIN et al. 1993).

Flow wave forms in grafted IMAs are significantly different from those in native vessels (Fig. 7.10), as shown in a study evaluating IMA coronary bypass grafts in 15 patients. Flow is less pulsatile with lower peak systolic velocities (31 cm/s in grafted IMAs vs 49 cm/s in native IMAs) (DEBATIN et al. 1993). There is greater flow during diastole, which frequently exceeds systolic flow. This flow pattern reflects simulation by the IMA graft of coronary flow dynamics, which are characterized by maximal flow during diastole and only little flow during systole. Flow in the IMA grafts varied considerably from 28 ml/min to 164 ml/min (mean = 80.3 ml/min) (DEBATIN et al. 1993). Through serial follow-up, PC flow mapping may aid in the early detection of graft disease, enhancing the ability to intervene successfully. The complex anatomy and close proximity of venous grafts to the moving heart have so far precluded their reliable quantitative evaluation.

Cardiac and respiratory motion, combined with a very small size, has made coronary arteries largely elusive to PC-based blood flow measurements. The potential of determining coronary flow reserve separately for each of the three major coronary arteries promises to serve as sufficient motivation to overcome the outlined problems in the near future.

Already success has been reported in measuring coronary sinus flow (VAN ROSSUM et al. 1992). PC measurements of mean blood flow in normal subjects are reported at 144 ml/min with a mean velocity of 2.1 cm/s and a mean cross-sectional vessel area of 1.2 cm^2. The flow pattern in the coronary sinus is described as biphasic: the first peak occurs in systole and the second peak in early diastole (VAN ROSSUM et al. 1992). At this time it is unclear whether coronary sinus flow measurements reflecting global left ventricular perfusion may become clinically useful in the assessment of coronary flow reserve (VAN ROSSUM et al. 1992).

7.3.6 Renal Arterial Blood Flow

Compromise of renal arterial blood flow has long been recognized as a cause of hypertension (HILLMAN 1989) and end-stage renal disease (RIMMER and GENNARI 1993). Despite considerable efforts, reliable identification of these patient subgroups has remained imperfect. Most imaging methods hitherto employed in the diagnosis of renovascular disease, including the accepted gold standard, contrast angiography, rely on the morphological assessment of renal arteries. These techniques are limited by their inherent inability to predict the hemodynamic importance of a particular arterial lesion. PC flow mapping promises to overcome this deficit by complementing morphological renal arterial MRI strategies (DEBATIN et al. 1991).

A prerequisite for any PC-based diagnosis of renovascular disease, however, is the ability to quantitate renal arterial flow accurately. In vitro and in vivo analysis has shown that accurate renal flow measurements are possible only if based upon breath-held data acquisitions (DEBATIN et al. 1994a). Relative to renal blood flow measurements obtained by means of para-aminohippurate clearance, non-breath-held cine-PC imaging unpredictably overestimates flow, reflecting artifactual enlargement of the apparent vessel size (Fig. 7.11). Accurate flow measurements, on the other hand, have been obtained with TRIADS-PC (time-resolved imaging with automatic data segmentation) (FREDRICKSON and PELC 1994), a view-order selection technique permitting the acquisition of six equally spaced PC frames in 37 s (Fig. 7.12).

In addition to facilitating the diagnosis of renovascular disease, PC flow analysis may prove valuable in assessing the effects of various vasoactive agents. Initial work with amino acids and dobutamine has clearly demonstrated the ability of breath-held renal PC measurements to characterize even subtle pharmacological effects (DEBATIN et al. 1994b).

Phase-contrast flow analysis can also be employed to quantitate the therapeutic effect of various interventions such as percutaneous transluminal angioplasty. Here, too, initial work in our laboratory has been rather promising.

7.3.7 Superior Mesenteric Artery

Mesenteric ischemia remains a frequently perplexing diagnostic dilemma. Cine-PC-based flow measure-

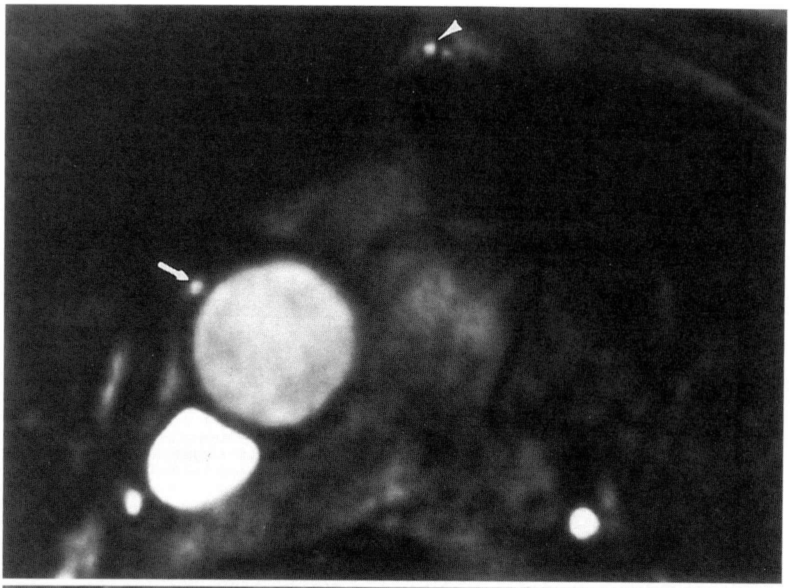

a

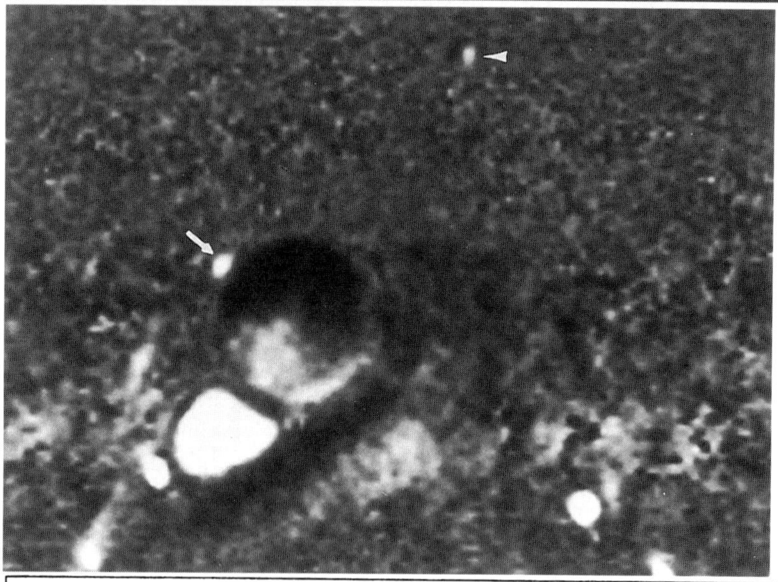

b

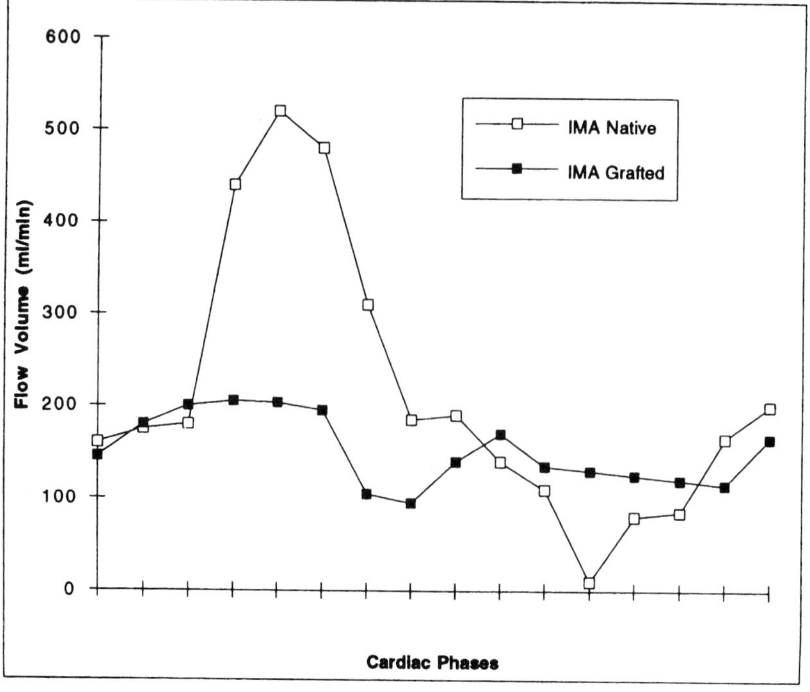

c

Fig. 7.10a–c. IMA bypass graft: axial cine-PC acquisition obtained at the level of the pulmonary artery bifurcation in a patient who had recently undergone surgery. **a** Magnitude and **b** SI encoded PC images of a single systolic frame demonstrate a right-sided IMA bypass graft in cross-section. **c** Flow volume plots compare the native right-sided IMA with the grafted vessel on the left. Peak systolic flow volume was lower in the graft than in the native vessel, reflecting assimilation of graft flow to coronary flow patterns. The patent IMA graft carried 110 ml/min to the left anterior descending coronary artery

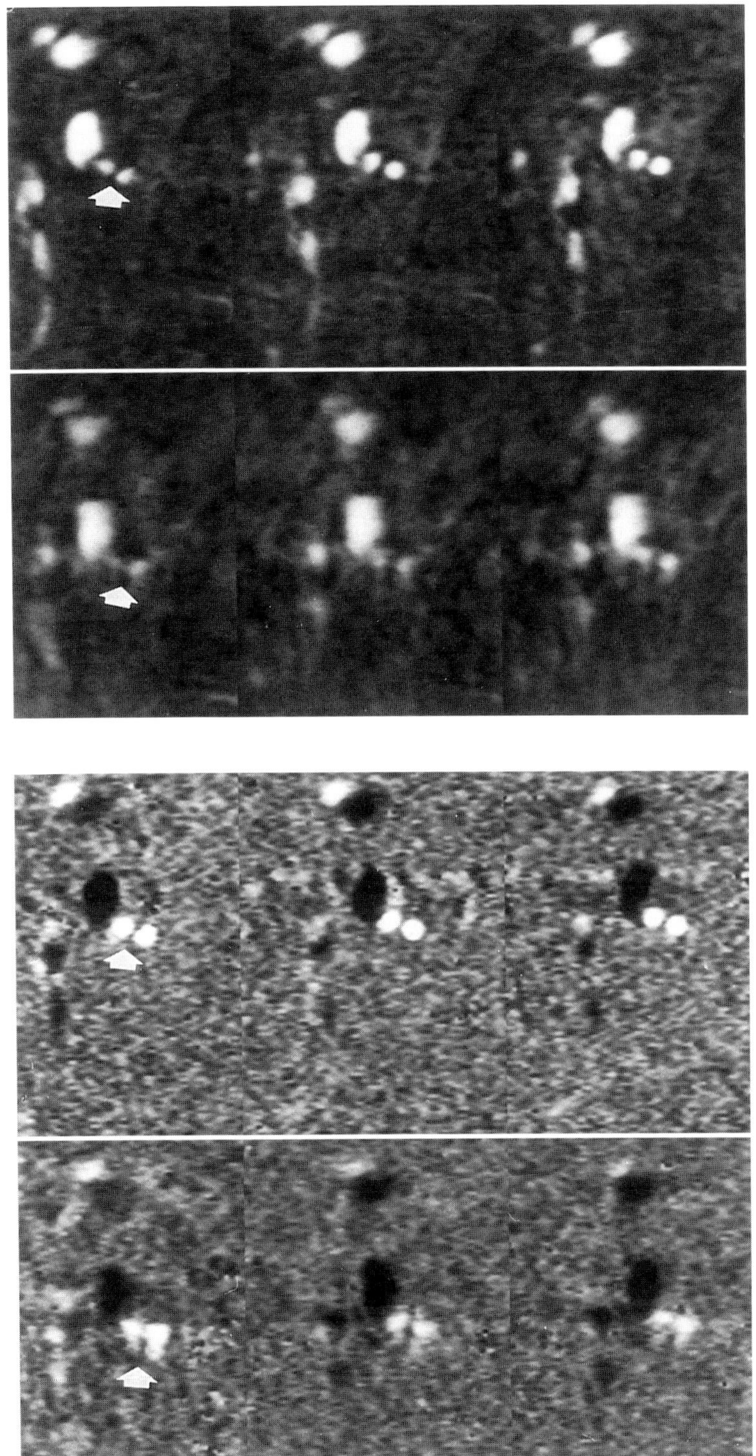

a

b

Fig. 7.11. **a** Magnitude and **b** PC images of three frames equally spaced throughout the cardiac cycle of breath-held TRIADS-PC (*upper rows*) and non-breath-held cine-PC (*lower rows*). The subject, who was evaluated in an initial volunteer study, is known to have two left renal arteries (*thick arrow*). Both vessels are clearly identified as such on both image sets. On the cine-PC images, the vessels are less well defined and characterized by blurred margins due to respiratory motion

Fig. 7.13b,c

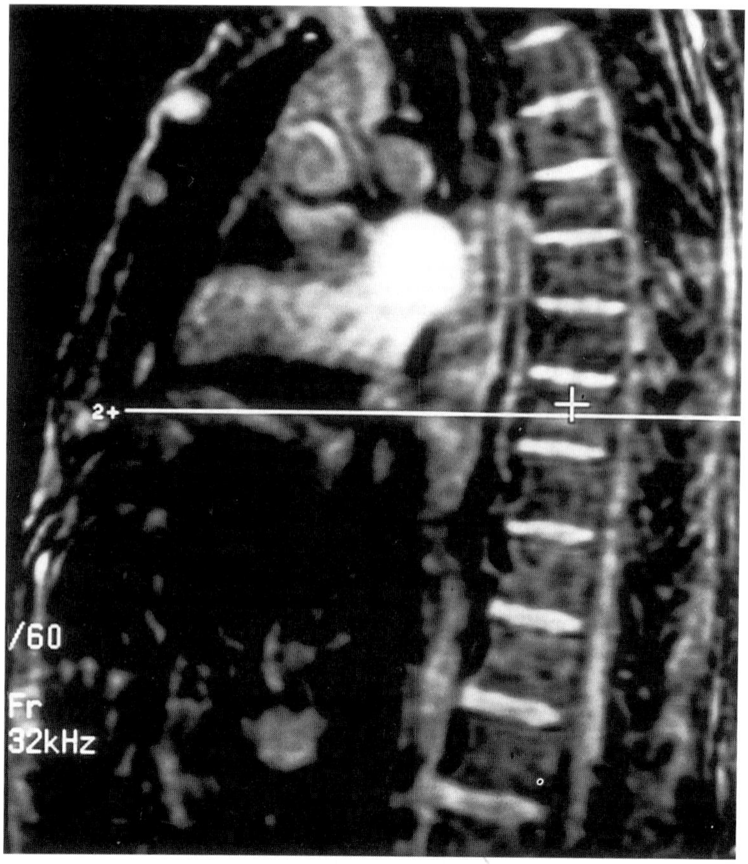

b

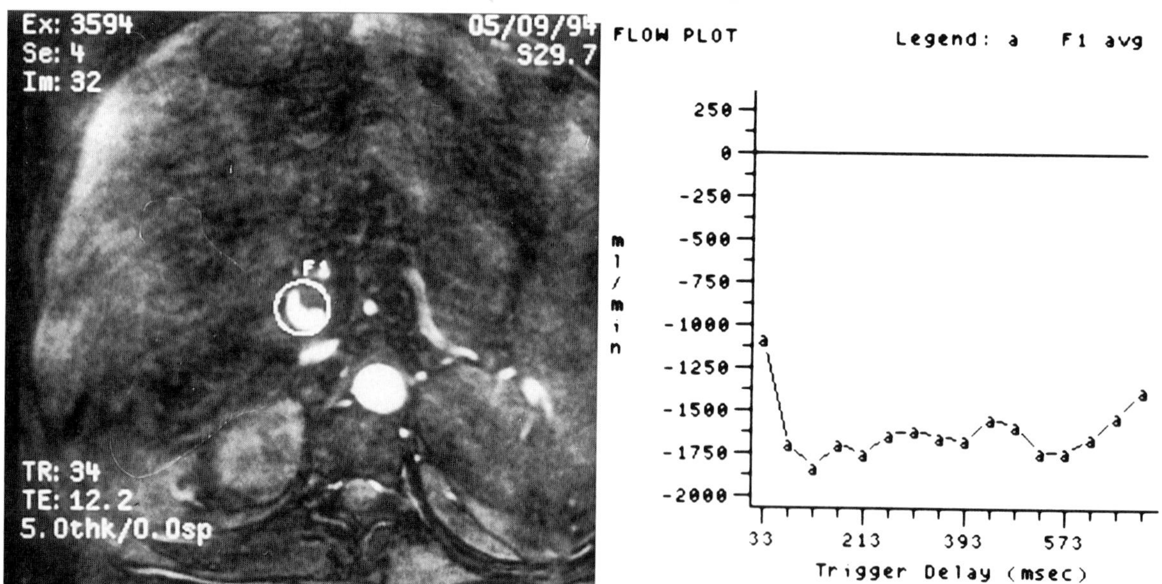

FLOW RATE / CYCLE (ml/min)

F1

Avg: -1611.9

FLOW DATA (ml/min)

TRIG DLY (msec)	F1 AVG
33	-1088.6
78	-1701.0
123	-1836.2
168	-1698.3
213	-1743.3
258	-1645.7
303	-1606.7
348	-1650.2
393	-1673.3
438	-1544.9
483	-1578.8
528	-1729.8
573	-1738.2
618	-1652.4
663	-1527.0
708	-1376.1

c

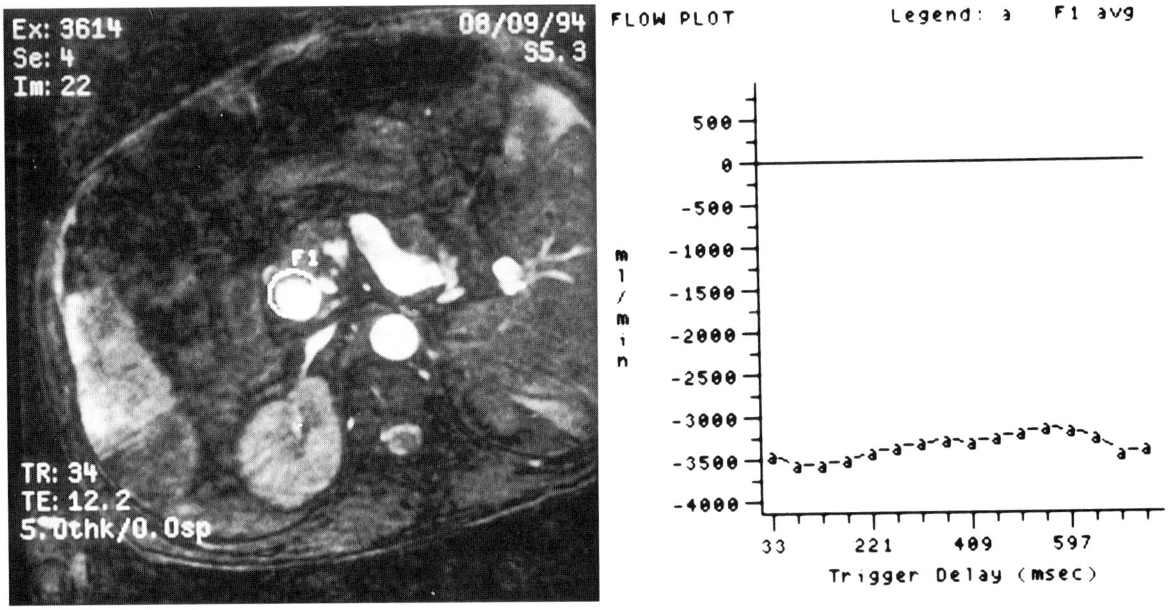

FLOW RATE / CYCLE (ml/min)

F1

Avg: -3346.7

FLOW DATA (ml/min)

TRIG DLY (msec)	F1 AVG
33	-3437.5
80	-3567.8
127	-3544.3
174	-3494.5
221	-3416.8
268	-3370.3
315	-3311.2
362	-3284.3
409	-3309.6
456	-3246.0
503	-3200.5
550	-3131.8
597	-3154.0
644	-3261.8
691	-3441.6
738	-3375.7

Fig. 7.13d

Phase-contrast flow mapping permits the quantification of the postprandial increase in portal flow (Thomsen et al. 1993). Similarly, changes induced by therapeutic interventions can be quantitatively assessed. In a recent study of 16 patients with portal hypertension, PC portal flow measurements were performed prior to as well as following transjugular intrahepatic portosystemic shunt (TIPSS) placement. While portal venous flow increased on average by 145% in the 14 patients with patent shunts, it changed by merely 5% in two patients with documented shunt occlusion (Figs. 7.13, 7.14).

Hemorrhage from gastroesophageal varices is the main complication in patients with portal hypertension (Galambos 1979). Since gastroesophageal collaterals drain into the azygous system, measurement of blood flow therein reflects the magnitude of blood flow in the collateral circulation (Mahl and Groszmann 1990). Since the magnitude of collateral flow appears to influence the course of variceal bleeding, azygous flow may be a predictor of variceal hemorrhage (Mahl and Groszmann). This was verified by Bosch et al. (1985), who, in a study encompassing 100 patients, demonstrated that azygous flow was markedly increased in patients with portal hypertension and revealed an association between the risk of variceal hemorrhage and highly elevated azygous flow. Similar results were obtained using the PC flow mapping technique in our laboratory. While normal azygous flow ranges between 3 and 29 ml/min, flow in 16 patients with portal hypertension was markedly increased, ranging from 87 to 735 ml/min (mean 336.3 ml/min). Following successful TIPSS placement in 14 patients, PC measurements revealed

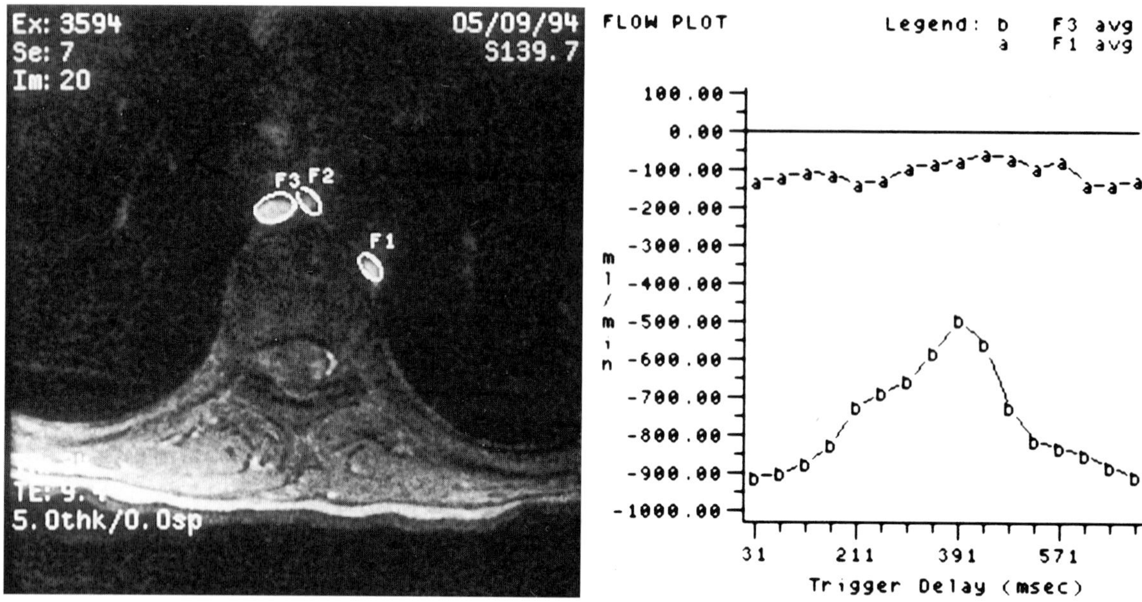

FLOW RATE / CYCLE (ml/min)

	F1	F3
Avg:	-104.8	-763.7

FLOW DATA (ml/min)

TRIG DLY (msec)	F1 AVG	F3 AVG
31	-133.2	-910.8
76	-117.5	-902.6
121	-107.0	-872.1
166	-111.5	-823.3
211	-140.6	-727.6
256	-127.8	-686.6
301	-92.9	-656.2
346	-79.7	-580.9
391	-74.3	-496.0
436	-58.7	-557.7
481	-68.3	-724.7
526	-91.8	-813.0
571	-77.1	-833.5
616	-136.6	-847.7
661	-135.3	-881.5
706	-123.7	-905.3

Fig. 7.13e

a decrease in azygous flow, on average by 59%, ranging from 21% to 88%. In one of two patients with occluded TIPSS, flow in the azygous venous system actually increased by 7% while a decrease of 2% was measured in the other patient (Marincek et al. 1994).

These data suggest that PC blood flow measurements of the azygous system might indeed reflect the degree of collateralization. In addition they certainly appear useful in the assessment of therapy effects.

7.3.9 Arterial System in the Lower Extremities

The flow pattern throughout the arteries of the lower extremities is typically triphasic. It is characterized by one systolic and two diastolic components (Abu Rahma and Diethrich 1988).

Normal flow volume in the lower extremities is highly variable. In a study of eight normal subjects cine-PC flow determinations of pedal flow, defined as the sum of flow in the trifurcation vessels measured at the level of the talotibial joint, ranged from 32 to 183 ml/min (mean 91 ml/min) (Debatin et al., to be published, b). Even the number of visualized vessels, usually three, may vary. Flow measurements over time in the same normal subjects showed considerable intrasubject variability. While it was significantly smaller than the intersubject variability, it can be as high as 100% in individual vessels (Debatin et al., to be published, b). This predominantly reflects nonstandardized subject and environmental factors, such as limb temperature and recent exercise, affect-

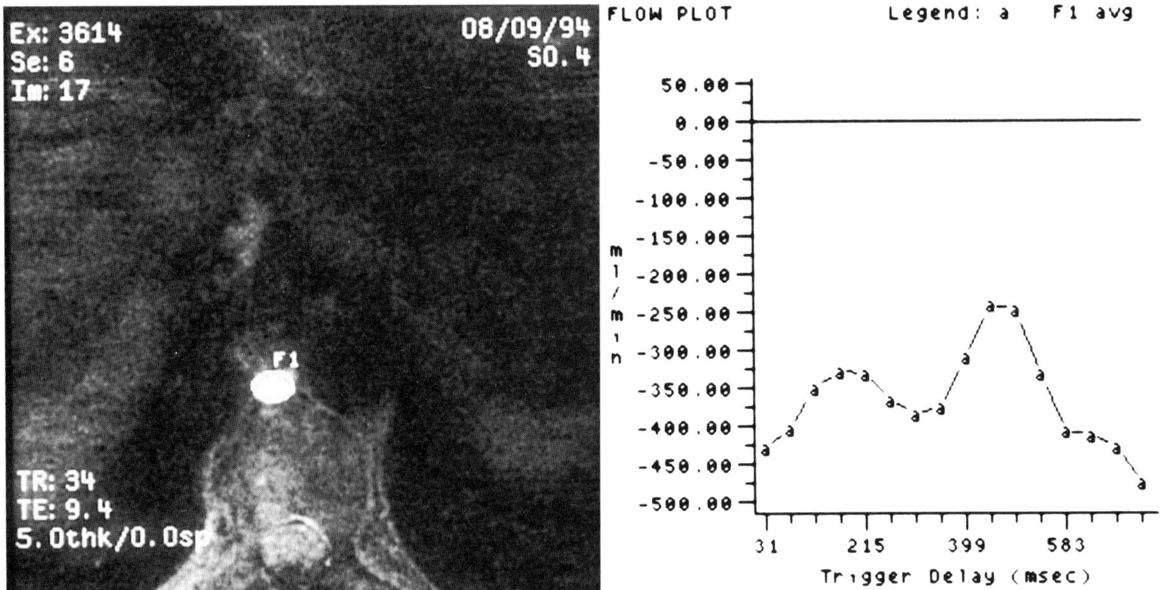

FLOW RATE / CYCLE (ml/min)

F1

Avg: -363.4

FLOW DATA (ml/min)

TRIG DLY (msec)	F1 AVG
31	-428.0
77	-403.5
123	-351.1
169	-327.6
215	-331.7
261	-366.8
307	-382.9
353	-376.3
399	-309.5
445	-239.3
491	-246.1
537	-330.5
583	-405.0
629	-412.7
675	-427.8
721	-475.1

Fig. 7.13f

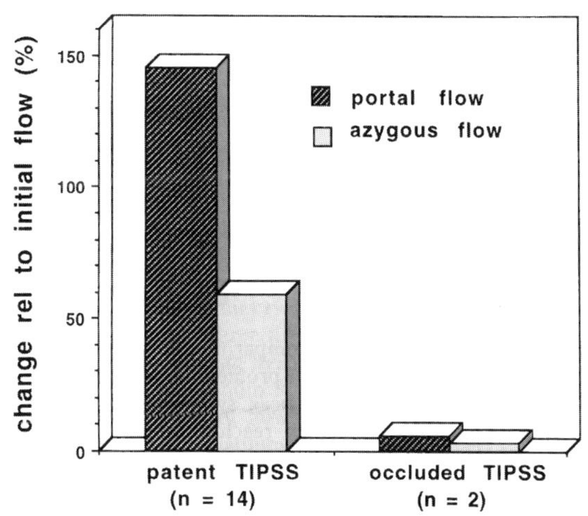

Fig. 7.14. Relative change of initial (i.e., before TIPSS placement) portal and azygous vein flow rates measured by PC flow mapping in patients with patent and occluded TIPSSs

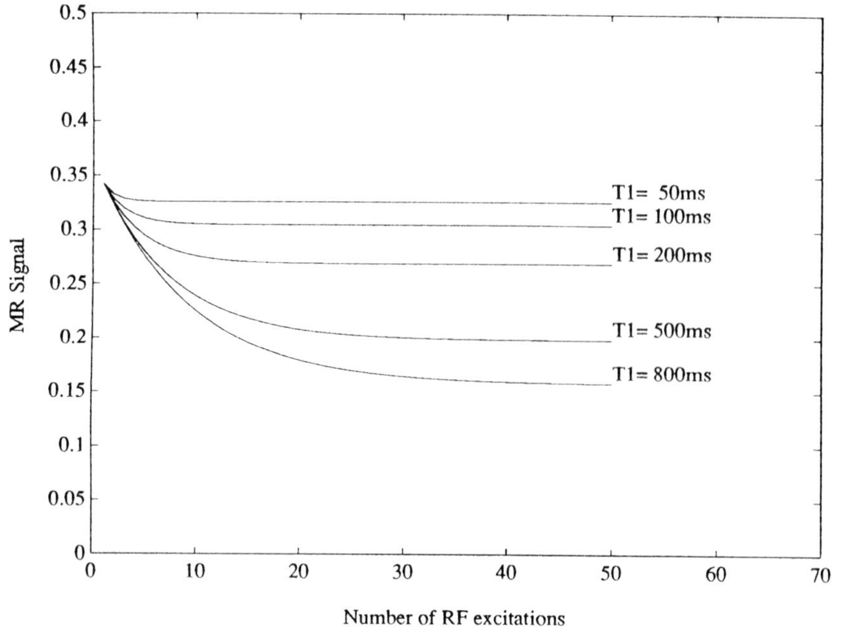

Fig. 8.1. Theoretical signal intensities as a function of the number of excitation pulses and the T1 of the blood (FLASH type sequence; TR/Fl, 40 ms/20°)

clinical routine are small molecules that rapidly diffuse out of the vessels into the large extravascular extracellular space. The half-life of the contrast in the plasma is approximately 12 min at 1 T and the relaxivity is $4.5\,l\,s^{-1}\,mmol^{-1}$.

In Fig. 8.2, the T1 values of blood samples, collected in a healthy volunteer after the injection of 0.3 mmol/kg gadolinium diethylenetriamine pentaacetic acid (Gd-DTPA), are plotted as a function of time. The initial T1 value of 490 ms decreases to 120 ms immediately after the injection and recovers monotonically to 200 ms after 10 min and to 390 ms after 1 h. In Fig. 8.3, MR angiograms obtained in a normal volunteer before and after the injection of 0.1 mmol/kg and 0.2 mmol/kg Gd-DTPA are compared. The injection of 0.1 mmol/kg Gd-DTPA clearly improves the vascular detail in comparison with that of an unenhanced MR angiogram. Increasing the dose from 0.1 to 0.2 mmol/kg further increases the signal intensity of the flowing blood. Follow-up studies have shown that this improved vessel detail already declines after the first contrast-enhanced acquisition (Bosmans et al. 1995).

Blood pool contrast agents that selectively remain in the vessels for a longer time are not yet available for clinical use, but some have been tested advantageously in animal models. Blood pooling agents consist of larger molecules or particles with strong T1 effects. The vascular half-life of these components can be many hours. This characteristic allows adequate angiograms to be obtained over a longer period and/or reduction of the dose to be injected. In Fig. 8.4, CE MRA acquisitions using Gd-DTPA and dextran-ferrioxamine (Dex-FeO) are compared (Bosmans et al. 1992). The latter contrast agent has a plasma half-life of 2 h and a relaxivity of $1.6\,l\,s^{-1}\,mmol^{-1}$ at 1 T and 35°. The MRA acquisitions have been performed in the lower legs of rabbits using a coronal 3D TOF acquisition. As the blood flow in these vessels is characterized by a high systolic flow component with short duration and an extremely low diastolic flow velocity, the net macroscopic motion of the blood is limited. The spins in the blood vessel are necessarily subjected to a large number of excitation pulses and the inflow-related signal enhancement is rapidly lost. In rabbits as well as in human subjects, this saturation effect strongly limits the quality of this type of plain MR angiogram (Fig. 8.4a). Immediately after the injection of 0.3 mmol/kg Gd-DTPA, The saturation is largely overcome (Fig. 8.4b). Twenty minutes after the injection, the effect of the contrast agent is already drastically reduced (Fig. 8.4c). The acquisition immediately after the injection of 0.145 mmol/kg Dex-FeO is shown in Fig. 8.4d. This acquisition provides angiograms that compare well with the higher dose acquisition of the Gd chelate. After 20 min, the image quality is still acceptable (Fig. 8.4e). Even 2 h after the injection, the major vessels are still well visualized.

Figure 8.5 shows the results of an MRA acquisition scheme that exploits the extremely short T1

Fig. 8.2. T1 values of the blood (ms) as a function of time (min) after injection of 0.2 mmol/kg Gd-DTPA in a healthy volunteer

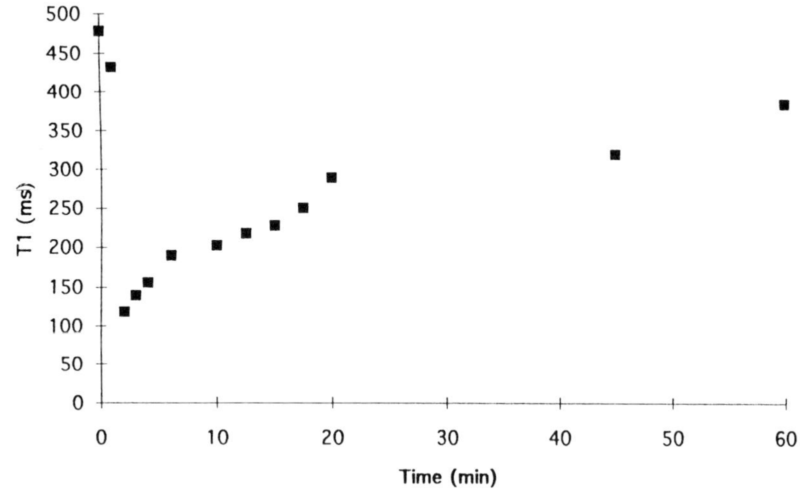

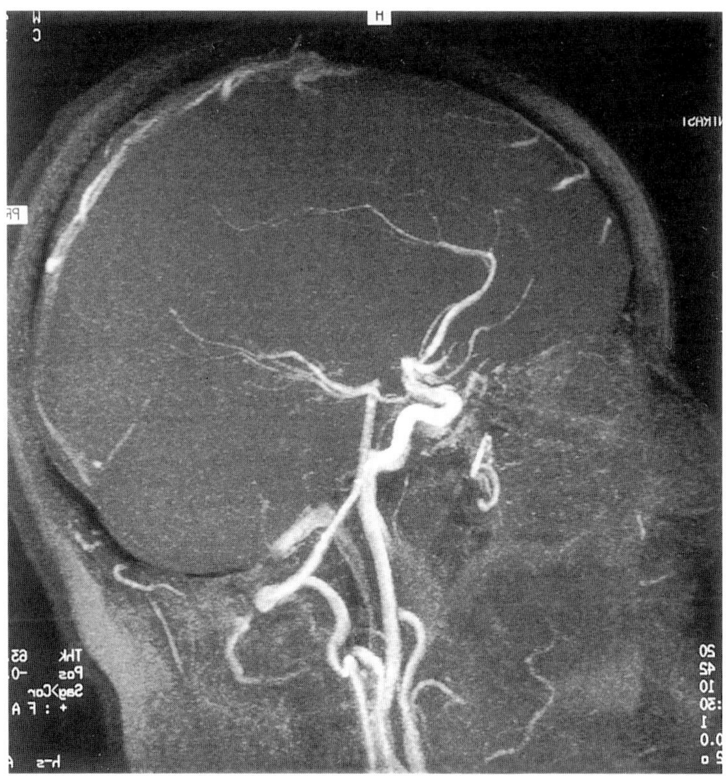

a

Fig. 8.3a–c. Comparative study in a normal volunteer on 2 separate days on the effects of dose of contrast material on quality of 3D TOF MR angiograms (TR/TE/Fl, 42 ms/10 ms/20°; magnetization transfer prepulse; acquisition time, 10 min). Images were obtained **a** before administration of contrast material, **b** immediately after the injection of 0.1 mmol/kg Gd-DTPA, and **c** immediately after the injection of 0.2 mmol/kg Gd-DTPA. The MR angiogram obtained after the injection of 0.2 mmol/kg Gd-DTPA provides the best image quality

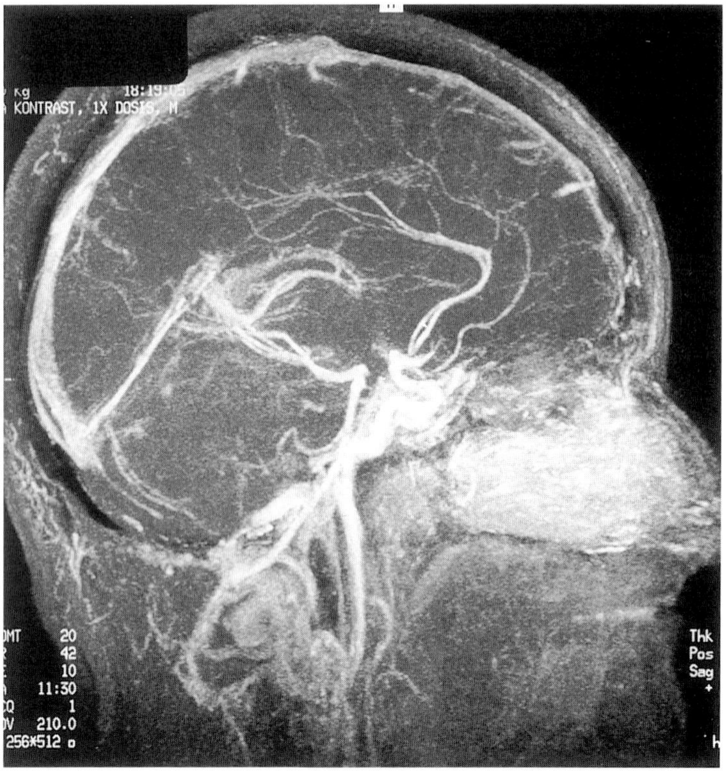
KONTRAST, 1X DOSIS, M
20
42
10
11:30
1
210.0
256*512
Thk
Pos
Sag

b

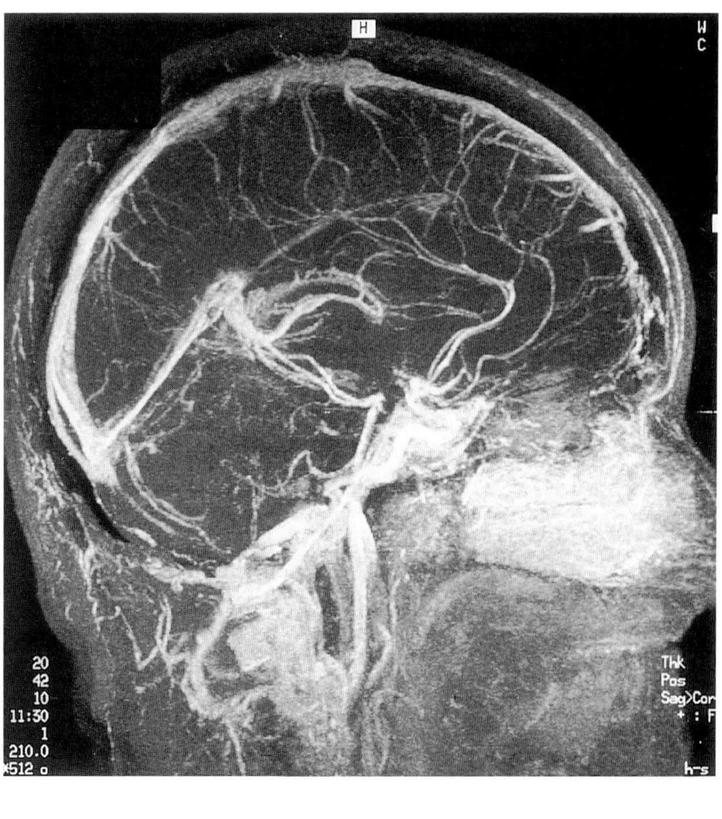
H
W
C
20
42
10
11:30
1
210.0
Thk
Pos

c

Fig. 8.3b,c

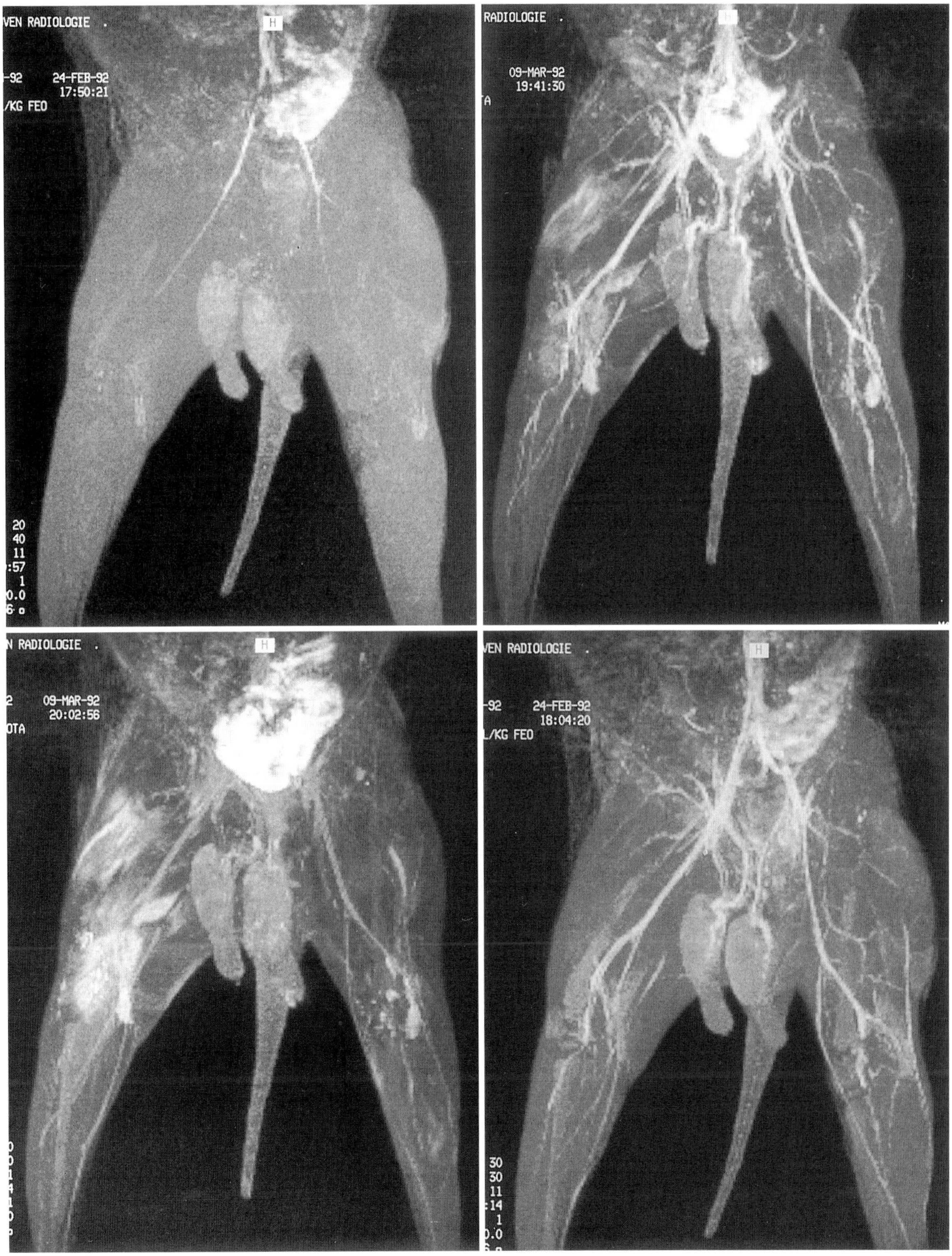

a b c d

Fig. 8.4a–e. Comparative study in the lower leg of a rabbit, using a 3D TOF acquisition (TR/TE/Fl, 40 ms/11 ms/20° before contrast injection; TR/TE/Fl, 30 ms/11 ms/30° after contrast injection; slab thickness/partitions, 56 mm/64 partitions; coronal acquisition plane; acquisition time, 8 min). **a** Plain TOF acquisition. **b** CE acquisition, immediately after 0.3 mmol/kg Gd-DTPA. **c** CE acquisition, 20 min after Gd-DTPA. **d** CE acquisition, immediately after 0.145 mmol/kg Dex-FeO. **e** CE acquisition, 20 min after Dex-FeO. Immediately after the injection of the contrast agent, Gd-DTPA- and Dex-FeO-enhanced MRA acquisitions provide comparable angiograms. Using the blood pooling agent, this effect remains over a longer time interval

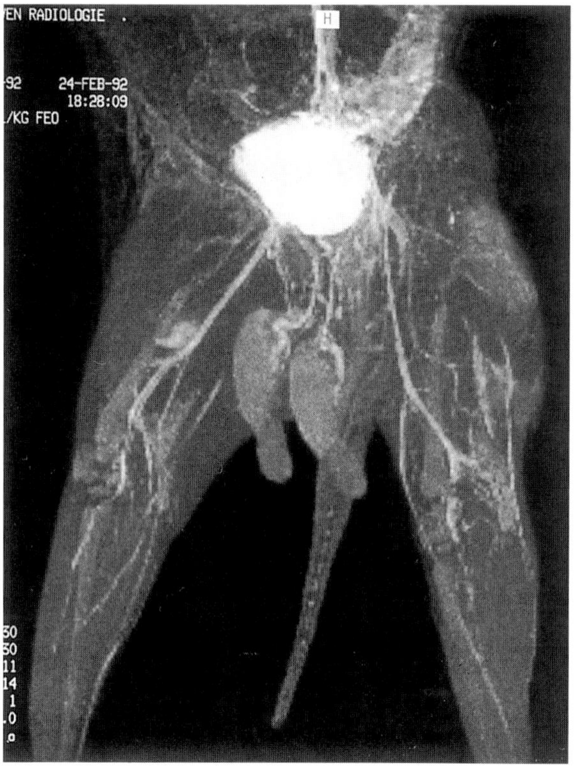

Fig. 8.4e

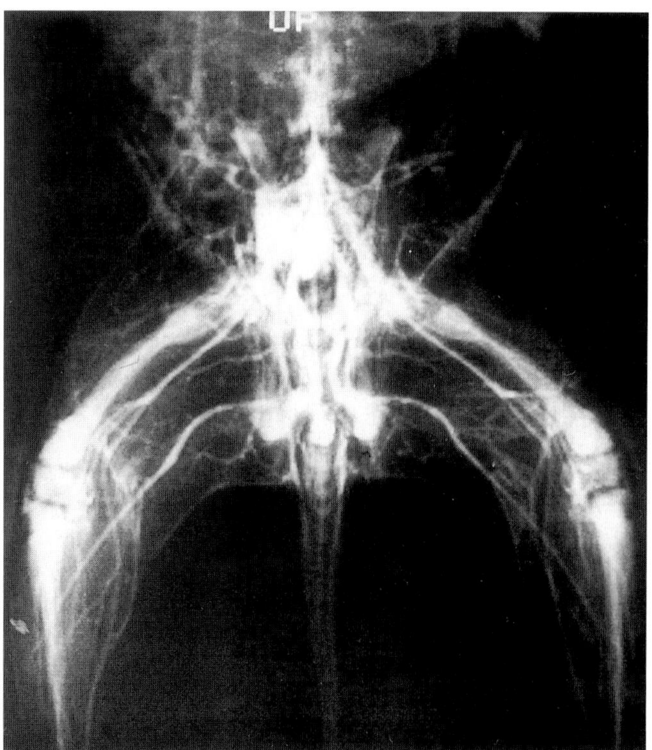

Fig. 8.5. Gd-DTPA polylysine-enhanced acquisition, using a dose of 0.5 mmol/kg. IR-based acquisition with 90° fat saturation prepulse. TR/TE, 100 ms/18 ms, and first-order flow compensation. This acquisition exploits the extremely short T1 of the blood. Total acquisition time, 26 s

values of the blood over the entire measurement. The contrast agent consists of Gd-DTPA that has been attached to a lysine chain. The plasma half-life is about 1 h and the relaxivity is 13 l s^{-1} $mmol^{-1}$ at 0.47 T and 39°. A total dose 0.5 mmol/kg has been injected. The applied technique is based on the inversion recovery (IR) phenomenon (MARCHAL et al. 1992). First, a 90° pulse is given to nullify the signal of fatty tissue. In the subsequent IR sequence, the IR time is set such that the soft tissue is nulled while the blood vessels have a hyperintense signal due to a significantly shorter T1. Due to the almost perfect suppression of the stationary tissues, no slice selection gradient is needed during the acquisition. A 2D projective image is acquired in less that 1 min.

From Fig. 8.6, it is obvious that phase-based MRA techniques also benefit from contrast enhancement. In the example, the lower legs of a rabbit have been examined with an MRA technique in which a flow-dephased acquisition is subtracted from a flow-rephased acquisition. Compared to a nonenhanced acquisition, the vascular detail is significantly increased after contrast enhancement.

Another contrast agent, which has been used by MOSELY et al. (1989) for MRA acquisitions in animals, is albumin Gd-DTPA with a relaxivity of 273 l s^{-1} $mmol^{-1}$. Recently, a new blood pool contrast agent consisting of a monomethoxy ether of poly (ethylene glycol) covalently attached to poly-L-lysine (PL), with PL serving as the carrier of Gd-DTPA, has been used successfully by BOGDANOV et al. (1993). This contrast agent has a relaxivity of 18 l s^{-1} $mmol^{-1}$ and a half-life of 14 h. The same group has also some experience with iron oxide-based T1 relaxation agents.

8.3 Optimization of Contrast-Enhanced MRA Acquisitions

In plain TOF sequences, the MR parameters have to be adjusted to provide optimal flow-related enhancement (HAACKE et al. 1990). This is no longer true in contrast-enhanced (CE) MR angiograms, where the quality is more related to the induced T1 shortening of the blood and the time evolution of this T1 value during the measurement. Optimal results would definitely be obtained if a sequence fast enough to catch the first pass could be designed. However, to cover a clinically useful volume of interest, the current MRA sequences typically last several minutes. Therefore, data are acquired when plasma concentrations of contrast material are significantly lower than the peak concentrations immediately after administration of a bolus of contrast material. As a first rule, CE MRA acquisitions should be performed as soon as possible after the injection of the contrast agent in order to take full advantage of the T1 shortening of the blood.

For a given TR and T1 of the blood, the optimal vessel signal is obtained at the Ernst angle. For a 5-min acquisition with a TR of 40 ms and a dose of 0.2 mmol/kg Gd-DTPA, a flip angle of about 45° is an appropriate choice. This is illustrated in Fig. 8.7, where CE TOF acquisitions in normal volunteers are compared. Imaging was performed in two different sessions after the injection of 0.2 mmol/kg Gd-DTPA. An acquisition with a flip angle of 45° yields a more detailed angiogram than one using a flip angle of 20° (Fig. 8.7a,b). Using a flip angle of 60° shows less vessels than using a flip angle of 45° (Fig. 8.7c,d). The optimal flip angle is lower than 45° when longer measurements are performed or when the acquisition is started later after the contrast injection.

The improved signal-to-noise ratio of flowing blood can theoretically be exploited to shorten the overall acquisition time. However, this can only be achieved without loss of vascular detail if the flip angle is adjusted accordingly. Unfortunately, acquisitions with short repetition time TR exclude the use of preparation pulses such as fat suppression or magnetization transfer that are commonly used to reduce the background signal (EDELMAN et al. 1992; LIN et al. 1993; ATKINSON et al. 1994).

The possible influence of the rapidly decreasing plasma concentrations of contrast material on the image quality can be evaluated by comparing a monophasic bolus injection with biphasic injections that produce a more uniform intravascular contrast concentration. In a recent study, these different types of injection did not influence the angiographic quality noticeably (BOSMANS et al. 1995, in press). A slow injection spread over the entire measurement time has been proposed by PRINCE et al. (1993) to improve the quality of 3D MR angiograms of the thoracic and abdominal arteries.

Other experiments that have been done to evaluate the influence of the progressive decrease of the contrast concentration in the blood during the acquisition have dealt with the k-space sampling schemes. It is well known that the lower frequency k-space lines add to the global image quality whereas higher frequency k-space lines provide the image detail. The possible effects are illustrated in Fig. 8.8a,b in which evaluation of three-dimensional time-of-flight MR angiography. Radiology 175:443–448

Marchal G, Michiels J, Bosmans H et al. (1992) Contrast-enhanced MRA of the brain. Technique and first clinical results. J Comput Assist Tomogr 16:25–29

Mosely ME, White DL, Wang SC et al. (1989) Vascular mapping using albumin-(Gd-DTPA), an intravascular MR contrast agent, and projection MR imaging. J Comput Assist Tomogr 13:215–221

Prince MR (1994) Gadolinium enhanced MR aortography. Radiology 191:155–164

Prince MR, Yucel EK, Kaufman JA, Harrison DC, Geller SC (1993) Dynamic gadolinium-enhanced three-dimensional abdominal arteriography. J Magn Reson Imaging 3:877–891

Wilms G, Bosmans H, Marchal G, Demaerel P, Baert AL (1995) Magnetic resonance angiography of supratentorial brain tumors. Neuroradiology, in press

Clinical Applications of Magnetic Resonance Angiography

9 Intracranial Vessels

G. MARCHAL, G. WILMS, and H. BOSMANS

CONTENTS

9.1 Introduction

X-ray angiography remains the gold standard for the study of intracranial vessels (BRANT-ZAWADZKI et al. 1983). For some time now, conventional cut film angiography has been replaced by digital subtraction techniques. After initial overenthusiasm for the intravenous application of digital subtraction angiography, the method was rapidly abandoned due to insufficient vessel contrast, superimposition of vascular territories, and the requirement of good patient cooperation (WILMS et al. 1983). Nowadays the intra-arterial selective and superselective study of the intracranial vessels with digital subtraction is routinely used for the visualization of the intracranial vessels (BRANT-ZAWADZKI et al. 1983).

The morphological information obtained from these studies offers exquisite anatomical detail with visualization of very small distal cortical or deep branches, especially if a matrix of 1024 × 1024 is used. The venous anatomy and drainage both of the normal brain and of pathological conditions is very well demonstrated. Besides the morphological information, some flow information is obtained due to the temporal character of the technique, with sequential image acquisition during the arterial, capillary, and venous phases. This allows visualization of local or generalized flow changes, such as increase or decrease, asymmetry between the hemispheres, and steal phenomena or venous shunting (BRANT-ZAWADZKI et al. 1983). In addition, the development of very small and flexible microcatheters now allows superselective catheterization of distal cerebral arteries, permitting detailed study of the contributing arteries to arteriovenous malformations or tumors (VINUELA et al. 1989).

Similarly, the treatment during interventional angiography of cerebral aneurysms, arteriovenous malformations, arteriovenous fistulas, and even stenotic disease based on vasospasm or atherosclerosis is part of the routine practice of many centers (VINUELA et al. 1989). Unfortunately, even in experienced hands, angiography remains an invasive method, with the possibility of severe neurological and even life-threatening complications (WAUGH and SACHARIAS 1992). The risk of a neurological event is estimated at 4%, with permanent deficit in 1% (HANKEY et al. 1990). Furthermore, the information obtained by angiography is mostly indirect, for both vascular and nonvascular diseases. The exact size of an aneurysm can be underestimated, hemorrhage in an arteriovenous malformation is not detected, etc. The absence of soft tissue information in angiography is a major drawback which necessitates additional cross-sectional techniques, such as computed tomography (CT) or magnetic resonance imaging (MRI).

G. MARCHAL, Professor, Dienst Radiologie, Katholieke Universiteit, Gasthuisberg, Herestraat 49, 3000 Leuven, Belgium
G. WILMS, Professor, Dienst Radiologie, Katholieke Universiteit, Gasthuisberg, Herestraat 49, 3000 Leuven, Belgium
H. BOSMANS, NMR Engineer, Dienst Radiologie, Katholieke Universiteit, Gasthuisberg, Herestraat 49, 3000 Leuven, Belgium

Transcranial Doppler, in particular in combination with color Doppler, is a valuable noninvasive alternative for the functional study of the cerebral vessels, but unfortunately with poor morphological performance (De Bray et al. 1988). It allows the detection and follow-up of flow alterations in the major cranial vessels and the evaluation of the intracranial vascular reserve during carotid compression, and can be performed even as a bedside examination. In comparison with angiography and ultrasonography (US) the most important advantage of newer techniques such as MRA (Blatter et al. 1992; Huston and Ehman 1993; Marchal et al. 1990, 1992; Pernicone et al. 1990, 1992; Ruggieri et al. 1992) and recently CT-angiography (CTA) (Schwartz et al. 1994) consists in the integrated display of both vessels and soft tissues. In addition MRA, like transcranial Doppler, allows the quantification of intracranial flow.

9.2 Technical Considerations

The MRA study of the intracranial vessels has received much attention in clinical testing. This is due to the favorable flow conditions in the brain vessels which are characterized by relatively high blood velocities providing a substantial inflow effect throughout the cardiac cycle. In addition, they are confined to a small volume of interest which can be studied with high-performance coils. Furthermore, the effects of the most common causes of MR artifacts, such as respiration, cardiac motion, and susceptibility changes, are minimal. Although the clinical experience with MRA of the intracranial vessels is rapidly increasing, there is still no single robust MRA sequence that can be used for all intracranial MRA applications. In addition, the imaging protocols have to be adapted to the type of suspected lesion. It is clear that the techniques will differ for the study of arterial or venous disease. The quality of venous MR angiograms is strongly determined by the saturation phenomenon. Arterial MR angiograms are less degraded by saturation but flow-induced dephasing may be highly undesirable. In the following sections we shall discuss the different techniques that can be used for visualization of intracranial vessels. A sequence protocol will be provided. The advantages of using multiple acquisitions to clarify unexpected signal behavior will be stressed.

9.2.1 Cerebral Arteriography

9.2.1.1 Time-of-flight Acquisitions

Although successful 2D MRA examinations of intracranial arterial diseases have been reported, the acquisition of true 3D data is preferred since they provide more detailed angiograms that can be viewed from any projection direction. Three-dimensional time-of-flight (TOF) techniques are often used to study intracranial arteries. An example of a 3D TOF acquisition in a normal young volunteer is shown in Fig. 9.1a. In clinical practice, the vessel detail is often inferior to this example. Saturation effects commonly arise in patients with a reduced cardiac output or lesions with impaired arterial flow. Therefore, the MR parameters should always be carefully adjusted (Ruggieri et al. 1989; Haacke et al. 1990). In addition, the many techniques developed to reduce saturation and to increase the signal-to-noise ratio between the blood vessels and the stationary tissues have to be applied whenever possible. This is especially true if high-resolution acquisitions are performed.

9.2.1.1.1 TONE. A first technique to overcome saturation in transverse MRA acquisitions of the brain is the use of "tilted optimized nonsaturating excitation" (TONE) RF pulses. These pulses provide flip angles that are variable along the slice selection direction. Typically, low flip angles are applied at the entrance side of the arteries in the volume of interest to delay the saturation phenomenon. The flip angle is then steadily increased along the cranial direction. Using increasing flip angles maximally recovers the remaining signal of the flowing blood (Tkach et al. 1992; Atkinson et al. 1994). In Fig. 9.1a and b, we compare the MR angiograms obtained with a regular RF pulse and a TONE RF excitation pulse (10° at the entrance side of the intracranial arteries and 30° near the top of the 3D volume). The peripheral branches are better visualized by using TONE excitation pulses. The saturation of the sagittal sinus is increased due to the relatively large flip angles at the entrance site of the cortical veins.

9.2.1.1.2 Prepulses. Another way to cope with saturation in the intracranial arteries with 3D TOF acquisitions is the suppression of the signal of the stationary tissues (Edelman et al. 1992; Lin et al. 1993; Atkinson et al. 1994). Fat suppression prepulses can be applied. However, this requires a

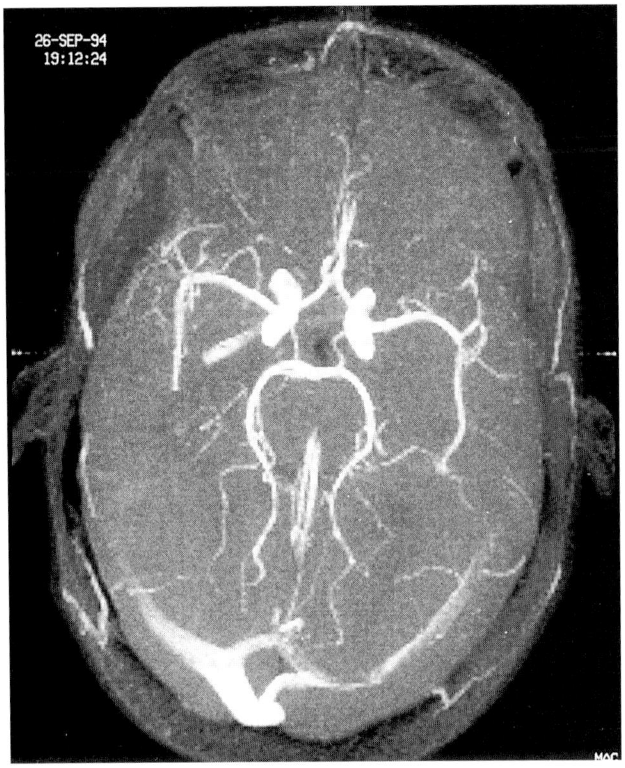

a

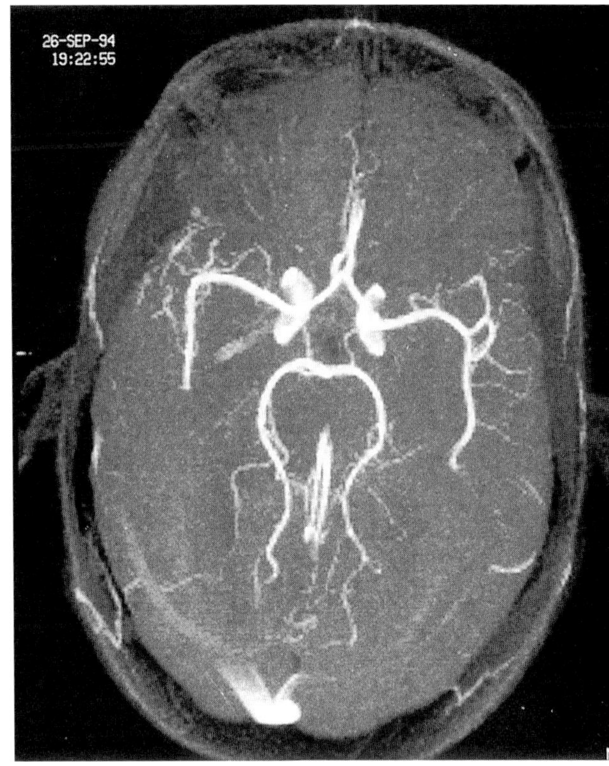

b

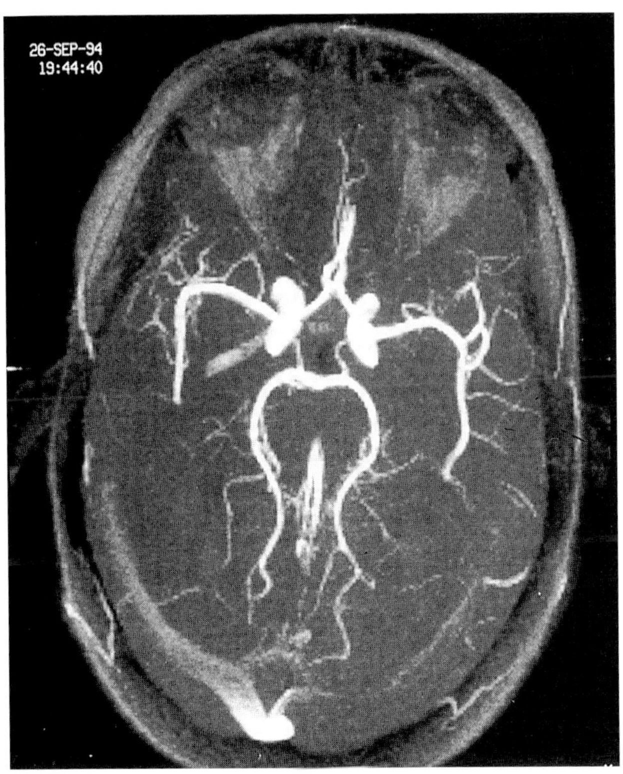

c

Fig. 9.1a–c. MRA of a normal young volunteer: 3D TOF, TR/TE/Fl 40 ms/10 ms/20°, rectangular FOV 158 × 210 mm, slab thickness 52 mm/64 partitions. **a** Transverse projection; **b** identical acquisition, TONE RF excitation pulse; **c** identical acquisition, TONE RF excitation and MT prepulse. As the different techniques are added, the quality of the MR angiograms steadily increases well-shimmed system that is free of eddy currents. If spins of the blood have a precessional frequency matching the resonance frequency of the fatty tissue before entering the volume of interest, they are saturated by the fat suppression prepulse. In addition, in the brain, the contribution of fatty signal is small. Superposition of fatty tissue in the maximum-intensity projections (MIPs) can largely be avoided by means of a targeted MIP, which is available on routine scanners. Therefore, this prepulse is not often used in routine MRA protocols for intracranial vessels.

The use of magnetization transfer (MT) prepulses is more widespread. In brain MRA, this technique mainly suppresses the signal of the brain tissue. The benefits of MT prepulses are more obvious than those of fat saturation pulses. Besides, this application does not place extreme requirements on the hardware of the system. An example of a 3D TOF with TONE RF excitation and MT prepulse is shown in Fig. 9.1c. The contrast-to-noise ratio between vessel and brain tissue is significantly improved and therefore small vessels are better visualized.

9.2.1.1.3 MOTSA (Multiple Overlapping Thin Slab Acquisition). Theoretically, all the intracranial vessels can be examined with a single transverse, sagittal, or coronal 3D TOF acquisition using a slab thickness of 128 mm/128 partitions. With this technique, however, there is obvious signal decrease in the blood vessels that are distal in the volume. In order to reduce the saturation, the number of excitations for the spins in the vessels has to be reduced. This can be done by acquiring multiple consecutive 3D TOF acquisitions restricted to a smaller part of the brain, using for example a slab thickness of 56 mm/64 partitions. If the acquisition volume is oriented transversely, the saturation of the blood is minimized. Due to the fact that the slab excitation profiles in the slice selection direction are never ideal, these multiple 3D acquisitions have to be mutually overlapping. MR angiograms are calculated from a single data set that is composed of images obtained during the different consecutive acquisitions (Marchal et al. 1990; Blatter et al. 1991; Lewin et al. 1991). If the MRA data are projected along an axis different from the slice selection direction, the transitions between the different slabs remain visible in the angiograms. This "Venetian blind artifact" can be reduced by increasing the slab overlap and reducing the number of partitions. Blatter et al. (1993) propose the performance of MRA acquisitions using 16 partitions, 50% slab overlap, and a postprocessing algorithm that combines the overlapping slabs from each adjacent slab by selecting on a pixel-by-pixel basis the brightest of the spatially coincident pixels.

In Fig. 9.2a, we show a preoperative TOF study in a patient with a brain tumor. The whole brain was imaged by means of four overlapping transverse 3D slabs, with a thickness of 32 mm/ 32 partitions. In the coronal projection, the transitions between the individual 3D volumes remain visible. Both in the veins and in the arteries with slow flow, the discontinuities at the transitions illustrate the persisting saturation effects (Wilms et al. 1995).

9.2.1.1.4 Contrast Enhancement. Another method to reduce saturation in a 3D TOF MRA is the intravenous injection of a contrast agent that shortens the T1 of the blood. Figure 9.2b illustrates the contrast-enhanced (CE) TOF study of the aforementioned patient with a brain tumor. A large 3D TOF acquisition with a slab thickness of 108 mm/128 partitions and MT prepulse was used. The veins and small arterial branches are better visualized on the CE TOF study than on the plain MRA study.

In the native CE images, arteries and veins are well visualized. In the MIP, however, the contrast-to-noise ratio of the vessels is poor. This is due to superposition of high signal intensities of enhancing subcutaneous tissues. A postprocessing procedure that nulls the signal of the subcutaneous tissues before projection provides angiograms with a better contrast-to-noise ratio (Fig. 9.2c). Theoretically, the same result could be obtained by targetted MIP. However, it is usually not possible to suppress all the enhancing soft tissues while preserving all the vessels, since most software for targetted MIP only offers box-shaped regions of interest.

Disadvantages of the use of contrast agents are the more invasive character of the technique and in increased cost. It is impossible to separate arteries from veins by means of saturation pulses. Finally, soft tissues may enhance and superimpose on the angiogram.

9.2.1.2 Phase-Contrast Acquisitions

Three-dimensional PC acquisitions are another means of overcoming the saturation (Pernicone et al. 1990). The choice of an appropriate VENC can be difficult in arterial disease as both extremely low and very rapid flow can be present, e.g., in arteriovenous malformations. A major advantage of the PC tech-

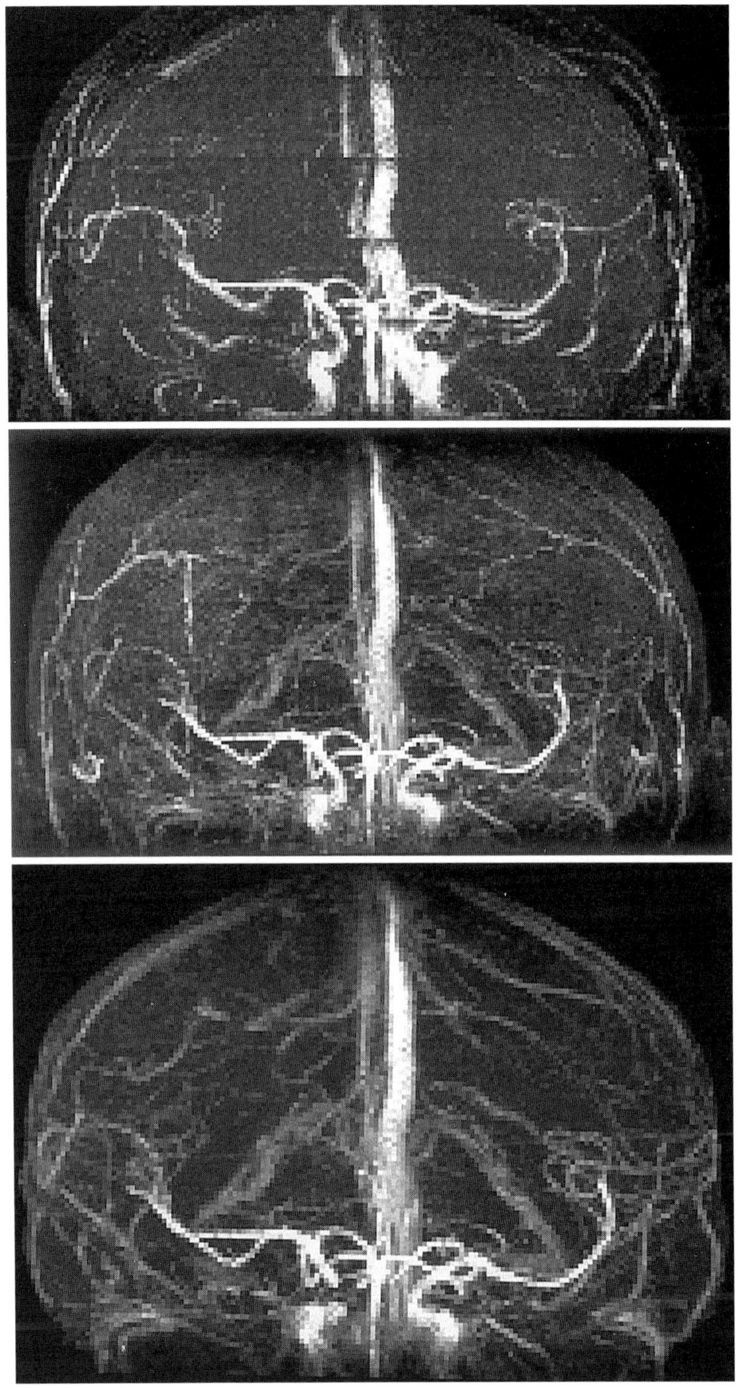
a
b
c

Fig. 9.2a–c. Preoperative TOF study (TR/TE/Fl 42 ms/10 ms/20°, FOV 210 mm, slab thickness 32 mm/32 partitions, matrix 192 × 256) in a patient with a brain tumor. **a** Coronal projection of four overlapping thin 3D slabs (MOTSA) with transverse orientation; total acquisition time 17 min. **b** Coronal projection of a single slab acquisition; 128 mm/128 partitions after the injection of 0.2 mmol/kg Gd-DTPA. **c** Coronal projection of the CE study after a postprocessing procedure that segments the brain prior to the MIP. The MOTSA technique provides a complete picture of the arteries. The CE acquisition provides additional information on the veins. The vessel contrast is significantly improved after filtering the data such that only the brain tissue and the vessels are left for the projection algorithm

nique is that the quality of the angiograms is largely independent of the orientation of the acquisition volume. In Fig. 9.3 we compare an MR angiogram calculated from the flow rephased acquisition of the PC sequence, which is in fact a regular TOF acquisition (Hausmann et al. 1991), with an MR angiogram calculated from the subtracted (PC) images in a patient with an arteriovenous malformation. In the 3D sagittal TOF acquisition, the saturation of the blood vessel signal is obvious and the resulting MR angiogram is incomplete. A transverse MOTSA acquisition had been preferable but at least three acquisition volumes would have been necessary to visualize the whole lesion, requiring a total acquisition time of at least 20 min. In the sagittal 3D PC acquisition, saturation is still present but the blood

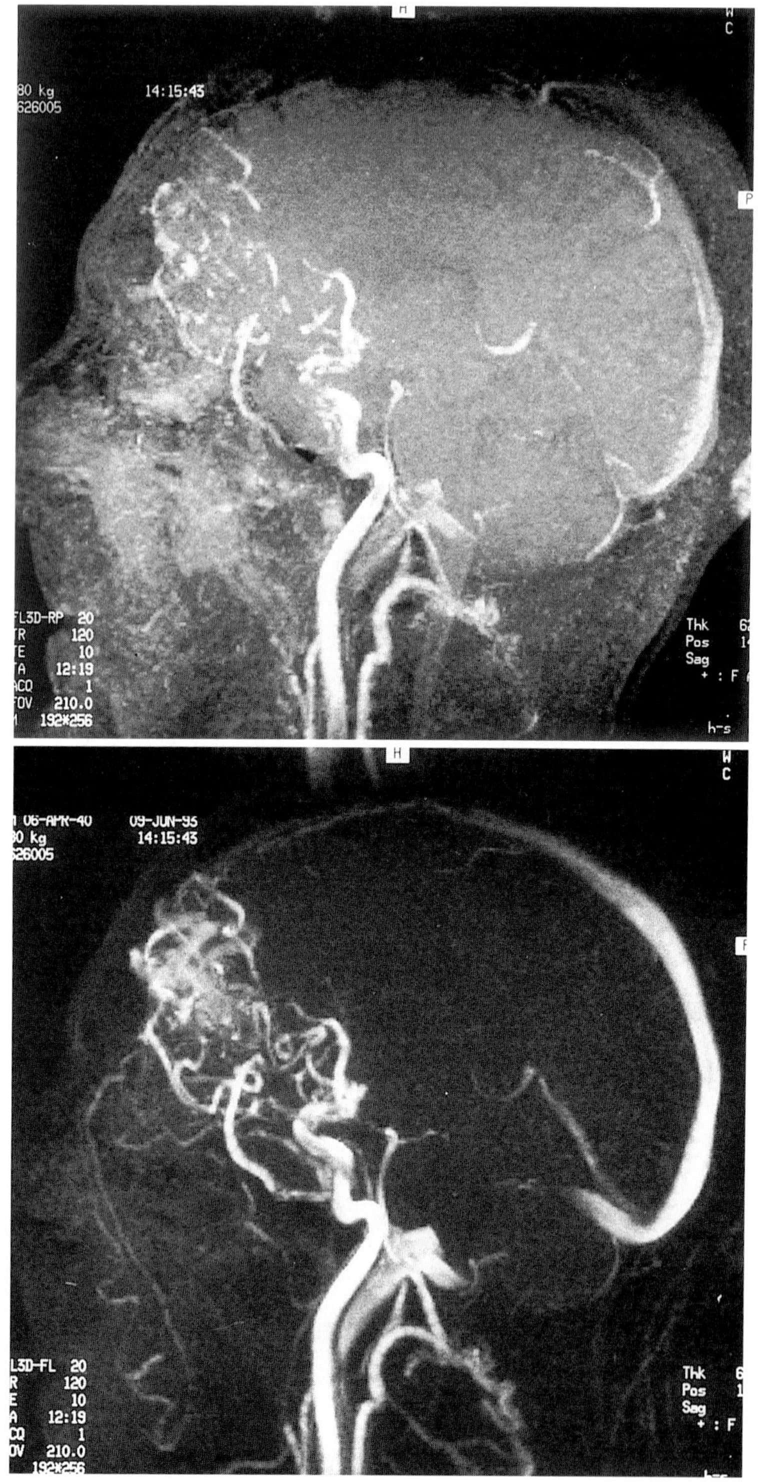

a b

Fig. 9.3a,b. PC acquisition in a patient with an AVM in the frontal lobe (TR/TE/Fl 30 ms/10 ms/20°, VENC 35 cm/s, FOV 210 mm, matrix 192 × 256, slab thickness 64 mm/32 partitions, sagittal acquisition volume, acquisition time of 12 min. **a** Projection of the flow rephased (TOF) acquisition. **b** Projection of the subtracted data. The TOF data are hampered by saturation effects. The lesion is incompletely shown. The PC acquisition shows the feeder arteries coming from both the internal and the external carotid artery, the nidus, and some draining veins

vessel signal remains significantly larger than the background signal. Using the PC acquisition, arteries and veins can be studied by the same procedure. The image detail is lower than that achieved with a CE (high-resolution) TOF acquisition but the latter suffers from superimposition of enhancing soft tissues. The use of PC acquisitions is also indicated in patients with acute thrombosis and bleeding associated with vascular malformations. In contrast to TOF acquisitions, all stationary tissue signals, including those from high signal intensity blood, are suppressed.

9.2.1.3 Flow-Induced Turbulence

Flow-induced dephasing is a very fundamental problem in both TOF and PC MRA. It consists in a signal drop out frequently seen in the knee of the middle cerebral arteries, in the carotid siphon, in hyperdynamic flow situations, and in aneurysms. It occurs with any bright blood acquisition technique that fails to refocus the flowing spins at the echo time. It cannot be overcome by techniques that increase the signal amplitudes of the spins, such as MOTSA and TONE. A first approach to reduce flow-induced dephasing is to use sequences in which the amplitude and the duration of the applied gradi-

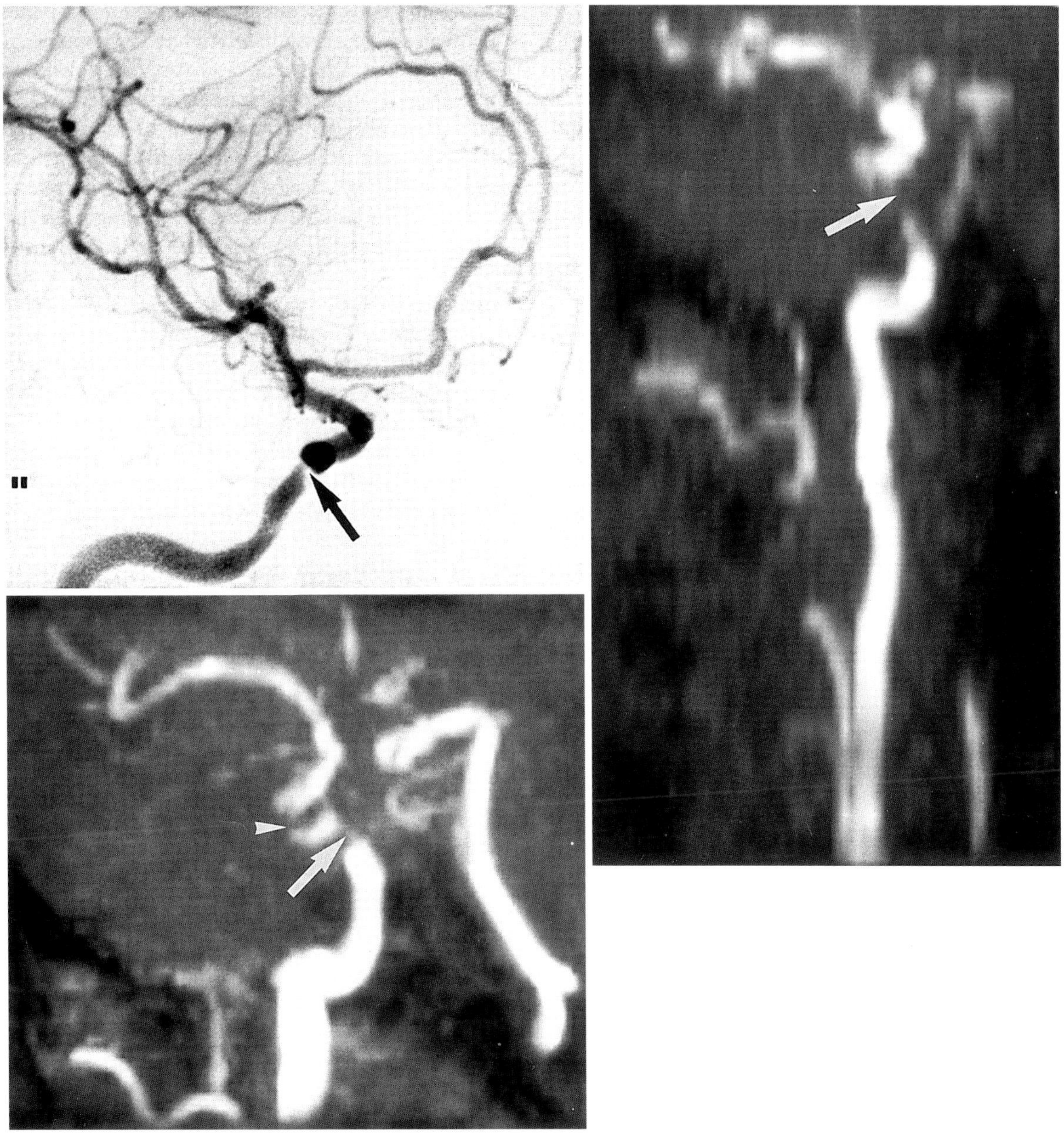

a b c

Fig. 9.4a–c. Comparative digital subtraction angiography (DSA) and TOF study in a patient with a stenosis in the cranial portion of the carotid artery. **a** DSA examination shows stenosis of the proximal carotid siphon (*arrow*). **b** 2D TOF acquisition, TR/TE/Fl 40 ms/10 ms/35°, FOV 200 mm, slice thickness 3 mm, venous presaturation, transverse slices. **c** 3D high-resolution TOF study, TR/TE/Fl 42 ms/11 ms/25°, FOV 200 mm, slab thickness 52 mm/64 partitions, matrix 192 × 512, transverse acquisition volume. The stenosis (*arrow*) is detected in both the 2D and the 3D TOF acquisition. The stenosis is overgraded in the 2D TOF study and accurately visualized in the 3D TOF study. Flow-induced signal voids in the carotid siphon are seen in the 3D study (*arrowhead*)

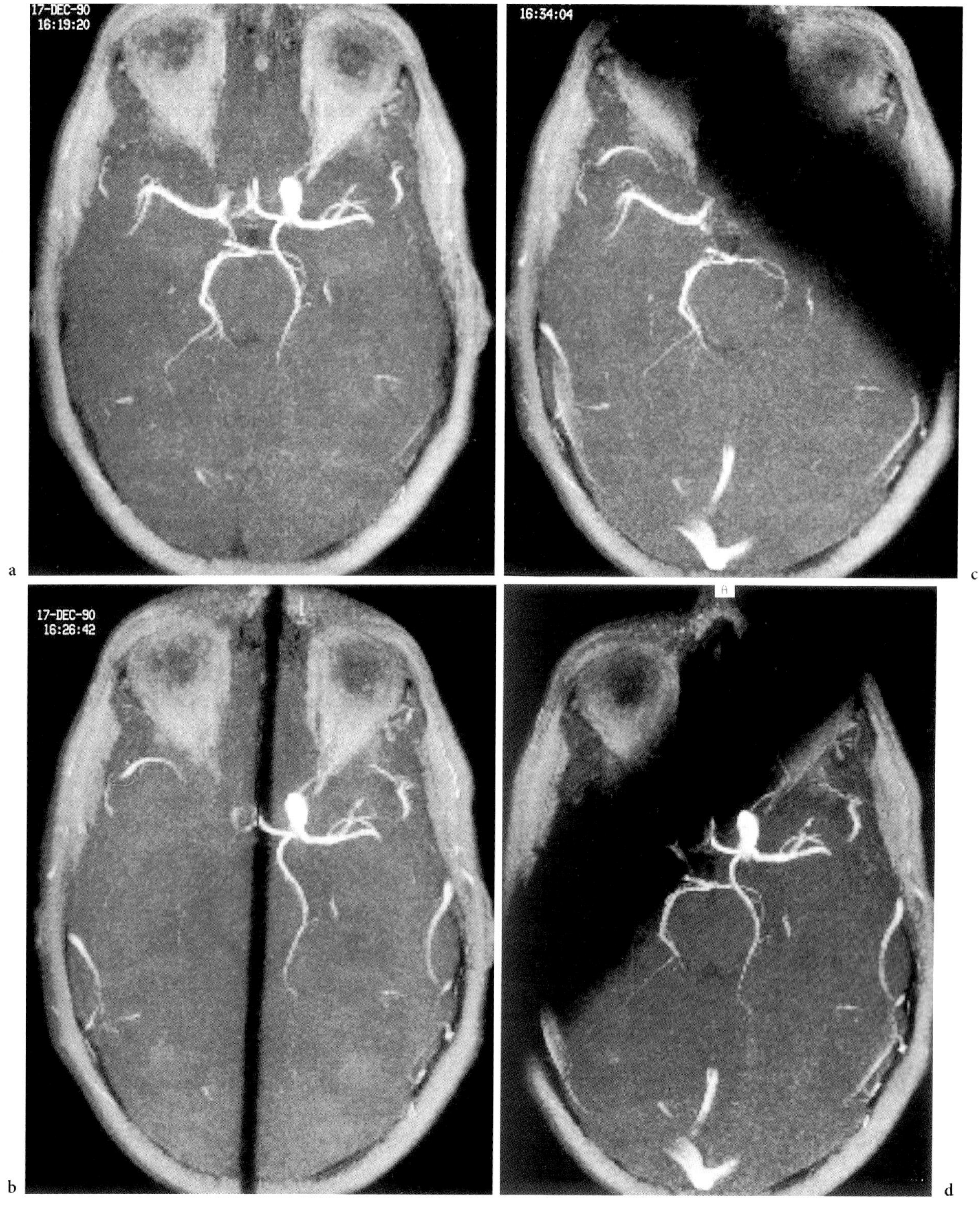

Fig. 9.6a–e. Selective MRA study (3D TOF, TR/TE/Fl 40 ms/7 ms/25°, FOV 220 mm, matrix 256 × 256, acquisition time of 5 min per slab) of a patient with a schwannoma at the bifurcation of the right carotid artery with total obstruction of the right internal carotid artery. **a** Reference image showing absence of signal in the right carotid siphon. **b** Selective saturation of the basilar artery in the sagittal plane. Absence of flow in the right middle cerberal artery and in the right posterior cerebral artery. **c** Selective saturation of the left carotid artery. This saturation pulse destroys the contributions of the left carotid artery. Note the disappearance of the left posterior cerebral artery which directly originates from the left middle cerebral artery and the persistence of the vascular signal in the right middle cerberal artery. **d** Selective saturation of the right internal carotid artery. The presence of flow signal in the right anterior communicating artery and the right posterior communicating artery indicates shunting over these vessels. **e** Selective saturation of the basilar artery in the coronal plane shows total disappearance of the vascular signal. This surprising result can be explained by simultaneous saturation of both the basilar artery and carotid arteries at a more caudal level, probably due to cervical lordosis

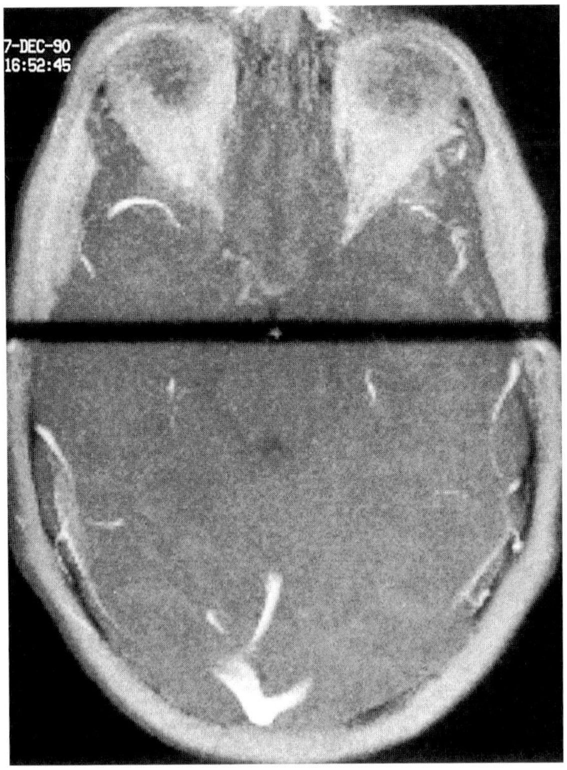

Fig. 9.6e

tions or when acquisitions with different VENC have to be performed. It should be stressed, however, that other slice or slab orientations can be chosen more freely than in 2D TOF. In the latter, the slices have to be perpendicular to the vessels and at least two acquisitions of, for example, 6 min (50 slices, TR 40 ms, 192 × 256 matrix) are needed to visualize the sagittal sinus. PC acquisitions can be applied along the axis of the vessel. A single sagittal acquisition volume with a total acquisition time of 12 min can be applied to visualize the sagittal sinus.

9.3 Normal Intracranial Vascular Anatomy

9.3.1 General Anatomy

9.3.1.1 The Internal Carotid Artery

The cervical internal carotid artery (ICA) enters the skull through the carotid canal in the petrous bone. After a short vertical course, the artery courses medioventrally towards the cavernous sinus. Here the course is vertical up to the sellar floor (presellar segment). There the artery shows an S-shaped configuration. Therefore this juxtasellar segment is called the "carotid siphon."

Small branches arising from the intracavernous position of the ICA are never seen in normal patients. At the level of the anterior clinoid process, the artery traverses the dura of the cavernous sinus and becomes the supraclinoid segment. Ventrally, the ophthalmic artery arises. More cranially, the posterior communicating artery originates and runs posteriorly to join the posterior cerebral artery. A few millimeters more distally, the anterior choroidal artery arises, courses around the brain stem, and enters the choroid plexus. This artery supplies important portions of the internal capsule. At the top of the supraclinoid segment the ICA bifurcates in the middle and anterior cerebral arteries.

9.3.1.2 Circle of Willis

The circle of Willis is a vascular ring connecting the internal carotid and vertebral arteries with each other. The anterior portion is formed by the horizontal segments of both anterior cerebral arteries, connected by the anterior communicating artery. This portion is connected with the posterior cerebral arteries by the posterior communicating arteries.

9.3.1.3 The Anterior Cerebral Artery

The anterior cerebral artery (ACA) first shows a horizontal segment called the CA1 segment. At this point medial lenticulostriate arteries arise, the largest of which is called the recurrent artery of Heubner. These small arteries supply parts of the internal capsule and of the basal ganglia. The two anterior cerebral arteries are joined by the anterior communicating artery.

The anterior cerebral artery first gives off two branches (the orbitofrontal and frontopolar arteries). Just at the level of the corpus callosum, the artery divides into the pericallosal artery, running around the corpus callosum, and the callosomarginal artery, running within the cingulate gyrus. Both arteries supply the medial aspect of the brain: the callosomarginal artery the frontal region and the pericallosal artery the parietal region. Cortical branches run over the cerebral hemisphere and anastomose with the middle cerebral artery (MCA) and posterior cerebral artery (PCA).

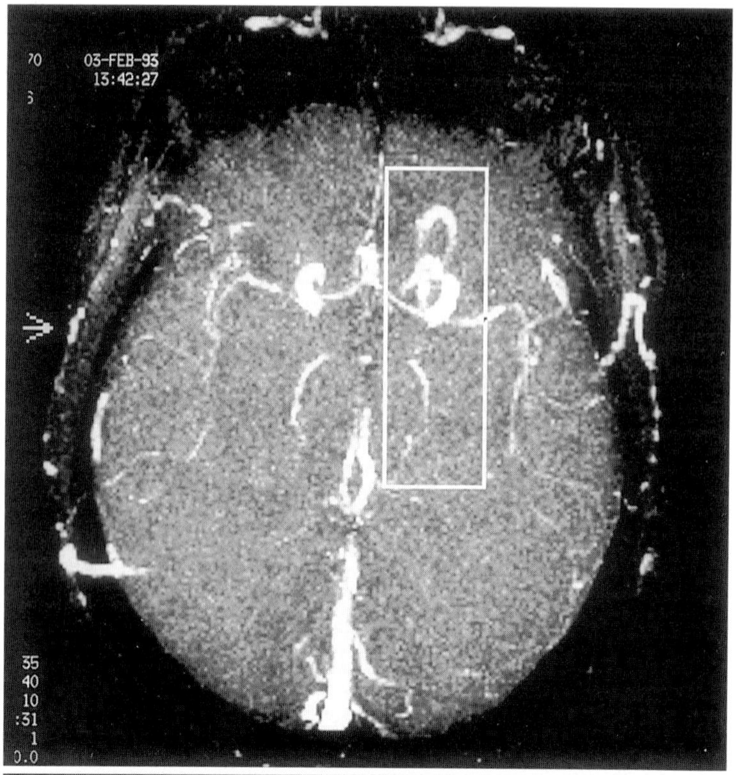

a

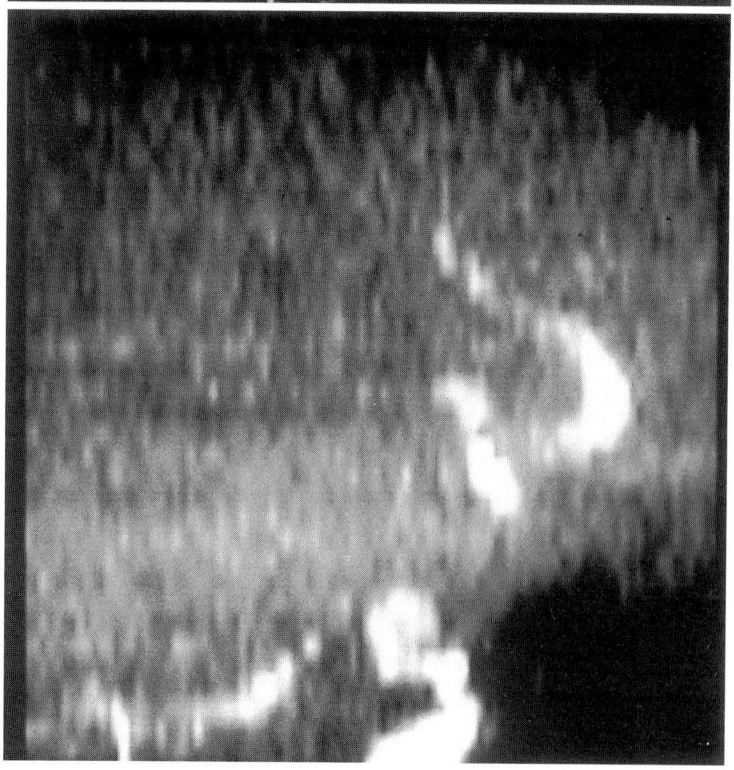

b

Fig. 9.7a–d. TOF study of a patient with a deep frontal venous anomaly. **a** Transverse projection of a 2D TOF, TR/TE/Fl 40 ms/10 ms/35°, slice thickness 3 mm, FOV 220 mm, matrix 256 × 256, transverse slices. **b** Sagittal projection of a small region of interest. **c** Transverse projection of a 3D TOF study after the injection of 0.1 mmol/kg Gd-DTPA. TR/TE/Fl 43 ms/10 ms/20°, slab thickness 60 mm/64 partitions, FOV 220 mm, matrix 256 × 512, transverse acquisition. **d** Sagittal projection of a small region of interest. The lesion was detected with the 2D TOF acquisition but the quality of the MR angiogram is poor. The MR angiogram of the CE 3D TOF study nicely shows the collector vein and the smaller medullary veins of the lesion

Fig. 9.7c,d

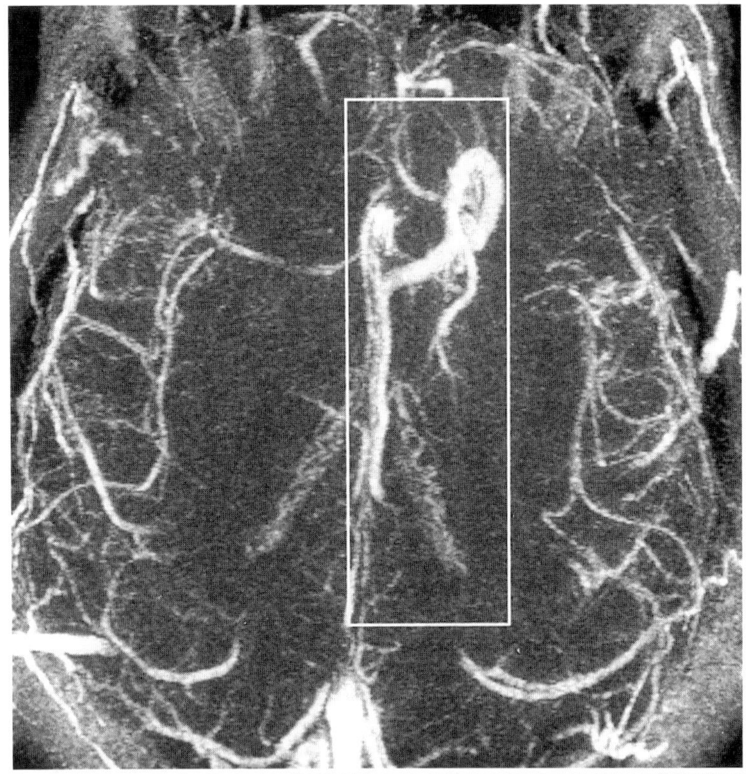

c

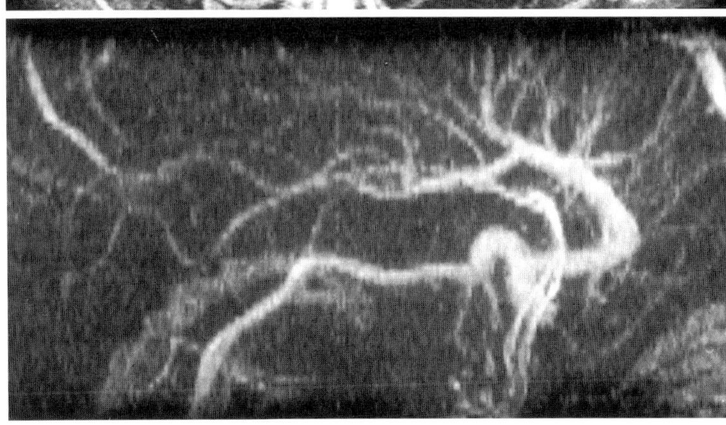

d

9.3.1.4 The Middle Cerebral Artery

The MCA takes a horizontal course under the frontal lobe. Here the lateral lenticulostriate arteries arise, supplying parts of the internal capsule and of the basal ganglia. Then the MCA courses around the island of Reil and extends into the fissure of Sylvius. The end arteries show an inward bend reaching to the top of the sylvian fissure. All these arteries are aligned along the sylvian line. Since the sylvian fissure runs posterosuperiorly, a triangle is described in the lateral view, called the sylvian triangle.

After giving off a small branch, the anterior temporal artery, the MCA usually bifurcates into an anterior and a posterior trunk. The arteries arising from the anterior trunk supply the rolandic sensorimotor region and Broca's area. The posterior trunk supplies the postrolandic parietal region and a large portion of the temporal lobe.

9.3.1.5 The Posterior Cerebral Artery

The PCA originates from the distal basilar artery and runs laterally around the brain stem. In the proximal part perforating arteries arise, supplying the thalamus. Of the ambient segment the most important branches are the medial and lateral posterior choroidal arteries. The former supplies the thalamus and the choroid plexus, while the latter supplies parts of the cerebral peduncle and the basal ganglia.

The terminal branches of the PCA supply the visual cortex (calcarine artery) and the surrounding parieto-occipital cortex.

9.3.1.6 The Vertebrobasilar System

The intracranial vertebral arteries enter the foramen magnum ventrally and converge in front of the medulla to form the basilar artery. Before the junction they give off the anterior spinal artery, which descends in front of the medulla oblongata, the posterior inferior cerebellar artery (PICA), which courses around the medulla and supplies the inferior vermis and parts of the cerebellum, and a posterior meningeal artery for the tentorium.

The basilar artery runs ventrally of the brain stem and bifurcates in the interpeduncular cistern. Smaller pontine branches are usually invisible even at angiography. The anterior inferior cerebellar artery (AICA) originates proximally and runs laterally over the pons to the cerebellopontine angle.

The superior cerebellar arteries run in the perimesencephalic cistern to supply the cerebellar hemisphere and the middle and superior cerebellar peduncle.

9.3.1.7 Normal Variants of the Arterial Anatomy

The internal carotid artery can be hypoplastic or absent. In these cases, the petrous and sellar segments are frequently absent. In other cases, the ascending pharyngeal artery takes over the cervical portion and joins the petrous portions of the carotid canal by means of intratympanic anastomoses. This is incorrectly called an aberrant intratympanic course of the ICA.

The most frequent variations of the circle of Willis are a "fetal" origin of the PCA, where the PCA arises directly from the carotid artery and the proximal PCA is hypoplastic (15% of cases), hypoplasia of one CA1 segment, with a hypoplastic anterior communicating artery joining the two cerebral arteries arising from the other side (25%), and hypoplasia of the posterior communicating artery (20%). Duplication of these arteries is described in about 5% of the cases.

Fetal remnants of anastomoses between the carotid arteries and the vertebral system can persist. The most frequent is the persistent trigeminal artery, originating from the junction of the presellar and juxtasellar ICA and joining the basilar artery through the cavernous sinus. The terminal branches of the ACA, PCA, and MCA are extremely variable.

9.3.1.8 Venous Anatomy

The most important veins of the brain are the dural venous sinuses, the superficial veins, and the deep cerebral veins. Bone and meninges are drained by diploic and meningeal veins.

The most important dural sinuses are: the superior sagittal sinus, the inferior sagittal sinus, the straight sinus, the transverse and sigmoid sinuses, and the cavernous and petrosal sinuses.

The superior sagittal sinus runs in the midline in a duplication of the falx. At the junction of the superior sagittal, straight, and transverse sinuses, a slight dilatation, the torcular herophili, is found.

The inferior sagittal sinus runs in the inferior free edge of the falx. At the junction of the vein of Galen and the inferior sagittal sinus, the straight sinus is formed and runs between the leaves of the joining falx and tentorium. The transverse sinus runs laterally over the tentorium. The sigmoid sinus then courses inferolaterally towards the internal jugular vein.

The inferior and superior petrosal sinuses extend from the cavernous sinus to the jugular bulb and the sigmoid sinus respectively. The sphenoparietal sinus is formed by meningeal and superficial middle cerebral veins and drains into the cavernous sinus. Both cavernous sinuses, lying next to the sphenoid bone and containing the ICA and cranial nerves III, IV, V1, V2, and VI, are connected via intercavernous sinuses.

The superficial or cortical veins drain the cortex. Two constant veins are seen. The vein of Labbé, draining posteriorly into the transverse sinus, and the vein of Trolard, draining into the superior sagittal sinus in the parietal region.

The deep venous system is formed by medial and lateral subependymal veins, the largest of which, the thalamostriate and septal veins, join to form the internal cerebral vein. The basal vein of Rosenthal, running along the brain stem, joins the internal cerebral vein behind the brain stem to form the vein of Galen. This vein courses around the splenium of the corpus callosum, where it joins the straight sinus in the incisura tentorii.

9.3.1.9 Normal Variants of the Venous Anatomy

Variants of the venous anatomy are very frequent. The most frequent anomaly is a hypoplasia of one of the transverse sinuses. The torcular can be connected to the jugular bulb by a midline occipital sinus. The superficial veins differ considerably in caliber. Sometimes the vein of Trolard is very large with a small vein of Labbé, while in other patients the opposite is true. The anatomy of the small subependymal vein is very variable.

9.3.2 MRA Display of the Anatomy and MRA Limitations

The major intracranial arterial branches are well visualised on MRA. Routinely the transverse projection is used for the display of the arteries (Figs. 9.8, 9.9a). Though this is a quite unusual projection in comparison to conventional angiography, it allows one to study the vessels of both hemispheres and/or of distal branches separately without superposition. Selective information for one hemisphere in the lateral projection can be obtained either by selective acquisition of one hemisphere or by projecting a single hemisphere out of the global angiogram by targeted MIP (Fig. 9.9b) (Mattle and Wentz 1992).

Routinely the major intracranial arteries are easily displayed: in their proximal course this includes the carotid siphon, the anterior and middle cerebral arteries, and their branches. The degree of visualization of the distal branches not only depends on the imaging parameters used but also varies widely among patients, depending on their cardiovascular status. Therefore MR angiograms are generally better in children than in adults. Very small arteries, such as the lenticulostriate arteries and the anterior choroidal arteries, are routinely visualized in children but are rarely visible even in young adults.

The application of the maximum intensity algorithm to extract an angiogram from a 3D data set further reduces the conspicuity of small vessels (Anderson et al. 1990). This is particularly true for the study of the circle of Willis, where the intensity of the small posterior communicating artery is further reduced by volume-averaging effects with the low-intensity CSF. Therefore if these are the focus of

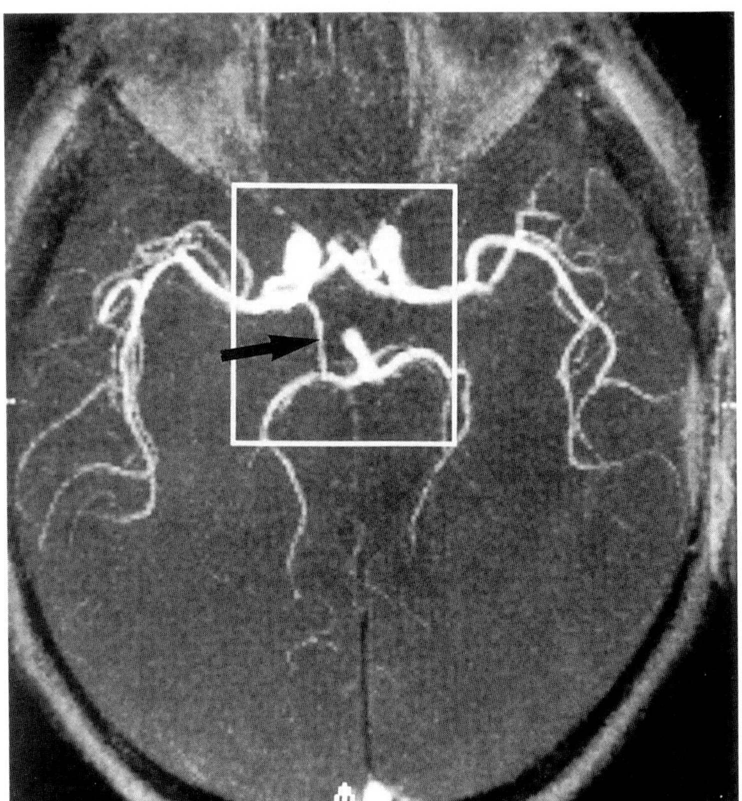

Fig. 9.8. Normal anatomy of the circle of Willis. Transverse MIP of 3D TOF MRA sequence. TR/TE/Fl 40 ms/10 ms/20°, magnetization transfer preparation pulse and TONE RF excitation, rectangular FOV 156 × 210 mm, slab thickness 52 mm/64 partitions, matrix 192 × 256, acquisition time 8 min, number of excitations (Nex) = 1. anterior part of the circle of Willis is formed by the horizontal segments of the anterior cerebral arteries. The right posterior communicating artery (*arrow*) is hypertrophic. The left posterior communicating artery is not visualized

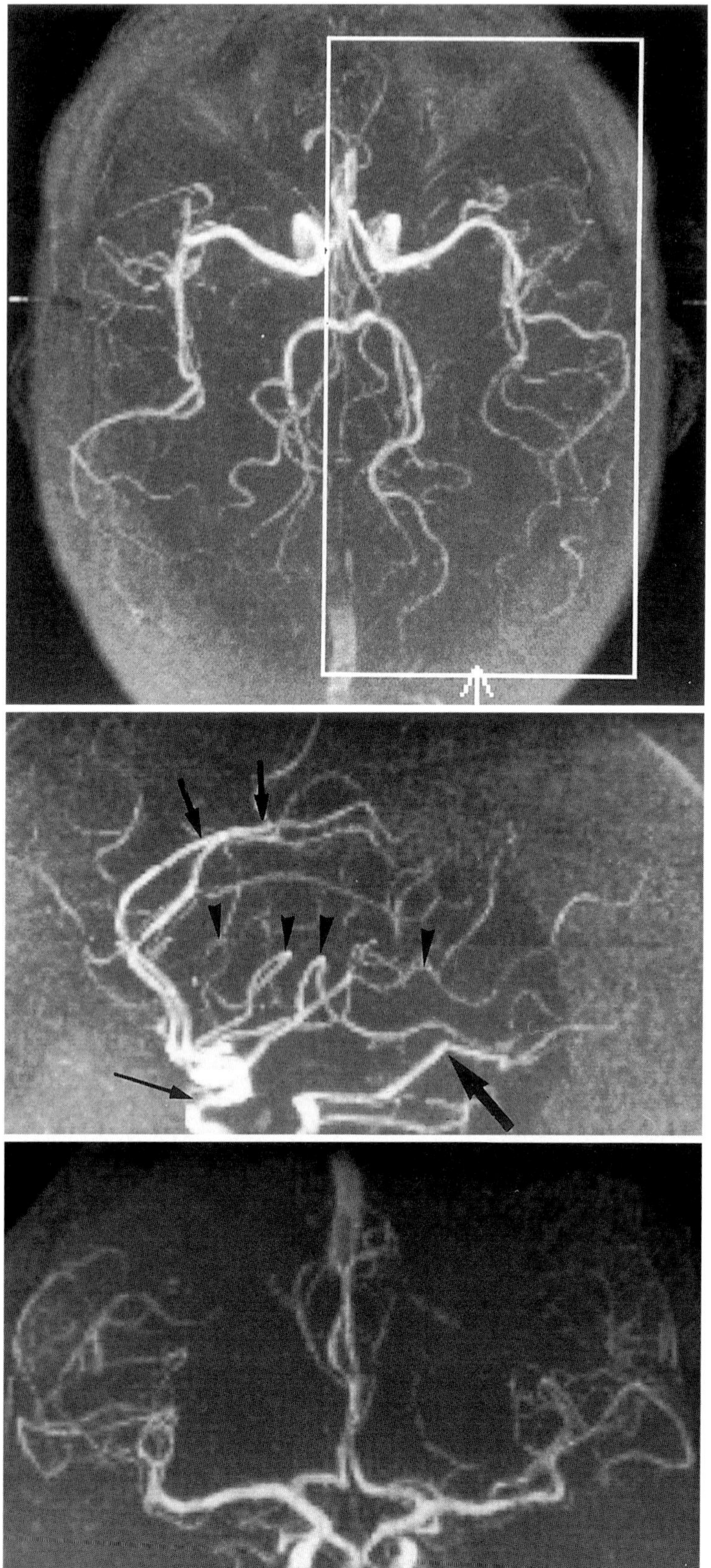

a b c

Fig. 9.9a–c. Normal anatomy of the brain vessels. **a** Transverse MIP of 3D TOF MRA sequence. TR/TE/Fl 40 ms/10 ms/20°, magnetization transfer preparation pulse, rectangular FOV 56 × 210 mm, MOTSA technique (two overlapping 3D acquisitions with slab thickness 56 mm/64 partitions an overlap of 4 mm), matrix 192 × 256, acquisition time 2 × 8 min, Nex = 1. Branches of the middle and posterior cerebral arteries are well visualized. **b** Sagittal targetted MIP of the left hemisphere of the same sequence. The anterior cerebral artery (*double arrow*) with the pericallosal and callosomarginal artery is well seen. The insular branches of the middle cerebral artery (*arrowhead*) are well visualized. The posterior cerebral artery (*arrow*) and its terminal cortical branches are also visualized. Note flow-related artifact in the carotid siphon (*thin arrow*). **c** Frontal MIP of same sequence. Excellent visualization of the normal anatomy the study, a careful review of the native images is mandatory.

The anterior communicating artery is best visible on coronal projections (Fig. 9.10). Persistent fetal branching of the posterior communicating artery from the supraclinoid portion of the carotid artery is a frequent finding and is always obvious on axial projections (Fig. 9.11). The respective contributions

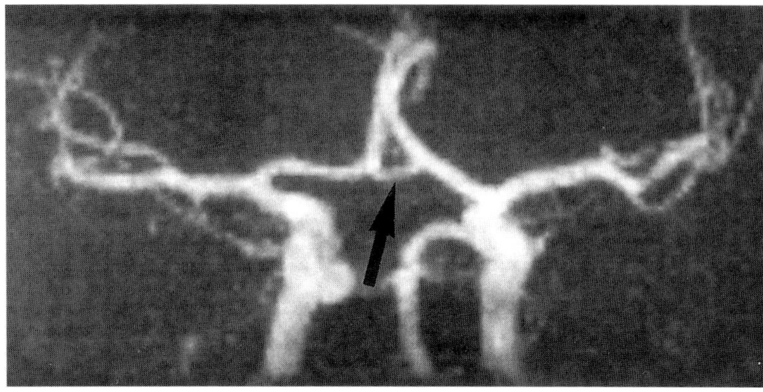

Fig. 9.10. Normal anatomy of the circle of Willis. Coronal MIP of 3D TOF MRA sequence. Note the excellent demonstration of the anterior communicating artery (*arrow*). TR/TE/Fl 40 ms/10 ms/20°, magnetization transfer preparation pulse and TONE excitation RF pulse, rectangular FOV 156 × 210 mm, slab thickness 52 mm/64 partitions, matrix 192 × 256, acquisition time 8 min Nex = 1

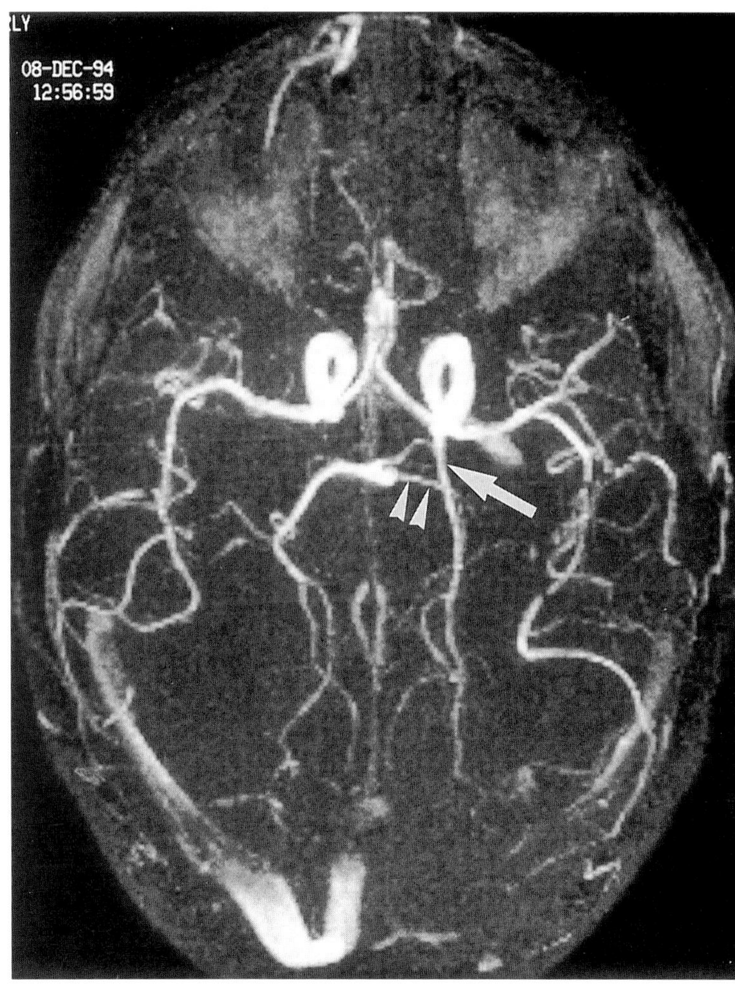

Fig. 9.11. Persistent fetal branching of the posterior communicating artery. Note the large posterior communicating artery (*arrow*) originating immediately from the supraclinoid portion of the internal carotid artery and giving rise directly to the posterior cerebral artery. There is hypoplasia (*arrowheads*) of the proximal posterior cerebral artery (P1 segment). Transverse MIP of 3D TOF acquisition. TR/TE/Fl 42 ms/10 ms/20°, FOV 210 mm, slab thickness 64 mm/64 partitions, matrix 256 × 256, acquisition time 10 min

of the different vessels to the cerebral territories can be deduced from selective MR angiograms with placement of spatial saturation volumes (see higher) (Mattle and Wentz 1992).

Coronal projections offer the best view of the anterior cerebral arteries, the anterior communicating artery, and the horizontal segment of the middle cerebral artery as well as the confluence of the vertebral arteries and the basilar artery (Fig. 9.9c). The carotid siphons are best seen on sagittal projections.

Spatial ambiguity is eliminated by stereoscopic viewing, closed vessel projection, and surface shaded rendering (Anderson et al. 1990).

The main dural venous sinuses are well visualized. Due to the different anatomical orientation the projection will have to differ according to the sinus being studied. The superior sagittal and straight sinuses are best evaluated in the sagittal projection (Fig. 9.12), and the transverse and sigmoid sinuses in the transverse (Fig. 9.13), frontal, or oblique projection. The smaller deep venous circulation is best visualized on gadolinium-enhanced venograms in the sagittal and axial plane (Fig. 9.12) (Marchal et al. 1992). The cavernous sinus can be studied in the transverse and the frontal projection.

Pseudoimages of thrombus in the dural sinuses can be caused by outpouching arachnoid granulations (Fig. 9.13).

9.4 Pathology

9.4.1 Arterial Stenosis and Occlusion

The most common manifestation of acute intracranial vascular obstruction is stroke, a sudden neurological deficit most commonly occurring in the sensorimotor or visual cortex. The symptoms can be limited or transient (transient ischemic attacks or reversible ischemic neurological deficits). However, very frequently cerebral infarction is massive and is the immediate cause of death. In case of survival severe neurological disability is common.

The most frequent cause of stroke is embolism originating in the extracranial neck vessels, with obstruction of distal branches or the main stem of the middle cerebral artery. The prognosis largely depends on the availability of the intracranial collaterals. Besides stenosis or obstruction of cerebral vessels, stroke can be due to other causes, such

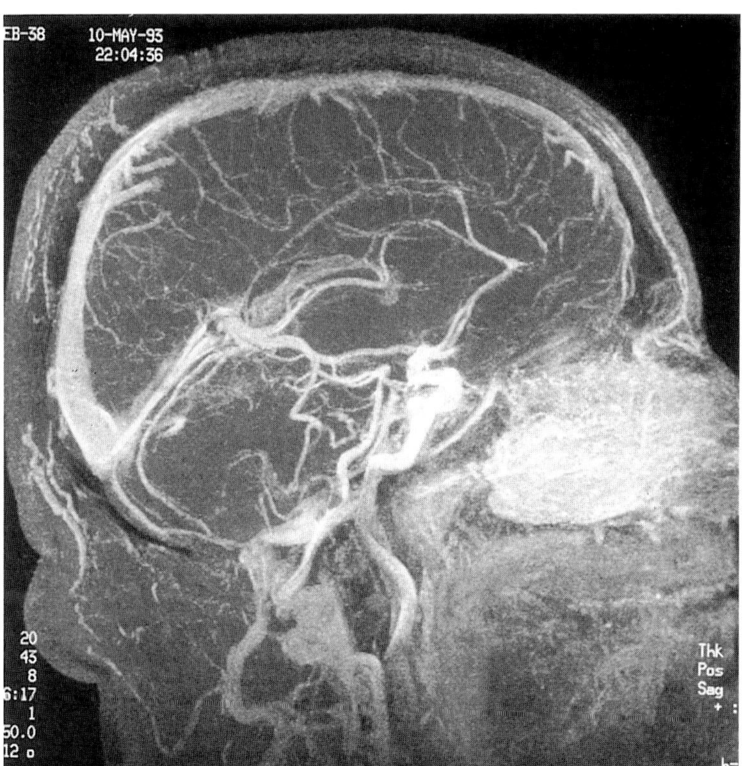

Fig. 9.12. Normal venous anatomy. Sagittal MIP of 3D TOF acquisition. TR/TE/Fl 42 ms/8 ms/20°, FOV 250 mm, matrix 384 × 512, slab thickness 64 mm/64 partitions, acquisition time 16 min. Gd-DTPA-enhanced study. There is excellent demonstration of deep and superficial cerebral veins

Fig. 9.13a,b. Venous aneurysm and venous filling defects caused by arachnoid (Paccioni's) granulation. **a** Selective intra-arterial DSA of the left vertebral artery. Anteroposterior transverse projection. A lateral venous aneurysm is seen on the right (*arrowhead*). Bilateral intraluminal filling defects (*arrows*) are present at the transition of the transverse and sigmoid sinuses, caused by arachnoid granulations. **b** Transverse MIP of 3D contrast-enhanced TOF MRA sequence. The venous aneurysm is well visualized (*arrowhead*). Filling defects (*arrows*) in the transverse sinus are well depicted on both sides. TR/TE/Fl 42 ms/10 ms/20°, magnetization transfer preparation pulse, rectangular FOV 152 × 210 mm, slab thickness 52 mm/64 partitions, matrix 192 × 512, acquisition time 8 min Nex = 1

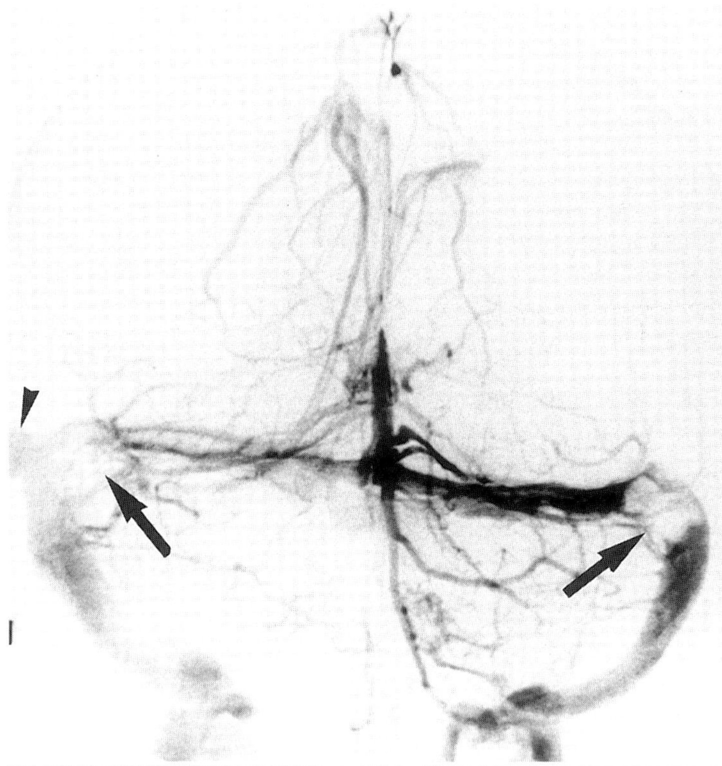

a

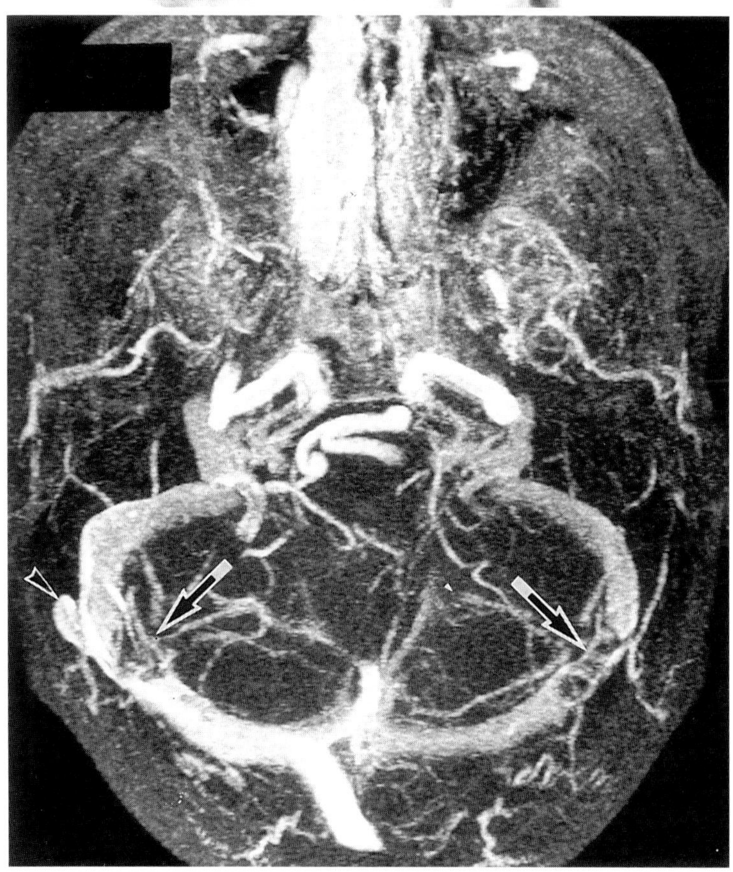

b

as intraparenchymal hemorrhage and subarachnoid hemorrhage. The detection of hemorrhage is important with respect to patient management and therapy.

Intracranial atheromatous stenoses mostly occur at the level of the carotid siphon, the basilar artery, or the main stem of the middle cerebral artery. Periphral stenoses are far less common. The importance of their detection increases if the extracranial vessels are normal. Failure of surgery of the neck vessels is commonly due to associated distal lesions.

In young patients the main causes of intracranial stenoses are infections or autoimmune-based arteritis (Heiserman et al. 1992), sickle cell anemia, and dissection or vasospasm (either traumatic or caused by aneurysmal rupture) (Goldberg 1992; Vogl et al. 1992; Zimmerman et al. 1992; Wiznitzer et al. 1990). In children, moya-moya disease and basal occlusion by osteopetrosis or achondroplasia have to be considered (Vogl et al. 1992; Lisovoski and Rosseaux 1993).

Total vascular obstruction is readily demonstrated by MRA. However, the impact of the demonstration of a vascular occlusion by MRA is still the object of debate since no effective thrombolytic treatment is yet available in these patients.

Low to moderate stenoses without any major effect on the local flow pattern are usually well demonstrated by MRA techniques (Pernicone et al. 1990, 1992). On the other hand stenoses which induce important local flow disturbances are generally overestimated both in diameter and in length because of incomplete flow rephasing in these areas (Fig. 9.14). Tight stenoses can simulate total

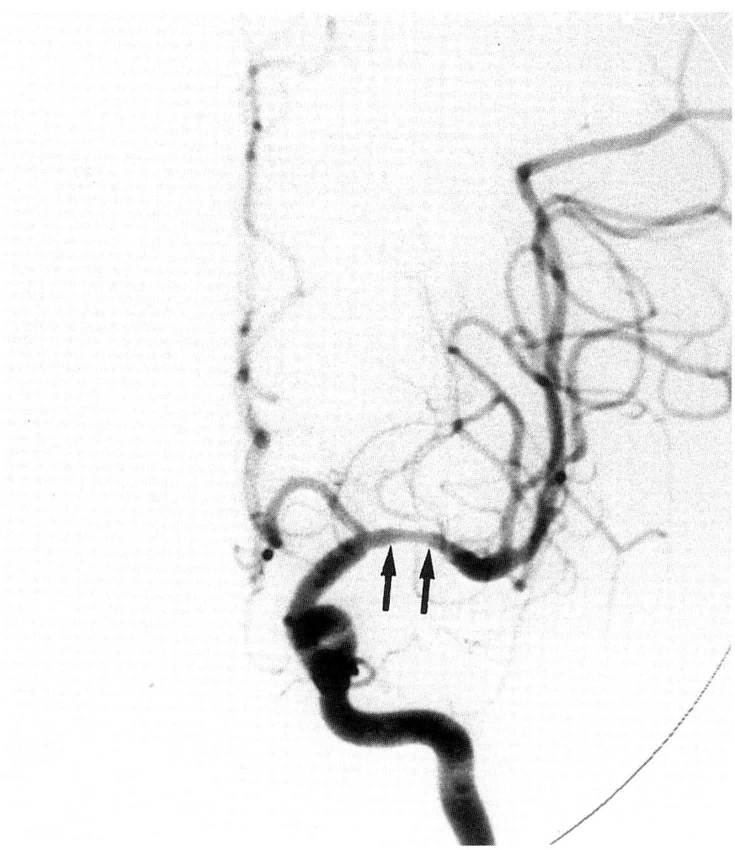

a

Fig. 9.14a–c. Middle cerebral artery stenosis. Presumed arteritis in a 3-year-old child. **a** Selective intra-arterial DSA of the left internal carotid artery. Anteroposterior projection. Moderate (50%–60%) stenosis (*arrows*) of the horizontal middle cerebral artery main stem. Note the involvement of the proximal 0.5 cm of a lateral lenticulostriate artery, originating at the site of maximal stenosis. **b** Transverse and **c** coronal MIP of 3D TOF MRA sequence. TR/TE/Fl 40 ms/10 ms/20°, magnetization transfer preparation pulse and TONE RF excitation, rectangular FOV 150 × 200 mm, slab thickness 52 mm/64 partitions, matrix 192 × 256, acquisition time 8 min Nex = 1. There is severe overestimation of the degree of stenosis (*arrows*), with total disappearance of the signal at the level of the lesion, due to turbulence. The presence of signal in the distal part of the middle cerebral artery indicates that the artery is not occluded. Stenosis of the lenticulostriate artery is not seen

Fig. 9.14b,c

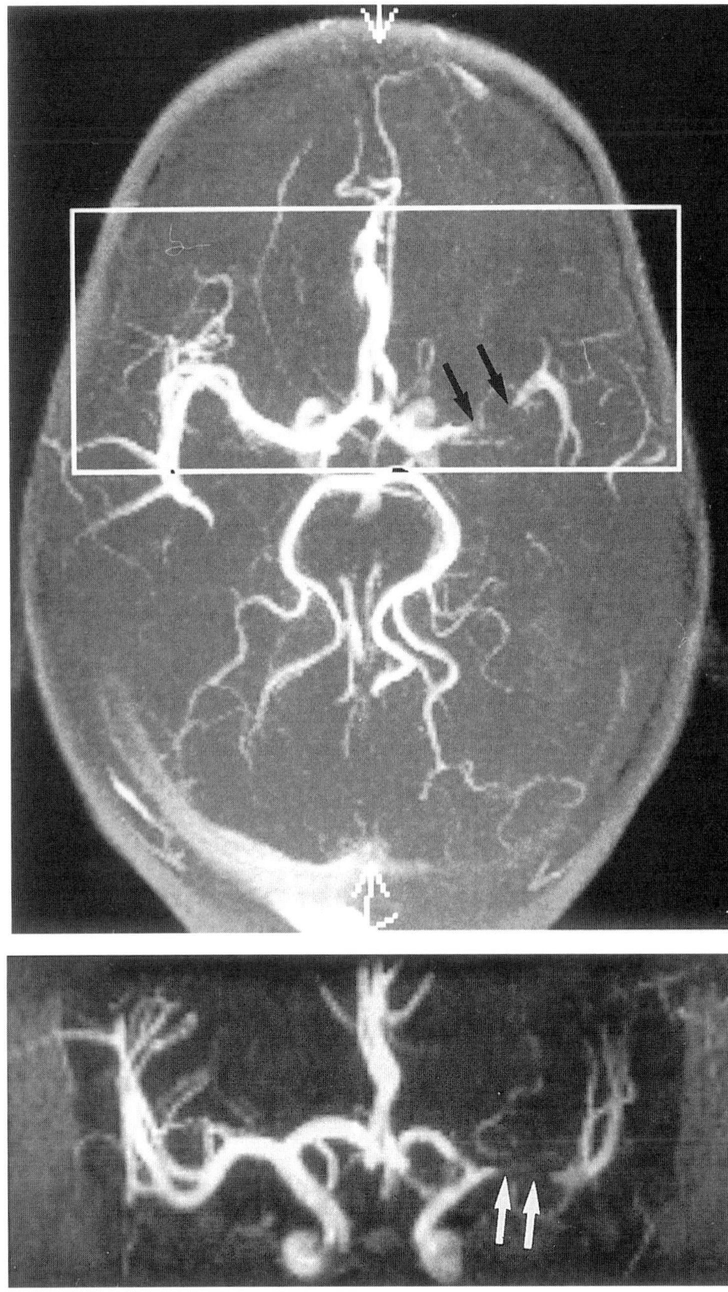

b

c

obstruction, particularly on 3D TOF MRAs, because of saturation of the slow-flowing blood distal to the stenosis (Fig. 9.15) (MASARYK et al. 1991; KATZ et al. 1989).

In general, saturation can be overcome by the injection of gadolinium since it makes the vascular signal less flow dependent (RUNGE et al. 1993; CREASY et al. 1990; MARCHAL et al. 1992). Another approach consists in the application of consecutive 2D acquisitions perpendicular to the poststenotic vessel (BOSMANS et al. 1992). One can conclude that total obstruction is present only if the previously mentioned methods fail to visualize any distal vessels (Fig. 9.15).

Besides the basal occlusions, the hypertrophic collaterals in the basal ganglia are the most striking feature of moya-moya disease (Fig. 9.16) (YAMADA et al. 1992). However, these collaterals together with associated infarcts in the cerebral parenchyma are often already visible on plain spinecho images.

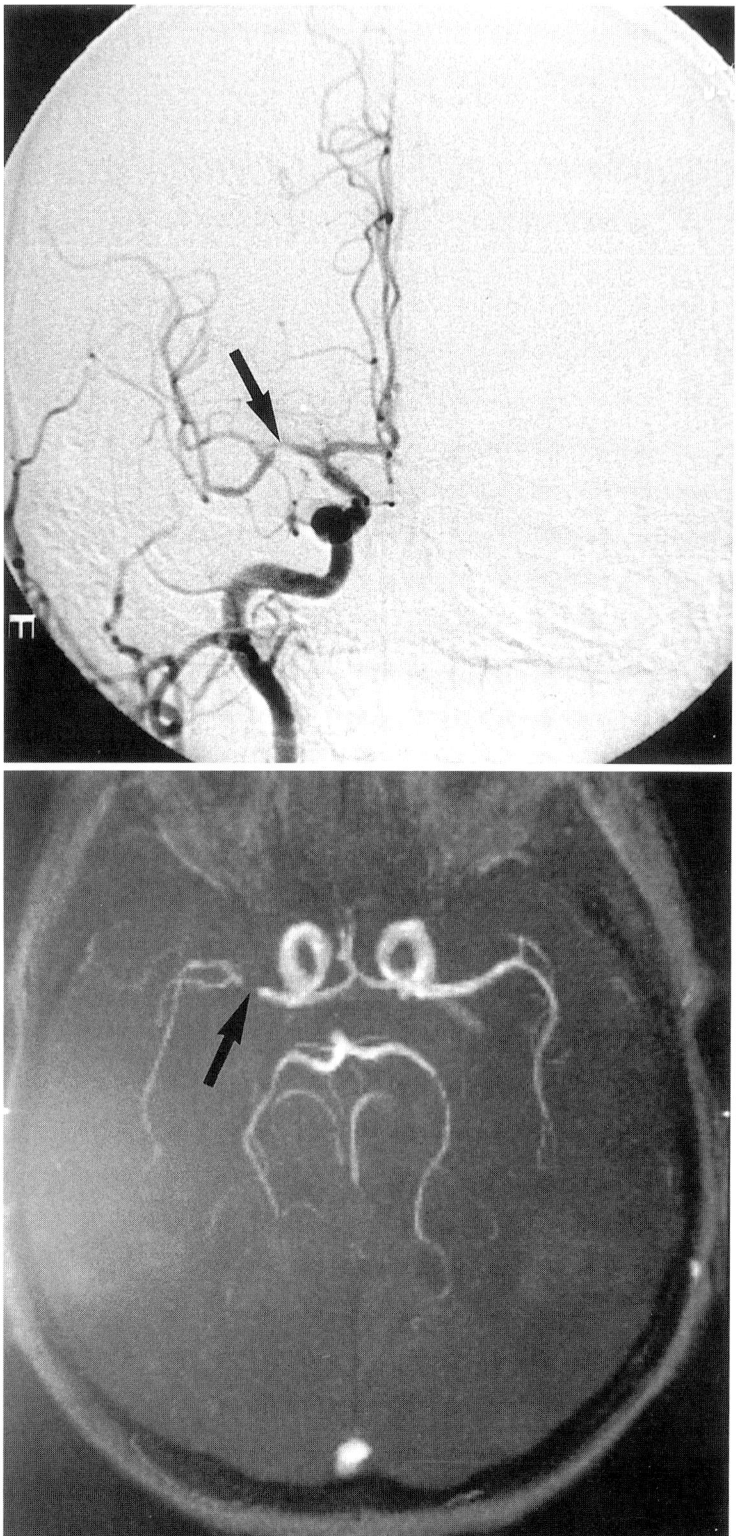

a

b

Fig. 9.15a–c. Middle cerebral artery stenosis in an 87-year-old patient with atheromatosis. **a** Selective intra-arterial DSA of the right common carotid artery. Anteroposterior projection. Severe (95%) stenosis (*arrow*) of the right middle cerebral artery is seen in the horizontal portion. **b** Transverse and **c** coronal MIP of 3D TOF MRA sequences. TR/TE/Fl 40 ms/10 ms/20°, magnetization transfer preparation pulse, TONE RF excitation, rectangular FOV 156 × 210 mm, slab thickness 52 mm/64 partitions, matrix 192 × 256, acquisition time 8 min Nex = 1. Stenosis (*arrow*) is overestimated, with total disappearance of the signal, simulating occlusion

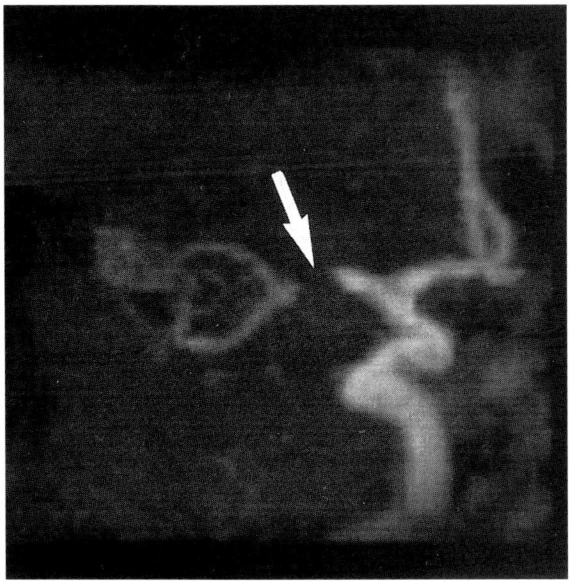

Fig. 9.15c

Intracranial dissection is usually an extension of an extracranial carotid dissection (GELBERT et al. 1991). The spontaneous hyperintense aspect of the thrombus can mimic flowing blood on the MIP. Therefore correlation with the spin-echo images or the applications of phase-contrast sequences is frequently needed (GELBER et al. 1991).

Although vasospasm can be detected by transcranial Doppler, morphological information can help to select those patients who could benefit from interventional neuroradiological procedures, such as angioplasty, papaverine infusion, or selective underlying aneurysm embolization, and to define the optimal time for intervention (MATHURIN et al. 1993).

As already mentioned, the prognosis of vascular occlusive disease of both the intra- and extracranial vessels largely depends on the available collateral circulation. Of the different possible collateral pathways, only the circle of Willis, which is close to the major vessels, can adequately be studied by MRA (MASARYK et al. 1989). Leptomeningeal collaterals

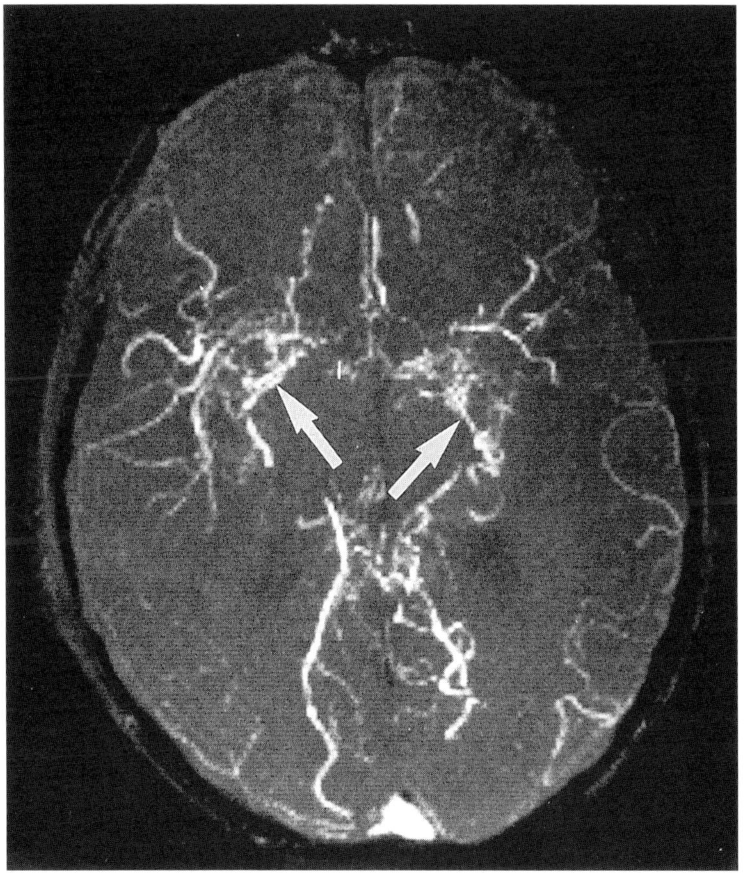

Fig. 9.16. Moya-moya disease in a 5-year-old child. Transverse MIP of 3D TOF MRA sequence (1.5-T MR system). TR/TE/Fl 40 ms/7 ms/20°, FOV 230 mm, MOTSA, slab thickness 32 mm/32 partitions, mutual overlap of 6 mm, total of five slabs; matrix 256 × 256, acquisition time 5 × 5 min, Nex = 1. The carotid siphon and the main stem of the middle cerebral anterior and posterior cerebral arteries are not visualized, pointing to basal occlusion. There is visualization of extensive collateral circulation (*arrows*) in the basal ganglia

In the diagnostic workup of a patient with a suspected intracranial aneurysm, angiography, by either catheter angiography, spiral CT, or MRA, has to answer several questions. The aneurysm has to be detected or absence of aneurysm has to be confirmed with certainty. It has to be kept in mind that aneurysms can be multiple (Fig. 9.17). The exact topographic location has to be determined, including identification of the vessel of origin, demonstration of the aneurysm neck, and clarification of relationship to neighboring vessels. The morphology and size must be demonstrated. In the case of multiple aneurysms it may be necessary to demonstrate which lesion has bled. Finally, vasospasm has to be ruled out.

Therapy for intracranial aneurysm is primarily neurosurgical and consists in clipping of the aneurysmal neck or, in the absence of a surgical neck, in "wrapping" of the aneurysmal dome.

Endovascular techniques can be used for certain locations that are less easily accessible to neurosurgery, such as the basilar tip or the cavernous sinus. The method can consist in parent vessel occlusion with a detachable balloon for giant aneurysms with a broad neck or selective embolization with GDCs (Guglielmi detachable coils).

Conventional spin-echo MRI can easily detect larger aneurysms. However, differentiation of small aneurysms from vascular loops and of signal void from dense bone from the anterior clinoid process can be difficult (Fig. 9.17).

Depending on the techniques used, MRA in combination with MRI has a detection rate of ca. 90%–95% for aneurysms as small as 3–4 mm (Fig. 9.18)

a

b

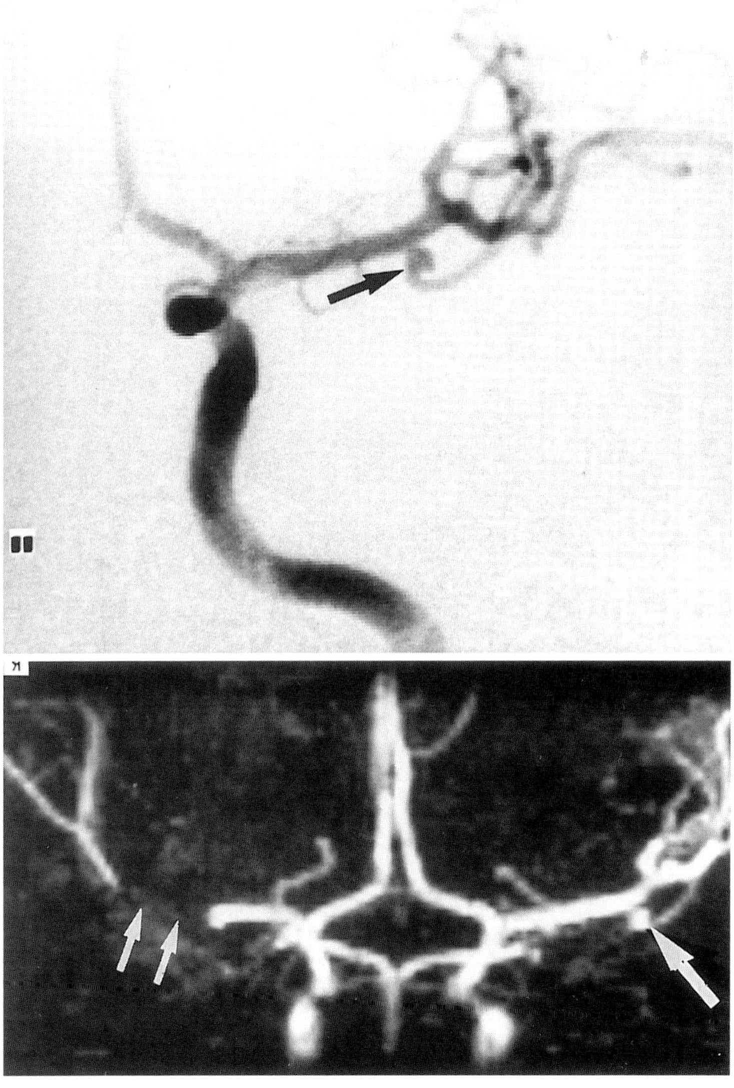

Fig. 9.18a,b. Small middle cerebral artery aneurysm. **a** Selective intra-arterial DSA of the left internal carotid artery. Anteroposterior submental projection. There is a small (diameter 3 mm) aneurysm (*arrow*) of the temporal branch of the left middle cerebral artery. **b** Coronal MIP of 3D TOF MRA sequence. TR/TE/Fl 40 ms/10 ms/20°, magnetization transfer preparation pulse and TONE RF excitation, FOV 153 × 210 mm, slab thickness 52 mm/64 partitions, matrix 192 × 256, acquisition time 8 min, Nex = 1. A small lesion (*arrow*) is well visualized. Note the signal drop out in the right middle cerebral artery due to a metallic clip (*double arrows*)

(CURNES et al. 1993; SEVICK et al. 1990; SCHUIERER et al. 1992; ROSS et al. 1990).

Since small aneurysms are easily missed because of superimposition of the carrier vessels a careful analysis of the MRA in multiple projections and of the native images is mandatory (Fig. 9.17).

Local turbulence often occurs in larger aneurysms and can simulate intraluminal thrombus or lead to underestimation of the severity of the lesion (Figs. 9.19, 9.20).

Saturation due to slow flow in larger aneurysms will reduce the signal in the lesion and the distal vessels (Fig. 9.21). Here, too, injection of gadolinium allows recovery of the vascular signal (MARCHAL et al. 1992). However, after gadolinium injection, the enhancement of the venous plexus in the region of the cavernous sinus can obscure intracavernous lesions. High signal from partial thrombosis or hemorrhage interferes with TOF angiograms, causing blurring of the aneurysm outline or overestimation of the patent lumen (Fig. 9.22) (GOULIAMOS et al. 1992). In these cases correlation with the spin-echo images and/or phase-contrast acquisitions will accurately demonstrate the exact size of the aneurysm (PERNICONE et al. 1990; HUSTON and EHMAN 1993). Besides the diagnosis of aneurysms, MRA, by virtue of its ability to yield multiple projections, allows very accurate evaluation of the anatomical implantation, the origin of the lesion, and the neck of the aneurysm. The transverse

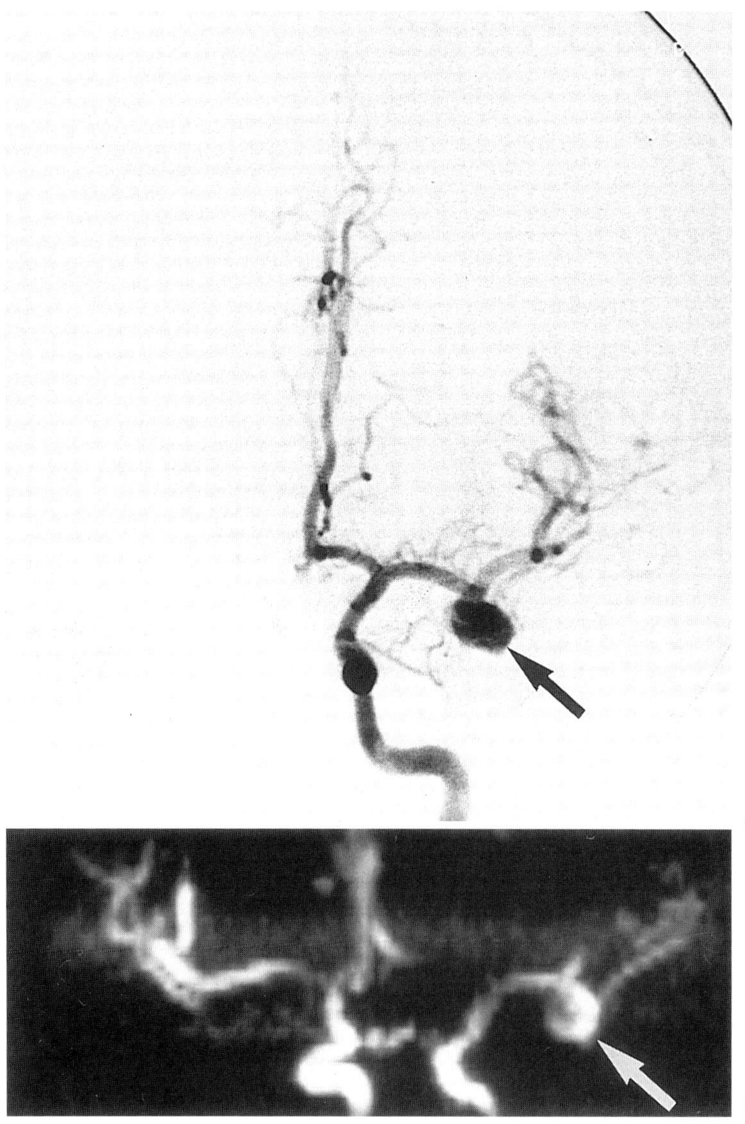

Fig. 9.19a,b. Middle cerebral artery aneurysm: signal loss due to turbulence. **a** Selective intra-arterial DSA of the left internal carotid artery. Anteroposterior projection. A large aneurysm (diameter 1 cm) (arrow) of the left middle cerebral artery is seen. **b** Coronal MIP of 3D PC MRA sequence, performed 1 week later. TR/TE/Fl 30 ms/10 ms/20°, FOV 210 mm, VENC 30 cm/s, slab thicknesss 40 mm/32 partitions, matrix 192 × 256, acquisition time 12 min, Nex = 1. There is central signal drop out in the aneurysmal sac (arrow) due to turbulence

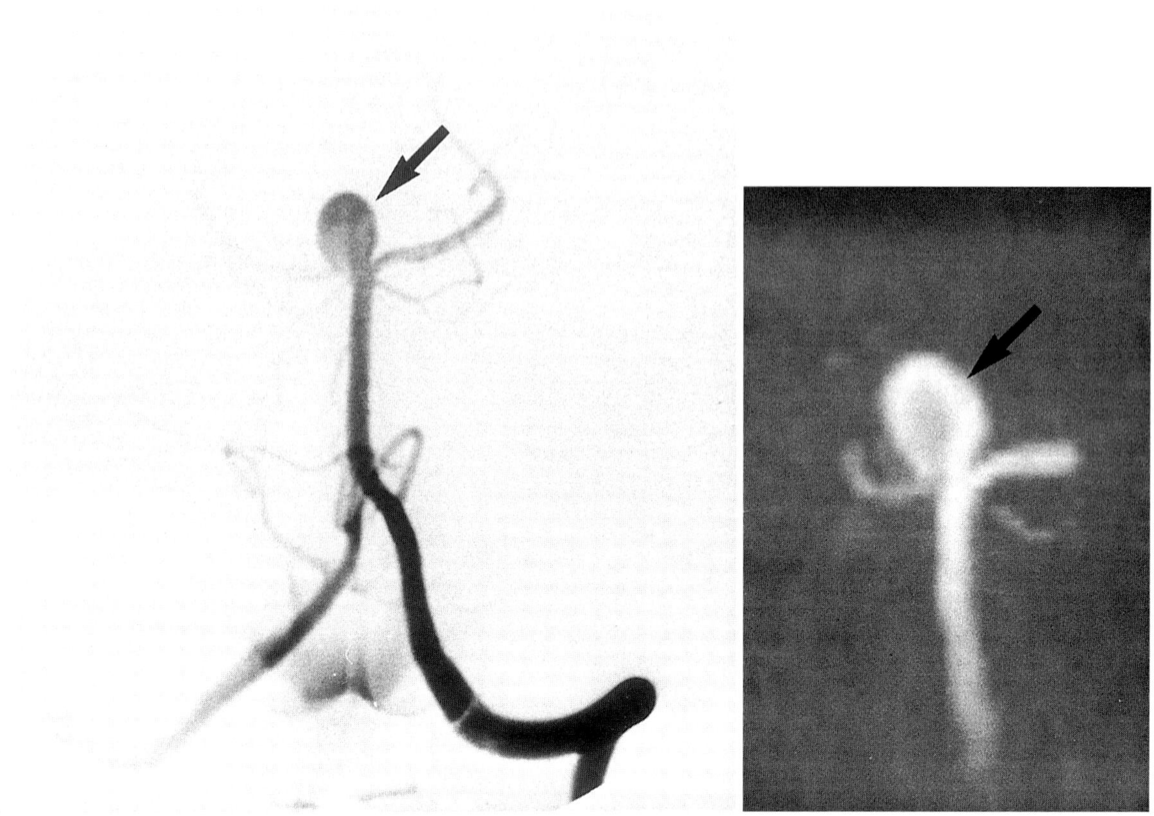

Fig. 9.20a,b. Basilar tip aneurysm: signal loss due to turbulence. **a** Selective intra-arterial DSA of the left vertebral artery. Anteroposterior projection, early arterial phase: a basilar tip aneurysm is shown (*arrow*). Turbulence within the aneurysmal sac is well depicted. **b** Coronal MIP of 3D TOF MRA sequence. TR/TE/Fl 40 ms/7 ms/20°, magnetization transfer preparation pulse and TONE RF excitation, rectangular FOV 165 × 220 mm, slab thickness 52 mm/64 partitions, matrix 192 × 256, acquisition time 8 min, Nex = 1. An aneurysm (*arrow*) of the basilar tip is very well visualized. Intra-aneurysmal turbulence causes signal loss

projections are of particular value in many cases of eccentrically developed aneurysms. As mentioned above, accompanying vasospasm is readily diagnosed and can easily be followed up by sequential MRA studies.

Given the high sensitivity of MRA in the detection of intracerebral aneurysms, one might propose screening of patients with this technique, with the performance of further angiographic investigations as necessary. After therapy with either surgical clipping or GDCs, it is impossible to follow up the lesion owing to susceptibility artifacts from the implanted material (Fig. 9.18). After endovascular balloon occlusion or parent vessel occlusion with detachable balloons, MRA can be of great value, provided balloons are used without a metal clip (Tsuruda et al. 1992). MRA is very well suited to the follow-up of patients with untreated aneurysms or after failure of endovascular therapy, to assess whether morphological changes in the aneurysmal configuration indicate new therapeutic approaches to be appropriate.

9.4.3 Arteriovenous Malformations

Intracranial arteriovenous malformations (AVMs) represent the most common type of cerebral vascular malformation. AVMs consist of a cluster or nidus of multiple arteriovenous shunts, fed by one or more hypertrophic arteries and drained by tortuous hypertrophic veins. Aneurysms which occur on the feeding arteries and draining veins as well as in the nidus are the source of frequently fatal hemorrhagic complications (Turjman et al. 1994). AVMs are usually large and already visible on plain spin-echo MRI (Graves and Duff 1990).

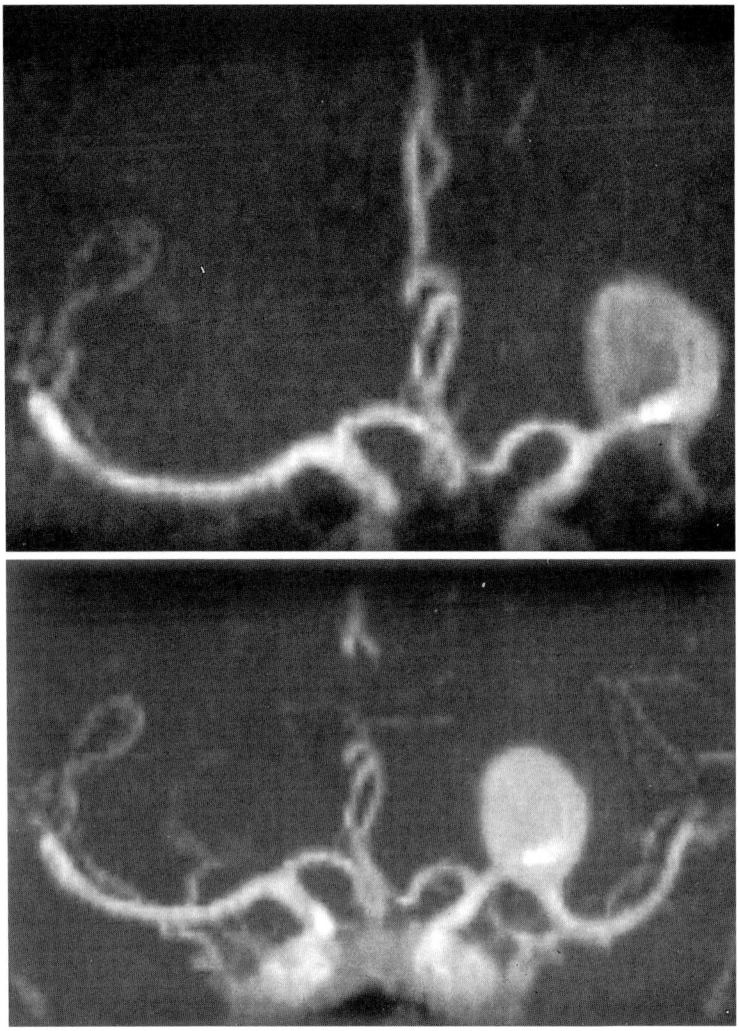

Fig. 9.21a,b. Middle cerebral artery aneurysm. Transverse MIP of 3D TOF acquisition, TR/TE/Fl 40 ms/10 ms/20°, FOV 220 mm, matrix 256 × 256, slab thickness 64 mm/64 partitions, acquisition time 10 min. **a** MIP of pre-Gd-DTPA study shows flow-induced turbulence and saturation in the middle cerebral artery. **b** MIP of contrast-enhanced TOF study shows filling of the aneurysm and good filling of the middle cerebral artery

The most severe clinical manifestation is hemorrhage, which has a high morbidity and mortality. However, many patients will develop substantial morbidity due to the lesion, even in the absence of hemorrhage. Most frequent manifestations are epilepsy or cranial nerve deficits. Considering the potential risk of bleeding even asymptomatic lesions are more aggressively treated nowadays.

Therapeutic possibilities include surgery, intravascular embolization, and radiotherapy. Combination of therapies is not uncommon. Angiography is not necessary for the diagnosis of cerebral AVMs since they are already well depicted on contrast-enhanced CT or spin-echo MR images. The choice of therapy can be made solely on the basis of topography and size.

Angiography has to demonstrate the number, size, and location of the feeding arteries, the size of the nidus, the presence of intranidal arteriovenous shunts, and the size and course of the draining veins. Aneurysms on the feeding arteries, in the nidus, or on the draining veins also have to be visualized, considering the important prognostic implications.

Because of the high flow in the feeding arteries, they are generally well visualized on 3D TOF MRA or PC MRA (Fig. 9.23) (Nüssel et al. 1991; Edelman et al. 1989). However, in some larger lesions, the hyperdynamic character of the flow in these feeding vessels can lead to severe local turbulence, leading to signal loss (Fig. 9.24). Therefore small aneurysms are frequently missed. On the other hand, in many cases both the nidus and the draining veins will be incompletely visualized due to saturation. Here again, the injection of gadolinium or the application of less saturating sequences (2D/3D MOTSA) can be very helpful (Blatter et al. 1991, 1992). Despite the spec-

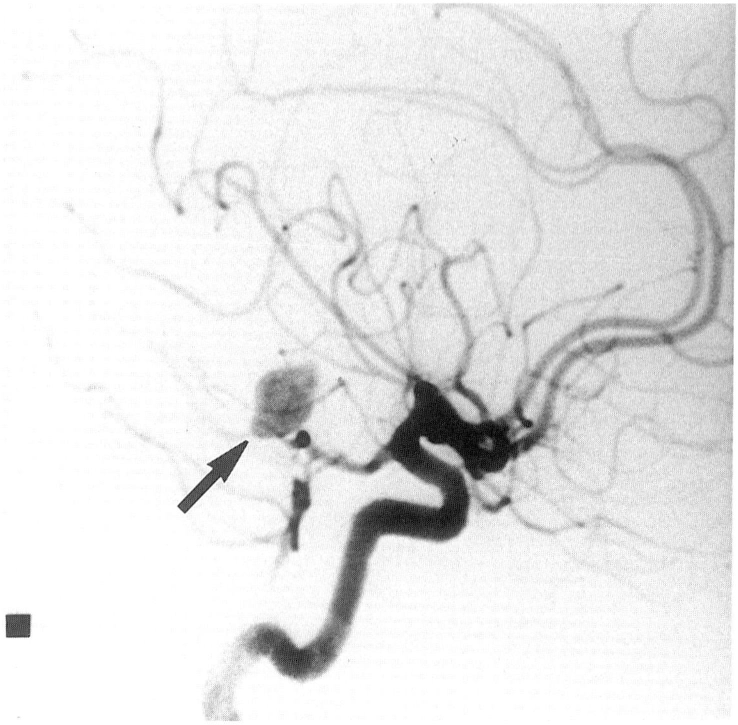

a

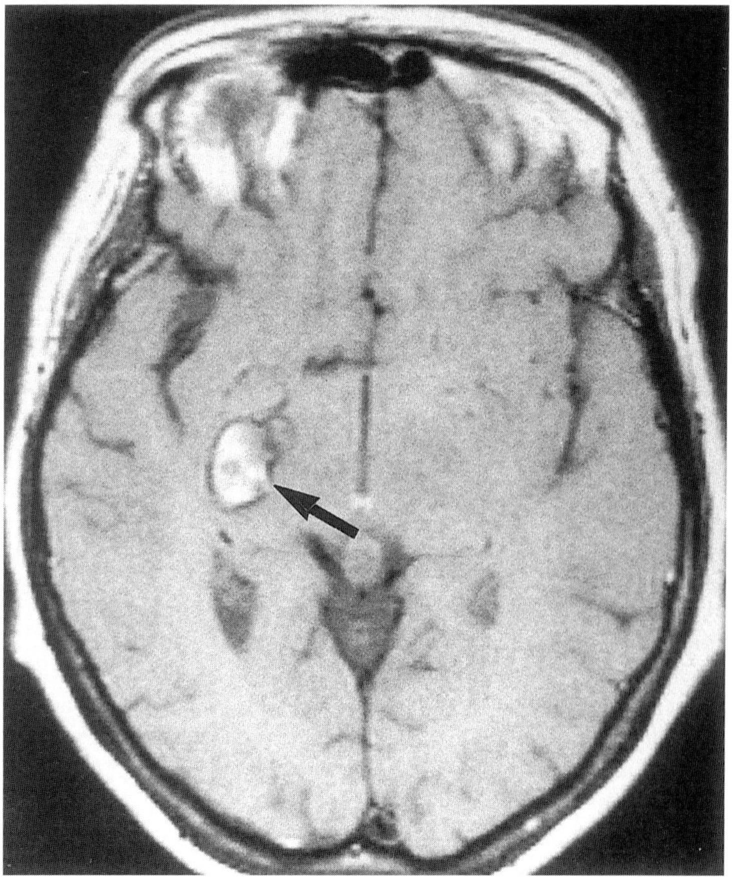

b

Fig. 9.22a–f. Posterior cerebral artery aneurysm, probably mycotic: superposition of recent hemorrhage. **a** Selective intra-arterial DSA of the right internal carotid artery. A large (diameter 8 mm) aneurysm (*arrow*) of the perimesencephalic portion of the right posterior cerebral artery is visualized. **b**, **c** Transverse T1-weighted (TR = 456 ms; TE = 40 ms, 1 acq.) spin-echo MR sequence. Hyperintense hemorrhage (*arrow*) is located posteriorly to the aneurysm (*arrowheads*), presenting signal void. **d** Sagittal and **e** transverse MIP of 3D TOF MRA sequence. TR/TE/Fl 43 ms/8 ms/20°, magnetization transfer preparation pulse and TONE RF excitation, FOV 165 × 220 mm, slab thickness 32 mm/32 partitions, matrix 256 × 256, acquisition time 6 min, Nex = 1. Both the aneurysm and the hemorrhage (*arrows*) are depicted as hyperintense. **f** Transverse MIP of 3D PC MRA sequence. TR/TE/Fl 45 ms/8 ms/20°, FOV 260 mm, VENC 30 cm/s, slab thickness 32 mm/32 partitions, matrix 192 × 256, acquisition time 9 min, Nex = 1. Stationary tissue of hemorrhage is subtracted. The size of the aneurysm is underestimated due to increased dephasing effects of flow turbulence in the PC MRA sequence

Fig. 9.22c,d

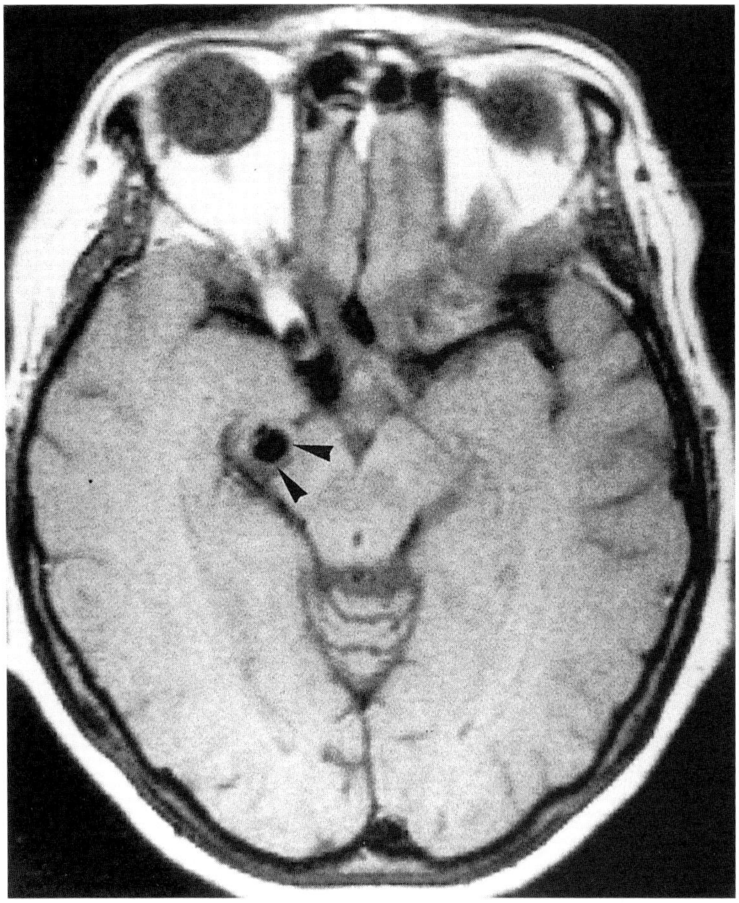

c

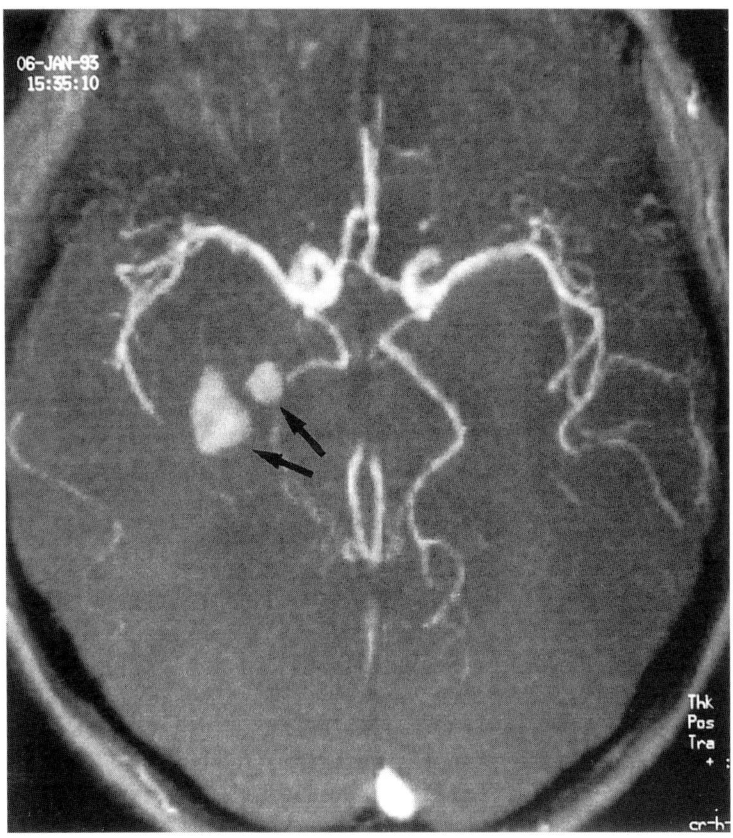

d

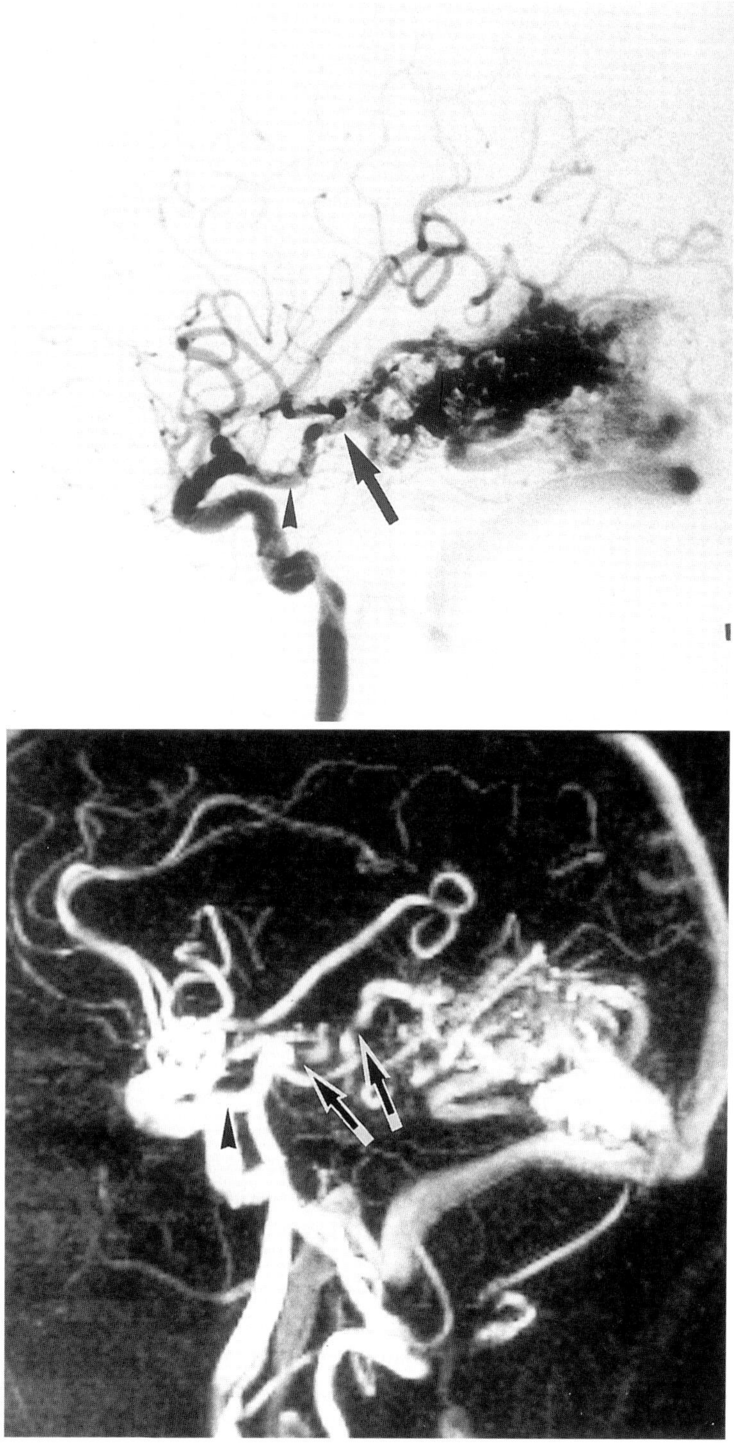

a

b

Fig. 9.24a,b. Left occipital AVM. **a** Selective intra-arterial DSA of the left internal carotid artery. Lateral projection. A huge nidus is present in the left occipital lobe. The lesion is fed by hypertrophic feeding arteries (*arrows*) arising from the posterior cerebral artery. There is embryonal branching of the left posterior cerebral artery of the left internal carotid artery via a hypertrophic "embryonal" posterior communicating artery (*arrowhead*). Venous drainage via hypertrophic cortical veins towards the transverse sinus. **b** Sagittal MIP of 3D PC MRA sequence. TR/TE/Fl 30 ms/10 ms/20°, FOV 210 mm, VENC 30 cm/s, two sagittal overlapping thin 3D acquisitions, slab thickness 48 mm/32 partitions, matrix 192 × 256, acquisition time 2 × 12 min Nex = 1. There is excellent visualization of the nidus. The posterior communicating artery is also well visualized (*arrowhead*). The perimesencephalic portion of the posterior cerebral artery and the proximal parieto-occipital artery show some interruption (*arrows*) due to turbulent flow. Venous drainage is well visualized

Fig. 9.25a–h. Left temporal glioblastoma. **a** Anteroposterior and **b** lateral projection of selective intra-arterial DSA of the left internal carotid artery. Elevation of the vascular branches of the middle cerebral artery is evident. Tumoral vessels with hypervascularization and tumoral blush are present in the temporal region. **c** Transverse, **d** coronal, and **e** sagittal MIP of unenhanced 3D TOF MRA sequence. TR/TE/Fl 43 ms/10 ms/20°, magnetization transfer preparation pulse and TONE RF excitation, FOV 200 mm, MOTSA, slab thickness 32 mm/32 partitions, mutual overlap of 6 mm at each side, matrix 256 × 256, acquisition time 5 × 6 min, Nex = 1. There is good visualization of vascular displacement of the left middle cerebral artery branches. Tumoral vessels are not visualized. **f** Transverse, **g** coronal, and **h** sagittal MIP of 3D TOF MRA sequence after gadolinium enhancement. TR/TE/Fl 43 ms/10 ms/20°, magnetization transfer preparation pulse and TONE RF excitation, single thick slab acquisition 120 mm/128 partitions, matrix 192 × 256, acquisition time 18 min, Nex = 1. The tumoral mass enhances and can be perfectly located. Enhancing tumor somewhat obscures vessels of the middle cerebral artery. Visualization of the tumoral vessels is questionable

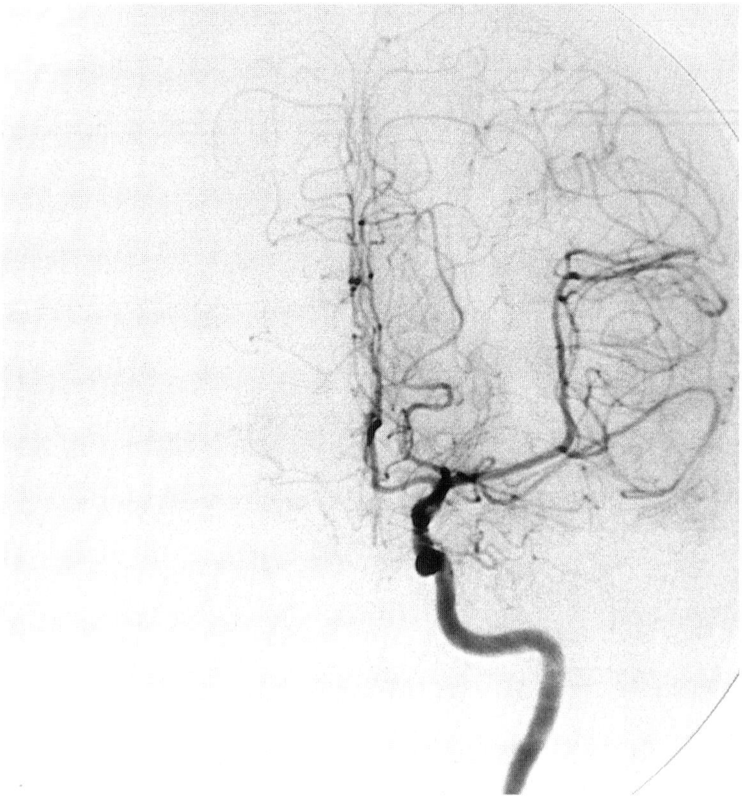

a

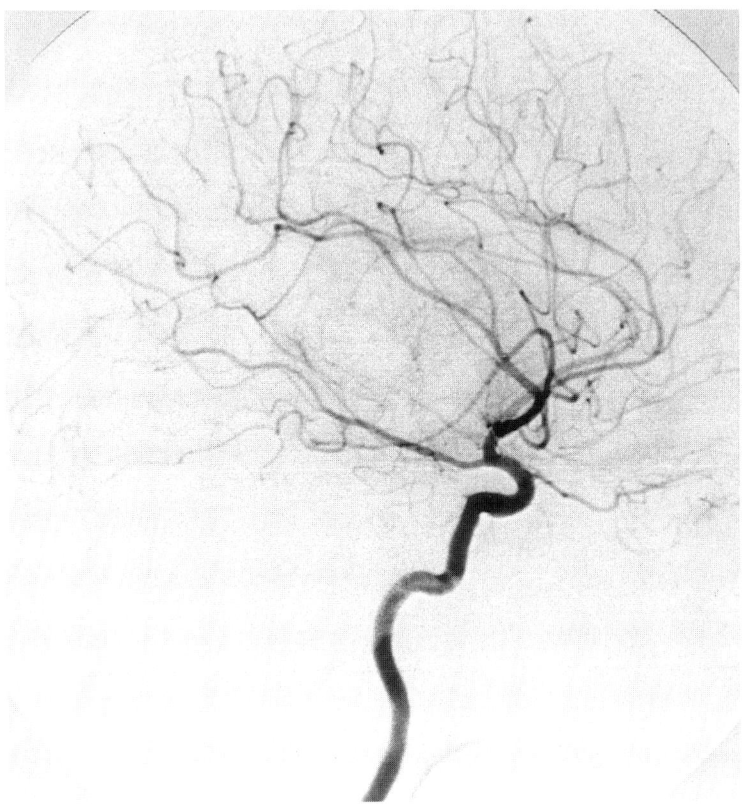

b

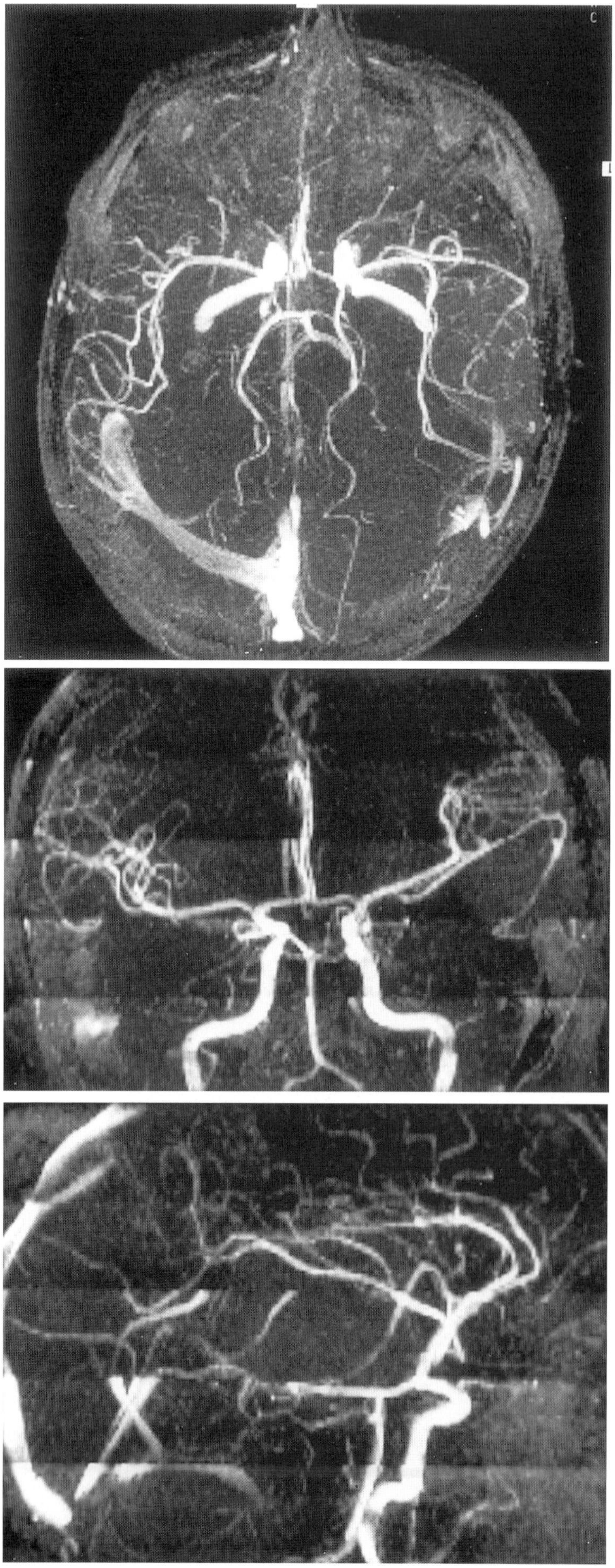

c

d

e

Fig. 9.25c–e

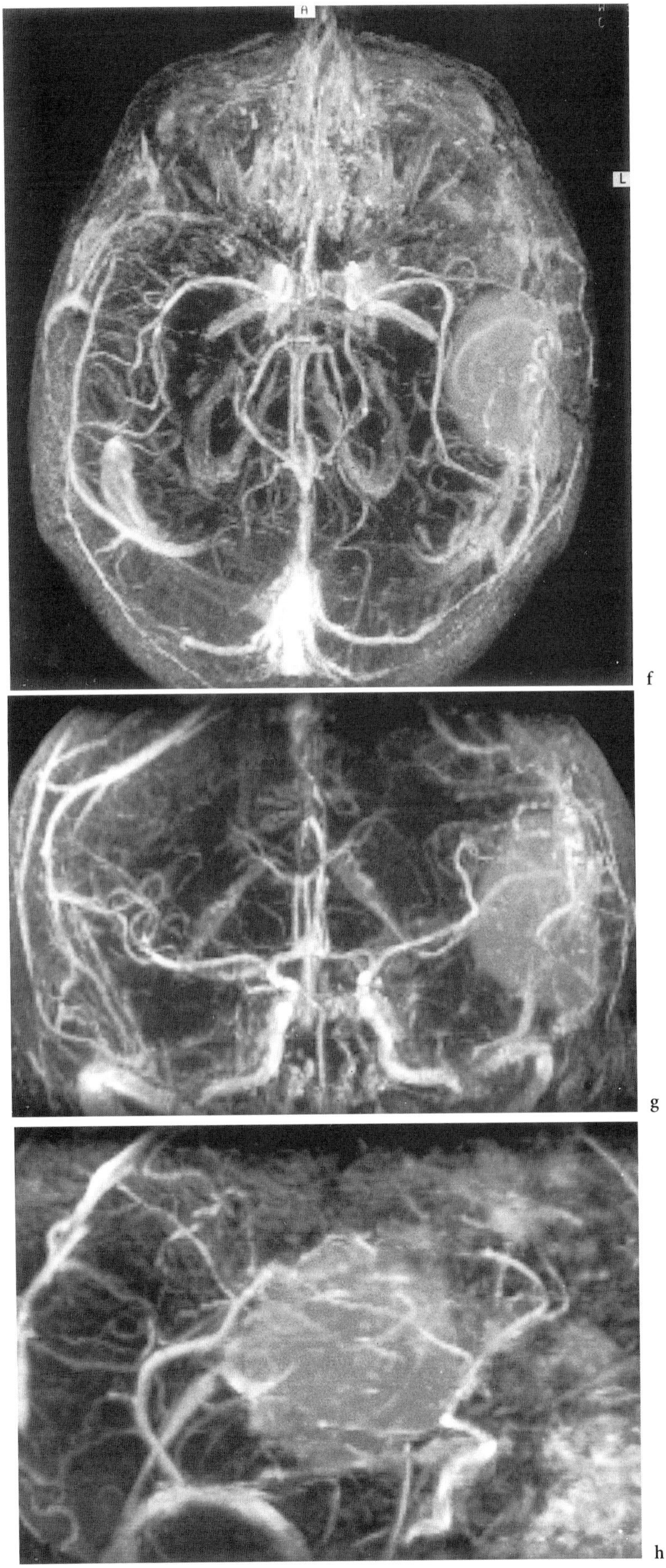

Fig. 9.25f–h

9.4.5 Venous Pathology

The clinical signs and symptoms of cerebral venous occlusive disease are often nonspecific, with acute neurological deterioration due to cortical venous infarction and petechial perivascular hemorrhages. It is most commonly due to spread of infection, dehydration, hematological disorders, coagulopathies, trauma, or tumors. The most frequently involved sinus is the superior sagittal sinus, followed by the transverse, sigmoid and cavernous sinuses. The superficial cortical veins that drain into the superior sagittal sinus are the most commonly occluded veins.

Many acute to subacute thromboses are already diagnosed on unenhanced T1 SE images by virtue of the hyperintense character of the fresh thrombus (Fig. 9.26c). The purpose of angiography is to confirm the CT or spin-echo MR diagnosis of cerebral venous occlusion and the evaluation of the extent of the venous occlusion.

In MRA the presence of a thrombus is seen as a filling defect within the lumen (NADEL et al. 1990; PADAYACHEE et al. 1991; FINN et al. 1993). High-intensity fresh thrombi can also cause false-negative images on TOF sequences (Fig. 9.26a). In these cases phase-contrast methods and the application of saturation pulses can be necessary (Fig. 9.26b,d). One has to be aware of possible false-positive results related to saturation in 2D TOF MRA sequences (Fig. 9.27). An appropriate choice of scan planes is therefore mandatory.

In Fig. 9.28 the compression of the sagittal sinus by a small meningioma is illustrated.

As mentioned above, giant arachnoid granulations can cause filling defects mimicking intraluminal thrombus in the dural venous sinuses (Fig. 9.13)

9.4.6 Vascular Compression

Vascular loops can cause compression of cranial nerves. The classical example is the compression of the facial nerve in the cerebellopontine angle by loops of the anterior inferior cerebellar artery (AICA), causing facial nerve palsy or neuralgia (FELBERT et al. 1992). Trigeminal neuralgia can be caused by an elongated basilar artery. The role of angiography consists in the visualization of the compressing vascular loops. With appropriate technique MRA is able to visualize the tortuosity of the basilar artery and the AICA. By correlation with the spinecho sequences the diagnosis of vascular compression can be made.

9.5 Conclusion

9.5.1 Imaging Protocols for Intracranial MRA of the Arteries

A 3D TOF acquisition through the region of suspicion is a quick and reliable first technique in many indications for intracranial MRA. However, frequently, unresolved questions remain that require additional approaches. A typical example is recent thrombus causing spontaneously hyperintense signal that interferes with the TOF angiogram. A T1-weighted (spin-echo) acquisition can be added to prove the presence of tissue with a short T1. Another approach is the use of a PC acquisition that subtracts the signal of the stationary tissues, including the blood clot, and therefore provides a more correct angiogram. In cases of suspected aneurysms with bleeding, it is even suitable to start the study with a PC acquisition.

A second common problem is the signal loss in blood vessels. An additional contrast-enhanced TOF or a 2D TOF study can then be performed to differentiate between saturation-induced signal loss (in which case the signal loss disappears) and flow-induced phase effects (in which signal loss is not overcome).

In the preoperative study of patients with brain tumors or in the workup of patients with AVMs it might be necessary to visualize all the vessels of the brain. The whole brain can be imaged by means of multiple overlapping small 3D TOF acquisitions. A second contrast-enhanced acquisition can be added to improve the visualization of the veins. PC acquisitions may be performed instead of TOF acquisitions. Though high-resolution PC angiography is time consuming, it can be advantageously used if the volumes of interest are limited. Occasionally, a PC study has to be repeated with different VENCs.

9.5.2 Imaging Protocols for Intracranial MRA of the Veins

In cases of suspected sagittal sinus thrombosis, a 2D TOF or a PC acquisition is an appropriate first MRA sequence. With 2D TOF, one must realize that both inflowing blood and tissues with a short T1 produce

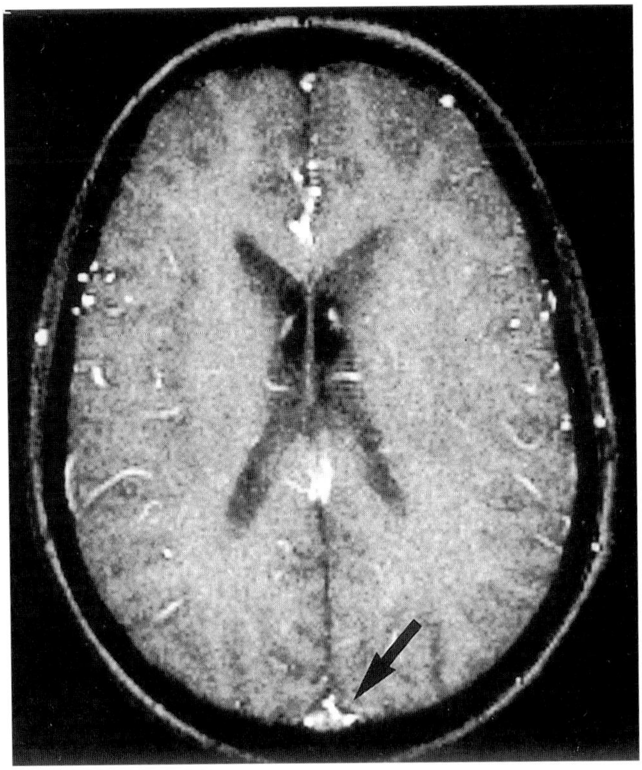

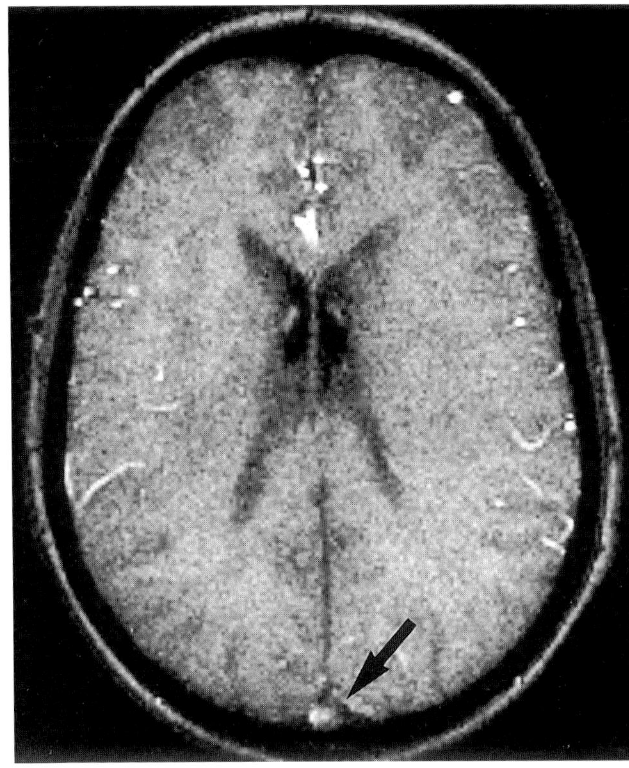

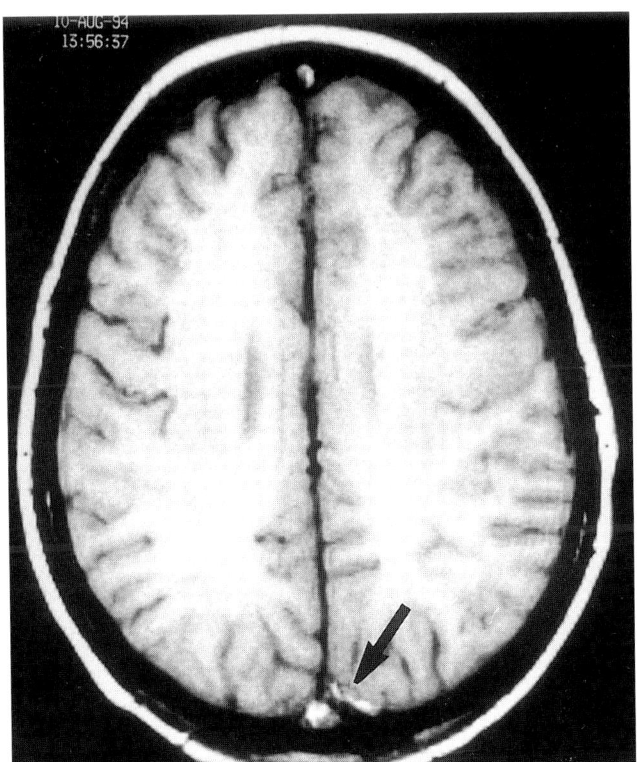

◀ c

▲ a,b

Fig. 9.26a–e. Sagittal sinus thrombosis. **a** Transverse sigle-slice 2D TOF MRA acquisition. TR/TE/Fl 40 ms/10 ms/90°, FOV 230 mm, slice thickness 4 mm, matrix 256 × 256, acquisition time 10 s, Nex = 1. Hyperintense signal (*arrow*) is present in arteries and in the superior sagittal sinus. **b** Identical acquisition with venous presaturation. Most of the hyperintense signal (*arrow*) in the superior sagittal sinus persists, pointing to venous thrombus. **c** Transverse T1-weighted spin-echo sequence at the same level. Thrombus is hyperintense (*arrow*). **d** Sagittal MIP of 3D PC MRA sequence. TR/TE/Fl 30 ms/10 ms/20°, FOV 210 mm, VENC 30 cm/s, slab thickness 45 mm/32 partitions, matrix 192 × 256, acquisition time 12 min, Nex = 1. The superior sagittal sinus (*arrows*) is only partially patent in the most distal third. **e** Selective intra-arterial DSA of the left internal carotid artery. Lateral projection; venous phase. The superior sagittal sinus is occluded. There is some patency (*arrows*) of the distal third, above the torcular herophili

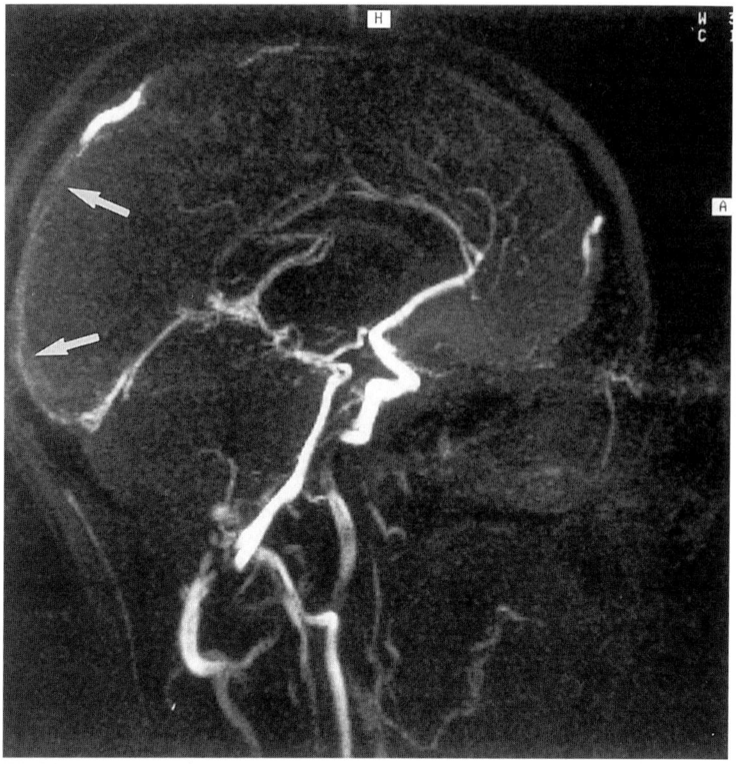

d

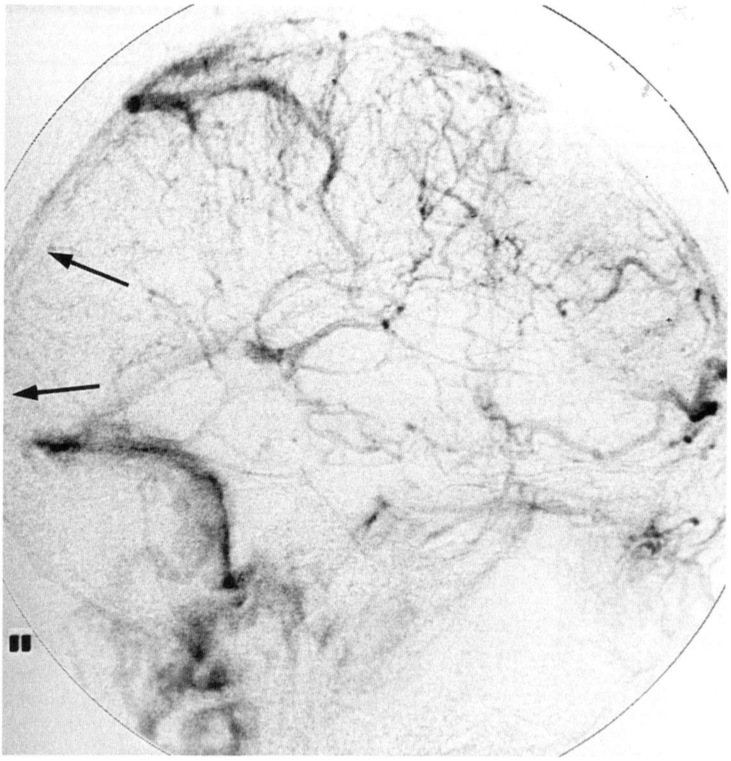

e

Fig. 9.26d,e

Fig. 9.27a–e. False-positive diagnosis of sinus thrombosis on 2D TOF MRA. **a** Transverse MIP of a parasagittal 2D TOF MRA acquisition. TR/TE/Fl 32 ms/10 ms/50°, FOV 230 mm, slice thickness 3 mm, matrix 256 × 256, acquisition time 8 s/slice, Nex = 1. **b** Parasagittal MIP of the same acquisition. **c** Coronal MIP of the same acquisition shows saturation phenomena in the TOF acquisition that cannot be differentiated from thrombosis of the sagittal sinus. **d** Coronal MIP of a transverse 2D TOF MRA acquisition. TR/TE/Fl 32 ms/10 ms/50°, FOV 230 mm, slice thickness 3 mm, matrix 256 × 256, acquisition time 8 s/slice, Nex = 1. The sagittal sinus is patent. **e** MIP of 3D PC acquisition. TR/TE/Fl 30 ms/10 ms/30°, FOV 220 mm, slab thickness 45 mm/32 partitions, matrix 192 × 256, VENC 30 cm/s, acquisition time 12 min, Nex = 1. The 3D PC acquisition provides a more reliable angiogram and does so independently of the projection direction

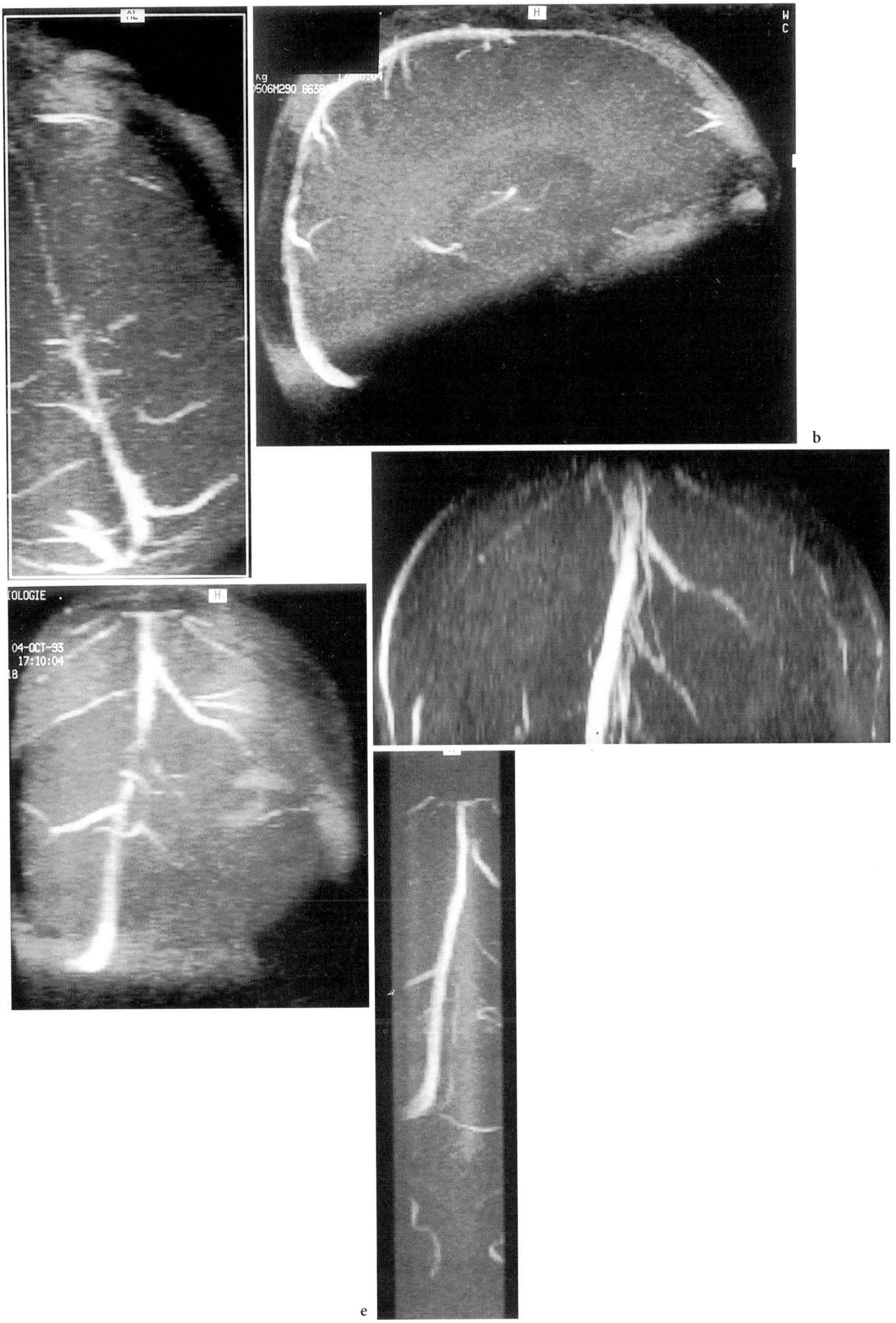

AL
H
W
C
04-OCT-93
17:10:04
a
b
c
d
e

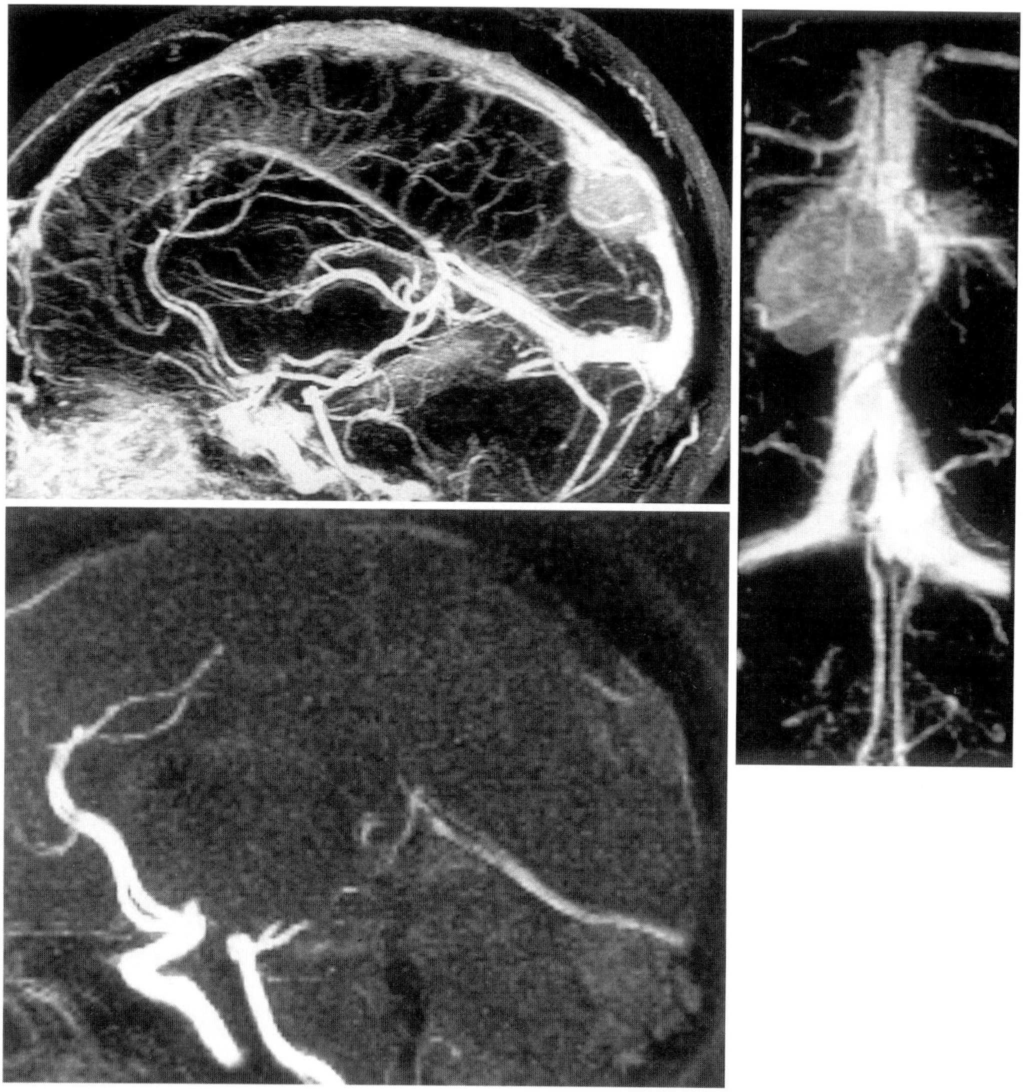

a b c

Fig. 9.28a–c. Meningioma with compression of the superior sagittal sinus. **a** Sagittal and **b** coronal MIP of 3D contrast-enhanced sagittal TOF MRA sequence. TR/TE/Fl 42 ms/11 ms/38°, magnetization transfer preparation pulse, FOV 230 mm, slab thickness 52 mm/64 partitions, matrix 192 × 512, Nex = 1, acquisition time 8 min. There is considerable narrowing of the superior sagittal sinus by the tumor. Collateral veins are visualized. **c** Sagittal MIP of 3D PC MRA sequence. Absence of signal in the superior sagittal sinus indicates reduced flow. TR/TE/Fl 30 ms/10 ms/20°, FOV 230 mm, VENC 30 cm/s, slab thickness 40 mm/32 partitions, matrix 192 × 256, Nex = 1, acquisition time 12 min

hyperintense signals. Furthermore slices have to be oriented as perpendicular as possible to the veins. A T1-weighted spin-echo acquisition is necessary to rule out thrombus with short T1 if the 2D TOF examination is negative. 3D PC acquisitions can be performed to visualize the suspected regions with better resolution. This acquisition offers the additional advantage that the signal of thrombi is suppressed. In a contrast-enhanced TOF acquisition, saturation is also overcome and improved blood vessel detail can be obtained. It must be remembered that extracranial tissues and lesions with an abnormal blood-brain barrier will also enhance.

Venous anomalies can also be demonstrated with MRA. 2D TOF or 2D PC acquisitions can be used as localizers. 3D PC with low VENC or contrast-enhanced 3D TOF acquisitions provide the most detailed angiograms.

Similarly, the best visualization of the venous drainage in larger intracranial lesions, including AVMs and tumors, is obtained with 3D PC imaging or 3D contrast-enhanced TOF acquisitions.

References

Anderson CM, Saloner D, Tsuruda JS, Shapeero LG, Lee R (1990) Artifacts in maximum-intensity-projection display of MR angiograms. Am J Roentgenol 154:623–629

Atkinson D, Brant-Zawadzki M, Gillan G, Purdy D, Laub G (1994) Improved MR angiography: magnetization transfer suppression with variable flip angle excitation and increased resolution. Radiology 190:890–894

Blatter DD, Parker DL, Robinson RO (1991) Cerebral MR angiography with multiple overlapping thin slab acquisition. I. Quantitative analysis of vessel visibility. Radiology 179:805–811

Blatter DD, Parker DL, Ahn SS et al. (1992) Cerebral MR angiography with multiple overlapping thin slab acquisition. II. Early clinical experience. Radiology 183:379–389

Blatter DD, Bahr AL, Parker DL, Robison RO, Kimball JA, Perry DM, Horn S (1993) Cervical carotid MR angiography with multiple overlapping thin slab acquisitions: comparison with conventional angiography. AJR 161:1269–1277

Bosmans H (1992) Optimization procedures for magnetic resonance angiography acquisitions. PhD Thesis, KU Leuven, Belgium, 1992

Bosmans H, Marchal G, Van Hecke P, Vandermeulen D, Suetens P (1990) Magnetic resonance angiography: techniques, prospects and limitations. Front Eur Radiol 7:69–86

Bosmans H, Marchal G, Van Hecke P et al. (1992) MRA review. Clin Imaging 16:152–167

Bosmans H, Marchal G, Lukito G et al. (1995) Contrast enhanced time-of-flight MR angiography of the brain: comparison of acquisition techniques in healthy volunteers. AJR (in press)

Brant-Zawadzki M, Gould R, Norman D, Newton TH, Lane B (1983) Digital subtraction cerebral angiography by intraarterial injection: comparison with conventional angiography. Am J Roentgenol 140:347–353

Chakeres DW, Schmalbrock P, Brogan M et al. (1991) Normal venous anatomy of the brain: demonstration with gadopentetate dimeglumine in enhanced 3-D MR angiography. AJR 156:161–172

Creasy JL, Price RR, Presbey T et al. (1990) Gadolinium-enhanced MR angiography. Radiology 175:280–283

Curnes JT, Shogry MEC, Clark DC et al. (1993) MR angiographic demonstration of an intracranial aneurysm not seen on conventional angiography. AJNR 14:971–977

De Bray JM, Joseph PA, Jeanvoine H et al. (1988) Transcranial Doppler evaluation of middle cerebral artery stenosis. J Ultrasound Med 7:611–616

Dumoulin CL, Souza SP, Walker MF (1989) Three-dimensional phase contrast angiography. Magn Reson Med 9:139–149

Edelman RR, Wentz KU, Mattle HP et al. (1989) Intracerebral arteriovenous malformations: evaluation with selective MR angiography and venography. Radiology 173:831–837

Edelman RR, Mattle HP, O'Reilly GV et al. (1990a) Magnetic resonance imaging of the flow dynamics in the circle of Willis. Stroke 21:56–66

Edelman RR, Mattle HP, Wallner B et al. (1990b) Extracranial carotid arteries: evaluation with "black blood" MR angiography. Radiology 177:45–50

Edelman RR, Ahn SS, Chien D et al. (1992) Improved time-of-flight MR angiography of the brain with magnetization contrast. Radiology 184:395–403

Evans AJ, Richardson DB, Tien R et al. (1993) Poststenotic signal loss in MR angiography: effects of echo time, flow compensation and fractional echo. AJNR 14:721–729

Felbert S, Birbamer G, Aichner F et al. (1992) Magnetic resonance imaging and angiography in hemifacial spasm. Neuroradiology 34:413–416

Finn JP, Zisk JHS, Edelman RR, Wallner BK, Hartnell GG, Stokes KR, Longmaid HE (1993) Central venous occlusion: MR angiography. Radiology 187:245–251

Gao J-H, Holland SK, Gore JC (1988) Nuclear magnetic resonance signal from flowing nuclei in rapid imaging using gradient echoes. Med Phys 15:809–814

Gelbert F, Assouline E, Hodes JE et al. (1991) MRI in spontaneous dissection of vertebral and carotid arteries: 15 cases studied at 0.5 Tesla. Neuroradiology 33:111–113

Goldberg HT (1992) Angiography of extra- and intracranial occlusive cerebrovascular disease. Neuroimaging Clin North Am 2:487–507

Gouliamos A, Gotsis E, Vlahos L et al. (1992) Magnetic resonance angiography compared to intra-arterial digital subtraction angiography in patients with subarachnoid haemorrhage. Neuroradiology 35:46–49

Guglielmi G, Vinuela F, Dion J, Duckwiler G (1992) Endovascular treatment of posterior circulation aneurysms by electrothrombosis using electrically detachable coils. J Neurosurg 77:515–524

Graves VB, Duff TA (1990) Intracranial arteriovenous malformations: current imaging and treatment. Invest radiol 25:952–960

Haacke EM, Masaryk TJ, Wielopolski PA et al. (1990) Optimizing blood vessel contrast in fast three-dimensional MRI. Magn Reson Imaging 14:202–221

Hankey GJ, Warlow CP, Sellar RJ (1990) Cerebral angiographic risk in mild cerebrovascular disease. Stroke 21:209–222

Hausmann R, Lewin JS, Laub G (1991) Phase-contrast MR angiography with reduced acquisition time: new concepts in sequence design. J Magn Reson Imaging 11:415–422

Heiserman JE, Drayer BP, Keller PJ, Fram EK (1992) Intracranial vascular stenosis and occlusion: evaluation with three-dimensional time-of-flight MR angiography. Radiology 185:667–673

Huston J III, Ehman RL (1993) Comparison of time-of-flight and phase-contrast MR neuroangiographic techniques. Radiographics 13:5–19

Katz BH, Quencer RM, Kaplan JO et al. (1989) MR imaging of intracranial carotid occlusion. AJNR 10:345–350

Kauczor HU, Engenhart R, Layer G et al. (1993) 3D TOF MR angiography of cerebral arteriovenous malformations after radiosurgery. J Comput Assist Tomogr 17:184–190

Laub GA, Kaiser WA (1988) MR angiography with gradient motion refocusing. J Comput Assist Tomogr 12:377–382

Lewin JS, Laub G, Hausmann R (1991) A direct comparison of three time-of-flight techniques. AJR 12:1133–1139

Lin W, Tkach JA, Haacke EM, Masaryk TJ (1993) Intracranial MR angiography: application of magnetization transfer contrast and fat saturation to short gradient-echo, velocity-compensated sequences. Radiology 186:753–761

Lisovoski F, Rosseaux P (1993) Cerebral infarction in young people, a study of 148 patients with early cerebral angiography. J Neurol Neurosurg Psychiatry 50:609–614

Marchal G, Bosmans H, Van fraeyenhoven L et al. (1990) Intracranial vascular lesions: optimization and clinical evaluation of three-dimensional time-of-flight MR angiography. Radiology 175:443–448

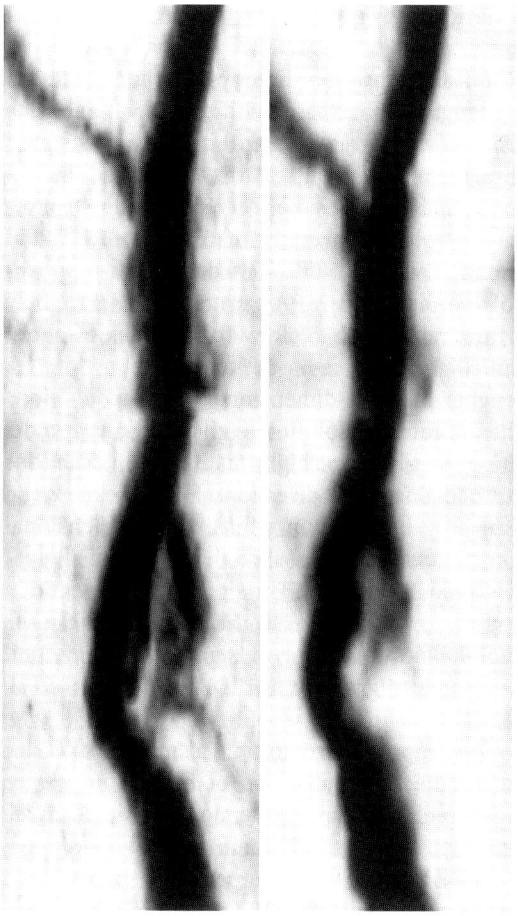

a b

Fig. 10.3. Direct comparison of a small-voxel 3D acquisition (**a**) and a larger voxel 2D acquisition (**b**) in a patient with a severe stenosis at the origin of the external carotid artery. In the 2D technique, the turbulence in the stenosis causes more severe signal voids than the 3D technique

usually destroys the vascular signal far below the skull base but also exaggerates poststenotic flow voids (Masaryk et al. 1991).

From a technical point of view, the limitation of the RF pulses to the investigated volume must be considered. In transaxial volume imaging, the RF power is restricted to the investigational volume. In coronal or sagittal imaging, RF is also supplied to all tissues above and below the imaging field – including the thorax and the heart. Presaturation of the blood in the thorax will affect the signal in the carotid arteries. Therefore, RF supply must be limited to the volume of interest. Use of dedicated neck coils with transmit and receive capacities will prevent the RF pulses from reaching the areas downstream of the carotid arteries, thereby preserving the entire longitudinal magnetization for the inflow MRA.

Nevertheless, the major problem in 3D time-of-flight MRA is signal saturation, which limits the investigation of the carotid artery to the bifurcational area. Parker et al. (1991) introduced the combination of multiple overlapping thin slice acquisitions (MOTSA) to carotid artery imaging (Fig. 10.4). Multiple, separately acquired 3D data sets are combined to yield a single reconstruction of the entire vessel. The volumes are acquired sequentially in order to prevent any saturation effect. Due to impairment of the slice profile at the boundaries of the volumes, the signal varies throughout each individual volume from the outer to the central slices. Therefore, a considerable overlap of one-third to one-quarter at each side of the volume is required. MOTSA techniques enable the investigation of the entire carotid artery without suffering from saturation effects. The disad-

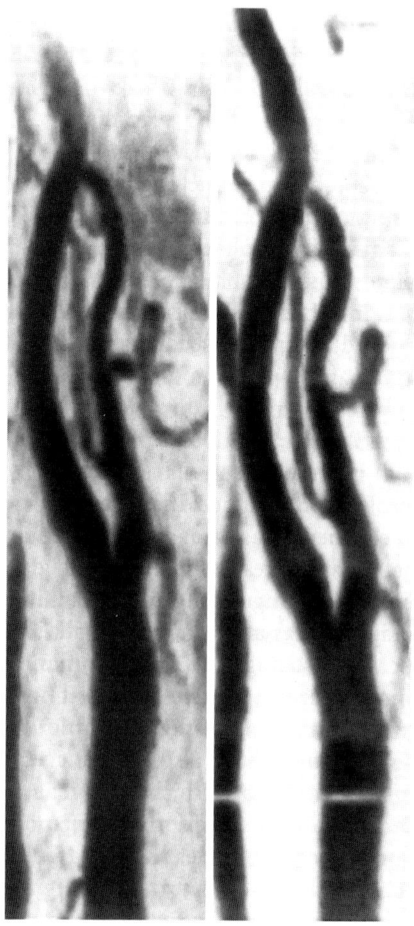

a,b

Fig. 10.4a,b. MRA of a normal carotid artery obtained in 3D acquisition. While inflow-saturation effects occur in the single-volume acquisition (**a**), the MOTSA technique (**b**) compensates the saturation effect and stabilizes the signal from the vascular origin to the distal part of the vessel delineated

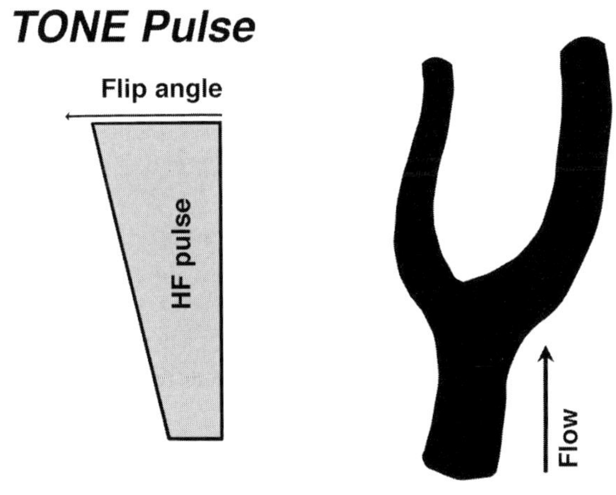

Fig. 10.5. Graphical depiction of the variation in gradient strength along the carotid flow direction in TONE applications. At the entrance side of the volume, less RF power is submitted to the system while at the distal end, the structures in the imaging field are exposed to higher RF values with stronger saturation effects. The slope of the gradient is related to the flow velocity. In slow flow conditions, the slope is steeper in order to compensate the saturation effects more efficiently

vantages are the problem of incorrect matching of the separate volumes due to gross patient motion and signal variation at the volume boundaries because of residual saturation effects.

Another technique to minimize saturation effects in transaxial volume time-of-flight MRA is the adjustment of the RF profile (and thus the flip angle) along the carotid flow direction. The strong signal at the entry side of the volume requires a smaller flip angle as compared to the very distal part of the carotid artery, where saturation already strongly affects the signal. The presaturated arterial magnetization can still deliver signal when the flip angle is sufficiently large. Thus, the RF profile for 3D acquisitions can be altered such that a small flip angle is used at the proximal carotid artery while a much larger flip angle recalls the weaker signal distally (Fig. 10.5). This technique is referred to as TONE (tilted optimized nonsaturating excitation) and was introduced by Purdy et al. (1992). The gradient of the TONE pulse (or ramped flip angle) directly depends on the velocity. In normal carotid flow, the gradient applied is 1:2 for a slab thickness of approximately 12 cm.

With recent hardware and software, in-plane matrices of 512 × 256 (or less) are most commonly used. The smaller voxel size shows benefits in areas of turbulence due to reduced intravoxel dephasing. Moreover, recognition of details like plaque morphology is improved.

Typical parameter settings for axial 3D time-of-flight carotid artery MRA are shown in Table 10.1.

10.2.3.2 2D Time-of-flight

The 2D alternative to volume imaging for carotid artery MRA shows improved sensitivity for inflowing magnetization based on minimization of the flow saturation effect: in sequential acquisitions each slice of the data set represents an entry slice with the strongest possible intravascular signal, enabling slightly higher flip angles and a consequent overall improvement in signal-to-noise ratio. Slice thickness varies around 2–3 mm, whereas 3D techniques enable use of a slice thickness of below 1 mm. The impact of a larger slice thickness and thus a larger voxel size on 2D MRA results is threefold: decrease in resolution, increase in dephasing effects for turbulent conditions, and increase of in-plane saturation effects.

For carotid artery imaging, these three effects must be taken into account with respect to different anatomical or pathological conditions.

1. *Resolution.* In 2D acquisitions the reconstruction of the investigated vessel requires a significant slice overlap to ensure a homogeneous vessel display. The decreased resolution affects the final result in terms of reduced detail visibility (plaque morphology, ulceration) and partial volume effects. Moreover, minor patient movements like swallowing affect the reconstruction by virtue of a small signal shift laterally. The MRA MIP image then shows the so-called venetian blind artifact, which can mimic or hide pathological alterations (Anderson et al. 1992).

2. *Larger voxel size – dephasing.* As previously mentioned, poststenotic, postbulbar, or postbifurcational turbulence causes spin dephasing that cannot be completely compensated by the imaging gradients. The total number of different possible phases

Table 10.1. Parameter settings for axial 3D time-of-flight MRA of the carotid arteries

Parameter	Single volume (TONE)	Multivolume (MOTSA)
TR	33 ms	33 ms
TE	8 ms	8 ms
Flip angle	25°	20°
Slab thickness	1 mm	1 mm
No. of partitions	5 × 32	128
Time	5 × 2.51 min	11.30 min

increases with voxel volume. With thicker slice imaging, signal voids increase due to dephasing problems (Bongartz et al. 1991).

3. *In-plane saturation effects.* Depending on both the larger voxel size and the higher flip angle applied, saturation effects are enhanced for flowing spins that run within an imaging plane instead of crossing it perpendicularly. In the case of the carotid and vertebral arteries, in-plane saturation is rarely expected with transaxial MRA applications. Only in cases of vascular kinking or coiling does a notable problem occur (Huston et al. 1993) (Fig. 10.6). In cases of severe coiling with reversed flow direction, the problem is even pronounced. In standard 2D sequential acquisitions, a travelling presaturation pulse is implemented into the sequence. This saturation pulse is placed parallel and directly superior to the imaging plane, nulling the cephalocaudad (venous) flow. If reversed flow occurs, the opposed arterial flow direction is affected simultaneously.

Two-dimensional MRA of the carotid system has two major benefits: first, sensitive display of slow flow in severe stenoses (improved inflow sensitivity) and second, improved signal-to-noise ratio of the original slice. This facilitates evaluation of the original data set prior to reconstruction. Information on the vascular wall and the nature of atheroma can be rendered by the cross-sectional 2D image (Litt et al. 1991; Heisermann et al. 1992) (Fig. 10.7).

Evaluation of the original slices is a demand for all MRA applications and can be based on 3D data sets, too. The improved signal-to-noise ratio in 2D acquisitions is due much more to the larger flip angle applied than to the larger slice thickness, which can be corrected for by the addition of several individual 3D thin slices.

Like the MOTSA technique, sequential 2D MRA allows imaging of the entire length of the carotid arteries. Typical parameter settings for sequential 2D time-of-flight MRA of the carotid arteries are given in Table 10.2.

10.3 Normal Anatomy/Variants

10.3.1 Carotid Arteries

The carotid arteries originate from the top of the aortic arch. The right artery has a common root

a

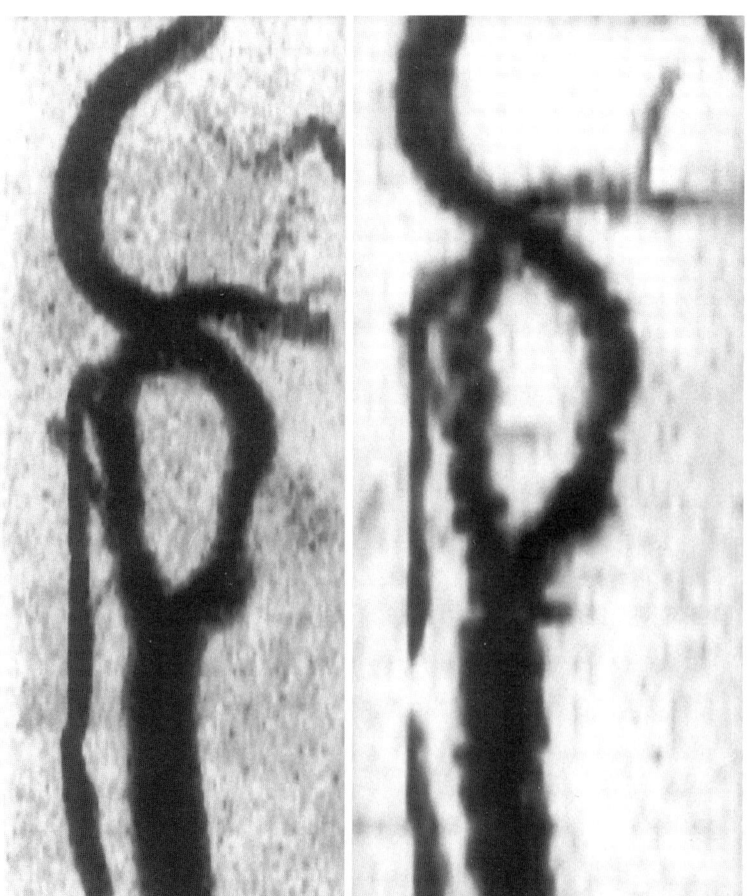

b

Fig. 10.6a,b. Mild postbulbar stenosis of the internal carotid artery in a patient with arteriosclerotic ischemic disease. The 3D technique readily displays all vasculature within the imaging volume (**a**), while the 2D approach (**b**) is handicapped by the motion-induced venetian blind artifact. Moreover, the horizontal partition of the external carotid artery is severely saturated in the 2D technique (in-plane saturation effect)

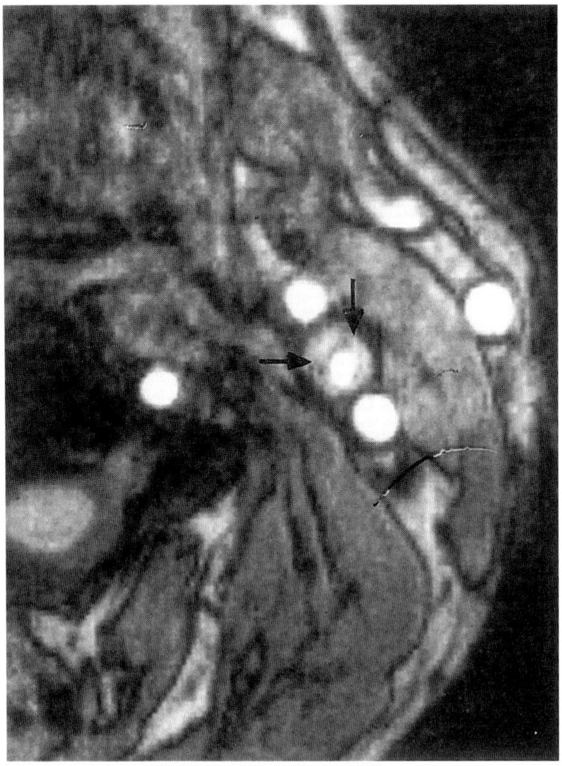

Fig. 10.7. Display of the original axial slice of a 3D MRA data set through the neck, the left internal carotid artery shows a moderate stenosis with a circular but asymmetric thickening of the vascular wall. The atheroma shows inhomogeneous signals with bright foci at the anterior aspect, suspicious for lipoid infiltration or fresh thrombotic material (*arrows*)

Table 10.2. Typical parameter settings for sequential 2D time-of-flight MRA of the carotid arteries

Parameter	Setting
TR	30 ms
TE	9 ms
Flip angle	35°
Slice thickness	3 mm (effective: 1.8 mm)
No. of slices	60
Time	5.14 min

with the right subclavian artery, called the brachiocephalic trunk, whereas the left subclavian artery originates in isolation. The most common aberrations are isolated origins of the right-sided arteries with absence of the brachiocephalic trunk, a common origin of the left carotid artery with the brachiocephalic trunk, a left-sided brachiocephalic trunk, and a variant origin of the right carotid artery from the aortic arch distal to the left subclavian artery (a. lusorica). The latter variant may cause a syndrome of dysphagia, because the misplaced right carotid crosses to the right side behind the esophagus and compresses it. The syndrome is called dysphagia lusoria.

The position of the carotid arteries' roots at the aortic arch guarantees a strong perfusion of the supra-aortic arteries, since centrifugal forces are pronounced at the outer curve of the arch. Physiologically, the flow accelerates with the diameter of the curve owing to the tendency to maintain a straight flow direction. Therefore a relatively large amount of the cardiac output is directly transported into the supra-aortic arteries for cerebral supply.

Depending on constitution and body size, the carotid arteries measure between 14 and 19 cm in an adult person, with an average of 17 cm between the aortic arch and the siphon of the internal carotid artery. In degenerative elongation of the vessel, tremendous variation may occur. For the purpose of MRA, the distance between the aortic arch and the carotid siphon is of greater importance than the actual length of the vessel.

The area surrounding the arteries of the neck consists of soft tissue, primarily muscle and fat. Both the carotid arteries are adjacent to osseous structures only at the skull base.

The carotid bifurcation projects itself onto the second or third cervical vertebra in conventional angiography, showing a symmetric position of both sides in about 60%. A variation of more than 5 cm cephalocaudad is often found upon clinical evaluation. At the bifurcation, the external carotid artery most often shows a smaller caliber than the internal carotid and is directed 45° frontomedially. The origin of the internal carotid is larger in diameter and expanded eccentrically, forming the carotid bulb. In most cases the carotid bulb points in the opposite direction to the external carotid artery. The carotid bulb may also include both arteries or extend solely into the external carotid artery. Evaluation of the bifurcational area can become complicated when the imaged area includes the bifurcation only, without demonstrating the peripheral vascular branches. The mainstem of the internal carotid artery preserves a straight upward direction without further truncations while the external carotid artery separates into several vessels which partially run horizontally. There is a much larger variation in the vessel's course as compared to the internal carotid (Kadir 1991).

10.3.2 Vertebral Arteries

The caliber of the vertebral arteries is normally small compared with the diameter of the carotid arteries.

Considerable variation in symmetry exists, the extreme case comprising a unilateral vertebral artery only. The supply of the posterior cerebral circulation by the basilar artery varies to a great extent, since there is also the embryonal type of blood supply via the posterior communicating arteries. As regards the basilar pathway only, the posterior perfusion is not affected by asymmetry of the cervical vessels or even the lack of one vertebral artery as long as the other supplies sufficient blood flow.

In the case of marked asymmetry of both vertebral arteries, the smaller one often ends in a minor artery of the cerebellar vascular system like the anterior inferior cerebellar artery (AICA) or the posterior inferior cerebellar artery (PICA). Moreover, the confluence of the vertebral arteries shows variations, such as an asymmetric course of the distal vertebral arteries or a fenestration of the basilar artery, e.g., a secondary separation of the basilar artery into two parallel vessels distal to the vertebral confluence (Wentz et al. 1994).

The vertebral arteries penetrate the transverse processes of the cervical vertebrae. The close relation of bone and vessel not only may give rise to vascular stenosis but also can cause problems in DSA or MRA.

In rare cases, primitive connections between the vertebral and carotid system can be found. These communicating vessels are usually unilateral. Up to four different connections are possible (Kadir 1991).

10.3.3 MRA Features in Normal Arteries of the Neck

The most important consideration in MRA of the supra-aortic vessels is the need for limitation of imaging to a certain region, e.g., to the bifurcation of the carotid arteries. Larger volumes are technically problematic and/or time consuming.

The delineation of the aortic origin of the carotids generally requires a faster acquisition method, as is described in greater detail in Chap. 11.

At the carotid bifurcation, the separation into two vessels alters the flow conditions (Blackshear et al. 1980). In the carotid bulb, laminar flow is converted to turbulent flow or inversed flow at the outer boundary. It remains unclear whether the bulb is actually formed by the local flow eddies or whether the turbulence is created by the bulb's geometry. The MRA display of the bifurcation area is affected by the typical flow conditions and by the MRA parameters. In any case, a dephasing problem occurs at the bifurcation that may cause misdiagnosis if not properly recognized.

In MRA applications which show a pronounced sensitivity to spin-dephasing, a local flow void mimics stenosis or plaque in the carotid bulb. Usually, the bulb is therefore displayed too small.

Branching of the external carotid artery and the transverse course of the secondary vessels also affect the MRA display. In transverse MRA acquisition techniques, the in-plane saturation prevents the visualization of these arterial parts. MRA of the external carotid artery is limited to the proximal and major branches (Bowen et al. 1994). Identical problems occur at the distal part of the internal carotid arteries or in cases of severe elongation or kinking of the vessels. The restricted quality of MRA in this area is worsened by susceptibility artifacts from the skull base.

10.4 Pathology

10.4.1 Stenosis and Occlusion

Atherosclerotic manifestations in the carotid system are a known problem in elderly patients and in those with metabolic disease. The exact pathological mechanism is unknown and therefore prevention is under discussion. The result of carotid atherosclerosis is either vascular stenosis or occlusion, usually at the bifurcation. Single stenosis is the most common finding, while tandem lesions distal to the carotid bifurcation occur in approximately 6%–8% of all cases (Akers 1987). If stenoses are found other than at the bifurcation, the intracranial portion of the internal carotid artery is most commonly affected. Besides stenotic disease of the internal carotid artery, atherosclerotic plaques of the proximal common carotid artery, directly at the aortic origin, are a known manifestation of the disease.

The clinical consequence of carotid stenosis is cerebral ischemia. Hemodynamic infarctions derive from significant stenosis of a major artery in cases of insufficient intracranial collateralization. Since the development of an atheromatous stenosis is a long process, collaterals can develop to prevent infarction. Hemodynamic ischemia is therefore usually a multifactorial process which results from multiple additive stenoses, cardiac dysfunction, and/or thrombotic alterations of collaterals. This form of cerebral infarction is not as common as the embolic stroke.

Embolic cerebral ischemia may originate from a huge variety of vascular areas including the heart, the venous system (patent ovale foramen), the aortic arch, and the carotid or vertebral arteries. Due to the known coincidence of stroke and carotid artery stenosis, various preventive regimens focus on the treatment of carotid atheroma. The atheroma in the vascular wall leads to secondary thrombosis, forming a prominent plaque that is able to organize and form fibrous tissue. Plaque morphology varies to a great extent, showing both smooth and uneven surfaces with sharp irregularities or central ulcerations. Either surface has an inherent risk of delivering particles to the bloodstream, causing emboli (Fig. 10.8). The emboli consist of the atheroma itself or of adherent thrombus. The latter seems to be pronounced in ulcerated plaques, although this is denied by several authors (Dixon et al. 1982).

To date, a single therapy does not exist that reliably prevents or cures atheromatous disease. Surgical treatment (endarterectomy or vasoplasty) and medical treatment are comparable in terms of a long-lasting effect on the outcome. From a clinical viewpoint, the distinction between symptomatic and asymptomatic manifestations is mandatory. A symptomatic stenosis leads to transient or permanent neurological deficits due to cerebral ischemia. Symptomatic patients present with a history of at least one ischemic cerebral insult while in asymptomatic patients, the atheroma has been found by diagnostic tests only.

In symptomatic patients, the risk of a central manifestation increases with the severity of the stenosis. In a recent trial performed by the North American Symptomatic Carotid Endarterectomy Trialist's Group (NASCET 1991), three stages or degrees of stenosis were distinguished according to their appearance on conventional angiography:

stage 1 – low grade stenosis: >30% stenosis
stage 2 – moderate stenosis: 31%–79% stenosis
stage 3 – severe stenosis: 71%–99% stenosis

The degree of the stenosis was defined by comparing the smallest diameter at the site of the pathology in the carotid bulb with the diameter of the internal carotid artery distal to the stenosis (Fox 1993).

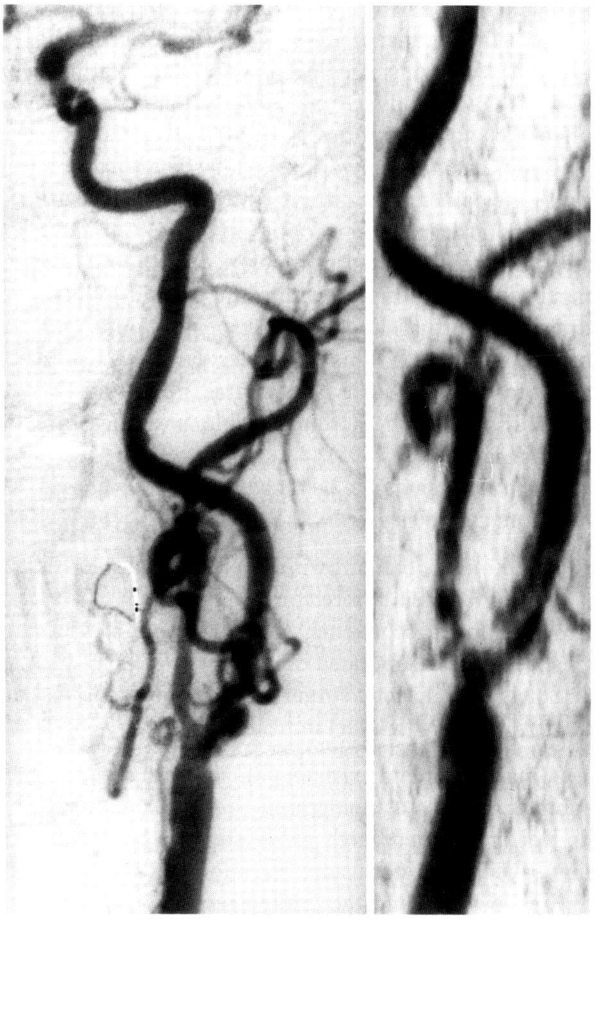

Fig. 10.8. Correlation of a DSA (**a**) and an MRA (**b**) investigation of a patient with severe atherosclerotic manifestations at the carotid bifurcation with stenotic alteration of both the internal and the external carotid artery. The additional plaque in the distal common carotid artery is visualized by either imaging modality. While DSA demonstrates the anatomical features only, MRA is influenced by the stenotic flow distortion

Beal M, Williams R, Richardson E, Fischer C (1981) Cerebral embolism as a cause of transient ischemic attacks and cerebral infarction. Neurology 31:860-869

Blackshear WM, Chikos DJ, Phillips PM (1980) Carotid artery velocity patterns in normal and stenotic vessels. Stroke 11:67-71

Bongartz G, Vestring T, Drews CI, Krings W, Peters PE (1991) Effect of slice orientation in 3D magnetic resonance angiography of the supraaortic arteries. Eur Radiol 1:158-164

Bongartz G, Behrendt J, Schuierer G, Reimer P, Peters PE (1994) MRA der Supraaortalen Arterien. Radiologe 34:430-436

Bowen BC, Quencer RM, Margosian P, Pattany PM (1994) MR angiography of occlusive disease of the arteries in the head and neck: current concepts. AJR 162:9-18

CASANOVA Study Group (1991) Carotid surgery versus medical therapy in asymptomatic carotid artery stenosis. Stroke 22:1229-1236

Collaborators North American Symptomatic Carotid Endarterectomy Trial (1991) Beneficial effect of carotid endarterectomy in symptomatic patients with high-grade carotid stenosis - (NASCET trial). N Engl J Med 325:445-453

Cumming MJ, Morrow IM (1994) Carotid artery stenosis: a prospective comparison of CT angiography and conventional angiography. AJR 163:517-523

Dillon EH, van Leuven MS, Fernandez MA, Eickelboom BC, Mali WPTM (1993) CT angiography: application to the evaluation of carotid artery stenosis. Radiology 189:211-219

Dixon A, Pais S, Raviola C (1982) Natural history of nonstenotic asymptomatic ulcerative lesions of the carotid artery: a further analysis. Arch Surg 117:1493-1495

Edelman RR, Mattle HP, Wallner B (1990) Entracranial carotid arteries: evaluation with "black blood" MR angiography. Radiology 177:45-50

Fox AJ (1993) How to measure a carotid stenosis. Radiology 186:316-318

Gryska U, Freitag J, Zeumer H (1990) Selective cerebral intraarterial DSA: complication rate and control of risk factors. Neuroradiology 32:296-299

Hankey GJ, Warlow CP, Sellar RJ (1990) Cerebral angiographic risk in mild cerbrovascular disease. Stroke 21:209-222

Heiserman JE, Drayer BP, Fram EK, Keller PJ, Bird CR, Hodak JA, Flom RA (1992) Carotid artery stenosis: clinical efficacy of two-dimensional time-of-flight MR angiography. Radiology 182:761-768

Hennerici M, Klemm C, Rautenberg W (1988) The subclavian steal phenomenon: a common vascular disorder with rare neurological deficits. Neurology 38:669-675

Hessel SJ, Adams DF, Abrams HL (1981) Complications of angiography. Radiology 138:273-281

Houser O, Mokri B, Sundt T (1984) Spontaneous cervical cephalic arterial dissection and its residuum: angiographic spectrum. AJNR 5:27-38

Hunink MGM, Polak JF, Barlan MM, O'Leary DH (1993) Detection and quantification of carotid artery stenosis: efficacy of various Doppler velocity parameters. AJR 160:619-625

Huston J, Lewis BD, Wiebers DO, Meyer FB, Riederer SJ, Weaver AL (1993) Carotid artery: prospected blinded comparison of two-dimensional time-of-flight MR angiography with conventional angiography and duplex US. Radiology 186:339-344

Kadir S (1986) Diagnostic angiography. Saunders, Philadelphia

Kadir S (1991) Diagnostische Radiologie. Thieme, Stuttgart

Kido DK, Moran CJ (1994) When should third-party payers provide reimbursement for MR angiography of the carotid arteries? AJR 162:19-20

Levy C, Laissy JP, Raveau V (1994) Carotid and vertebral artery dissections: three dimensional time of flight MR angiography and MR imaging versus conventional angiography. Radiology 190:97-103

Li W, Kramer J, Kleefield J, Edelman RR (1992) MR angiography of the extracranial carotid arteries using a two slab oblique 3-D acquisition. AJNR 13:1423-1428

Litt AW, Eidelman EM, Pinto RS, McLachlan SJ, Schwartzenberg S, Weinreb JC, Kricheff II (1991) Diagnosis of carotid artery stenosis: comparison of 2DFT time-of-flight MR angiography with contrast angiography in 50 patients. AJNR 12:149-154

Masaryk AM, Ross JS, DiCello MC, Modic MT, Paranandi L, Masaryk TJ (1991) 3DFT MR angiography of the carotid bifurcation: potential and limitations as a screening examination. Radiology 179:797-804

Masaryk TJ, Obuchowski NAS (1993) Noninvasive carotid imaging: caveat emptor. Radiology 186:325-331

McNeill D, Dreisbach J, Marsden RJ (1980) Spontaneous dissection of the internal carotid artery: its conservative management with heparin sodium. Arch Neurol 37:54-55

Mittl RL, Broderick M, Carpenter JP (1994) Blinded reader comparison of magnetic resonance angiography and duplex ultrasonography for carotid artery bifurcation stenosis. Stroke 25:4-10

Parker DL, Yuan C, Blatter D (1991) MR angiography by multiple thin slab 3D acquisition. Magn Reson Med 17:434-451

Patel MR, Klufas RA, Kim D, Edelman RR, Kent KC (1994) MR angiography of the carotid bifurcation: artifacts and limitations. AJR 162:1431-1437

Peters PE, Bongartz G, Drews CI (1990) Magnetresonanztomographie der hirnversorgenden Arterien. Fortschr Röntgenstr 152:528-533

Polak JF, Bajakian RL, O'Leary DH, Anderson MR, Donaldson MC, Jolesz FA (1992) Detection of internal carotid artery stenosis: comparison of MR angiography, color Doppler sonography and arteriography. Radiology 182:35-40

Polak JF, Kalina P, Donaldson MC, O'Leary DH, Whittemore AD, Mannick JA (1993) Carotid endarterectomy: preoperative evaluation of candidates with combined doppler sonography and MR angiography. Work in progress. Radiology 186:333-338

Purdy DE, Cadena G, Laub GA (1992) The design of variable tip angle selection (TONE) pulses for improved 3D MR angiography. Society of Magnetic Resonance in Medicine, Book of Abstracts, p 882

Schild H (1994) Angiographie - angiographische Interventionen. Thieme, Stuttgart

Schwartz RB, Jones KM, Chernoff DM (1992) Common carotid artery bifurcation: evaluation with spiral CT. Radiology 185:513-519

Steinke W, Kloetzsch C, Hennerici M (1990) Carotid artery disease assessed by color Doppler flow imaging: correlation with standard Doppler sonography and angiography. AJR 154:1061-1068

The North American Symptomatic Carotid Endarterectomy Trialist's Group (1991) North American symptomatic carotid endarterectomy trial. Methods, patient characteristics and progress. Stroke 22:711-720

Urchuk SM, Plewes DB (1992) Mechanisms of flow-induced

signal loss in MR angiography. J Magn Reson Imaging 2:453–462

Warlow C, Group European Carotid Surgery Trialists' Collaborative (1991) MRC European carotid surgery trial: interim results for symptomatic patients with severe carotid stenosis and with mild carotid stenosis. Lancet 337:1235–1255

Wentz KU, Rother J, Schwartz A, Mattle HP, Suchalla R, Edelman RR (1994) Intracranial vertebrobasilar system: MR angiography. Radiology 190:105–110

11 The Aortic Arch

J.S. LEWIN

CONTENTS

11.1 Introduction

Since the introduction of MRA, much attention has been paid to the cervical and intracranial segments of the carotid and vertebral arteries, with significantly less attention to the region of the aortic arch. Fortunately, current time-of-flight (TOF) MRA methods are well suited to the evaluation of the aortic arch and proximal brachiocephalic arteries (EDELMAN et al. 1989; LEWIN et al. 1991). Though a much less common cause of neurological symptoms than cervical carotid disease, disease of the aortic arch branches may arise from a variety of etiologies, including atherosclerosis, inflammatory arteritis, trauma, fibromuscular dysplasia, neoplasm, and radiation (PROVAN 1989). MRA can provide accurate noninvasive evaluation of stenosis and occlusion from these causes, and can allow for confirmation of vessel patency following surgical intervention. This chapter will discuss MRA methods applicable to evaluation of the aortic arch, with particular emphasis on 3D TOF techniques, followed by a discussion of pathological entities which may be encountered.

J.S. LEWIN, MD, Assistant Professor, Department of Radiology, University Hospitals of Cleveland, Case Western Reserve University, 2074 Abington Rd., Cleveland, OH 44106, USA

11.2 Imaging Techniques

Optimal aortic arch image acquisition with either 2D or 3D TOF techniques is best performed with a surface coil. In many patients, a Helmholtz neck coil placed as low on the neck as possible allows for adequate signal to noise through the level of the aortopulmonary window and above. Alternatively, specialized head and neck MRA coils have been developed which allow an adequate field-of-view through the aortic arch and provide transmit-receive capabilities (ANDERSON et al. 1990). Unfortunately, some patients may be too large to allow satisfactory placement of a Helmholtz neck coil, and may require use of the body coil for transmission and reception. This may significantly degrade the quality of the aortic arch MR angiogram. The use of specialized body Helmholtz coils may help circumvent this limitation in the future.

11.2.1 3D Time-of-flight

Despite initial concerns regarding the sensitivity of 3D acquisition schemes to image degradation from the degree of physiological motion encountered in the upper thorax, the 3D TOF method appears well suited to evaluation of the aortic arch and its branches. The 3D technique applied in imaging of the aortic arch is similar to that used for the cervical carotid and intracranial circulation and comprises a 3D fast imaging with steady-state precession (FISP) gradient-echo pulse sequence using a low flip angle and first-order flow compensation in the readout and volume select directions. Typically, the echo time is reduced to 5–7 ms through use of asymmetric echo sampling, typically centering the echo at point 64 of a 256 point sampling interval. The optimal flip angle and repetition time are dependent on the thickness of the volume of excitation, and are described in sample protocols below. Typically, a 90° axially oriented spatial presaturation pulse is placed superior to the imaging volume to suppress signal from inflowing venous spins (Felmlee and Ehman 1987).

As in the application of 3D TOF MRA techniques in other regions of the body, the best image quality is achieved through a combination of optimization of flow-related enhancement and reduction of phase-related signal loss (Haacke et al. 1990). Flow-related enhancement is maximized through use of an imaging volume orientation perpendicular to the inflowing blood and optimized repetition time and flip angle, wherease signal loss from intravoxel phase dispersion is reduced by use of first-order flow compensation gradients and a short echo time, and by minimizing voxel size through the extremely thin partitions available with 3D techniques. Since the signal loss from phase cancellation results from the variance of phase across an imaging voxel, the use of the smaller voxel results in a narrow effective range of velocity-induced phase shifts per voxel and minimizes the resulting phase cancellation and signal loss.

The primary limitation of TOF methods arises from the effect of progressive spin saturation, which is much more evident in 3D methods but is present in all inflow-based techniques. This results from saturation from the inflowing spins as blood traverses the volume of excitation, resulting in progressively diminishing contrast with a total loss of vascular conspicuity after the flowing spins have been present within the imaging volume for a period on the order of magnitude of the longitudinal relaxation time (T1) of blood. The distance into the imaging volume in which contrast remains sufficient to visualize a blood vessel depends upon the blood velocity, repetition time, flip angle of excitation, and T1 of blood. Thus, with a thick acquisition volume or slab, only rapid arterial blood retains sufficient flow-related enhancement to be visualized throughout the volume, with more slowly moving venous or pathologically slowed arterial blood flow unable to span the entire volume before becoming sufficiently saturated to prevent adequate visualization.

Several strategies can be used to counter progressive saturation of inflowing spins in TOF MRA of the aortic arch. These include (a) decreasing the flip angle and increasing the repetition time, (b) limiting the length of vessel imaged by decreasing the thickness of the imaging slab to allow more slowly inflowing blood to cross the entire the volume before becoming saturated, and (c) altering the flip angle across the volume of excitation through use of a ramped volume excitation (TONE) pulse (Lewin and Laub 1991; Atkinson et al. 1994).

11.2.1.1 Thick-Volume Approach

The 3D TOF technique initially described for imaging of the aortic arch used one or multiple 64 partition imaging volumes, from 75 to 80 mm in thickness, to cover the aortic arch and brachiocephalic vessels (Fig. 11.1a) (Lewin et al. 1991; Oneson et al. 1992). While this technique works well in younger patients with good cardiac output, the method suffers from the effects of progressive spin saturation in the more distal aspects of the vessels, which is most notable in the subclavian arteries (Figs. 11.2a, 11.3a). However, with a relatively low flip angle (20°), this method does allow visualization of the primary branches of the aortic arch in younger individuals (Fig. 11.4). When available, this technique should be replaced with one using a ramped volume of excitation, as discussed below.

11.2.1.2 Multiple Overlapping Thin-Volume Approach

A technique that takes advantage of the effect of decreasing the volume thickness applies sequential acquisition of multiple thin 3DFT volumes (with 16–32 partitions each) to cover the region of interest (Figs. 11.1b, 11.2b, 11.3b) (Lewin and Laub 1991; Lewin 1992; Parker et al. 1991). The diminished effect of inflowing spin saturation allows visualization of more slowly flowing blood, while the advantages of increased spatial resolution, small voxel size, and improved signal-to-noise ratio of the 3D TOF technique are retained. The acquisition of multiple thin volumes also allows for a large anatomical coverage. The overlap of volumes with this technique is necessary to compensate for the decrease in signal intensity in the 3D partitions at the edges of the volume of excitation that result from a nonrectangular volume excitation profile. This overlap results in redundant data and an increase in imaging time when compared to the single thick 3D TOF or volume excitation pulse techniques. Insufficient overlap may result in loss of vascular information at the junction of volumes, while greater overlap further increases the total examination time (Lewin and Laub 1991; Parker et al. 1991). Typical parameters for this technique include flip angle of 20°, TR/TE of 24/6, 300 mm FOV, 256 × 192 matrix, and 36 mm 32 partition imaging volume with 12–18 mm overlap depending on the rectangularity of the volume excitation RF pulse profile.

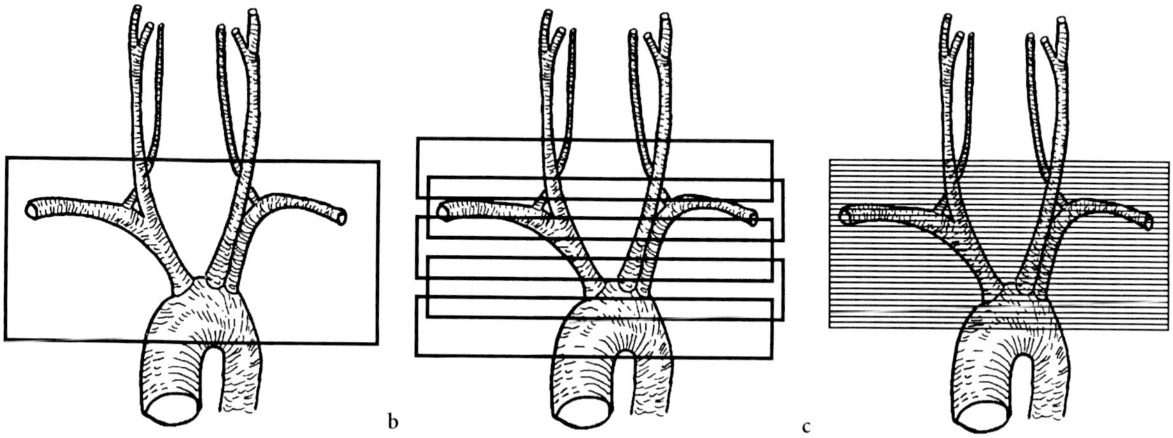

a b c

Fig. 11.1. **a** Thick-volume 3D TOF MRA. One or multiple large 64 partition imaging volumes, from 75 to 80 mm in thickness, are used to cover the region of interest. The volume is oriented in an axial plane, to be approximately perpendicular to the aortic arch branches, and is placed with its inferior edge at the level of the aortopulmonary window to minimize saturation of blood within the pulmonary arteries. **b** The multiple overlapping thin-volume approach applies multiple axial thin 16–32 partition slabs with 30%–50% overlap to cover the region of interest, with five overlapped volumes used in this diagram. As with the thick-volume approach, the inferior aspect of the lowermost slab is placed at the level of the aortopulmonary window to minimize saturation effects. This diagrammatic representation demonstrates 33% overlap, which is achievable when the volume excitation RF profile is relatively rectangualr. As the excitation profile worsens, greater degrees of overlap are necessary. **c** The 2D TOF method results in minimum possible spin saturation. Best inflow is achieved with axial orientation, as diagrammed. However, in practice coronal, sagittal, and oblique planes are useful to cover the areas of interest in a smaller number of slices

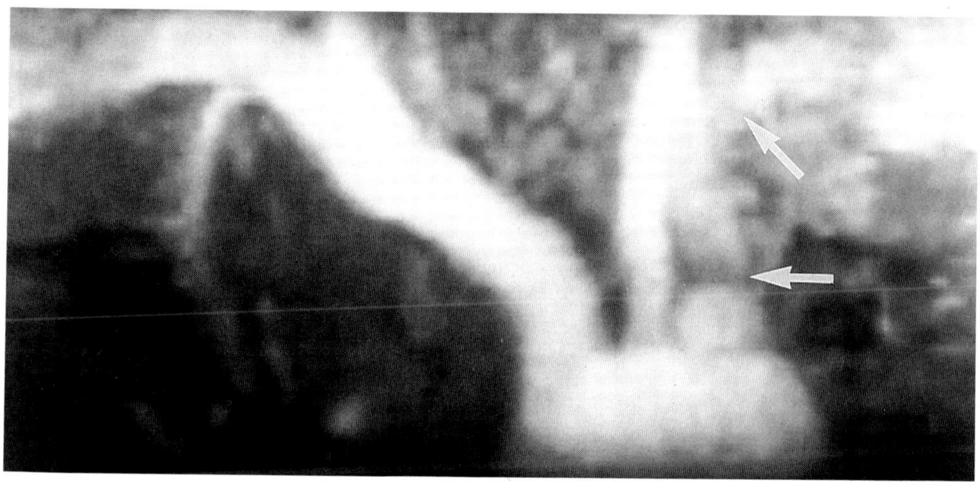

a

Fig. 11.2a–d. Technique comparison in a healthy older subject. **a** Thick-volume 3D TOF method using a 75 mm-thick imaging volume. In a healthy older subject, adequate vascular contrast can be achieved with this method for the origins of the brachiocephalic and left common carotid arteries. However, even with a normal subject the left subclavian artery may be difficult to evaluate due to the effects of progressive saturation (*arrows*). **b** Multiple overlapping thin-volume acquisition significantly reduces saturation effects and allows better visualization of the left subclavian artery at the expense of an increased imaging time and junctional artifact. **c** Single thick-volume acquisition with use of a 10°–30° varying flip angle TONE pulse allows visualization of a large region of interest (80 mm) with minimal saturation effects. Note improvement in left subclavian artery visualization (arrows). **d** TONE pulse with 13°–27° flip angle variation and 80 mm thickness is sufficient to visualize the brachiocephalic vessels with minimal saturation effects when the arch itself is outside the volume of origin, and yields a more uniform vascular and background signal intensity than the 10°–30° TONE pulse

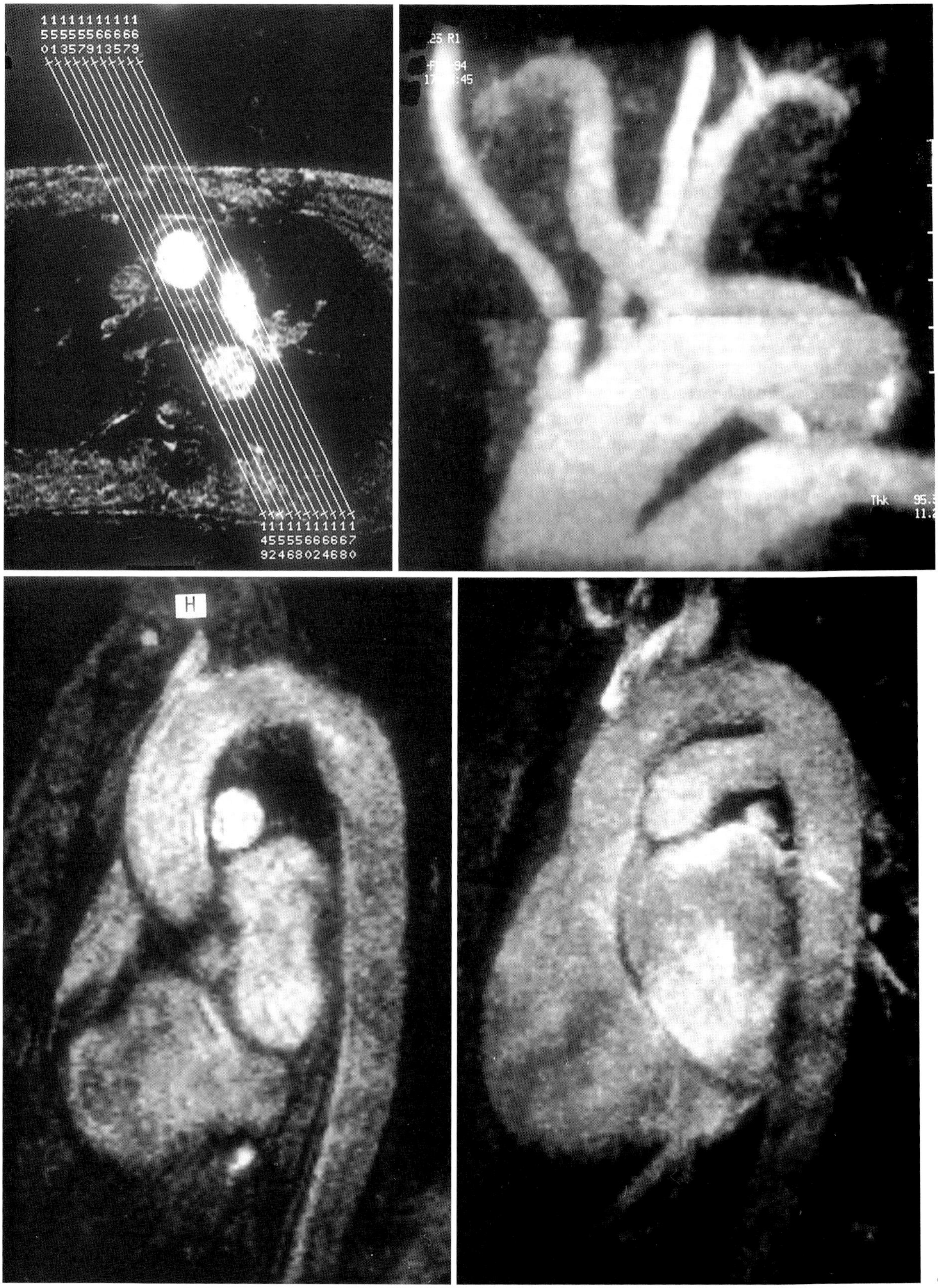

a d b c

Fig. 12.3a–d. Normal individual: Static bright blood MRA of the thoracic aorta (2D TOF GRE FLASH sequence, cardiac-gated segmented k-space acquisition, 1.5 T, body coil, TR/TE = 15.8/8.4 ms, slice thickness = 5 mm with 1 mm of overlap, field of view = 350 mm, matrix = 210 × 256, 14 segments with 15 lines/segment). Using a transverse localizer slice (**a**), an oblique sagittal acquisition plane can be selected for imaging of the entire thoracic aorta (**b** = single 2D section, **c** = 2D MR-projection angiogram). Atypical supra-aortic branch origin is demonstrated on a 3D TOF MIP MR angiogram (**d**) (GRE FISP sequence, 1.5 T, body coil, TR/TE = 35/6 ms, flip angle = 15°, TONE, slice thickness = 1.2 mm, volume = 75 mm with 64 partitions, double-slab, matrix = 192 × 256)

(Friedman 1983). As a result of an abnormality of the media, this narrowing is located at or adjacent to the insertion of the ligamentum arteriosum. In approximately 50% of cases AC is an isolated lesion, while in the other 50% it is associated with congenital cardiac lesions. AC can be classified according to its morphology, localization, and hemodynamic function. More than 95% of the stenoses are segmental and can be operated on without technical problems. More recently, percutaneous angioplasty has been successfully applied. In rare cases a long area of narrowing due to hypoplasia can be observed, necessitating prosthetic operative techniques. In 1/200 cases of AC the stenosis has an atypical location, such as the aortic arch, the descending aorta, or the abdominal aorta.

The prognosis of an untreated AC is poor. Life expectancy has been calculated at not more than 35 years (Wood 1956). The preferred therapeutic regimen in infancy is conservative medical therapy or dilatation of the patent ductus arteriosus as a means of maintaining systemic pressure, with elective surgical correction after 1 year of age and before adulthood. The optimal patient's age for operation ranges from 8 to 14 years, when the aorta has reached more than 50% of its definitive lumen (Heberer et al. 1963). In hemodynamically significant AC the typical clinical symptom is arterial hypertension which develops proximal to the zone of stenosis, confined to the head and upper extremities. In the lower extremities there is a relative or absolute hypotension. Upper extremity hypertension, almost invariably present in adults, may be absent or relatively slight in infants. Diagnostic imaging should be able to detect the presence and location of AC, to characterize its features, and to identify collateral pathways in order to demonstrate the hemodynamic effect of the stenosis. A significant pressure gradient before and behind the stenosis can safely differentiate between true coarctation and pseudo-coarctation, which does not require surgery.

Noninvasive two-dimensional echocardiography with simultaneous pulsed Doppler studies, which has been advocated as the sole preoperative imaging modality in AC, is not generally accepted. Traditionally the definitive preoperative diagnostic procedure has been aortography, enabling determination of the site and grade of the stenosis, arterial collaterals, and possible cardiac malformations, as well as pressure gradients across the AC (Figs. 12.4a, 12.6a,b). Oblique sagittal vascular MRI employing a SE sequence can not only accurately depict the site and localization of the lesion but also assess the severity and type of AC (Amparo et al. 1984; Bank et al. 1987; von Schulthess et al. 1986) (Fig. 12.4b). Dynamic evaluation of the lesion is possible using GRE cine techniques (Simpson et al. 1988), including in the oblique sagittal plane. Normal blood flow appears with bright signal (Fig. 12.4c) whereas in regions of turbulent or high velocity flow, signal void areas can be observed within the blood pool. In significant stenosis a turbulent dark jet occurs just behind the AC at peak systole (Fig. 12.4d) but not during diastole. This technique has been shown to be useful in pre- and postoperative assessment and in pre- and postangioplasty management (Bank et al. 1987; Rees et al. 1989). Furthermore, GRE cine studies permit evaluation of aortic valve stenosis, which is frequently associated with AC (Fig. 12.5a–c). Measurement of poststenotic jet velocities on the basis of a field-even-echo rephasing (FEER) velocity mapping sequence has been reported in aortic valve stenosis and in AC. Although experience with this technique is limited, the findings seem to agree well with Doppler US measurements (Kilner et al. 1991). In addition to dynamic cine imaging, static MRA of the thoracic aorta on the basis of 2D and 3D TOF GRE sequences has proved useful in visualizing anatomically representative images and in evaluating AC features such as dilatation of the ascending aorta, aortic stenosis, collateral pathways (Fig. 12.6c–e), and abnormalities of the aortic arch and its branches (Fig. 12.4d).

12.3.2.2 Aortic Trauma

Aortic laceration or rupture predominantly results from sudden horizontal deceleration injuries of the thorax. In more than 80% of cases the laceration is located at the aortic isthmus just distal to the insertion of the ligamentum arteriosum. A further site of laceration is the proximal ascending aorta. The prognosis of patients with aortic rupture is poor. The mortality initially is approximately 80%; of the early survivors, 50% die within 24 h and 90% within 4 months. When typical signs are present on chest radiography (Seltzer et al. 1981) following blunt thoracic trauma, it is imperative that aortography is performed. On the arteriogram a traumatic aneurysm can be observed most frequently, and may show penetration. Furthermore, intimal tears may occur in 5%–50%, dissection in up to 10%, and posttraumatic coarctation in rare cases. Because of the emergency situation in most patients, MR examination currently does not play a role in the

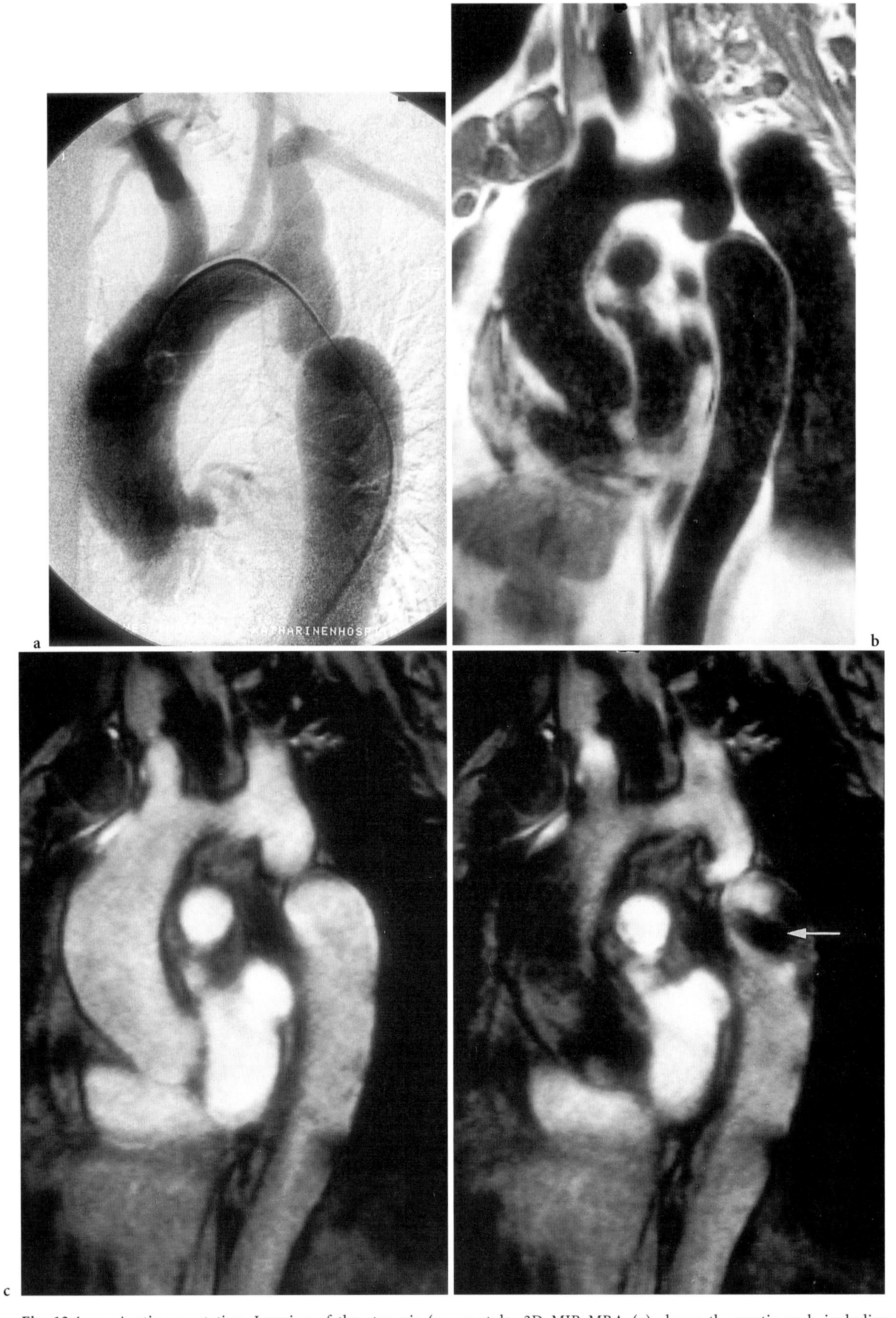

Fig. 12.4a–e. Aortic coarctation. Imaging of the stenosis (**a** intra-arterial digital subtraction angiography; **b** cardiac-gated T1 SE sequence **c,d** GRE cine study during diastole/systole). Note the turbulent dark jet (*arrow*) seen only during peak systole. 3D MIP MRA (**e**) shows the aortic arch including dilatation of subclavian arteries and the brachiocephalic trunk

Fig. 12.4e

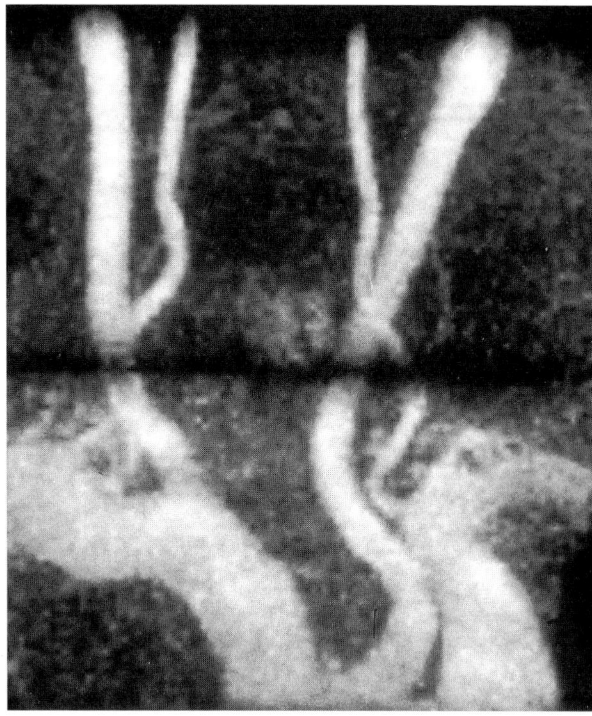

initial diagnostic evaluation. Therefore, experience in this field is not available. However, vascular MRI has proved of value in the evaluation of chronic posttraumatic pseudoaneurysms (MOORE et al. 1984) and complications following aortic surgery.

12.3.2.3 Thoracic Aortic Aneurysm

Aortic aneurysms are found in approximately 10% of autopsies; the main cause is atherosclerosis. Aneurysms can be classified into true lesions (all layers of the wall intact), false lesions (all layers of the wall disrupted), and pseudo-aneurysms (one or two layers of the wall intact). Atherosclerotic thoracic aortic aneurysms (TAAs) tend to have a fusiform configuration, only a small number being saccular. They are located in the ascending or descending aorta. Saccular TAAs, mostly of mycotic or posttraumatic etiology, are most frequently located in the ascending aorta, as are aneurysms due to syphilis, Marfan syndrome, Takayasu's arteritis, degenerative processes of the media, and dilatation associated with valvular disease. In patients with ascending aortic aneurysm it is imperative to exclude pericardial effusion or hemipericardium. To evaluate the prognosis the true diameter of the vessel has to be measured, because aneurysms greater than 6 cm in diameter have an increased prevalence of rupture and require surgical intervention. Delineation of the aneurysmal shape can be helpful in determining its etiology. A TAA should also be characterized with respect to its location, extent, and relationship to the aortic branch vessels. The aortic wall should be evaluated for possible aortitis, mural thrombus, and periaortic hematoma.

For MR examination of TAAs, multiplane SE imaging is initially useful. Coronal, sagittal oblique, and transverse T1-weighted slices provide complete information, including aneurysmal location and size and the nature of intramural changes (Fig. 12.7a–d). In a study reported by YUCEL et al. (1990) it was found that SE MRI of the aortic wall is superior to angiography in depicting the extent of atheromatous plaques and intramural thrombus. Furthermore, the authors reported that MRI is better than CT in differentiating acute intramural hematoma, which shows increased signal in T1- and T2-weighted images, from atherosclerotic plaque and chronic intraluminal thrombus, in which intermediate to low signal occurs. Hemopericardium appears with an intermediate signal on T1-weighted images and a dark signal on T2-weighted images. Subacute pericardial processes have a bright signal on both T1- and T2-weighted images. Abnormalities of blood flow in TAA can be evaluated using GRE sequences. In cine studies slow flow (Fig. 12.7e–h), thrombotic deposits,

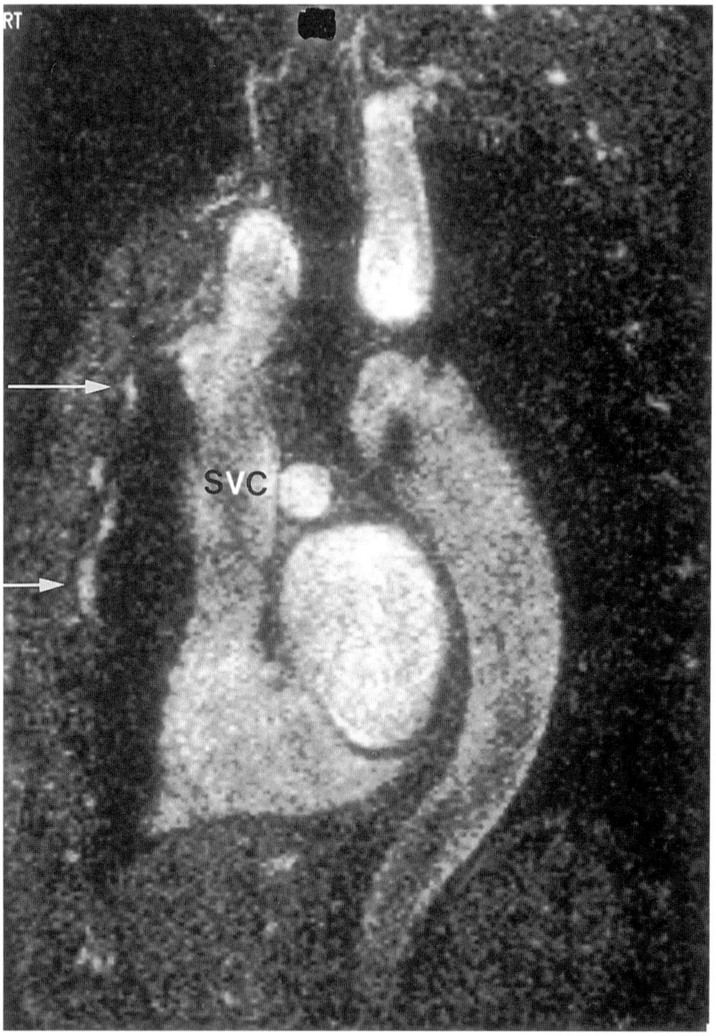

Fig. 12.6d

and the aortic valve status can be assessed over the cardiac cycle. 3D TOF projection angiograms have proved useful in accurately evaluating ostial supra-aortic branch status when involvement is suspected. 2D TOF projection angiograms acquired in the oblique sagittal plane can be rotated along the long axis of the body, visualizing the entire thoracic aorta and the abnormality (Fig. 12.7i).

12.3.2.4 Thoracic Aortic Dissection

Aortic dissection predominantly occurs in the sixth and seventh decades of life. It is more frequent in males than females and is found in 1/400 autopsies. In most cases the etiology is a degenerative process of the media which results in a subintimal hemorrhage, subsequent intimal tear, and extension of a false lumen. The disease is accelerated by hypertension, preexisting mycotic aneurysm, or aortic laceration (Kadir 1986).

For prognostic and therapeutic reasons thoracic aortic dissection (TAD) has been classified independently of the origin of the intimal flap. In the most frequently used classification system, namely the Stanford classification (Dailey et al. 1970), type A TAD [i.e., type I and II of the DeBakey classification (DeBakey et al. 1965)], which includes the ascending aorta, can be differentiated from type B (i.e., type III of the DeBakey classification), in which only the descending aorta is affected. Type I dissection (29%–34%) includes the ascending aorta, the aortic arch, and the descending aorta. The lesion usually begins close to the aortic root in the anterolateral ascending aorta and continues into the aortic arch superiorly along the convexity and the posterolateral descending aorta. Type II dissection (12%–21%) is localized in the ascending aorta. Type

Fig. 12.6e

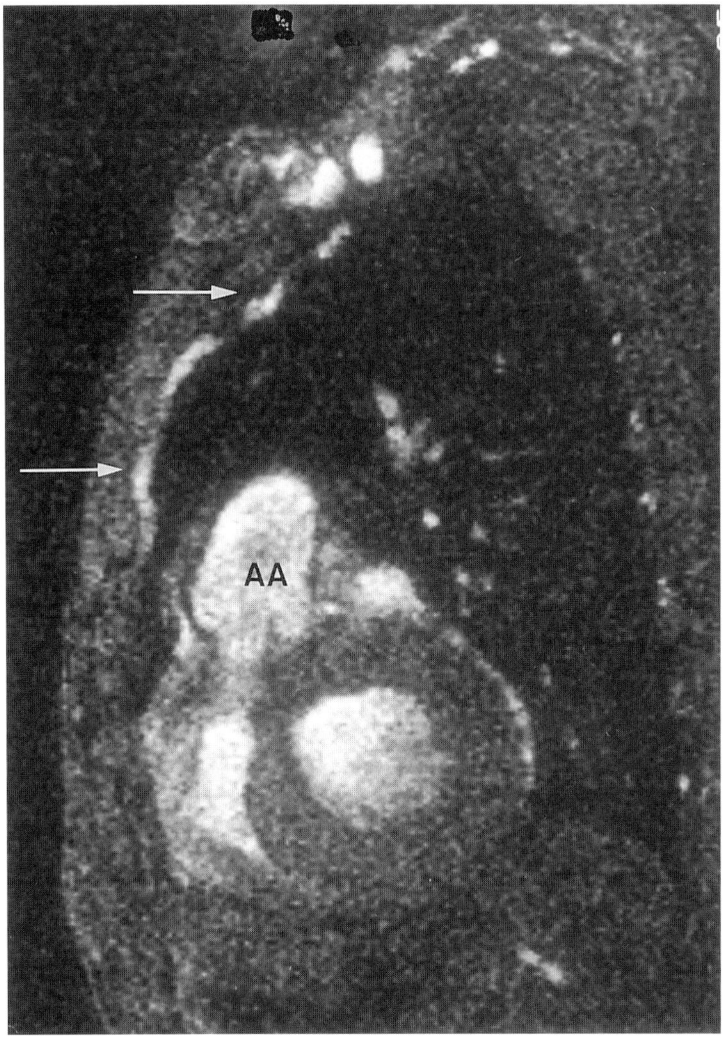

I and II dissection may be associated with aortic valve insufficiency, pericardiac hemorrhage, and dissection of the supra-aortic branches. In type III dissection (approx. 50%) the entry is usually located close to the isthmus distal to the origin of the left subclavian artery, and the lesion involves the descending aorta and may extend into the abdominal aorta and its branches.

In the majority of cases, two lumina exist. The false lumen may develop either antegrade or retrograde from the origin of the entry. Reentries can be observed in the aortic arch, in the mid-descending aorta, at the level of the abdominal aortic branches, at the aortic or iliac bifurcation, and at the femoral bifurcation in rare cases. Usually blood flow in the false channel is slower than in the true lumen and the true channel may appear narrow due to compression by the wide false channel. Multiple flaps can develop, and when reentries occur (as happens in ca. 87% of cases), blood flow in both patent channels demonstrates the typical "double-barrel" aorta. If no reentry develops (ca. 13% of cases), spontaneous thrombosis of the false channel may occur. Untreated, acute dissection has a high mortality of 30% within 24h, 60% at 3 weeks, and 80% at 3 months (HARRIS et al. 1967).

In dissection of the ascending aorta, immediate surgical management is mandatory due to the life-threatening complications which frequently occur, as mentioned above. Operative repair includes the placement of a graft in the ascending aorta with or without prosthesis of the aortic valve. Risk of surgical complications is 10%–15% for treatment of aortic dissection when aortic arch replacement is performed, and 5%–10% for treatment of ascending aorta and arch aneurysms (ROFSKY et al. 1993). In contrast, chronic dissection of the descending aorta usually can be managed conservatively when

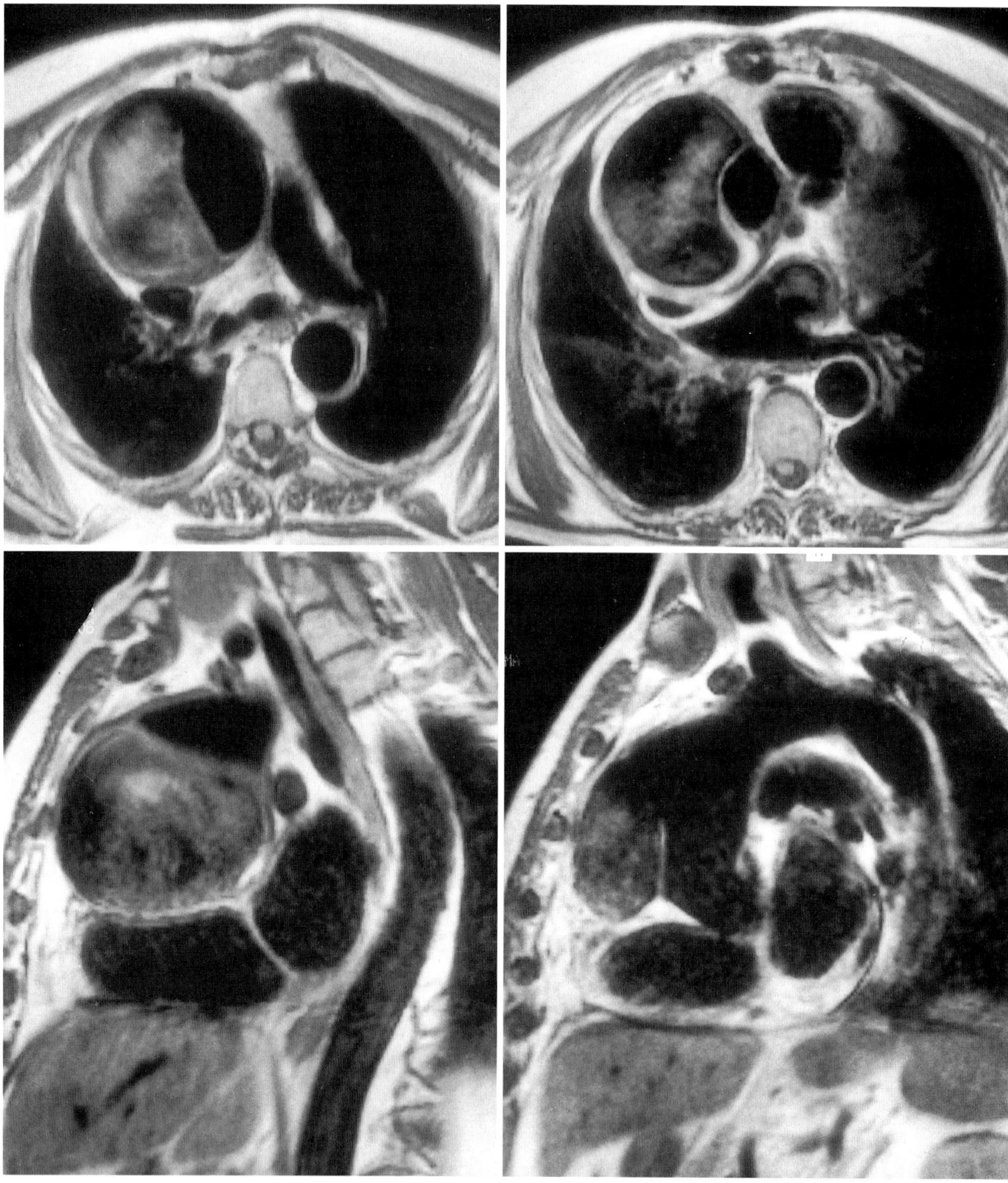

a c

b d

Fig. 12.7a–i. Aneurysmatic dissection of the supravalvular ascending aorta (type I). Transverse (**a**, **c**) and oblique sagittal (**b**, **d**) cardiac-gated T1 SE images show a dark true lumen and inhomogeneous bright signal in the aneurysmatic false lumen. In transverse (**e**, **g**) and oblique sagittal (**f**, **h**) GRE cine studies (2D TOF GRE FISP) bright signal within both channels during diastole (**e**, **f**) demonstrates patency, whereas during systole a turbulent signal void can be identified within the true lumen (**g**, **h**). On a 2D MR-projection angiogram (TOF GRE FLASH) (**i**) the entire thoracic aorta is visualized with bright signal in both channels. *LV*, left ventricle; *LA*, left atrium

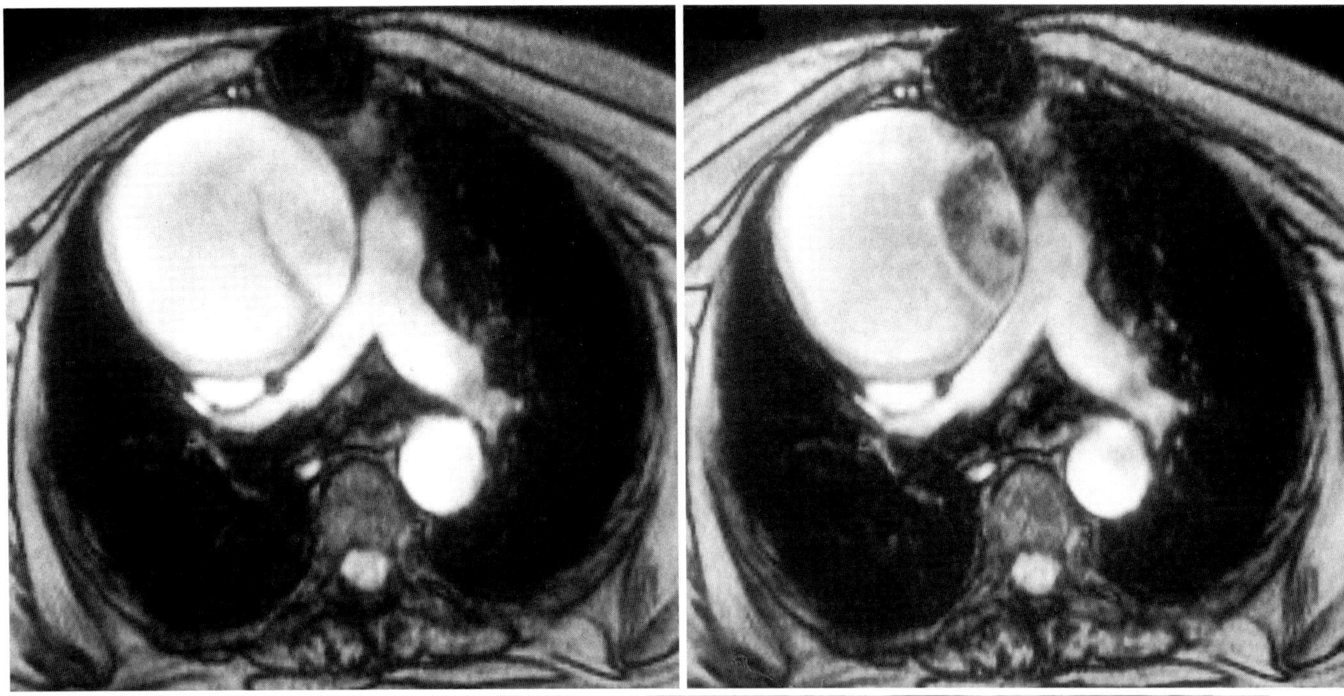

▲ e,g

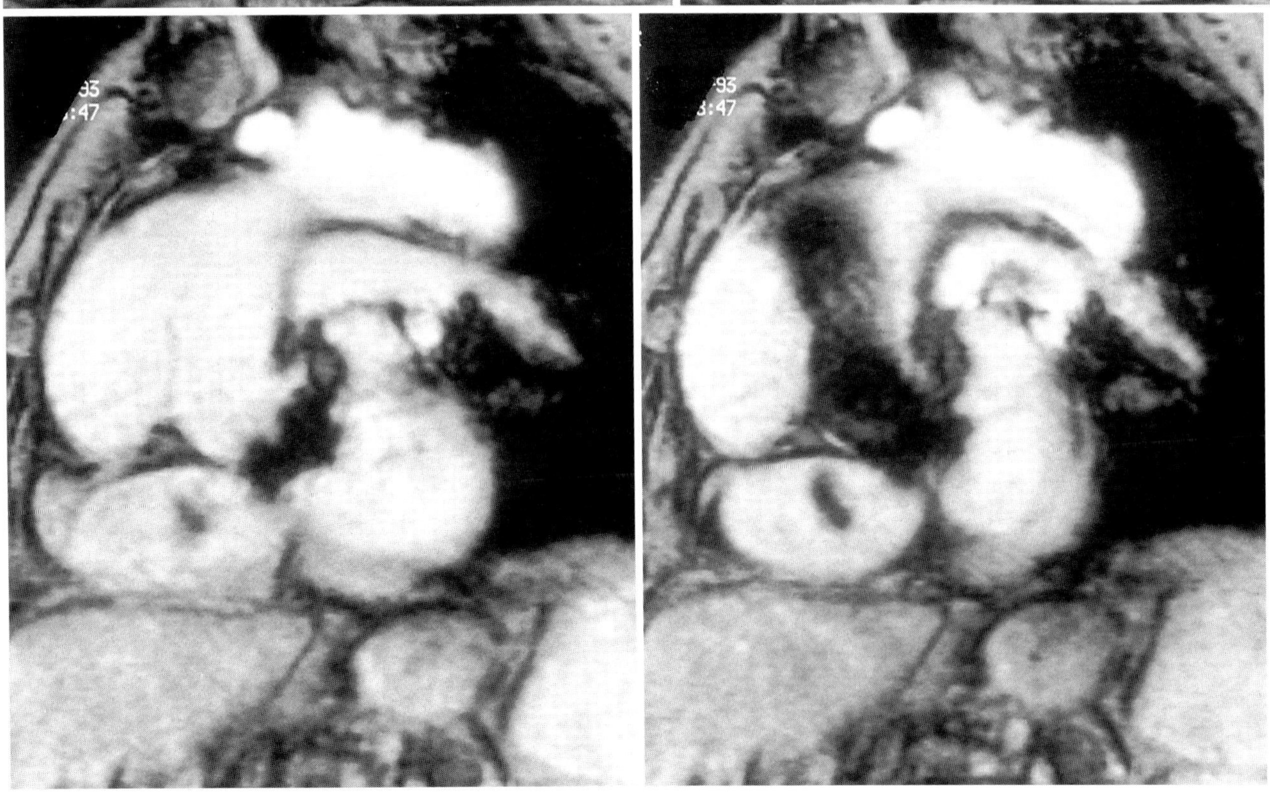

f,h

Fig. 12.7e–h

a reentry exists and no ischemic complications become evident. However, surgical intervention will be necessary in cases in which the dissection acutely occludes a spinal artery, the mesenteric artery, renal arteries, or iliac arteries. Diagnostic information which is required in patients with TAD includes accurate identification and location of the aortic dissection, extent of the dissection, sites of entry and reentries, thrombus in the false lumen, supra-aortic and abdominal branch vessel involvement,

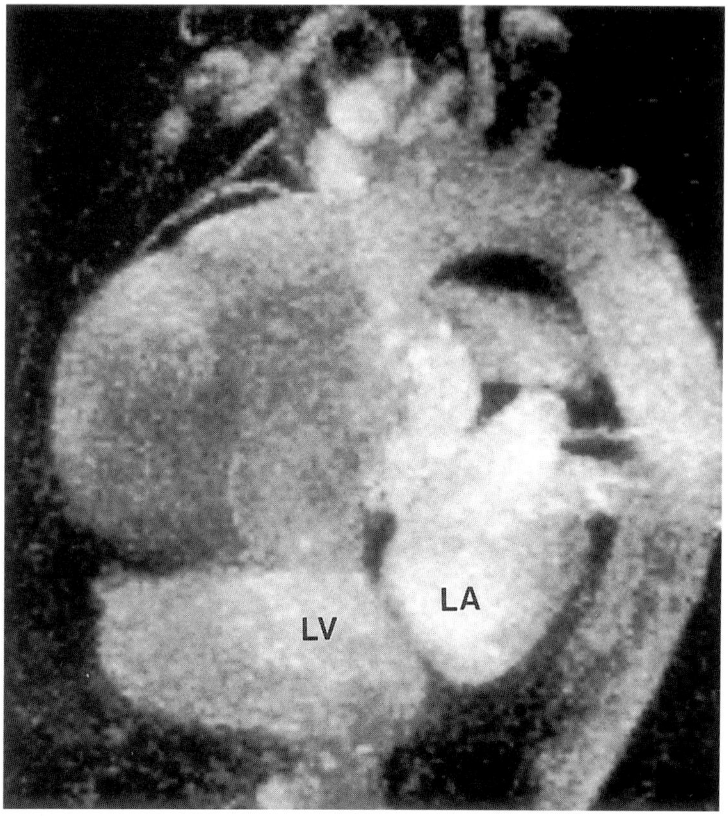

Fig. 12.7i

extravascular rupture, aortic insufficiency, coronary artery involvement, and hemopericardium.

Aortography has long been considered indispensable in the diagnosis and evaluation of TAD. It can be performed with minimal risk by experienced personnel. In a review article, CIGARROA et al. (1993) evaluated the diagnostic accuracy of different imaging methods. Sensitivity and specificity of angiography were reported to be 88% and 94%, respectively. False-negative results may occur due to thrombosis of the false lumen, simultaneous opacification of both channels overlapping the intimal flap, and unusual location of tears. With the introduction of noninvasive echocardiography, CT, and MRI, the role of angiography has changed. Bedside transthoracic echocardiography (TTE) or transesophageal echocardiography (TEE) rapidly enables the diagnosis of TAD. However, the examination is highly dependent on the operator's experience, and the examination area is limited. In their article CIGARROA et al. (1993) reported a sensitivity and specificity of 59%–85% and 63%–96%, respectively, for TTE, 99% and 97%, respectively, for TEE, and 83%–100% and 87%–100%, respectively, for contrast-enhanced CT. The introduction of ultrafast CT (HAMADA et al. 1992) and helical CT (COSTELLO et al. 1992) has increased the accuracy in assessing the site of tear and the location of reentries whereas the number of known CT pitfalls due to motion artifacts and mistimed opacification may have decreased. Moreover, using this method, involvement of supra-aortic branches may be evaluated in 3D reconstructions (PROKOP et al. 1993). In a comparative diagnostic study, SOMMER et al. (1995) reported a sensitivity and specificity of 100% for the detection of both TAD and the involvement of the aortic arch vessels. The major disadvantage of CT is the inability to depict abnormalities of the aortic valve and coronary vessels, and the necessity of using large volumes of contrast material, which may limit use of the method particularly in patients with impaired renal function. Hence, in the future noninvasive MRI may gain more acceptance and provide more information on TAD than CT, utilizing the advantages of obtaining images of the thoracic aorta in transverse, sagittal oblique, and coronal planes (Fig. 12.8), of potential coronary artery imaging, and of functional cine imaging.

Typical patterns seen with MRI include the intimal flap separating true and false lumens, and sites of entry and reentries. However, atypical patterns are described by WOLFF et al. (1991), such as

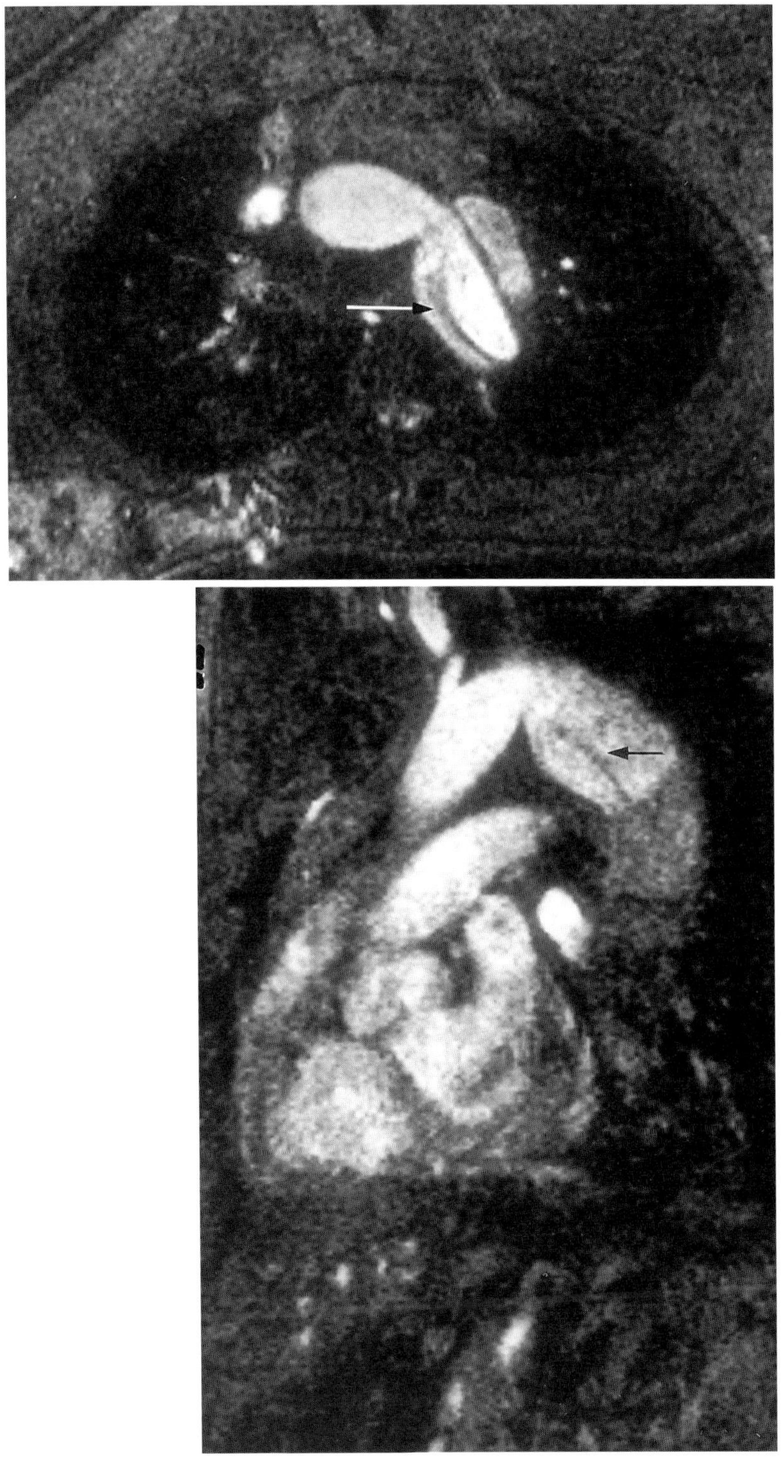

a

b

Fig. 12.8a–f. Thoracic aortic dissection of type III. On transverse and coronal individual MR slices (2D GRE FLASH, segmented k-space acquisition) the intimal flap can be delineated accurately (*arrows*). In addition, the site of entry (**a**, **b**) and both the true and false channels (**c**, **d**) can be evaluated, with higher signal in the smaller true lumen and lower signal in the larger false lumen. Reconstruction of MIP MR angiograms is possible from individual transverse (**e**) and oblique sagittal (**f**) sections

aortic wall thickening, which may occur more frequently in acute (up to 30% of cases than in chronic dissection. On T1 SE images rapidly flowing blood in the true and false channels appears devoid of signal, and the intimal flap can be delineated as a linear structure of intermediate signal. In MRI studies by KERSTING-SOMMERHOFF et al. (1988) and NIENABER et al. (1992) TAD was detected with a 95%–100% sensitivity and a 90%–100% specificity. TEE and MRI showed similar sensitivities, whereas the specificity of MRI was superior in respect of lesions localized in the ascending aorta. For MR identification of the site

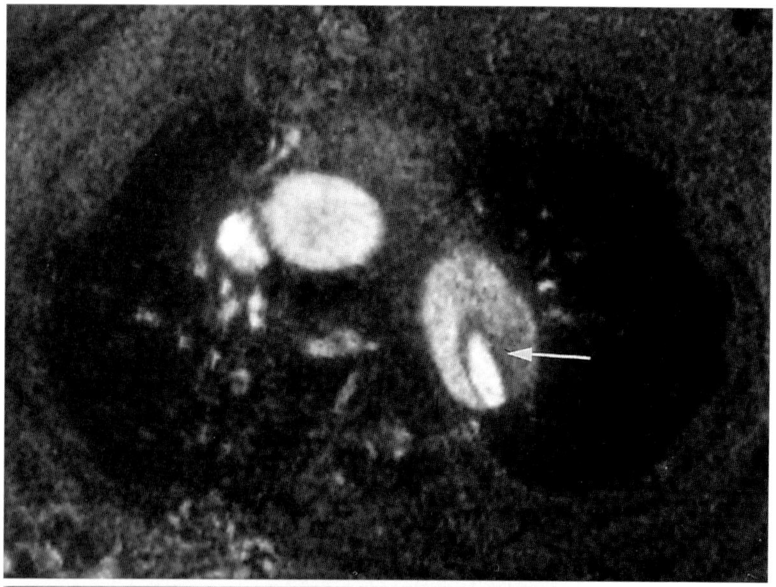

c

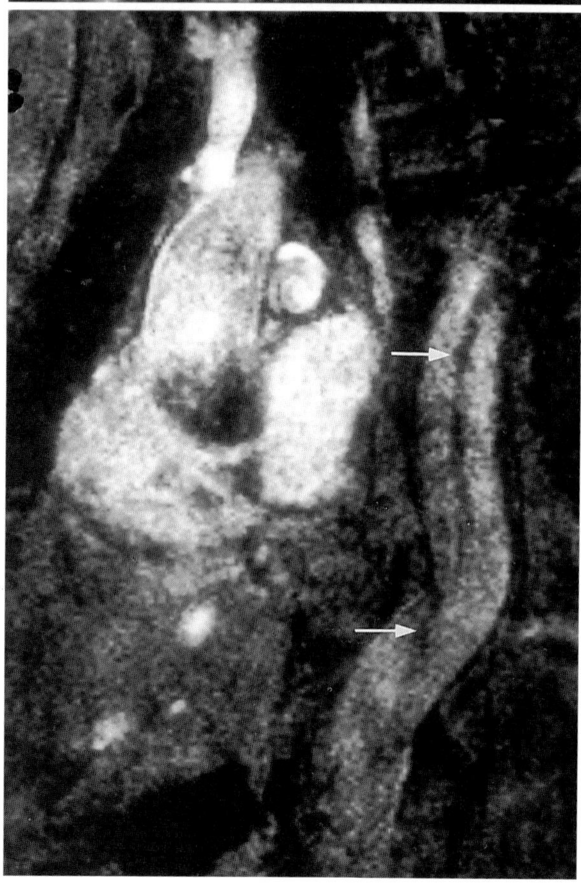

d

Fig. 12.8c,d

of entry, a sensitivity of 85% and a specificity of 100% were reported. In their article NIENABER et al. (1992) identified aortic arch thrombus and pericardial effusion with a 100% sensitivity and specificity. MRI has proved of value not only for initial detection of acute dissection but also for postoperative follow-up to exclude complications (WHITE et al. 1988; ROFSKY et al. 1993).

The major advantage of SE imaging results from the high contrast between the black blood pool, the intermediate signal of vascular wall, and the high signal of adjacent fat tissue. A limitation of SE

Fig. 12.8e,f

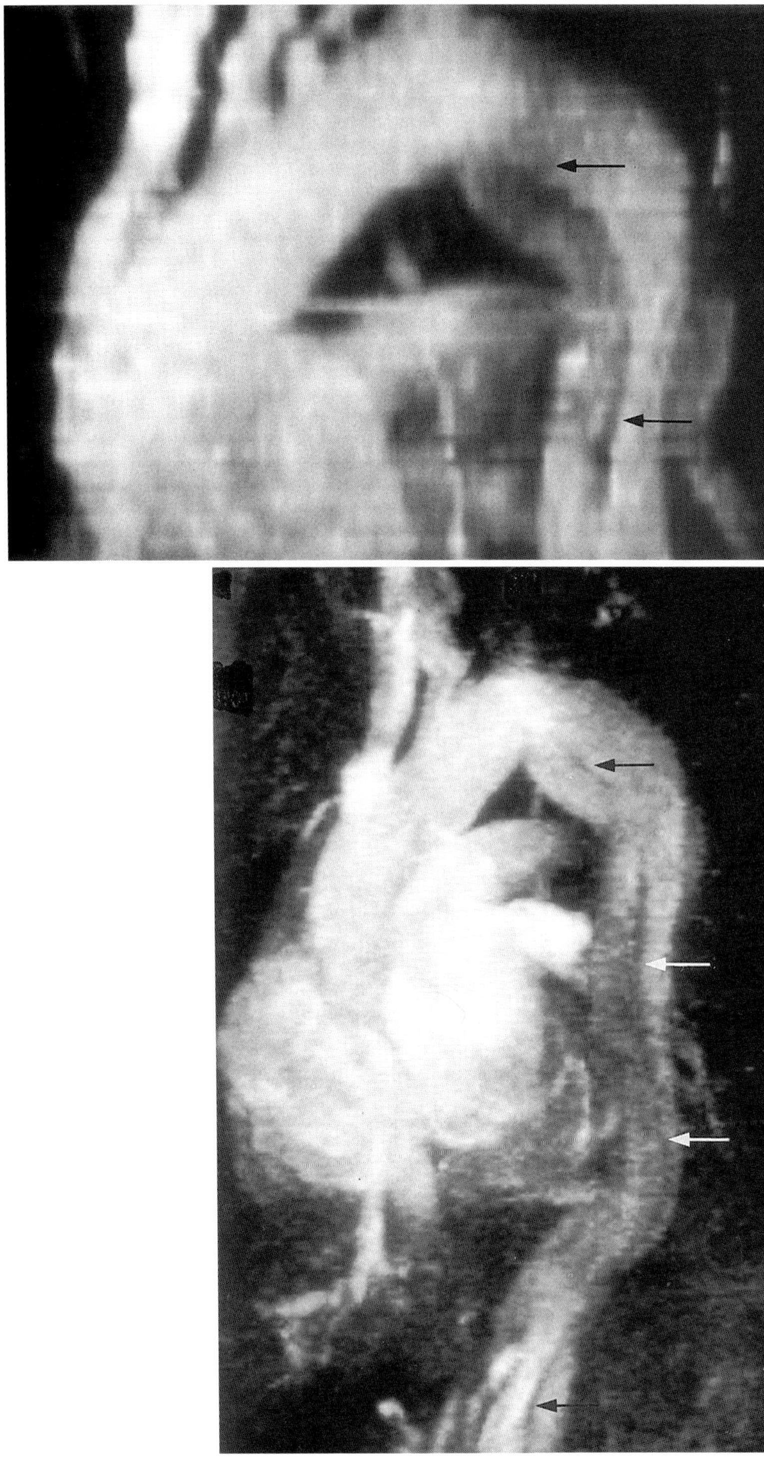

e

f

imaging is the fact that aortic dissection can be mimicked by artifacts due to fat shift, adjacent anatomical structures such as the left brachiocephalic vein or superior pericardial recess, and ghosting in the phase-encoding direction caused by respiratory, cardiac, and pulsatile vessel motion, particularly in the ascending aorta (LOTAN et al. 1989; SOLOMON et al. 1990). Furthermore, on SE images the signal of slow flowing blood in the false lumen may mimic thrombus material. Therefore, in these cases dynamic GRE cine studies (FRUEHWALD et al. 1989) should be applied by which means the true and false lumens can be accurately evaluated and flowing blood can be clearly distinguished from intimal flap and thrombus material (Figs. 12.2c,d, 12.7e–h). In addition, such studies provide information that facilitates localization of the entry and further communications between the channels, detection of aortic leaks into a contained pseudo-aneurysm, and evaluation of aortic valve insufficiency. By means of velocity-encoded cine MR studies, blood flow velocity has been quantified within the true and false lumens, and flow direction has been determined (CHANG et al. 1991).

Besides GRE cine studies, static bright blood MRA employing 2D and 3D TOF GRE sequences is able to assess accurately the anatomical and hemodynamic conditions in TAD, including evaluation of the site of entry and delineation of intimal flap structures (Fig. 12.8a–d). Breath-hold 2D TOF GRE FLASH sequences have been shown to be most informative in the oblique sagittal orientation (WALLNER et al. 1991; WHITE et al. 1994) or when transverse slices are acquired utilizing high signal of perpendicular inflow. Low signal from the dissection membrane can be delineated accurately from bright signal in both channels when no thrombus exists. Projection angiograms can be established from both transverse and sagittal oblique sections (Fig. 12.8e,f). In the article of WHITE et al. (1994) the sensitivity of conventional 2D TOF MRA was reported to be high in the ascending and descending portions, ranging from 89% to 100%; specificity was reported to range from 62% to 100% in the ascending and arch portions, whereas aneurysmal or dissecting lesions could be detected more accurately in the descending portion. In our experience, use of the k-space technique in 2D GRE acquisition may further improve diagnostic accuracy in the ascending and descending portions of the aorta. The aortic arch portion can be evaluated most effectively using 3D TOF MRA (LEWIN et al. 1991). Thus, this technique is a helpful adjuvant to 2D TOF MRA in cases in which the dissection is suspected to extend into the aortic arch and supra-aortic branches.

Overall, GRE sequences are far more reliable than the SE technique in providing the necessary preoperative information in type I dissections regarding the status of the aortic valve, the proximal coronary arteries, the site of intimal tear, and involvement of the supra-aortic branches. However, each of the imaging techniques has certain advantages as well as limitations in the evaluation of suspected TAD (LINK and LESKO 1992). Therefore, a combination of SE imaging and GRE MRA has been recommended (HARTNELL et al. 1994). Although echocardiography, CT, and MRI have greatly enhanced our ability to evaluate the thoracic aorta by noninvasive means, aortography was until relatively recently considered the examination technique of choice, especially when surgical intervention was being considered (PETASNICK 1991). Currently, echocardiography is considered the first method to be employed in the diagnostic workup of suspected acute aortic dissection, whereas in patients with stable or chronic dissection both MRI and MRA are increasingly accepted as the gold standard for defining aortic anatomy and blood flow conditions due to their excellent image quality and high accuracy (CIGARROA et al. 1993). In emergency patients, however, spiral CT-angiography has been shown to be the preoperative method of choice due to its high diagnostic accuracy and short examination time.

12.4 Conclusion

Because MRI combines the major attributes of angiography, echocardiography, and CT, its role in the evaluation of the thoracic aorta is steadily increasing. It is generally accepted that cardiac gated SE sequences are extremely useful in the evaluation of chest anatomy including depiction of the aortic lumen, aortic wall, and adjacent mediastinal structure, due to their high resolution and contrast. Compared with SE imaging, MRA of the thoracic aorta provides additional information, usually gained only with echocardiography or arteriography. On dynamic GRE cine studies, flowing blood generates a signal that allows the visualization of blood as it pulsates through the aorta. Cine MRA enables reliable differentiation between normal flow, slow flow, and thrombus. Turbulent flow generates a signal void, thereby allowing the detection and qualitative assessment of the pathophysiological consequences of anatomical abnormalities. Breath-hold non-cine MRA techniques allow easier evaluation of aortic arch branch vessels and provide further accurate anatomical information which is relevant to any type of aortic thoracic disease. With phase-mapping techniques, blood velocity can be measured in a variety of aortic disorders and results can be used for calculation of pressure gradients. Recent advances in the field of 2D and 3D MRA will greatly enhance the overall role of MRI in the evaluation of thoracic aortic lesions by allowing determination of blood flow conditions and assessment of branch vessels. Thus, utilizing the transverse and oblique sagittal planes, MRI and MRA have become the procedures of choice for evaluating congenital anomalies of the thoracic aorta, which includes aortic arch anomalies and aortic coarctation. Furthermore, MRI and MRA are indicated for the evaluation of aortic aneurysms, thrombotic and inflammatory wall lesions, and aortic dissection, and for postoperative follow-up to exclude graft occlusion, anastomotic aneurysm, aneurysm of the false channel in TAD, and compression of the true channel which may produce cerebral, renal, or mesenteric ischemia. However, MRI and MRA of the thoracic aorta currently still may be of limited value in the detection of small focal intimal flap lesions due to motion artifacts and reduced spatial resolution. Furthermore, problems may occur in emergency conditions due to the relatively long examination time and uncomfortable monitoring of the patient. Therefore, in clinically symptomatic patients with acute aortic disease who do not have contraindications to iodinated agents, contrast-enhanced helical CT-angiography is preferred as the imaging method of choice at present.

Acknowledgment. The authors thank Mrs. E. Griepeutrog and Mrs. I. Meurer for assistance with data acquisition and preparation of image material.

References

Amparo EG, Higgins CB, Shafton EP (1984) Demonstration of coarctation of the aorta by magnetic resonance imaging. AJR 143:1192–1194

Anderson CM, Edelman RR, Turski PA (1993) Clinical magnetic resonance angiography. Raven Press, New York

Bank ER, Aisen AM, Rocchini AP, Hernandez RJ (1987) Coarctation of the aorta in children undergoing angioplasty: pretreatment and posttreatment MR imaging. Radiology 162:235–240

Chang JM, Friese K, Caputo GR, Kondo C, Higgins CB (1991) MR measurement of blood flow in the true and false channel in chronic aortic dissection. J Comput Assist Tomogr 15:418–423

Chien D, Edelman RR (1991) Ultrafast imaging using gradient echos. Magn Reson Q 7:31–56

Cigarroa JE, Isselbacher EM, DeSanctis RW, Eagle KA (1993) Diagnostic imaging in the evaluation of suspected aortic dissection: old standards and new directions. N Engl J Med 328:35–43

Costello P, Ecker CP, Tello R, Hartnell GG (1992) Assessment of the thoracic aorta by spiral CT. AJR 158:1127–1130

Dailey PO, Trueblood HW, Stinson EB, Wuerflein RB, Shumway NE (1970) Management of acute aortic dissection. Ann Thorac Surg 10:237–247

DeBakey ME, Henly WS, Cooley DA, Morris GC, Crawford ES, Beall AC Jr (1965) Surgical management of dissecting aneurysms of the aorta. J Thorac Cardiovasc Surg 49:130

Edelman RR, Mattle HP, Wallner B et al. (1990) Extracranial carotid arteries: evaluation with black blood MR angiography. Radiology 177:45–50

Edelman RR, Chien D, Kim D (1991) Fast selective black blood imaging. Radiology 181:655–660

Friedman WF (1983) Congenital heart disease. In: Petersdorf RG, Adams RD, Braunwald E, Isselbacher KJ, Martin JB, Wilson MD (eds) Principles of internal medicine. McGraw-Hill, New York

Fruehwald FXJ, Neuhold A, Fezoulidis J, Globits S, Mayr H, Wicke K, Glogar D (1989) Cine-MR in dissection of the thoracic aorta. Eur J Radiol 9:37–41

Hamada S, Takamiya M, Kimura K, Imakita S, Nakajima N, Naito H (1992) Typa A aortic dissection: evaluation with ultrafast CT. Radiology 183:155–158

Harris PD, Malm JR, Bigger JT Jr, Bowman FO (1967) Follow-up studies of acute dissecting aortic aneurysms managed with antihypertensive agents. Circulation 35 (suppl 1):183–187

Hartnell GG, Finn JP, Zenni M, Cohen MC, Dupuy DE, Wheeler HG, Longmaid HE (1994) MR imaging of the thoracic aorta: comparison of spin-echo, angiographic, and breath-hold techniques. Raidology 191:697–704

Heberer G, Rau G, von Buch KG, Gehl H (1963) Die chirurgische Behandlung der Coarctatio aortae (Aortenisthmusstenose) im höheren Lebensalter und bei zusätzlichen Herz- oder Gefäßanomalien. Dtsch med wochenschr 88:773–787

Kadir S (1986) Diagnostic angiography. Saunders Philadelphia

Kersting-Sommerhoff BA, Higgins CB, White RD, Sommerhoff CP, Lipton MJ (1988) Aortic dissection: sensitivity and specificity of MR imaging. Radiology 166:651–655

Klipstein RH, Firmin DN, Underwood SR, Rees RSO, Longmore DB (1987) Blood flow patterns in the human aorta studied by magnetic resonance. Br Heart J 58:316–323

Lewin JS, Laub G, Hausmann R (1991) Three-dimensional time-of flight MR angiography: applications in the abdomen and thorax. Radiology 179:261–264

Link KM, Lesko NM (1992) The role of MR imaging in the evaluation of acquired diseases of the thoracic aorta. AJR 158:1115–1125

Lotan CS, Cranney GB, Doyle M, Pohost GM (1989) Fat-shift artifact simulating aortic dissection on MR images. AJR 152:385–386

Mayo JR, Culham JAG, MacKay AL, Atkins DG (1989) Blood MR signal suppression by preexcitation with inverting pulses. Radiology 173:269–271

Mirowitz SA, Lee JKT, Gutierrez FR, Brown JJ, Eilenberg SS (1990a) Normal signal-void pattern in cardiac cine MR images. Radiology 176:49–55

Mirowitz SA, Lee JKT, Gutierrez FR, Brown JJ, Eilenberg SS (1990b) "Pseudocoarctation" of the aorta: pitfall on cine MR imaging. J Comput Assist Tomogr 14:753–755

Moore EH, Webb WR, Verrier ED, Broaddus C, Gamsu G, Amparo E, Higgins CB (1984) MRI of chronic posttraumatic false aneurysm of the thoracic aorta. AJR 143:1195–1196

Nienaber CA, Spielman RP, von Kodolitsch Y et al. (1992) Diagnosis of thoracic aortic dissection: magnetic resonance imaging versus transesophageal echocardiography. Circulation 85:434–437

Petasnick JP (1991) Radiologic evaluation of aortic dissection. Radiology 180:297–305

Prokop M, Schaefer C, Kalender WA, Polacin A, Galanski M (1993) Vascular imaging with spiral CT. Radiologe 33:694–704

Rees S, Somerville J, Ward C, Martinez J, Mohiaddin KH, Underwood R, Longmore DB (1989) Coarctation of the aorta: MR imaging in late postoperative assessment. Radiology 173:499–502

Rofsky NM, Weinreb JC, Grossi EA, Galloway AC, Libes RB, Colvin SB, Naidich DP (1993) Aortic aneurysm and dissection: normal MR imaging and CT findings after surgical repair with the continuous-suture graft-inclusion technique. Radiology 186:195–201

Seltzer SE, D'Orsi C, Kirshner R, DeWeese JA (1981) Traumatic aortic rupture. Plain film findings. AJR 137:1011–1014

Simpson IA, Chung KJ, Glass RF, Sahn DJ, Sherman FS, Hesselink J (1988) Cine magnetic resonance imaging for evaluation of anatomy and flow relations in infants and children with coarctation of the aorta. Circulation 78:142–148

Solomon SL, Brown JJ, Glazer HS, Mirowitz SA, Lee JK (1990) Thoracic aortic dissection: pitfalls and artifacts in MR imaging. Radiology 177:223–228

Sommer T, Fehske W, von Smekal A et al. (1995) Spiral computed tomography, multiplane transesophageal echocardiography and magnetic resonance imaging in the diagnosis of thoracic aortic dissection ROFO 162:104–111

von Schulthess GK, Higashino SM, Higgins SS, Didier D, Fisher MR, Higgins CB (1986) Coarctation of the aorta: MR imaging. Radiology 158:469–474

Wallner B, Edelman RR, Kim D, Finn JP (1991) Assessment of thoracic and abdominal aortic aneurysms using MR angiography. ROFO 154:11–16

White RD, Ullyot DJ, Higgins CB (1988) MR imaging of the aorta after surgery for aortic dissection. AJR 150:87–92

White RD, Obuchowski, NA, VanDyke CW et al. (1994) Thoracic aortic disease: evaluation using a single MRA volume series. J Comput Assist Tomogr 18:843–854

Wolff KA, Harold CJ, Tempany CM, Parravano JG, Zerhouni EA (1991) Aortic dissection: atypical patterns seen at MR imaging. Radiology 181:489–495

Wood P (1956) Diseases of the heart and circulation, 2nd edn. Eyre and Spottiswoode, London

Yucel EK, Steinberg FL, Egglin TK, Geller SC, Waltman AC, Athanasoulis CA (1990) Penetrating aortic ulcers: diagnosis with MR imaging. Radiology 177:779–781

13 Thorax – Coronary Arteries

B. SIEWERT, R.R. EDELMAN, and W.J. MANNING

CONTENTS

13.1 Introduction

Modalities for imaging the coronary arteries can be divided into noninvasive (radiological or nuclear medicine) and invasive studies (contrast angiography). Noninvasive imaging methods include nuclear medicine studies, such as single-photon emission computed tomography (SPECT) and positron emission tomography (PET), sonographic procedures like two-dimensional (2D) echocardiography, and magnetic resonance imaging. Even though SPET and PET are able to demonstrate perfusion abnormalities of the myocardium, these procedures cannot be used to image the coronary arteries directly and to visualize underlying stenoses. In addition, direct measurements of coronary artery blood flow are not possible.

B. SIEWERT, MD, Department of Radiology, Beth Israel Hospital, Harvard Medical School, 330 Brooklyn Avenue, Boston, MA 02215, USA
R.R. EDELMAN, MD, Department of Radiology, Beth Israel Hospital, Harvard Medical School, 330 Brooklyn Avenue, Boston, MA 02215, USA
W.J. MANNING, MD, Departments of Radiology and Medicine (Cardiovascular Division), Beth Israel Hospital, Harvard Medical School, 330 Brooklyn Avenue, Boston, MA 02215, USA

Using 2D echocardiography, the coronary arteries can be imaged directly. Transthoracic echocardiography has variable success depending on the location of the vessel, due to interference of the chest wall and pulmonary parenchyma. The best results obtained are for the left main coronary artery, which can be seen in more than 80% of patients. The usefulness of sonography is somewhat limited since only 5%–10% of patients with angina demonstrate left main coronary artery stenosis on conventional angiography (ROBERTS et al. 1983). Furthermore, information about the distal portion of the vessel cannot be obtained. The imaging of the remaining coronary arteries is even more difficult: the proximal left anterior descending artery can be seen in 53% of patients, the left circumflex artery in 34%, and the right coronary artery in 56% (CHEN et al. 1980; OGAWA et al. 1980; DOUGLAS et al. 1988). Encouraging results have been reported for transesophageal echocardiography in the detection of left main coronary artery stenosis (YOSHIDA et al. 1990), but it does carry the risk of death, esophageal perforation and laceration, aspiration, hypoxia, and arrhythmias (DANIEL et al. 1991).

The "gold standard" for diagnosis of coronary artery disease is still contrast angiography, which, for several reasons, is not an ideal screening test. Conventional angiography is an invasive procedure and may be complicated by death (0.12%–0.20%), cerebrovascular accident (0.03%–0.20%), or myocardial infarction (0.0%–0.25%). Minor related problems include vascular complications, local infection (0.57%–1.6%) and arrhythmias (0.30%–0.63%) (DAVIS et al. 1979; KENNEDY and the Registry Committee of the Society for Cardiac Angiography 1982; WYMAN et al. 1988). Surprisingly, even though this is not a test that should be performed routinely, up to 20% of requested angiograms in patients with suspected cardiovascular disease do not show significant disease (JOHNSON et al. 1989). In addition, these are very expensive procedures, with 1.8 billion dollars per year spent in the United States on angiography alone (WITTELS et al. 1990).

duces a phase shift that is proportional to velocity (VAN DIJK 1984; WALKER et al. 1988). Therefore, two images with different flow sensitivities are acquired and subtracted to eliminate phase shifts from sources other than flow. The first image is insensitive to flow, whereas the second is sensitive to the shift in the phase of the MR signal of flowing blood. Subtraction of these images produces a phase difference image in which the phase shift is proportional to the flow velocity along the direction of slice selection. Image subtraction eliminates background phase shifts unrelated to blood flow.

13.2.4.2 Flow Measurements in Coronary Arteries

For flow measurements in coronary arteries, a segmented turboFLASH sequence, in which two pairs of segmented turboFLASH images are acquired in a single breath-hold (EDELMAN et al. 1993), can be used (Fig. 13.1). The gradients are calibrated to permit flow velocities up to 150 cm/s. In the EDELMAN et al. study, the matrix size was 96×256 with pixel dimensions of 1.4×0.8 mm. Four images were reconstructed by this sequence comprising magnitude and phase reconstruction of the flow insensitive and sensitive data. With this technique coronary artery flow was measured in 16 subjects at rest, four of whom had additional intravenous injection of adenosine. Coronary flow velocities increased at least fourfold with the administration of the drugs. These results demonstrate that MRA can measure the velocity of coronary artery flow noninvasively. It may be useful for determining the physiological significance of coronary artery stenoses.

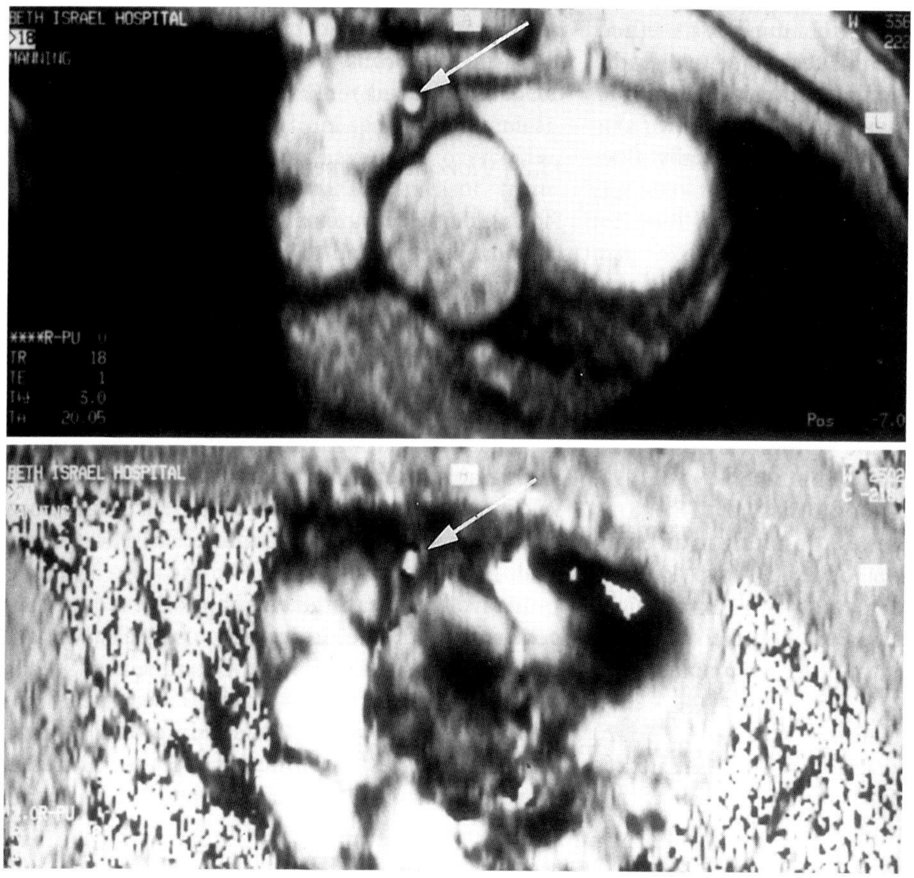

a

b

Fig. 13.1a,b. Flow measurements using breath-hold phase contrast segmented turboFLASH. **a** Axial magnitude reconstruction. Right coronary artery (*arrow*) in cross-section. **b** Phase image. The pixel intensity is related to the phase shift, which itself is dependent upon flow velocity. Right coronary artery (*arrow*)

13.3 Normal Anatomy

13.3.1 General Anatomy

Blood is supplied to the heart by the right and left coronary artery which arise from the right and left sinuses of Valsalva, respectively. The right coronary artery (RCA) travels in the anterior atrioventricular groove to the crux of the heart. When it arises from the RCA, the conus artery is the first branch. The next branch is the sinoatrial node artery. Traversing the perimeter of the heart, the RCA gives off acute marginal branches to the right ventricle. From its distal portion, proximal to or just at the crux, the RCA gives off the posterior descending artery, which supplies the posterior one-third of the ventricular septum and the diaphragmatic portion of the left ventricle.

In general, the RCA supplies the right ventricle and the inferior portion of the left ventricle, with the left coronary artery supplying the remainder of the left ventricle. In right dominant systems, however, the right coronary artery continues in the posterior atrioventricular groove around the back of the heart as the posterolateral segment artery, giving off a variable number of posterolateral branches. This region may also be supplied by the left circumflex artery.

The left coronary artery (LCA) originates as the left main coronary artery from the left sinus of Valsalva, and courses slightly posteriorly, at first, and then bends more anteriorly. After a short distance it bifurcates into the left anterior descending (LAD) and left circumflex (LCX) arteries. The most frequent variation is trifurcation of the left main coronary artery into the LAD, the LCX, and an artery between called the ramus intermedius artery (Green and Elliott 1989).

The LAD runs in the anterior interventricular sulcus, directly above the ventricular septum, giving off septal and diagonal branches. The latter course over the anterolateral wall of the left ventricle. The LAD may extend completely around the apex or stop short of it, with the difference being made up by the posterior descending artery (coming from the RCA).

The LCX runs in the atrioventricular groove towards the posterolateral segment artery. It gives off an atrial circumflex artery and obtuse marginal branches which supply the lateral wall of the left ventricle. In a left dominant system, the RCA is congenitally short and the LCX continues as a left atrioventricular groove artery.

13.3.2 MRA Display of Anatomy and MRA Limitations

Using MRA, the proximal portions of the left main, left anterior descending, and right coronary arteries can be reliably imaged (Fig. 13.2). Their more distal portions, up to the diaphragm, can often be visualized. This has previously not been possible and represents a major improvement in coronary MRA.

Limitations of the technique arise from specific anatomical and physiological constraints. Stenoses may be missed because of inadequate spatial and temporal resolution, improper timing of the acquisition within the cardiac cycle, irregular breathing and misregistration between breath-holds, partial voluming of tortuous vessels and overlapping adjacent anatomical structures, reverse flow past an area of stenosis (collateral flow), and the unknown signal intensity and contrast characteristics of vessel wall and plaque on coronary MRA sequences. To date, the most important drawbacks for coronary MRA seem to be its spatial resolution of 0.5–1.5 mm, which does not yet match that of contrast coronary angiography, and its dependence on a regular cardiac rhythm (see also Sect. 13.2.1). Irregular breathing and misregistration between breath-holds may not be a problem in volunteers, but have been found to be of major significance in patients (Duernickx et al. 1994).

The imaging of the portion of a vessel distal to a severe stenosis is still a major problem; due to its poor flow, distinction from an occlusion may be difficult (Fig. 13.3). The use of intravenous contrast media, especially of blood pool contrast agents which are currently under development, may be helpful in improving image quality.

13.4 Pathology

13.4.1 General Pathology

Coronary artery disease is a major cause of morbidity and mortality in the United States and in most of the industrialized world. Myocardial infarction is still the leading cause of death in the United States (Yang et al. 1988), resulting in approximately 500 000 deaths in 1994. Risk factors for coronary artery disease include hypertension, hyperlipidemia, cigarette smoking, positive family history, and diabetes (Kannel et al. 1971).

The most important clinical clue to coronary artery stenosis is the presence of anginal discomfort

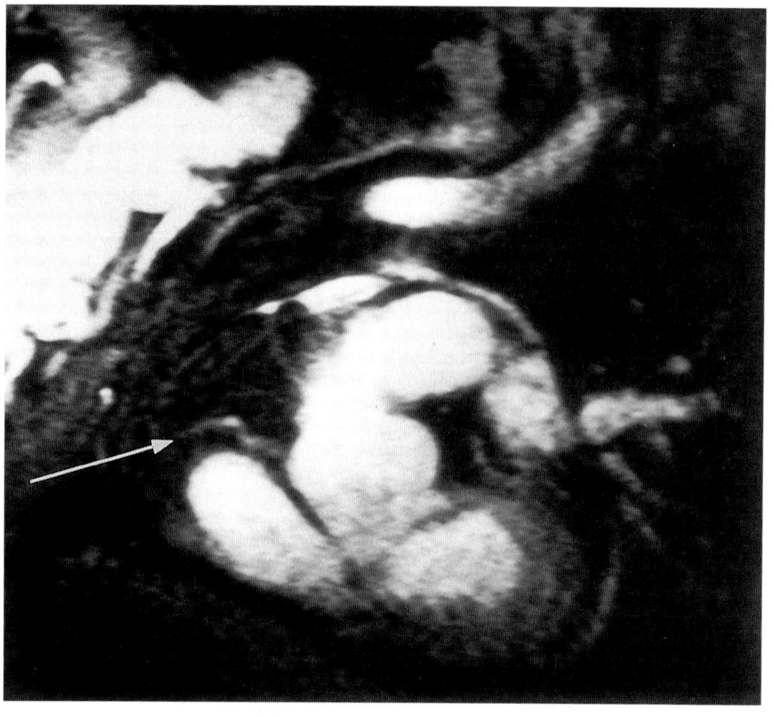

a

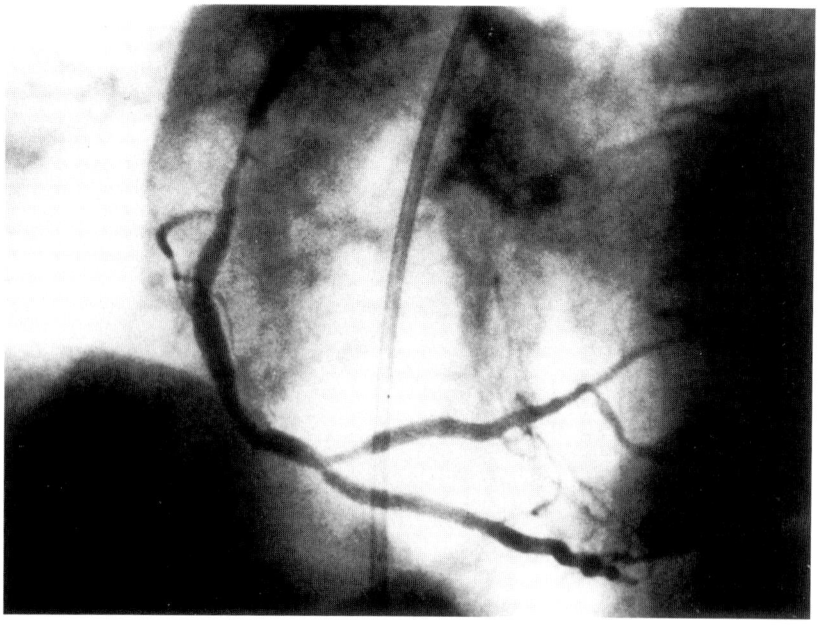

b

Fig. 13.3a,b. Severe stenosis of the right coronary artery. **a** Coronary MRA (oblique ECG-gated segmented turboFLASH) suggests an occlusion (*arrow*). Because of very slow flow distal to the stenosis there is insufficient flow contrast in this portion of the vessel. **b** Coronary catheterization demonstrates severe stenosis

tiguous sections as a "cine loop," for more accurate diagnosis (Fig. 13.4). Stenosis appears as a narrowing of the predicted vessel lumen, similar to conventional angiography. High-grade stenosis may be mistaken for occlusion, as described above.

13.4.3 Comparison of MRA with Alternative Imaging Modalities

Studies comparing coronary MRA and conventional angiography have revealed different sensitivities and

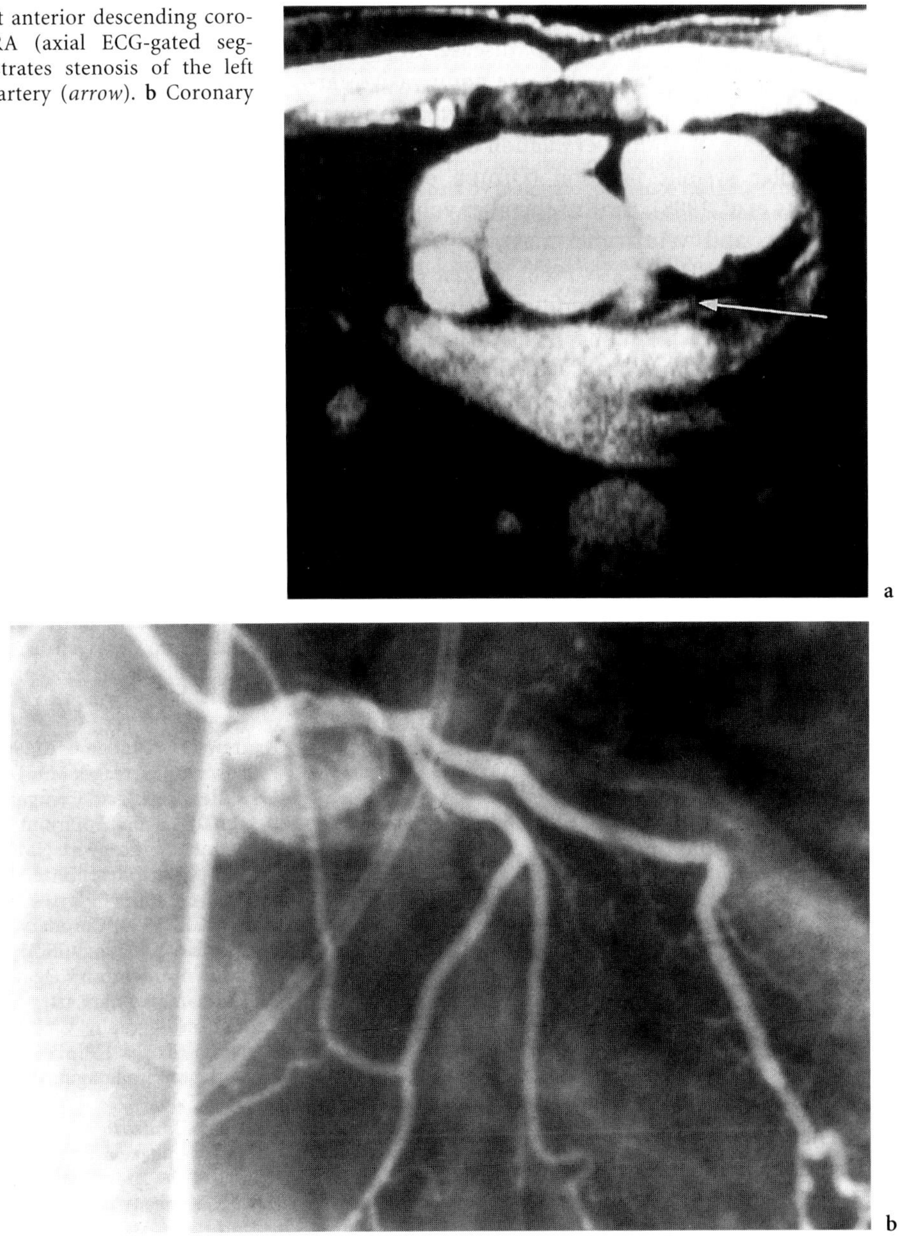

Fig. 13.4a,b. Stenosis of the left anterior descending coronary artery. **a** Coronary MRA (axial ECG-gated segmented turboFLASH) demonstrates stenosis of the left anterior descending coronary artery (*arrow*). **b** Coronary catheterization

specificities for the detection of coronary artery stenosis. The initial report by MANNING et al. (1993b) demonstrated a sensitivity of 90% and a specificity of 92% in the detection of hemodynamically significant (>50% luminal reduction) proximal coronary artery stenosis, these figures being comparable to those achieved with thallium scintigraphy. Subsequently, PENNELL et al. (1994) reported a sensitivity of 88% for a patient group with similar types of coronary artery stenosis. However, this degree of accuracy could not be reproduced by others: using the same technique (segmented k-space), DUERINCKX and URMAN (1994) demonstrated a sensitivity of 62%, while POST et al. (1994a) were able to demonstrate a sensitivity of only 41%. These investigators were unable to detect those stenoses of less than 50% in diameter (POST et al. 1994b).

13.5 Conclusion

With its completely noninvasive depiction of coronary artery anatomy and its cost-effectiveness, coro-

Table 14.1. Indications for pulmonary angiography in patients with suspected pulmonary embolism [derived from KELLEY et al. (1991) and STEIN et al. (1993)]

V/P scan probability	Doppler ultrasound for DVT	Clinical suspicion for PE
Low	Negative	High
Intermediate	Negative	Intermediate-high
High	Negative	Low-intermediate

V/P, ventilation-perfusion; DVT, deep venous thrombosis; PE, pulmonary embolism

Magnetic resonance angiography has been used to study pulmonary embolism. Other pulmonary diseases in which MRA could prove helpful include pulmonary hypertension, arteriovenous fistulae, pulmonary varix, and pulmonary sequestration. Since flow velocities can be measured with MRI, additional applications may include the evaluation of patients undergoing lung transplantation or pneumonectomy.

14.2 Technical Considerations

14.2.1 Challenges of Pulmonary MRA

Pulmonary MRA must overcome two major difficulties. Imaging time must be short to minimize artifact from the movement of the pulmonary vessels during breathing. Another challenge is the presence of magnetic susceptibility artifacts in MR evaluation of the lungs produced by the air/lung interface, resulting in rapid dephasing of the MR signal and decreased signal to noise and resolution. However, high spatial resolution is mandatory to allow for the detection of blood clots in small vessels when diagnosing pulmonary embolism.

Two-dimensional (2D) and three-dimensional (3D) techniques have been applied to overcome some of the problems mentioned above and each modality has been found to have drawbacks and advantages for pulmonary MRA; depending on the individual question that has to be answered, one or the other is advantageous. The best results have been obtained using time-of-flight techniques (FOO et al. 1992; WIELOPOLSKI et al. 1992).

14.2.2 Two-dimensional Time-of-flight Techniques

With 2D techniques, images can be acquired during a single breath-hold, so that there is less blurring from respiratory motion (HATABU et al. 1989, 1992), and high-resolution single slices can be obtained. However, studies using breath-hold techniques require the acquisition of multiple images to include the entire lung and therefore they are prone to misregistration on the projection images from variant locations during breath-holding. In addition, long breath-holds are not practical in patients with pulmonary embolism, who may be in respiratory distress.

Vascular enhancement in 2D scans is superior to that in 3D scans due to inflow of unsaturated spins (KELLER et al. 1989). If cardiac triggering is used, maximum arterial inflow enhancement is possible and movement of pulmonary vessels is reduced in cardiac cycles. Flow contrast to noise is maximal at end-systole (VON SCHULTHESS et al. 1985). However, saturation effects in vessels running parallel to the imaging plane, especially at end-diastole, may mask the presence of an embolus.

Attempts to improve vascular signal have included the use of intravascular contrast agents which reduce the T1 relaxation of blood and allow for maximum vessel enhancement independent of inflow effects and provide much better blood/thrombus contrast (RUBIN et al. 1992; DOUEK et al. 1993).

However, 2D sequences are prone to susceptibility artifacts because of the close relation of pulmonary vessels and the bronchial tree. This problem can be partially overcome by the use of short echo times, thus allowing less signal dephasing to occur.

14.2.3 Three-dimensional Time-of-flight Techniques

Three-dimensional techniques are not as prone to signal loss from susceptibility and offer advantages in resolution and image quality. 3D acquisitions were not used initially for pulmonary angiography because of their higher sensitivity to motion artifacts. Initial attempts were made with a single breath-hold but signal to noise was low when using the body coil, requiring multiple averaging. Other approaches have employed multiple breath-holds to reduce motion artifacts (such as ghosting) and to increase signal to noise. However, the exact location of the breath-hold is difficult to reproduce, thus resulting in image blurring. Therefore, breath-holding is not suitable for 3D imaging.

Two different approaches, pseudogating and respiratory gating, have been developed to suppress motion artifact from breathing. With respiratory gating, the respiratory cycle is monitored and phase-encoding steps are acquired at the same point of the respiratory cycle. Since only a few phase-encoding steps can be acquired during one respiratory period, this is a slow process leading to extremely long scanning times. Thus, this technique has rarely been applied. With pseudogating (WIELOPOLSKI et al. 1986, 1992; HAACKE et al. 1987), each line of data is acquired synchronized with the respiratory period. If multiple lines are collected in the part of the respiratory cycle with least motion, scanning times can be reduced considerably. This has been employed by WIELOPOLSKI et al. (1992) with a syncopated FLASH 3D technique.

Data analysis of 3D acquisitions allows multiplanar reconstructions and maximum intensity projections. The pulmonary vessels can then be displayed in any orientation, reducing vessel overlap, which makes the interpretation of the images less ambiguous. This is a major advantage over conventional angiography, where only a limited number of views are obtained. Scanning time has been reduced to 8–13 min with sufficient signal to noise and resolution to visualize up to the fifth- and sixth-order branches of the pulmonary arterial tree.

14.2.4 Current Techniques

At present, there is still some controversy about the optimal parameters for 2D or 3D imaging of the pulmonary vessels when no contrast medium is applied. We use a 3D gradient-echo technique (syncopated 3D FLASH) without cardiac gating, respiratory triggering, or breath-holding (Fig. 14.1). A short echo time is recommended to reduce dephasing due to the high susceptibility between vessels and surrounding air and to avoid flow compensation. Imaging time is 10–13 min for the entire pulmonary vasculature

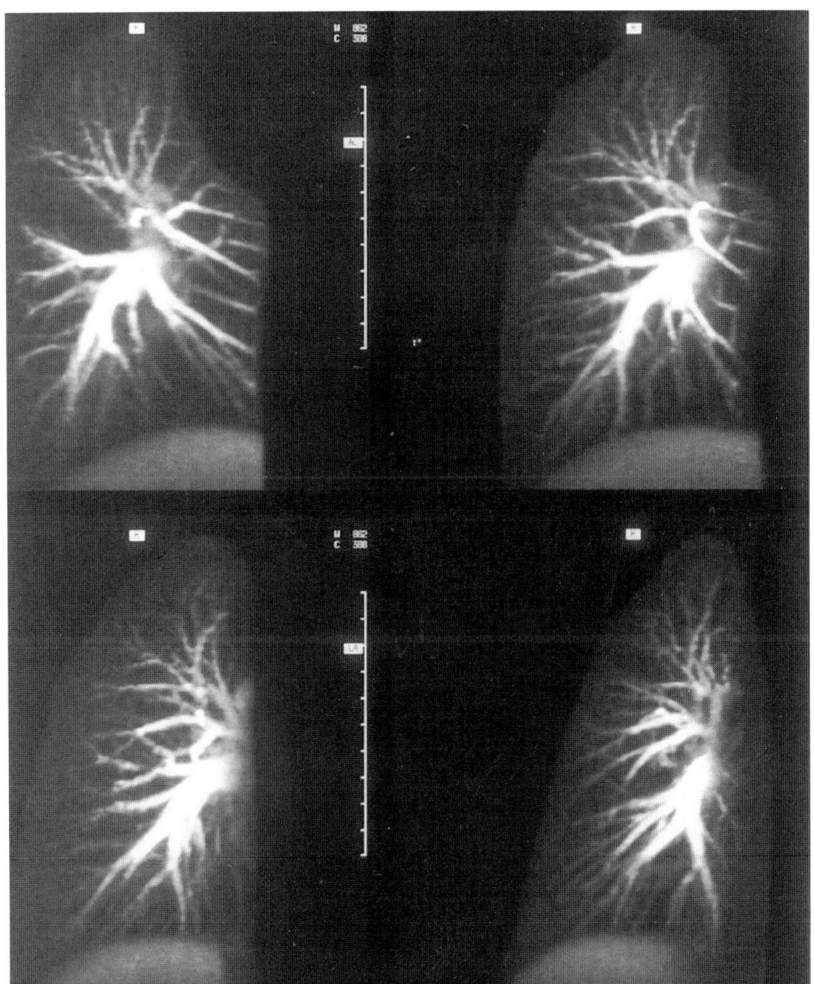

Fig. 14.1. 3D time-of-flight MRA. Sagittal acquisition in a healthy volunteer clearly demonstrating pulmonary vasculature in different maximum intensity projections

using six to eight averages in order to reduce respiratory motion artifacts. A thick sagittal slab is preferred since arteries benefit from inflow effects while veins are saturated because blood remains inside the imaging volume, thus reducing overlapping and ambiguity between arteries and veins. When respiratory gating is used, the signal from arteries and veins is comparable due to the long inflow times between each data collection that permit both arteries and veins to recover their full magnetization. Pulmonary arteries and veins can be imaged selectively by applying saturation slabs to suppress signal from veins or arteries, respectively. For thicker slices, these might be best applied outside the imaging volume; for thinner slices, they can be applied as traveling saturation pulses.

14.2.5 Signal to Noise

Signal to noise is low when a body coil is used. Several techniques have been introduced using various types of surface coils to increase signal to noise (Akins et al. 1986; Kulkarni et al. 1986; Hatabu et al. 1989; Hayes and Roemer 1990). Phased-array multiple receiver coils collect data simultaneously from two coils placed anteriorly and posteriorly over one lung (Hatabu et al. 1992; Hayes and Roemer 1990). Signal to noise for this approach increases up to 80% in comparison to the body coil (Hayes and Roemer 1990). In addition to these considerable improvements in signal to noise, surface coils permit the visualization of the pulmonary arteries down to sixth- to seventh-order branches, making it possible to acquire images with higher in-plane resolution in short imaging times (Hatabu et al. 1989). This can be advantageous in the detection of small peripheral thrombi, for which spatial resolution has been limited thus far.

14.3 Normal Anatomy

14.3.1 General Anatomy

The pulmonary arterial system begins at the pulmonary valve with the main truncus, which divides into the right and left branches within the pericardium. While the left ramus is a continuation of the main truncus, the right ramus is a side branch. Right and left rami then divide into the right and left arteries in the hila and travel with the bronchial tree into the lung. The right artery divides into an anterior truncus that supplies the upper lobe and a larger inferior truncus that supplies both the lower and middle lobes. The portion of the inferior artery that lies between the middle and inferior lobes is called the interlobar artery. On the left, the ramus divides into an upper and a lower truncus that supply the upper lobe and lower lobe/lingula, respectively. The intrapulmonary ramifications of the pulmonary arteries travel in the center of the lobules along with the bronchial tree (Greene 1989).

Unlike the pulmonary arteries, which travel at the center of the lobules, pulmonary veins travel at the margins along the septal and periacinar planes. Pulmonary veins may be distinguished from arteries by two characteristics: veins extend towards the left atrium and they do not accompany bronchi. Therefore, veins tend to take a more horizontal course in the lower lobes and a more vertical course in the upper lobes.

14.3.2 MRA Display of Anatomy and MRA Limitations

The pulmonary vascular tree, from the main to the subsegmental pulmonary arteries, has been demonstrated accurately with MRA in healthy volunteers (Julsrud et al. 1989; Wielopolski et al. 1992). However, there are certain limitations of pulmonary MRA in patients: Vascular enhancement with either 2D or 3D time-of-flight angiography depends on rates of blood flow and may be limited in patients with vascular occlusion, pulmonary hypertension, or low cardiac output, resulting in poor contrast between thrombus and flowing blood. Further contrast reduction comes from interfering respiratory motion which blends the signal from blood and thrombus. Unfortunately, this is particularly common in patients with acute pulmonary embolism who are very ill and dyspneic and are therefore unable to cooperate with the MR study. In interpreting 3D MIP reconstructions, it is important to be aware that clots evident on the individual slices may be obscured on standard MIP reconstructions.

MRA findings are sometimes difficult to interpret. A variety of circumstances can lead to signal loss mimicking embolism. These include saturation effects in vessels running in-plane, turbulent or slow flow, and susceptibility artifacts. The differentiation of clot from slow flow is sometimes possible using cine MRI, where flow can be demonstrated at some point during the cardiac cycle, or by using the SPAMM (spatial modulation of magnetization) technique (Axel and Dougherty 1989a,b). The SPAMM technique is a tagging technique in which stripes from tagging are maintained in stationary clot, while they are dispersed in flowing blood (HATABU et al. 1994). Using this technique, an accuracy of greater than 90% has been reported for the differentiation from slow flow.

14.4 Pathology

14.4.1 Pulmonary Embolism

14.4.1.1 General Pathology

Pulmonary embolism is an extremely common disorder and the most common pulmonary emergency in hospitalized patients (KELLEY et al. 1991). It accounts for approximately 50000 deaths per year in the United States (WESSLER 1976). Even though early detection is crucial, the diagnosis is missed in 60%–70% of fatal cases, as shown by autopsy studies.

Pulmonary emboli generally originate in the deep veins of the leg. The incidence of lower extremity thrombosis among patients with thromboemboli ranges from 80% to 100%, but it is clinically evident in only 50% (KISTNER et al. 1972). It is important to recognize that emboli can also arise from the right heart and the pelvic veins and, less often, from axillary, heptic, and renal veins. The formation of thrombus is related to stasis of slow-flowing blood in the extremities, damage to the vessel wall, and alterations in the coagulation mechanism. Stasis is quite common, resulting from bed rest in the postoperative period, congestive heart failure, or even prolonged periods of sitting during ground or air travel.

No single symptom, combination of symptoms, or sign is diagnostic of pulmonary embolism. In a series of 327 patients, the following symptoms were observed: chest pain (88%), pleuritic in 74% of cases, dyspnea (84%), apprehension (59%), and cough (53%). Hemoptysis (30%), diaphoresis (27%), and syncope (13%) were present in only a minority of patients. The most common sign was tachypnea (92%), followed in frequency by rales (58%), increased pulmonic second sound (53%), tachycardia (44%), and fever (43%). Gallop rhythms (34%), phlebitis (32%), edema (24%), murmur (23%), an cyanosis (19%) were less common. The classic triads of hemoptysis, cough, and diaphoresis; hemoptysis, chest pain, and dyspnea; or dyspnea, chest pain, and apprehension were uncommon (BELL et al. 1977).

An integral part of the workup of pulmonary embolism is the evaluation of the deep veins of the lower extremities (KELLEY et al. 1991; STEIN et al. 1993) since most therapy is directed toward preventing recurrent emboli, rather than treating the clot already present in the lung (except in patients with large clots causing cardiopulmonary compromise). Therefore, the presence of deep venous thrombosis is more important for treatment than the detection of small emboli in the lung. Initial results reported a sensitivity of 100% and a specificity of 90% for clot detection in deep venous thrombosis using MR (SPRITZER et al. 1990). In the future, it may even be possible to combine an MR evaluation of the lung with a scan of the deep venous system.

Anticoagulant treatment for pulmonary embolism is highly effective and known to improve outcome, by reducing mortality from 30% to less than 10% (DALEN and ALPERT 1975). Management is determined by the degree to which the circulation has been compromised. Initial therapeutic strategies involve heparinization when the diagnosis is first suspected (SASAHARA et al. 1992). If the diagnosis is established, the treatment can be continued or changed to thrombolytic agents; the latter were demonstrated to be superior to heparin alone when clot resolution was estimated by pulmonary angiography (SASAHARA et al. 1972; Urokinase Streptokinase Pulmonary Embolism Trial 1974). Patients with massive embolism should receive thrombolytic therapy for 24h. In case of contraindication or failure to respond to thrombolysis and/or vigorous pressor support, emergency surgery with pulmonary embolectomy and IVC interruption should be performed (GOLDHABER 1985). In patients with contraindications to anticoagulation, or in whom recurrent emboli develop despite adequate anticoagulation, the insertion of an IVC filter should be considered.

14.4.1.2 MRA Findings

On "bright blood" MRA sequences, pulmonary emboli are usually seen as low-intensity regions along the arterial tree compared to nearby flowing blood. Central emboli can be depicted easily, if of such low signal intensity (Fig. 14.2a). But signal intensity from thrombus is variable and is a major problem in pulmonary MRA. Depending on its age, thrombus may appear as a complete signal void or as iso- or hyperintense compared to the surrounding blood. If acute emboli appear bright, they can be confused

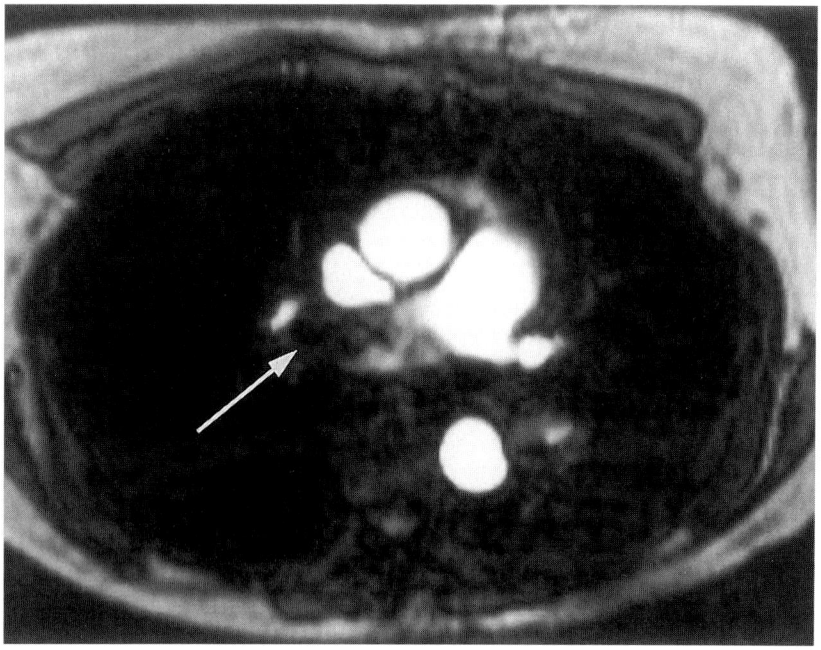

a

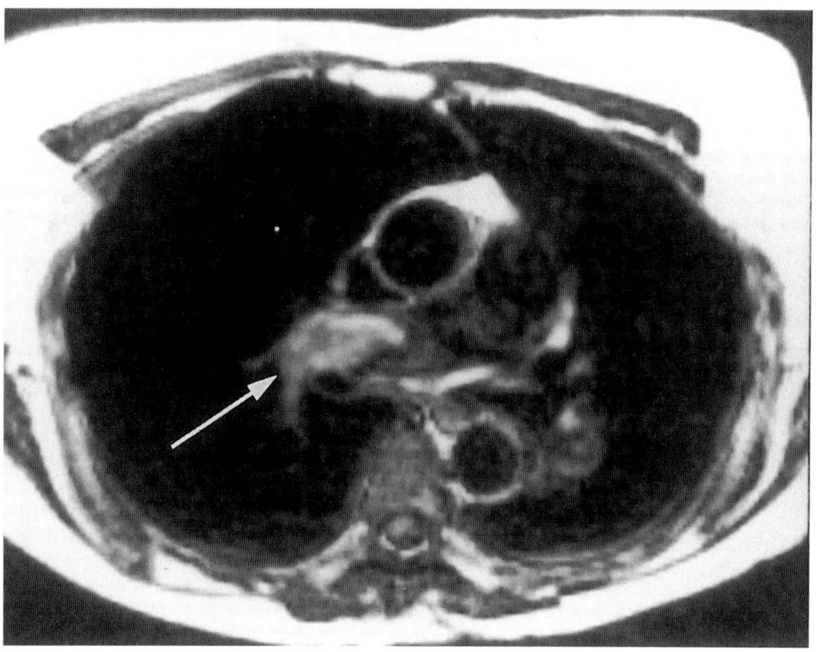

b

Fig. 14.2a,b. Pulmonary embolism. **a** Cine scan demonstrating signal loss in the right pulmonary artery due to pulmonary embolism (*arrow*). **b** T1-weighted spin-echo image. Embolus appears as a region of higher signal intensity (*arrow*) compared to flowing blood

with flowing blood on "bright blood" MRA sequences. On the other hand, on spin-echo sequences, pulmonary emboli appear bright, in contrast to flowing blood, which itself produces a signal void (Fig. 14.2b).

By adding an inversion pulse to a 3D "bright blood" angiographic sequence, strong T1 weighting can be provided to help distinguish thrombi from surrounding tissue and blood. With this technique, an inversion time TI can be chosen such that signal from flowing blood is reduced, while signal from thrombus may appear enhanced (Fig. 14.3). These images can then be subtracted from 3D syncopated FLASH images in order to depict the thrombus as a more conspicuous filling defect (Wielopolski 1993).

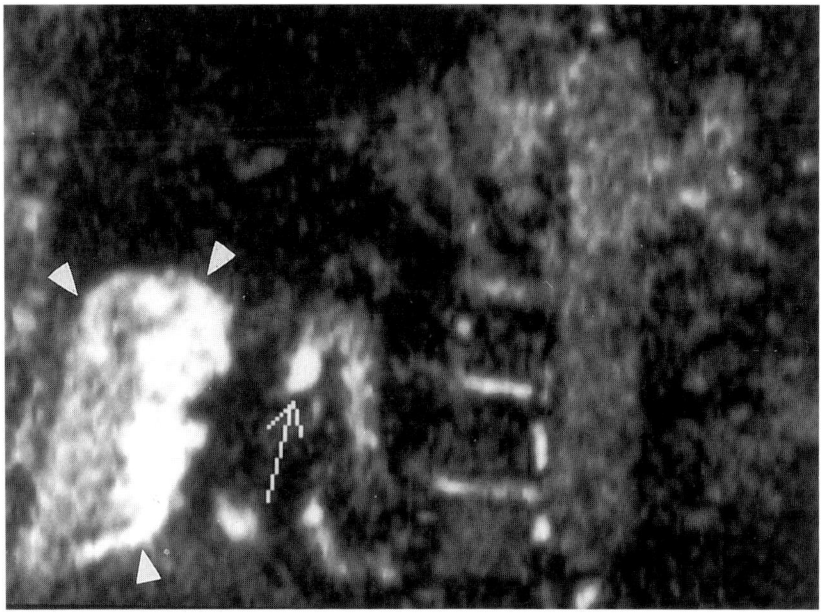

Fig. 14.3. 3D IR FLASH. Small focus of high signal intensity in the right descending pulmonary artery due to pulmonary embolism (*arrow*). Large region of increased signal intensity, laterally, caused by pleural effusion (*arrowheads*)

THAKUR et al. (1990) described the use of ferrosome, a superparamagnetic agent, for the detection of pulmonary embolism in an animal experiment. This substance eradicates the signal from lung by shortening the T2 of blood more than that of clot, making pulmonary embolism more discernible.

The differentiation of acute from chronic pulmonary embolism is also problematic. POSTERANO et al. (1989) reported encouraging results using a cine pulse sequence with limited flip angle, in combination with a standard spin-echo sequence for this purpose. Diagnostic criteria for acute emboli included intraluminal filling defects or abrupt vessel cutoffs with curvilinear capping of the high signal intensity blood column. Chronic embolism was diagnosed when abrupt vessel cutoffs without capping or webs or coarctations were present.

14.4.1.3 Comparison of MRA with Other Imaging Modalities

The sensitivity of MRA for detection of acute central pulmonary emboli varies between 92% and 100% (with pulmonary angiography as the "gold standard"); the specificity was 62% for a 2D time-of-flight sequence (GRIST et al. 1993). The specificity of MRA is thus intermediate between pulmonary angiography (100%, by definition) and ventilation-perfusion scintigraphy (10% at 100% sensitivity level, 52% at 82% sensitivity level) (The PIOPED Investigators 1990). The low specificity of MRA is due to frequent false-positive results in patients with atelectasis or previous pulmonary embolism. Spurious results may also be produced by signal loss from turbulent or slow flow, saturation effects in vessels running parallel to the imaging plane, or magnetic susceptibility artifacts mimicking thrombus (as discussed in Sect. 14.3.2).

Cine MRI techniques have a reported 82% sensitivity, 92% specificity, and 80% accuracy in the detection of chronic, central emboli (GEFTER et al. 1990). Computed tomography (CT) has been demonstrated to be superior to pulmonary angiography in the detection of chronic, proximal clot (REMY-JARDIN et al. 1992), whereas angiography is more sensitive in the detection of vascular stenoses and distortions. With the availability of ultrafast and spiral CT, CT examination of the pulmonary vessels are possible in one breath-hold. These images can be reformated in any desired orientation.

Both techniques have their advantages and drawbacks. Advantages of CT over MRI are faster scanning times, improved contrast to noise, higher spatial resolution, the ability to evaluate pulmonary parenchyma, and easier patient monitoring (GEFTER et al. 1993). MRA, on the other hand, does not require the administration of intravenous contrast media, allows selective imaging of the pulmonary arteries, and may incorporate the evaluation of the deep venous system of the legs into a single examination.

14.4.2 Other Entities

Other indications for which pulmonary MRI has proved useful include: evaluation of the pulmonary arteries in patients with congenital heart disease (Gomes et al. 1990), arteriovenous malformations (AVMs; Dinsmore et al. 1988), sequestration (Oliphant et al. 1987), scimitar syndrome (Baxter et al. 1990), and pulmonary hypertension. AVMs cause a flow-related signal void on spin-echo images which is indistinguishable from the surrounding lung. With MRA, these lesions appear as high signal intensity lesions due to the presence of flowing blood. Pulsatility of the lesion may be found on dynamic cine images during the cardiac cycle (Dinsmore et al. 1988).

In MRA of pulmonary hypertension the following findings are seen: dilatation of the central pulmonary vessels with attenuation of peripheral vessels, and loss of normal systolic distension and diastolic collapse. The pulmonary tree demonstrates tortuous vessels with peripheral pruning. Signal intensity in the pulmonary artery normally varies depending on the point in the cardiac cycle. These variations in signal intensity between systole and diastole are not found in patients with pulmonary hypertension, due to decreased distensibility of the central vessels (Gefter et al. 1990; Bogren et al. 1989), and intravascular signal is high in the proximal pulmonary arteries during systole (von Schulthess 1985). Flow patterns can be investigated using velocity-encoded cine MRI as a method for estimating pulmonary vascular resistance. In patients with pulmonary hypertension, flow has lower peak systolic velocity and there is more retrograde flow compared to healthy volunteers (Bogren et al. 1989; Kondo et al. 1992).

14.5 Conclusion

Magnetic resonance angiography has been used to investigate the pulmonary vasculature with some success. However, the evaluation of pulmonary MR angiograms requires considerable experience. Particularly in the setting of pulmonary embolism, differentiation between artifacts or vascular abnormalities related to chronic lung disease and acute thrombus may be difficult. In addition, small emboli may be indistinguishable from mucous plugs or collapsed lung because the arteries run parallel to the bronchi. Though preliminary results for the detection of central emboli are encouraging, the overall results of MRA for acute pulmonary embolism do not yet establish this technique as a screening test.

With new contrast agents the acquisition of MR-based ventilation/perfusion scans has become possible. Mismatched perfusion defects have been demonstrated in animal models using polylysine-gadolinium-DTPA as intravenous and gadopentetate dimeglumine as bronchial contrast media (Berthezene et al. 1992). Such studies may prove useful adjuncts to directed MRA in patients with regional abnormalities on standard ventilation-perfusion scan.

References

Akins EW, Fitzsimmons JR, Mancuso AA et al. (1986) Double loop receiver coil for MR imaging at 0.15 T. J Comput Assist Tomogr 10:1083–1088

Axel L, Dougherty L (1989a) MR imaging of motion with spatial modulation of magnetization. Radiology 171:841–845

Axel L, Dougherty L (1989b) Heart wall motion: improved method of spatial modulation of magnetization for MR imaging. Radiology 172:349–350

Baxter R, McFadden PM, Gradman M, Wright A (1990) Scimitar syndrome: cine magnetic resonance imaging demonstration of anomalous pulmonary venous drainage. Ann Thorac Surg 50:121–123

Bell W, Simon TL (1976) A comparative analysis of pulmonary perfusion scans with pulmonary angiograms. From a national cooperative study. Am Heart J 92:700–706

Bell WR, Simon TL, DeMets DL (1977) The clinical features of submassive and massive pulmonary emboli. Am J Med 62:355–360

Berthezene Y, Vexler V, Clement O et al. (1992) Contrast-enhanced MR imaging of the lung: assessment of ventilation and perfusion. Radioloy 183:667–672

Biello DR, Matter AG, McKnight RC et al. (1979) Ventilation-perfusion studies in suspected pulmonary embolism. AJR 133:1033–1037

Bogren HG, Klipstein RH, Mohaiaddin RH et al. (1989) Pulmonary artery distensibility and blood flow patterns: a magnetic resonance study of normal subjects and of patients with pulmonary arterial hypertension. Am Heart J 118:990–999

Dalen JE, Alpert JS (1975) Natural history of pulmonary embolism. Prog Cardiovasc Dis 17:259–270

Dinsmore BJ, Gefter WB, Hatabu H, Kressel HY (1988) Pulmonary arteriovenous malformations: diagnosis by gradient refocused MR imaging. J Comput Assist Tomogr 12:560–564

Douek PC, Loubeyre P, Delignette A et al. (1993) Ultrafast MR imaging of the major pulmonary vessels with Gd-DTPA. In: Society of Magnetic Resonance Imaging 1993, San Francisco, p 346

Foo TKF, MacFall JR, Hayes CE et al. (1992) Pulmonary vasculature: single breathhold MR imaging with phased array coils. Radiology 183:473–477

Gefter WB, Hatabu H, Dinsmore BJ et al. (1990) Pulmonary vascular cine MR imaging: a noninvasive approach to dynamic imaging of the pulmonary circulation. Radiology 176:761–770

Gefter WB, Krishanu BG, Holland GA (1993) MR, CT enhance diagnosis of pulmonary emboli. Diagn Imaging Int 38–43

Goldhaber SZ (1985) Strategies for management. In: Goldhaber SZ (ed) Pulmonary embolism and deep venous thrombosis. Saunders, Philadelphia, pp 121–134

Gomes AS, Lois JF, Williams RG (1990) Pulmonary arteries: MR imaging in patients with obstruction of the right ventricular outflow tract. Radiology 174:51–57

Greene RE (1989) Anatomical and functional basis of imaging the respiratory system. In: Taveras JM, Ferrucci JT (eds) Radiology: diagnosis-imaging-intervention, vol 1. Lippincott, Philadelphia, chapter 39

Grist TM, Sostman HD, MacFall JR et al. (1993) Pulmonary angiography with MR imaging: preliminary clinical experience. Radiology 189:523–530

Haacke EM, Lenz GW, Nelson D (1987) Pseudogating: elimination of periodic motion artifacts in MRI without gating. Magn Reson Med 4:162–174

Hatabu H, Gefter WB, Kressel HY et al. (1989) Pulmonary vasculature: high resolution MR imaging. Radiology 171:391–395

Hatabu H, Gefter WB, Listerud J et al. (1992) Pulmonary MR angiography utilizing phased-array surface coils. J Comput Assist Tomogr 16:410-417

Hatabu H, Gefter WB, Axel L et al. (1994) MR imaging with spatial modulation of magnetization in the evaluation of chronic central pulmonary thromboemboli. Radiology 190:791–796

Hayes CE, Roemer PB (1990) Volume imaging with MR phased arrays. Magn Reson Med 16:181–191

Julsrud PR, Ehman RL, Hagler DJ, Ilstrup DM (1989) Extracardiac vasculature in candidates for Fontan surgery: MR imaging. Radiology 173:503–506

Keller PJ, Draywe BP, Fram EK et al. (1989) MR angiography via 2D acquisitions but yielding a 3D display: a work in progress. Radiology 173:527–532

Kelley MA, Carson JL, Palevsky HI et al. (1991) Diagnosing pulmonary embolisms: new facts and strategies. Ann Intern Med 114:300–306

Kistner RL, Ball JJ, Nordyke RA, Freeman GC (1972) Incidence of pulmonary embolism in the course of thrombophlebitis of the lower extremities. Am J Surg 124:169–176

Kondo C, Caputo GR, Takayuki M et al. (1992) Pulmonary hypertension: pulmonary flow quantification and flow profile analysis with velocity-encoded cine MR imaging. Radiology 183:751–758

Kulkarni MV, Patton JA, Price RR (1986) Technical considerations for the use of surface coils in MRI. AJR 147:373–378

Mill SR, Jackson DC, Older RA et al. (1980) The incidence, etiologies, and avoidance of complications of pulmonary angiography in large series. Radiology 136:295–299

Oliphant L, McFadden RG, Carr TJ, Mackenzie DA (1987) Magnetic resonance imaging to diagnose intralobar pulmonary sequestration. Chest 19:500–502

The PIOPED Investigators (1990) Value of ventilation/perfusion scan in acute pulmonary embolism: results of the Prospective Investigation of Pulmonary Embolism Diagnosis (PIOPED). JAMA 263:2753–2759

Posterano RH, Sostman HD, Spritzer CE, Herfkens RJ (1989) Cine-gradient-refocused MR imaging of central pulmonary emboli. AJR 152:465–468

Remy-Jardin M, Remy J, Wattinne L, Girand F (1992) Central pulmonary thromboembolism: diagnosis with spiral volumetric CT with a single-breath-hold technique: comparison with pulmonary angiography. Radiology 185: 381–387

Rubin GD, Herfkens RJ, Napel SA et al. (1992) Breath-hold pulmonary MR angiography: comparison of imaging strategies (abstract). Radiology 185:217

Sasahara AA, McIntyre KM, Belko JS, Hyers TM (1972) A national cooperative trial of thrombolysis in pulmonary embolism: phase I results of urokinase therapy. J La State Med Soc 124:130–136

Sasahara AA, McIntyre KM, Cella G, Palla A, Sharma GVRK (1992) Venous thromboembolism and chronic venous disorders. In: Loscalzo J, Creager MA, Dzau VJ (eds) Vascular medicine. Little, Brown and Co, Boston, pp 1049–1074

von Schulthess GK, Fisher MR, Higgins CB (1985) Pathologic blood flow in pulmonary vascular diseases as shown by gated magnetic resonance imaging. Ann Intern Med 103:317–323

Simon M (1989) Radiology of pulmonary hemodynamic disturbances and thromboembolism. In: Taveras JM, Ferrucci JT (eds) Radiology: diagnosis-imaging-intervention, vol 1. Lippincott, Philadelphia, chapter 71

Spritzer CE, Sostano HD, Wilkes DC, Coleman RE (1990) Deep venous thrombosis: experience with gradient-echo MR imaging of 66 patients. Radiology 177:235–241

Stein PD, Hult RD, Saltzman HA, Pirnov G (1993) Strategy for diagnosis of patients with suspected acute pulmonary embolism. Chest 103:1553–1558

Thakur ML, Vinitski S, Mitchell DG el al. (1990) MR imaging of pulmonary parenchyma and emboli by paramagnetic and superparamagnetic contrast agents. Magn Reson Imaging 8:625–630

Urokinase Pulmonary Embolism Trial Study Group (1970) Urokinase pulmonary embolism trial: phase I results. A cooperative study. JAMA 214:2163–2172

Urokinase streptokinase pulmonary embolism trial: phase II results (1974) JAMA 229:1606–1612

Wessler S (1976) Venous thrombosis: scope of problem. In: Prophylactic therapy of deep vein thrombosis and pulmonary embolism. DHEW Publication No. 76-866, pp 1–10

Wielopolski PA (1993) Pulmonary arteriography. MRI Clin North Am 1:295–313

Wielopolski PA, Haacke EM, Adler LP (1986) Evaluation of the pulmonary vasculature with three-dimensional magnetic resonance imaging. Med Phys 13:794–805

Wielopolski PA, Haacke EM, Adler LP (1992) Three dimensional MR pulmonary vascular imaging: preliminary experience. Radiology 183:465–472

15 The Abdominal Aorta and Its Branches

I.P. ARLART and L. GUHL

CONTENTS

15.1 Introduction

The abdominal aorta and its branches are responsible for transporting blood into the arteries of the pelvis and lower extremities and maintaining the parenchymal function of the liver, spleen, pancreas, adrenals, and kidneys. Angiography visualizes in a highly specific way not only primary vascular lesions of different anatomical areas of the abdomen but also parenchymal lesions of different etiology. Thus, conventional x-ray angiography was considered the method of choice for many years before newer noninvasive technologies became available such as ultrasonography (US) and computed tomography (CT). These methods were introduced successfully into the diagnotic armamentarium for evaluating parenchymal and perivascular disorders as well as vascular abnormalities, particularly in vessels with larger diameters. Contrast-enhanced CT and color Doppler US have in fact revolutionized the diagnostic approach in abdominal vessel disease. Due to the high accuracy of US and CT in these areas, indications for angiography have been reduced and changed significantly. Today, abdominal angiography still plays a prominent role as a highly accurate diagnostic procedure in vaso-occlusive disease, renovascular disease, and gastrointestinal bleeding that cannot be localized by endoscopy, particularly in cases in which therapeutic interventions can be performed during the same procedure by the radiologist (i.e., recanalization by administration of fibrinolytic agents or transluminal angioplasty or implantation of vascular stents, embolization therapy using different agents and materials, and localized administration of chemotherapeutics). Furthermore, angiography has proved useful for preoperative vascular mapping in order to demonstrate abnormal vascular supply or intraparenchymal vasculature prior to local tumor resection.

Noninvasive MRA has been introduced into the abdominal area to a very restricted extent compared with its intracranial application. Although like angiography, CT, and US, MR sequences can yield intraluminal flow information and visualize the perivascular space, MRA for abdominal arteriography has achieved only limited acceptance: its application has been considered not clinically useful due to low spatial resolution, severe artifact problems, and the lack of consensus concerning pulse sequences.

15.2 Technical Considerations

In the abdomen, problems may arise with the use of vascular MRI due to susceptibility artifacts and motion artifacts caused by respiration, pulsatile vessels, and bowel movement. Furthermore, dephasing effects and saturation effects can be observed which reduce image quality and flow signal. Concentration of different vascular structures in the area of interest usually causes arteriovenous overlap, which may impair the diagnostic accuracy of arterial MRA. In contrast to MR examination of the thoracic aorta, in the abdomen flow-encoded SE sequences play only a secondary role. Although it has been demonstrated that in T1-weighted SE images aortic wall structures and perivascular space can be evaluated accurately

---

I.P. ARLART, Professor, Dr., Ärztlicher Direktor, Radiologisches Institut, Katharinenhospital, Kriegsbergstraße 60, 70174 Stuttgart, FRG
L. GUHL, Dr., Radiologisches Institut, Katharinenhospital, Kriegsbergstraße 60, 70174 Stuttgart, FRG

due to high resolution and contrast (Anderson et al. 1993) (for typical sequence parameters see Fig. 15.8), the value of this sequence is limited particularly by the unclear findings which may occur when using bright blood gradient-echo (GRE) MRA. With regard to the use of bright blood MRA in the abdomen it has become obvious that the application of a sole standard type of pulse sequence is not able to yield satisfactory diagnostic images of the different vascular structures. Thus, a variety of examination protocols have to be employed on the basis of two-dimensional and three-dimensional time-of-flight (TOF) and phase-contrast (PC) techniques which can be used alone or in combination.

15.2.1 Abdominal Aorta and Iliac Arteries

Currently, the most useful MR technique for imaging the entire abdominal aorta seems to be a breath-hold 2D TOF GRE fast low-angle shot (FLASH) sequence by which most vessels of interest are covered in a coronal acquisition plane by a relatively small number of slices within a short acquisition time (Arlart et al. 1992; Edelman et al. 1989). Following a transverse localizer slice through the abdominal area of interest, a series of GRE images will be acquired with a conventional multislice technique at the end of expiration after a breath-hold using a first-order (velocity) flow compensation in frequency- and section-encoding directions. It is recommended that TR should be as short as possible, bearing in mind that an appropriate balance should be achieved between TR and the flip angle. A 30–40° flip angle has been shown to be most useful in order to optimize flow signal and to limit pulsatility "ghosts" in the phase-encoding direction when using 1.5-T equipment. T1 relaxation times are shorter at low field strength, necessitating use of larger flip angles than at 1.5 T to achieve satisfactory signal suppression for stationary tissues.

Optimal parameters will vary according to field strength and machine. Typical sequence parameters used in our institution can be taken from Fig. 15.1. Usually up to 20 sections should be acquired with a 20%–30% overlap, and projection angiograms will be established from individual slices using a maximum-intensity projection (MIP) algorithm. Eighteen projections can be reconstructed with rotation angles of 3° along the long axis of the body. Signal void in the lower part of the abdominal aorta may occur due to in-plane saturation. Thus, acquisition of sections parallel to the aortic flow direction should be avoided by selection of image planes in a 10° coronal-axial angle. Additional tranverse sections are recommended for accurate evaluation of aortic diameter, intraluminal flow utilizing maximization of inflow, and aortic wall structures. In coronal plane acquisition, presaturation of venous flow has been shown to be of only limited value in the abdomen due to the complexity of the vessel anatomy. However, when tranverse acquisition is used for imaging of renal arteries or iliac arteries, venous flow signal can be eliminated by selective presaturation pulses and traveling saturation pulses, respectively, which are incorporated in the sequence. Thus, for examination of iliac arteries and the aortic bifurcation, a flow-compensated single-slice non-breath-hold 2D TOF GRE FLASH sequence can be recommended. The integrated traveling saturation pulse should be positioned 5–10 mm below the imaging section and moves at a constant distance with the imaging section during acquisition. Transverse slice acquisition utilizes an arterial inflow direction nearly perpendicular to the section. Typical sequence parameters are shown in the legend to Fig. 15.3. Usually, 50–70 consecutive slices cover the iliac arteries that will be used for reconstruction of projection angiograms. Diagnosis should be made from both the MIP angiograms and the individual slices.

It has to be recognized that signal void may occur due to in-plane saturation under turbulent or slow blood flow conditions as well as in tortuous iliac arteries. In these cases acquisition of thin slices perpendicular to the flow direction, i.e., oblique axial images, or slice acquisition during systolic peak flow is an alternative means of improving intravascular signal intensity. Furthermore, image quality can be impaired by motion artifacts due to respiratory and bowel movement. Breath-holding may be useful when imaging of infrarenal parts of the abdominal aorta and the bifurcation is required. In order to exclude misdiagnosis it should be borne in mind that focal signal void can be observed when surgical clip material is present, and that false-positive signal may occur due to perivascular fat, ascites, methemoglobin within a thrombus, and extravascular hematoma.

15.2.2 Abdominal Aortic Branches

The ideal angiographic MR technique for imaging the abdominal aortic branches, in particular visceral arteries and renal arteries, does not currently exist. Thus, different 2D and 3D MRA modalities may be applied on the basis of TOF or PC techniques, both of which have advantages and drawbacks. None of the sequences is able to visualize peripheral vascular structures or intrarenal arterial branches with satisfactory quality.

15.2.2.1 2D/3D TOF MRA Sequences

The number of renal arteries can be determined best by use of a coronal flow-compensated breath-hold 2D TOF GRE FLASH multislice sequence, whereas for the evaluation of flow within the renal arteries, transverse slice acquisition has been shown to be superior (Kim et al. 1990). The recommended sequence design is similar to that used for MRA of the abdominal aorta. Slice thickness should be 3–5 mm with 1 mm of overlap (for typical sequence parameters, see the legend to Fig. 15.11). Both systolic peak flow acquisition and the segmented k-space technique using a cardiac-gated breath-hold sequence seem able to increase flow signal and signal-to-noise ratio. In order to diminish overlapping of renal arteries by venous flow signal, the application of selective presaturation pulses in an oblique coronal direction through both kidneys and the inferior vena cava (IVC) is mandatory. However, passage of the presaturation band through parts of the heart should be avoided. Residual venous signal intensity can be overcome by rotation of the MIP angiograms and reduction of the number of individual sections included in the projection angiogram (targeted MIP). Alternatively, venous signal can be eliminated by application of a traveling saturation pulse which is included in the sequence (see above).

Compared with 2D TOF MRA, 3D TOF MRA displays a higher spatial resolution and less intravoxel dephasing. For imaging of the renal arteries a flow-compensated 3D GRE fast imaging sequence with steady-state precession (FISP) (Arlart et al. 1992; Loubeyre et al. 1993) can be recommended in which a tilted optimized nonsaturation excitation (TONE) technique should be included in order to reduce in-plane saturation effects (Purdy et al. 1992). Following a coronal localizer slice, transverse acquisition of a 75-mm slab with 64 partitions has proved useful within an examination time of 14 min (for typical sequence parameters see legend to Fig. 15.11). When larger volumes are requested, a double slab can be acquired. The upper border of the slab should be localized just above the origin of the renal artery to gain optimal signal from inflowing spins. Infrarenal presaturation bands should be localized through the IVC to eliminate venous flow signal and additionally through the anterior abdominal wall to reduce motion artifacts. Reconstructed MIP angiograms can be rotated along the long or cross axis of the body.

Major disadvantages of 3D TOF MRA are its sensitivity to motion artifacts, the fact that strong signal from stationary tissue with short T1 reduces vascular contrast, and the occurrence of progressive inflow saturation in a thick volume due to the long T1 of the blood. The latter phenomenon can be observed particularly in the presence of slow aortic blood flow due to low cardiac output or aneurysmal disease. The signal intensity of the abdominal aorta and renal arteries, which is reduced under these conditions, may be enhanced using a multiple overlapping thin slab acquisition (MOTSA) technique (Parker et al. 1991). Recently, 3D TOF MR angiograms of high contrast and quality have been presented using a sequence that includes selective inversion recovery (SIR) rapid acquisition with gradient-echo (RAGE), fat saturation, cardiac gating, and presaturation of venous flow (Li et al. 1994). In this sequence acquisition of blood inflow is maximized during systole, and intravoxel dephasing and pulsatile flow artifacts are minimized by collecting data during systole. As a result, the authors reported a reduction of vessel boundary blurring and ghosting artifacts due to background motion which will improve arterial lumen definition. Furthermore, the length of visualized renal arteries has been shown to be greater than with a conventional 3D TOF pulse sequence. More recently, a 3D subtraction TOF MRA technique has been published which includes slice acquisition during a single breath-hold of 18–30 s using signal targeting with alternating radiofrequency (STAR) and a scheme that combines echo planar imaging (EPI) readouts and k-space segmentation (Wielopolski et al. 1995). Preliminary results demonstrate excellent visualization of the renal arteries to the level of the intrarenal branch vessels.

15.2.2.2 2D/3D PC MRA Sequences

PC MRA has several advantages compared with TOF methods (Dumoulin et al. 1990). The source of contrast in phase-sensitive methods is velocity and does not rely on entry of relaxed spin magnetization into the imaged volume. The ability to encode for variable velocities allows the detection of slow flow in a large field of view. Flow velocities can also be quantitatively measured from 2D PC imaging, which may be of increasing interest for a number of clinical applications, including the determination of flow reduction behind a stenosis (AMANUMA et al. 1992; DEBATIN et al. 1994). Further advantages of PC MRA are excellent suppression of stationary background tissue and the ability to generate phase images in three different flow-encoding directions, facilitating separation of arteries from veins.

PC methods are only sensitive to the component of flow in the direction of the applied flow-encoding pulse. Hence, to obtain a total flow image, three acquisitions are needed. This feature can be used to distinguish overlapping vessels with orthogonal flow. 2D PC MRA provides fast images acquired within a single breath-hold following an SE localizer slice. Renal arteries can be examined in both axial and coronal planes. MIP angiograms are established along the slice direction. The major advantage of 3D PC MRA compared with 2D PC MRA is its superior image quality with respect to resolution and signal-to-noise ratio. Additionally, a retrospective view from any projection angle is possible, including the examination of subvolumes. For renal artery imaging, acquisition of a 10- or 20-cm slab including 60 and 128 partitions, respectively, is recommended using TR/TE of 23/10 ms, a flip angle of 20°, and a velocity-encoding gradient optimized on 30–80 cm/s. Acquisition time varies from 14 to 18 min depending on the selected matrix (RICHTER et al. 1993). MIP angiograms can be established in transverse and coronal oblique projections. The major limitations of 3D PC MRA are the long acquisition time and increased sensitivity to signal void from intravoxel spin phase dispersion due to turbulent flow, which may induce overestimation of stenotic disease.

15.2.2.3 Contrast-Enhanced MRA

In contrast-enhanced MRA the T1 of blood can be made short compared with that of fat, muscle, and other background tissues by the administration of a paramagnetic contrast agent. Since gadolinium-enhanced MRA does not depend on the inflow of unsaturated spins, the saturation problem that complicates routine TOF imaging is eliminated. Turbulent flow is not a problem. It is possible to image in any plane without concern over saturation. The image volume can be orientated to achieve optimal coverage of the vessels of interest with a minimum of imaging time. PRINCE (1994) reported that renal artery imaging is directly possible with excellent contrast, high signal-to-noise ratio, and high resolution in a three-dimensional Fourier-transform acquisition optimized for the shortened T1 of blood. With a gadolinium-chelate dose of 0.2 mmol/kg and a 3- to 5-min infusion during the image time, the arterial blood T1 is predicted to be 150–200 ms; in fact it will be shorter since the recirculation time is less than 3–5 min. Image time and infusion timing have a significant effect on image quality and on the arterial-to-venous enhancement ratio. Following a coronal T1-weighted localizer image for renal MRA a 3D spoiled GRE volume that is centered on the arteries of interest should be acquired during contrast infusion. Recommended imaging parameters include coronal acquisition plane, 9- to 12-cm sections with 60 partitions covering the arteries of interest, a TR/TE of 25/6.9 ms, a flip angle optimized at 40°, and first-order flow compensation (PRINCE 1994). No presaturation or breath-holding has to be performed. The resulting images are described as similar to those obtained with conventional angiography. Analysis should be carried out on MIP angiograms and on raw data or reformatted images.

15.3 Application of MRA

15.3.1 *Normal Anatomy and Variants*

The main sources of blood supply to the esophago-gastrointestinal tract are the celiac artery, the superior mesenteric artery (SMA), and the inferior mesenteric artery (IMA). In approximately 65% of persons the celiac artery divides into three branches, the splenic, left gastric, and hepatic arteries. In about 10% the right hepatic artery is replaced by a branch of the SMA. As regards the arterial supply of the kidneys, in about two-thirds of individuals a single right and single left renal artery can be found. Multiple renal arteries occur unilaterally in 32% and bilaterally in 12%. An independent superior renal pole artery can be observed in 7%, and an inferior pole artery in about 5%. Aberrant renal vascular supply may be expected in those with renal malformations or horseshoe kidneys. Atypical accessory supply via collaterals is seen in proximal renovascular occlusion or hypervascularized renal tumors. At the level of L4 the aorta bifurcates into the right and left common iliac arteries. The two common iliac arteries divide into the external and internal iliac arteries. Branches of the external iliac artery are the inferior epigastric artery and the deep iliac circumflex artery. Main branches of the internal iliac artery can be separated into vesical arteries, obturator artery, internal pudendal artery, iliolumbar arteries, and gluteal, uterine, and prostatic arteries (KADIR 1986).

Currently, the diagnostic value of arterial MRA in the abdomen and pelvis is still limited by a variety of problems which have not been adequately resolved. Major problems include a reduced spatial resolution, a reduced signal intensity in peripheral arteries with a small diameter and slow blood flow, in-plane saturation effects when using TOF techniques, susceptibility artifacts, and motion artifacts (ANDERSON et al. 1993). Because of these known drawbacks, peripheral or smaller tortuous abdominal arteries and intraparenchymal arteries cannot be visualized satisfactorily with MRA at present. Thus, the major clinical application of MRA has resided in the imaging of vessels with a larger lumen, i.e., the abdominal aorta, the main iliac arteries, the renal arteries, and the proximal visceral arteries. The abdominal aorta can be imaged in its complete course, including bifurcation and iliac arteries, whereas renal arteries can be visualized satisfactorily only up to the segmental bifurcation. For the examination of the entire abdominal aorta and iliac arteries, 2D TOF MRA sequences have been shown to be most useful; for imaging of aortic branches, 2D and 3D TOF techniques are still in competition with 2D and 3D PC techniques.

15.3.2 Pathology

Diagnostic information regarding the abdominal aorta and iliac arteries is required in arterial occlusive disease due to atherosclerosis, trauma, or embolization. Further indications for vascular imaging are aneurysms and aortic dissection. Imaging of renal arteries is of particular interest in hypertensive patients in order to ascertain whether high arterial blood pressure is of renovascular origin.

15.3.2.1 Aorto-iliac Occlusive Disease

Aorto-iliac occlusive disease is mainly caused by atherosclerosis, which can be observed particularly in males over 50 years of age. Cigarette smoking, arterial hypertension, and metabolic disorders such as diabetes mellitus accelerate atherosclerotic disease. Clinical symptoms such as intermittent thigh and buttock claudication, impotence, paresthesia, ischemic neural pain, or limb weakness are usually the background to the diagnostic examination. Progression of chronic occlusive atherosclerosis is slow and permits the development of extensive collaterals. Acute occlusions may be caused by trauma, thromboembolism, aneurysmal thrombosis, in situ thrombosis following atherosclerotic stenosis, metastatic tumors, coagulation disorders, and iatrogenic lesions due to catheter manipulations or surgical interventions. The prognosis of aortoiliac occlusive disease without management is poor due to its continuous progression. Surgical management includes thrombendarterectomy and diverse bypass procedures, in particular the implantation of aortoiliac or aortofemoral bifurcation grafts. Of major therapeutic importance today are percutaneous angioplasty and endovascular stenting. Diagnostic examination of occlusive or stenotic arterial disease requires characterization of the lesion and exact evaluation of the site and extent, the stenotic grade, the number of collaterals, and the status of adjacent vascular branches proximal and distal to the lesion.

Noninvasive ultrasonography (US) is generally accepted as a fast and simple method for detecting aortic lesions with high accuracy. Reliable localization of aortic stenosis or occlusion is possible, particularly when color Doppler is used. However, accurate US information in respect of the iliac arteries is obtained in only ca. 50% of cases. Limitations of US are related to body habitus, bowel gas, examination experience, and technical factors. Aortography employing a catheter is considered the gold standard due to its high accuracy; although it is an invasive procedure, the risk to the patient is reduced when a transfemoral approach is selected. However, when a transaxillary approach is necessary due to complete aortic occlusion, risks may increase. The major advantage of angiography is the possibility of immediate percutaneous management upon proof of occlusive disease. Since the introduction of non-invasive MRA for imaging of the abdominal region, vascular imaging problems have been greatly reduced by application of a flow-compensated 2D TOF GRE FLASH sequence in which sections are acquired within a breath-hold in the oblique coronal plane. Using this sequence, successful evaluation of occlusive aortic disease and postoperative follow-up has been reported (ARLART et al. 1992a; KIM et al. 1990) (Figs. 15.1, 15.2). In contrast, for imaging of the aortic bifurcation and iliac arteries, transverse slice acquisition with consecutive saturation of venous flow has been suggested to be superior since the slice

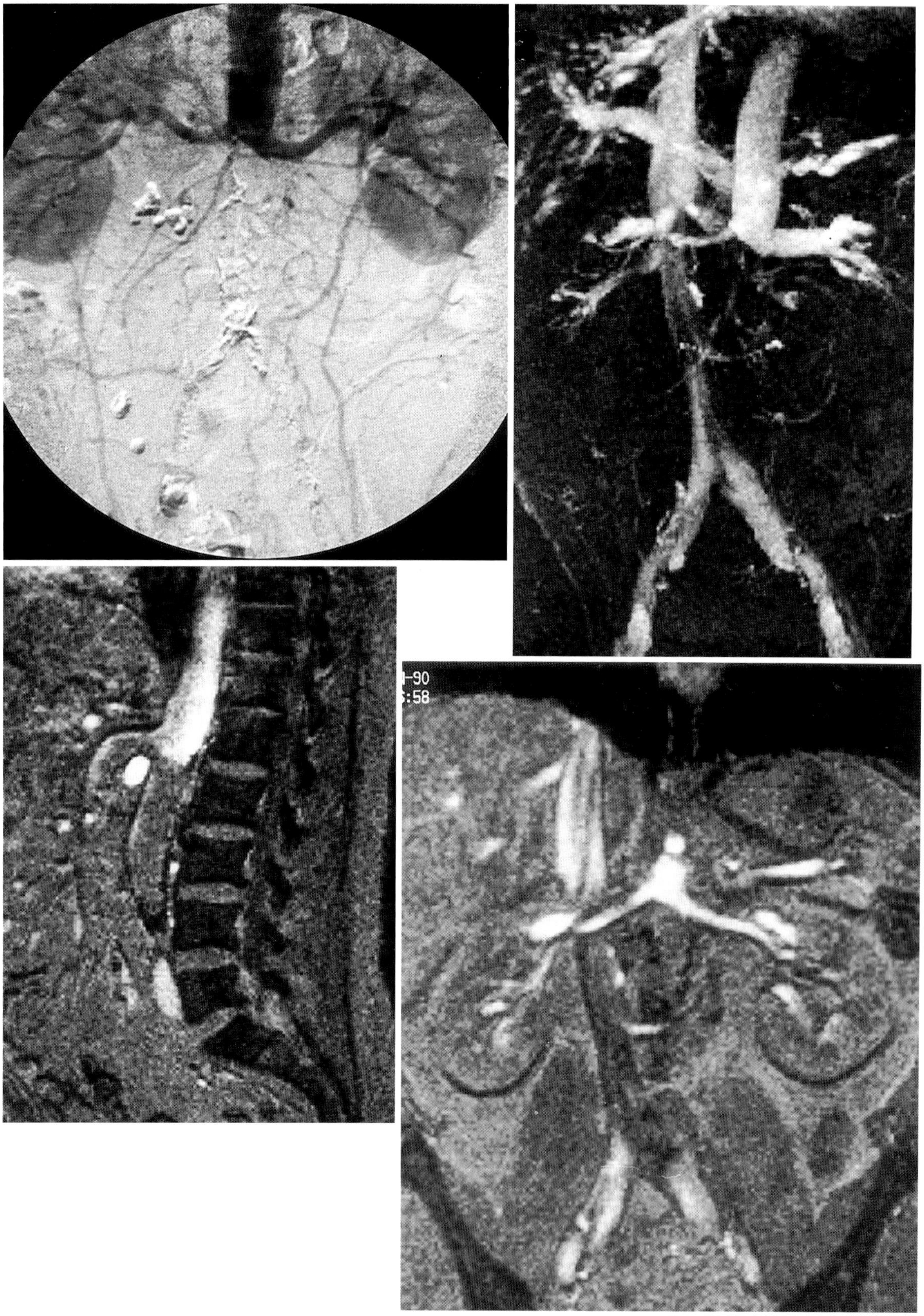

a b c d

Fig. 15.1a–d. Complete thrombotic occlusion of the infrarenal abdominal aorta. The diagnostic information on a DSA image (**a**) and a MR-projection angiogram (**b**) is similar (flow-compensated multislice 2D TOF GRE FLASH sequence, 1.5 T, body coil, coronal slice acquisition, three slices within a breath-hold of 15 s, TR/TE = 51/10 ms, flip angle = 30°, slice thickness = 5 mm with 1 mm of overlap, field of view = 420 mm, matrix = 256 × 256). In coronal (**c**) and sagittal (**d**) individual sections bright flow signal in both renal arteries and the SMA can be identified, as well as intermediate signal of the occlusive material within the aortic lumen

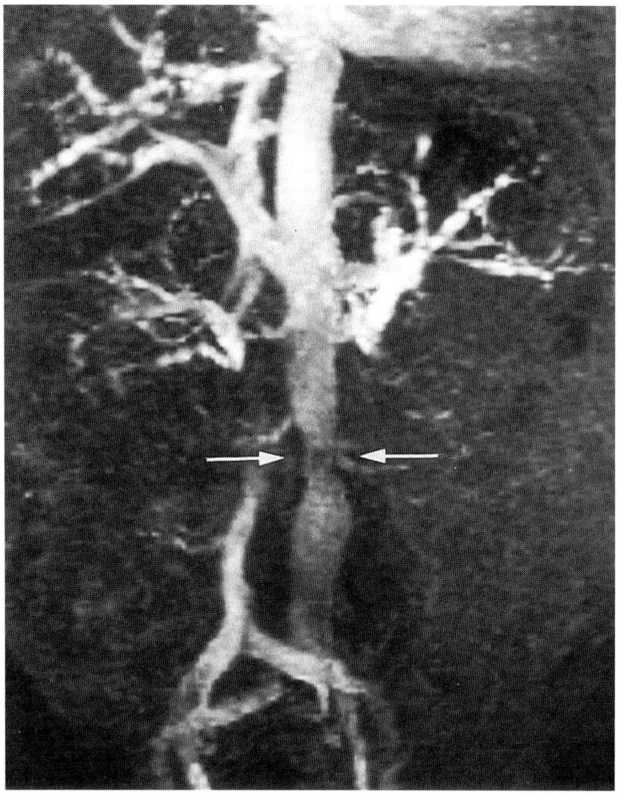

Fig. 15.2. Infrarenal aortic stenosis due to atherosclerosis. On a coronal 2D TOF MR projection angiogram the entire abdominal aorta can be evaluated, including the stenotic area (*arrows*) between renal arteries and the aortic bifurcation. Note the reduced signal distal to the stenosis

orientation is nearly perpendicular to the inflow direction and complete elimination of overlapping venous signal is achieved (Owen et al. 1993; Yucel et al. 1993b) (Figs. 15.3–15.5). On individual GRE images, plaque material usually appears dark and can be easily distinguished from the high signal of flowing blood (Fig. 15.1b). However, in cases with thrombus material containing methemoglobin, signal may be high and may mimic a patent lumen. This equivocal finding can be clarified by additional SE imaging. Yucel et al. (1993b) reported a sensitivity and specificity of 92% and 83%, respectively, in delineating iliac artery stenosis >50%, whereas in complete occlusions sensitivity may increase to 100%. When signal void occurs due to in-plane saturation in tortuous iliac arteries and slow or turbulent blood flow, for instance in aneurysmal disease, thin slice acquisition during systolic peak flow can be recommended to improve flow signal (Fig. 15.5c). One reported limitation of MRA, particularly when employing projection angiograms for the purpose of diagnosis, lies in the differentiation of severe stenosis from short occlusion. In order to reduce the risk of misdiagnosis, transverse individual sections have to be included in the analysis in such cases. Nevertheless, the accuracy of grading of iliac artery stenosis sometimes may be limited. In our experience there is a risk of overestimating stenotic disease when using a non-cardiac-gated sequence as well as a risk of underestimating it when using a systolic peak flow acquisition. Further limitations of iliac arterial MRA include signal void, which may occur due to motion artifacts and surgical clip material. Far more reliable MR angiograms of the iliac arteries can be obtained when a gadolinium-enhanced technique is used in that arterial signal can be acquired rapidly within some minutes and is independent of inflow (Prince et al. 1994). Figures 15.9c and d demonstrate excellent arterial imaging comparable to conventional angiograms.

15.3.2.2 Abdominal Aortic Aneurysm

Abdominal aortic aneurysm (AAA) can be proved in 2%–6% of autopsy studies. Atherosclerosis is responsible for more than 90% of AAAs. In more than 90% the infrarenal part of the aorta is involved, and in 30% the aneurysmal extension includes the iliac arteries (Kadir 1986). Aneurysms occur most frequently over the age of 60 years, and males are affected 10 times more frequently than females.

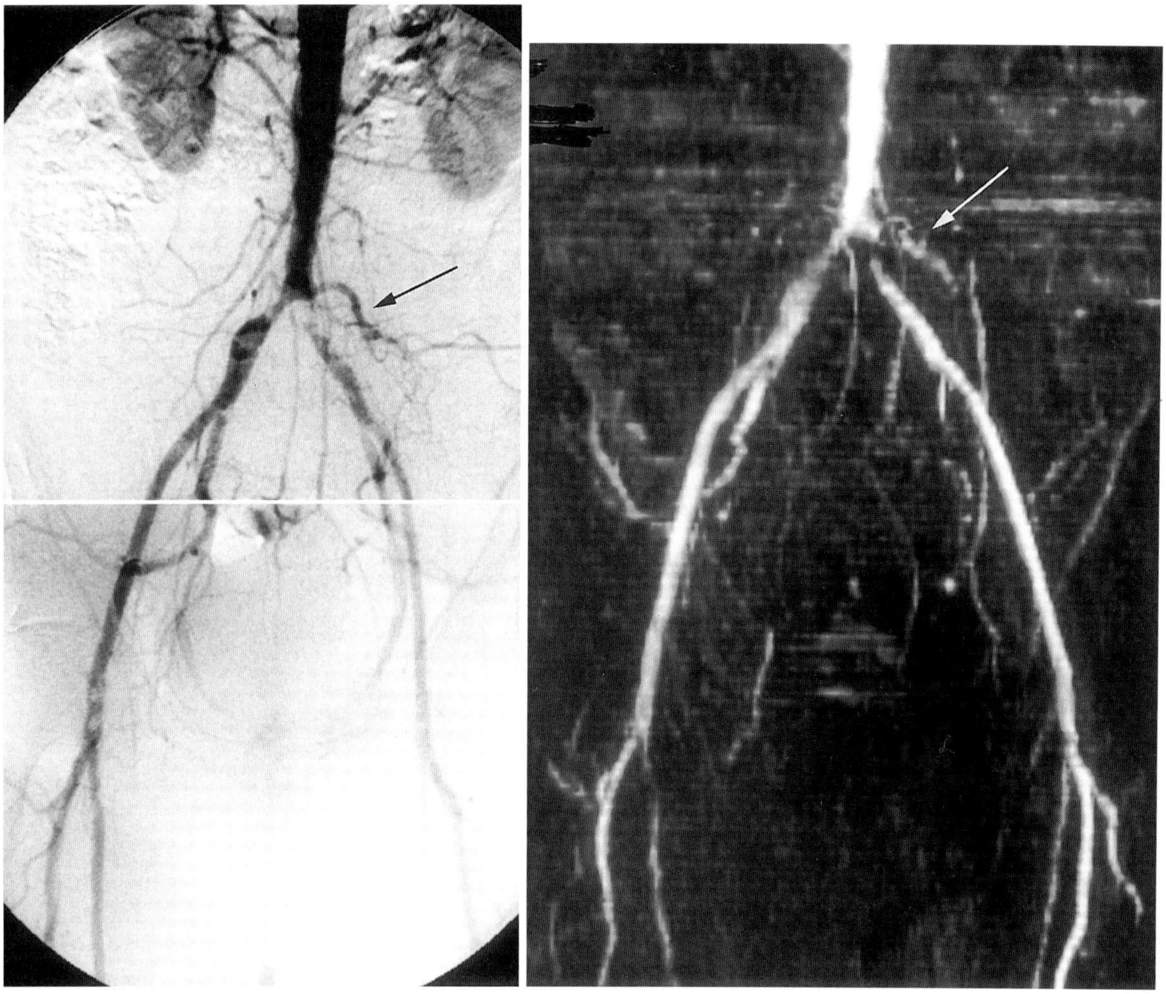

a b

Fig. 15.3a,b. Stenosis of the right common iliac artery; short occlusion of the left common iliac artery. Information on the DSA image (**a**) is identical to that on a coronal MR-projection angiogram (**b**) (transverse slice acquisition, flow-compensated single-slice non-breath-hold 2D TOF GRE FLASH sequence, 1.5 T, body coil, TR/TE = 33/9 ms, flip angle = 40°, slice thickness = 3–5 mm with 1 mm of overlap, field of view = 300 mm, matrix = 256 × 256, elimination of venous signal by a traveling saturation pulse). Note the collateral pathway via the left Vth lumbar artery (*arrow*), clearly visualized by both techniques

Hypertension accelerates the disease. AAA is defined as a localized widening of the lumen of more than 2.5 cm and appears fusiform in most cases. Whereas in fusiform or saccular aneurysms all layers of the arterial wall are dilated, in saclike false aneurysms the layers are disrupted and surrounded by periarterial tissue. The growth rate of smaller aneurysms is 2–3 mm per year, while larger aneurysms may grow by up to 4 mm per year. In autopsy studies the frequency of rupture is related to the aneurysmal size and ranges from 9.5% in aneurysms less than 4 cm in diameter to 60.5% in those of more than 10 cm in diameter (Darling et al. 1977). Clinical symptoms can be observed in only 40% of patients and may indicate rapid growth or imminent rupture.

Surgical repair of AAA is commonly recommended when the diameter of the aneurysm exceeds 5–6 cm, or when, in follow-up studies, rapid progression is proved. Because operative mortality has progressively decreased to 2%–3% due to improvement of surgical techniques and elective interventions, prophylactic graft replacement is advocated in patients in good health. In contrast, aneurysmal rupture is a life-threatening event with a mortality of more than 60%. Surgical techniques depend on the aneurysmal localization and extent and include endoaneurysmorrhaphy or segmental resection, and the implantation of tube or bifurcation grafts (Vollmar 1985). For the planning of surgical intervention, preoperative information is required

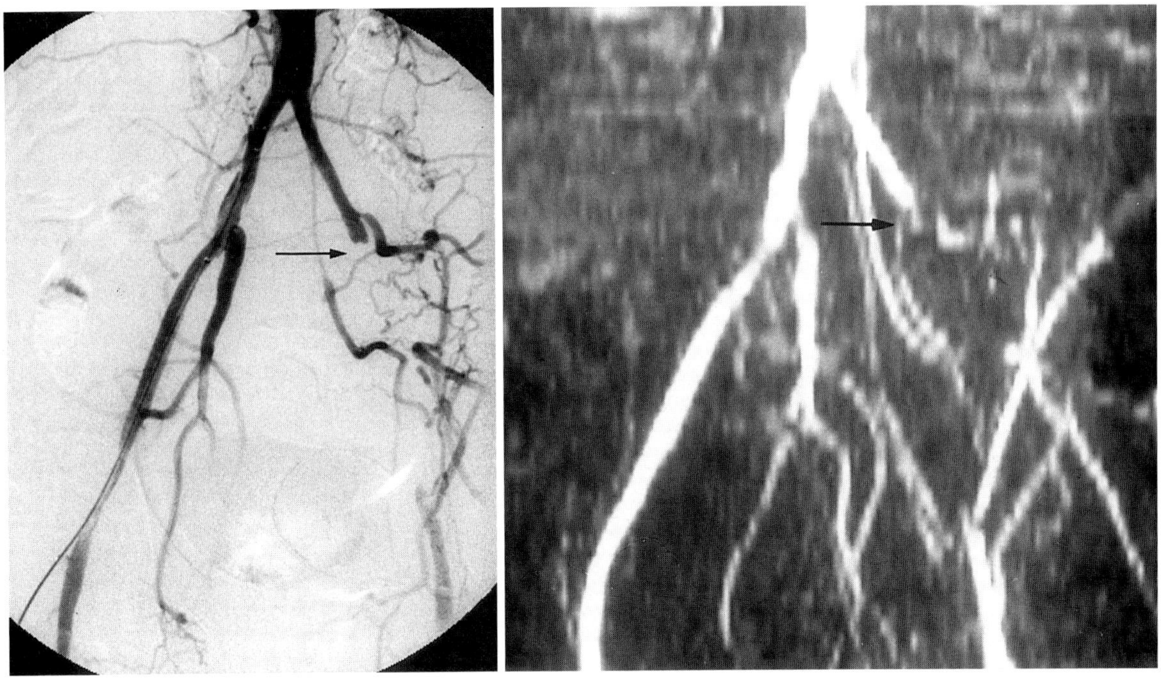

Fig. 15.4a,b. Complete occlusion of the left external iliac artery. The occlusion (*arrow*) can be identified similarly on a DSA image (**a**) and a coronal 2D TOF MR-projection angiogram (**b**) (transverse slice acquisition)

on the type and localization of the aneurysm, its diameter and craniocaudad extent, the aneurysmal wall structures and periaortic space, the relationship to renal arteries, the status of the mesenteric circulation, and the presence of iliac aneurysm or stenosis. The detection of accessory renal arteries which arise from the aneurysm is of major importance for the prevention of segmental renal infarction following surgery.

Inflammatory aneurysms of the abdominal aorta account for 5%–23% of all aneurysms. The definition of inflammation includes mural thickening, fragmentation and attenuation of the internal elastic lamina, loss of smooth muscle in the media, dense connective tissue with neovascularity surrounding the aortic wall, and infiltration by lymphocytes, plasma cells, and histiocytes. Adherence of various visceral structures can be observed, e.g., ureter, duodenum, vena cava, and sigmoid colon (CULLENWARD et al. 1986). Opinions as to the etiology of such inflammatory aneurysms differ; some have suggested an immunological basis while others have proposed that they form part of the spectrum of inflammatory aortitis or bear a relationship to retroperitoneal fibrosis. Although the frequency of rupture is no higher than in atherosclerotic AAA, indications for surgery may be more restricted because repair is associated with higher morbidity and mortality. Preoperative diagnosis allows improved surgical planning and permits the institution of preoperative steroid treatment to reduce the inflammatory reaction.

Ultrasonography has emerged as a highly valuable tool for the detection of suspected AAA, with an accuracy of 100% (AMPARO et al. 1985), and is being introduced for screening of the elderly. US is useful and accurate for sequential measurement of the diameter of a known asymptomatic AAA, and it can detect intraluminal thrombus as well as perianeurysmal fibrosis or perivascular masses in ruptured cases. Color Doppler US may improve the evaluation of renal artery involvement. However, due to known limiting factors the primary role of US is to identify an AAA, whereas other examination techniques are more useful to stage AAA for surgery.

Contrast-enhanced CT has been proved to be the method of choice in the evaluation of aneurysmal extent, diameter, thrombus, wall calcifications, and perivascular space, and in the management of patients with suspected imminent or acute aneurysmal rupture. Inflammatory aneurysms can be accurately identified by CT showing an enhancement of perianeurysmal fibrosis and can be clearly differentiated from perianeurysmal hematoma. In addition, CT is sensitive in the evaluation of postoperative perigraft fluid collections, anastomotic pseudoan-

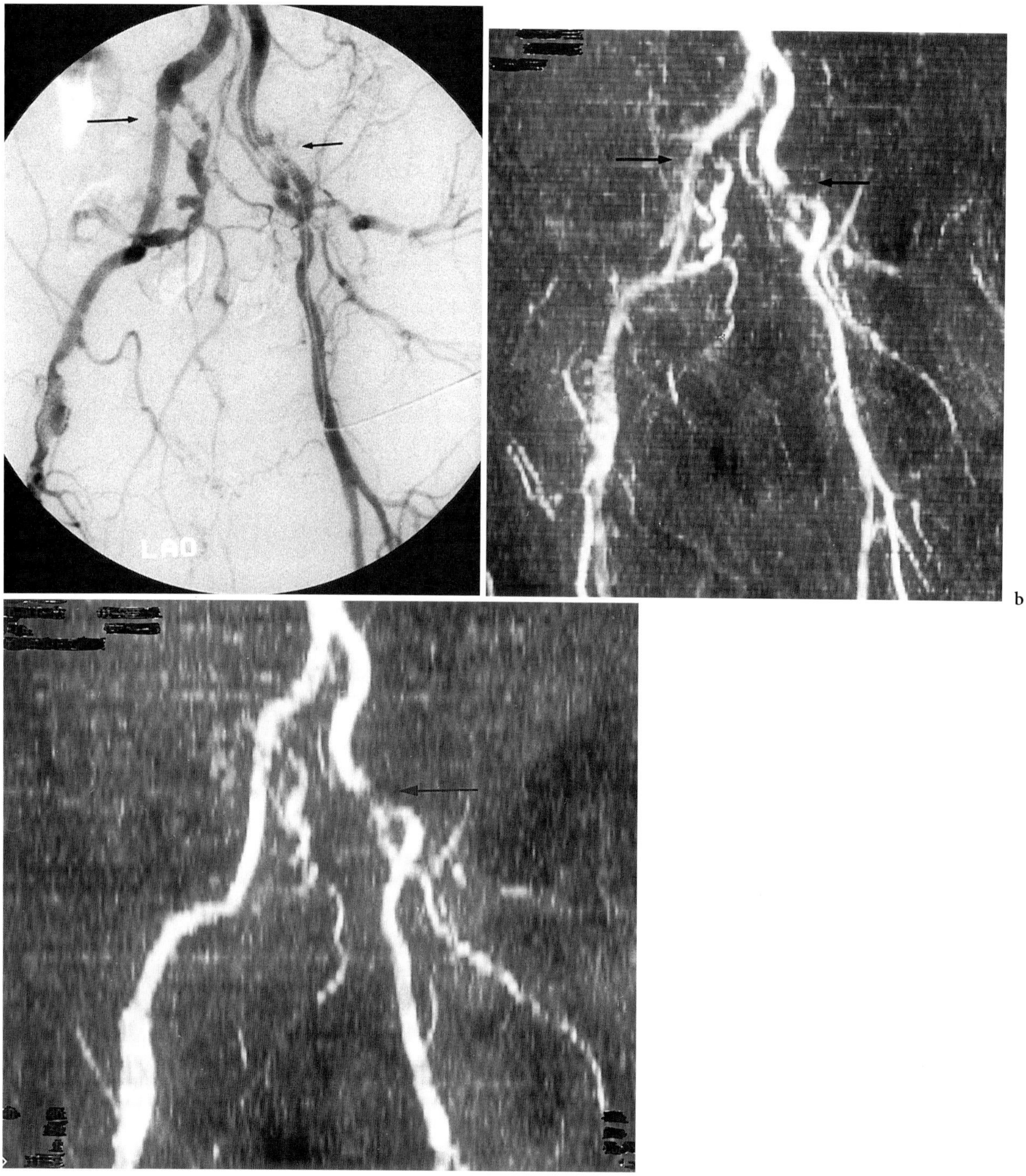

Fig. 15.5a–c. Eccentric stenosis at the origin of the right external iliac artery and at the left iliac bifurcation (*arrows*). On the DSA image (**a**) an additional eccentric calcified stenosis can be seen at the right femoral bifurcation which was operated on between the DSA and the MRA study. 2D TOF MIP MRA was performed without (**b**) and with (**c**) cardiac gating (acquisition during peak systole). Signal void of the left stenosis appears similar with both techniques, whereas in the right iliac artery signal void disappears during peak systolic acquisition

eurysms, aortoenteric fistulas, and abscess formations (SIEGEL and COHAN 1994).

Using currently developed helical CT angiography (CTA), direct visualization of the proximal part of renal and mesenteric arteries or iliac arteries involved in the aneurysm is possible in 3D reconstructions (PROKOP et al. 1993). Thus, current CT technology seems to be the ideal preoperative tool and may substitute for conventional aortography, long considered the gold standard for the staging of AAA. However, spiral CT studies demand larger volumes of intravenously administered contrast agent than does arteriography. This may be considered a limitation of CT in the group of patients with AAA due to the progressive atheroslerosis that frequently is associated with impaired renal function. Further limitations of helical CT given current technology include (a) the fact that only small portions of the aorta can be evaluated due to narrow collimation, (b) long processing and reconstruction times, (c) vessel overlapping in MIP images, and (d) problems in identifying internal morphology and calcifications within the vessel or vessel wall. Thus, spiral CTA may show promise in evaluating AAA, but it is not without problems and therefore remains investigational at the moment (SIEGEL and COHAN 1994).

Although angiography frequently may underestimate aneurysmal extent and diameter, particularly in cases with mural thrombus, it is highly accurate in the evaluation of renal, visceral, and iliac arteries. As a noninvasive method and given its ability to image in multiple planes, vascular MRI has certain advantages over CT. In detecting AAA, demonstrating its size and craniocaudad extent, and differentiating intraluminal thrombus from patent lumen without contrast material, SE imaging has been assessed as being of similar value to conventional CT and angiography (AMPARO et al. 1985; EVANCHO et al. 1985; KANDARPA et al. 1992; RUFF and WATSON 1988). However, limitations of SE imaging have been observed, such as low sensitivity in detecting accessory renal arteries, inability to demonstrate patency of renal or iliac arteries, and inability to accurately differentiate slow blood flow from thrombus (EVANCHO et al. 1985; KOSLIN et al. 1988). MRA techniques employing GRE sequences have been shown to improve diagnostic accuracy in AAA since low to intermediate signal of thrombus material could be differentiated safely from high signal of flowing blood (ARLART et al. 1992a; ECKLUND et al. 1994; KIM et al. 1990) (Fig. 15.6). Oblique coronal sections can be recommended for evaluating aneurysmal extent in a large field of view and for identifying the number of renal arteries. However, misdiagnosis when employing coronal slice acquisition may result from methemoglobin within the thrombus that mimics turbulent flow in the aneurysmal sac (Fig. 15.7). In these cases transverse slice acquisition perpendicular to the flow direction may significantly enhance the inflow effect, resulting in a brighter signal within the patent aortic lumen. Additional application of SE sequences as well as GRE cine studies (GROSS-FENGELS et al. 1991) may also clarify equivocal findings.

Aortic wall structures and their pathology can be evaluated most accurately on sections acquired in the transverse plane. Inflammatory changes of the aortic wall, in which thickening of the middle and outer layers appears with low signal intensity, can be clearly identified and distinguished from the inner layer and the bright signal of intraluminal flowing blood on individual GRE slices (ARLART et al. 1992a) (Fig. 15.8). Furthermore, MRA techniques can be successfully applied for evaluation of the relationship of renal and iliac arteries to the aneurysm or atherosclerotic occlusive disease. Promising initial results have been obtained using a sequential slice acquisition in the transverse plane on the basis of a 2D TOF GRE sequence (Figs. 15.3b, 15.6f, 15.10c–e) which includes selective presaturation pulses to eliminate venous flow signal. Limitations of MRA in renal artery imaging are discussed in the following section. Nevertheless, noninvasive MRA as the sole method of investigation can provide information on aneurysmal anatomy and flow characteristics of similar quality to that obtained with conventional contrast-enhanced CT, and can evaluate renal and iliac arteries in many cases as effectively as angiography (ECKLUND et al. 1994).

Recently, contrast-enhanced 3D TOF imaging has emerged as a rapid and sensitive method of imaging the aortoiliac system (PRINCE 1994). Gadolinium is dynamically injected intravenously and images are acquired in a coronal plane. Scan times depend on the slab volume selected and usually do not exceed 5 min. Not only is 3D TOF imaging faster, but artifacts are far less of a problem than when 2D TOF imaging is used. In his study, PRINCE (1994) reported 3D TOF to have a sensitivity and specifity of 100% in the evaluation of aortoiliac aneurysm. Furthermore, the majority of accessory renal arteries could be identified by contrast-enhanced MRA. These data are promising, and gadolinium-enhanced MRA seems to be on its way to becoming superior to conventional TOF or PC technique (Fig. 15.9), and ulti-

Fig. 15.6a–f. Atherosclerotic abdominal aortic aneurysm. DSA image (**a**) and 2D TOF MR-projection angiogram (**b**) (coronal slice acquisition). On the MR angiogram displacement of the IVC by the aneurysmal sac can be seen. On individual coronal (**c**) and tranverse (**d**) slices bright flow signal of the patent lumen can be clearly distinguished from the intermediate to low signal of mural thrombus (*arrows*). **e** Arteriovenous overlap of the aortic bifurcation and iliac arteries on coronally acquired MIP angiogram (*A*, artery; *V*, vein; *UB*, urinary bladder). By transverse acquisition of slices, venous signal can be eliminated by a traveling saturation pulse which is integrated in the sequence: on the MIP angiogram reconstructed in a coronal view (**f**) stenotic disease of the right common iliac artery can be identified (*arrow*)

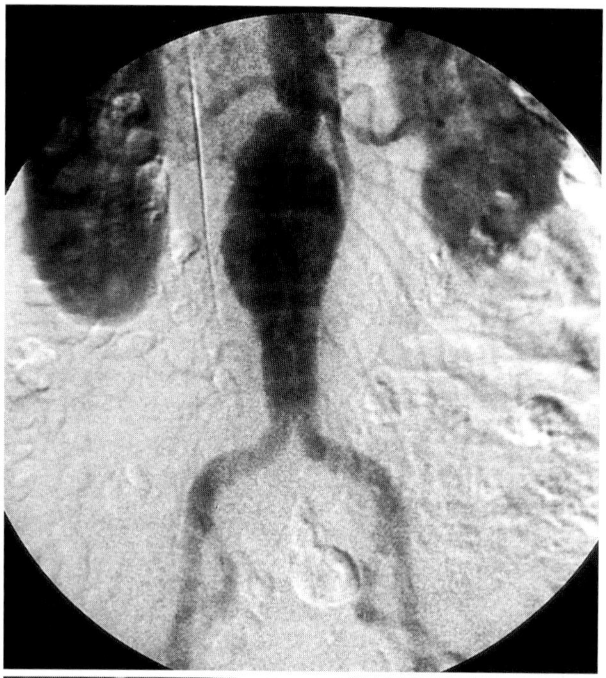

a

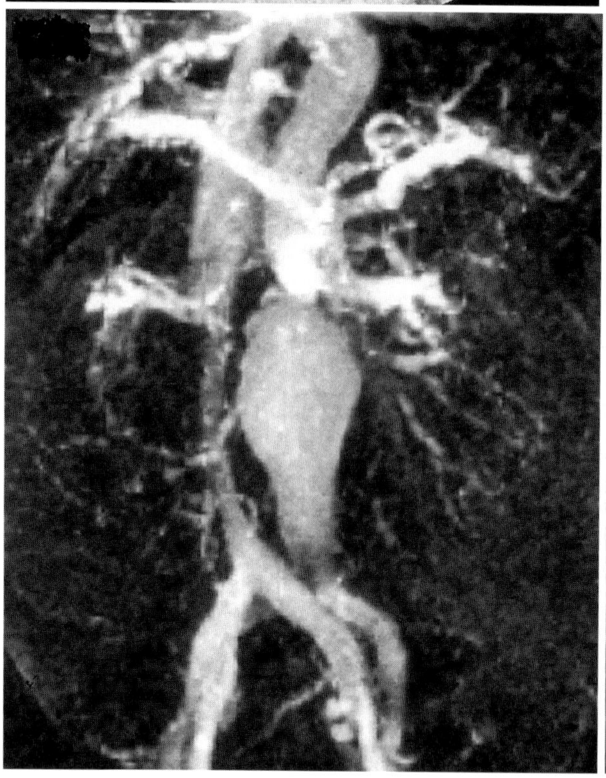

b

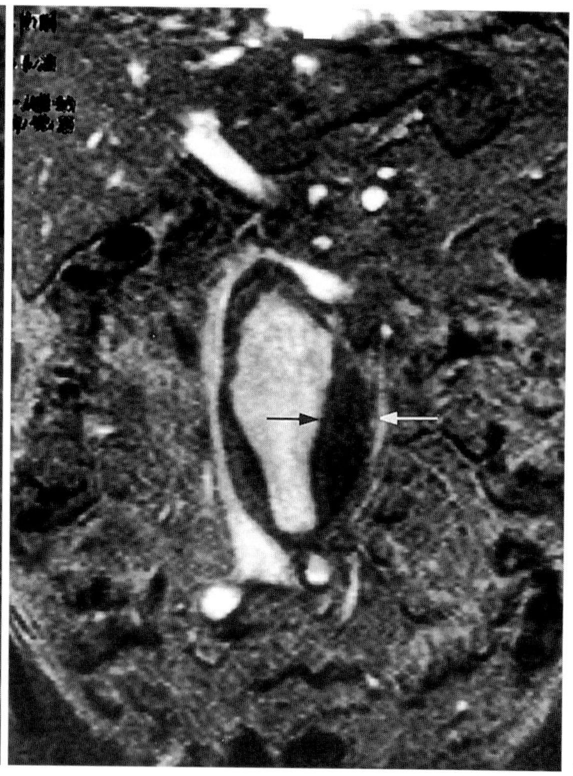

c

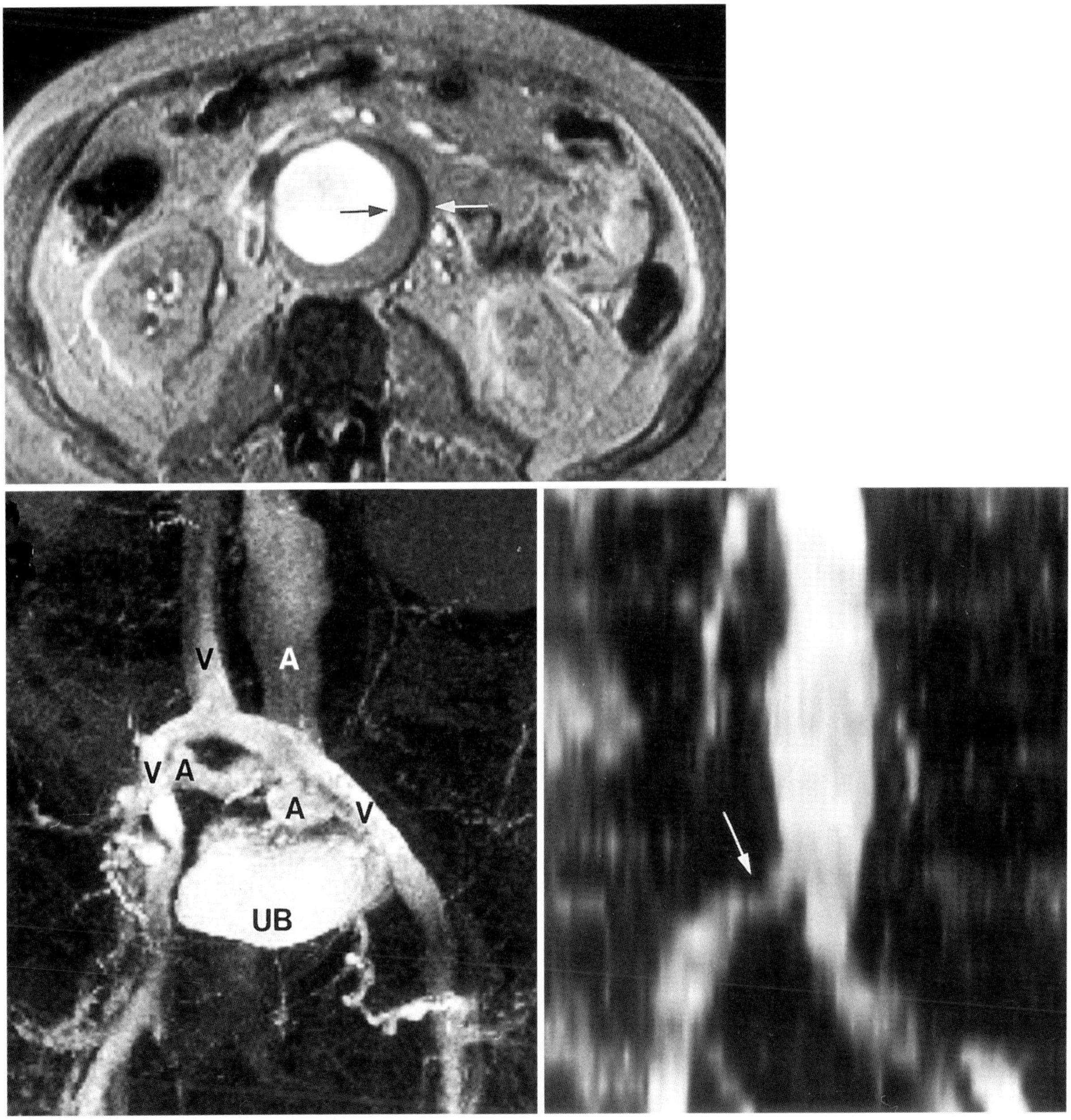

d

e

f

Fig. 15.6d–f

mately may provide the high accuracy which is reported with helical CTA.

15.3.2.3 Abdominal Aortic Dissection

In most cases abdominal aortic dissection (AAD) has its origin in the thoracic aorta (type I or III dissection, DeBakey classification). AAD may involve the aortic branches, the iliac arteries, and in rare cases the femoral arteries. A local aortic or iliac dissection is usually caused iatrogenically, for instance by arterial catheterization or aortoiliac endarterectomy. In most cases of AAD two patent channels can be found which appear as the typical "double barrel" aorta when distal reentries exist at the level of the aortic or iliac bifurcation. Blood supply to the celiac trunk, the renal arteries, and the mesenteric arteries may come from either the true or the false lumen or both depending on channel location and site of reentries. Chronic dissection with patency of both channels is usually managed conservatively, including diagnos-

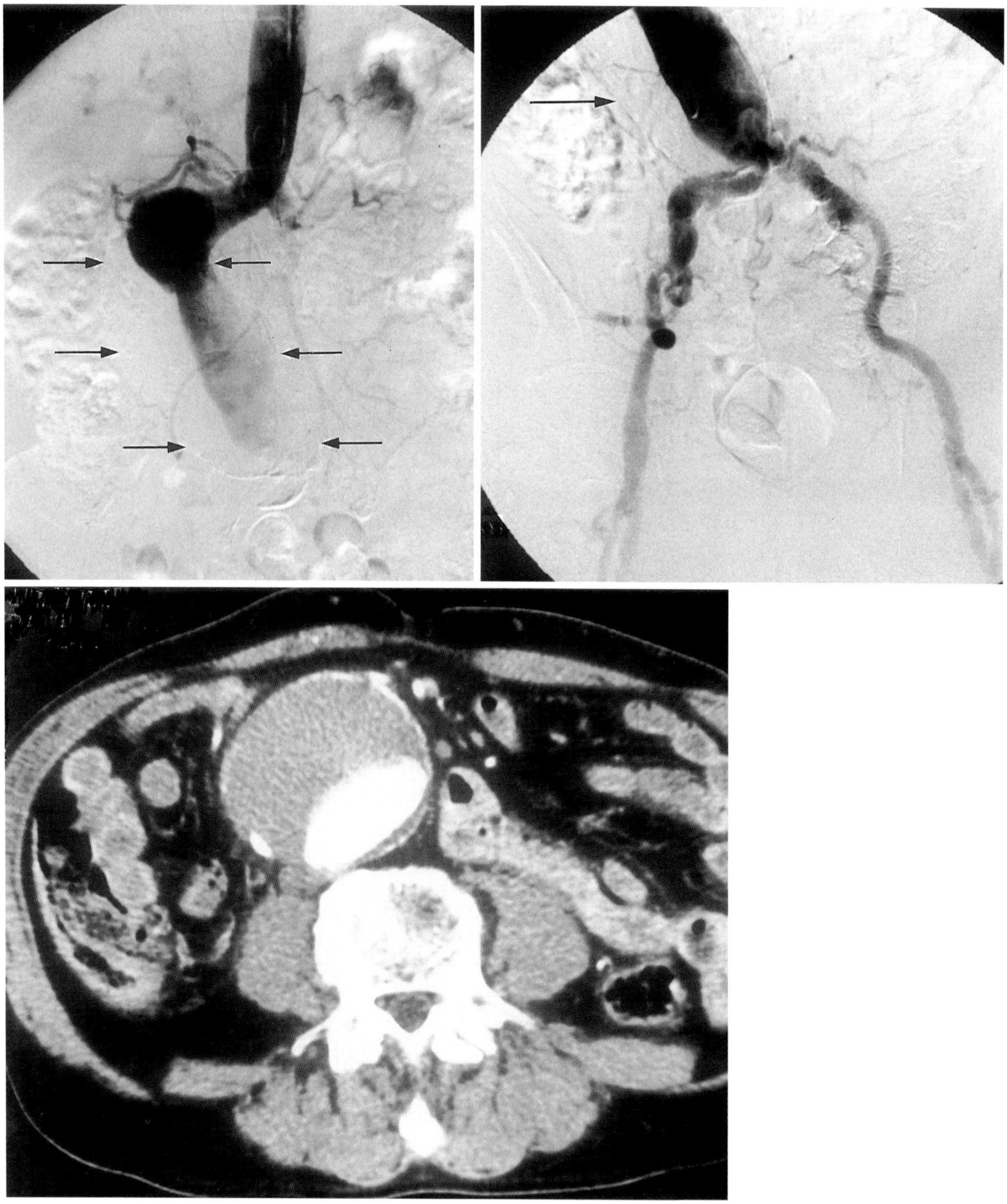

a b c

Fig. 15.7a–e. Atherosclerotic abdominal aortic aneurysm with extensive mural thrombus. On the DSA images (**a, b**) thrombus can be estimated by the distance between patent lumen and aortic wall calcification (*arrows*), and is confirmed by contrast-enhanced CT (**c**). On a 2D TOF MR-projection angiogram (**d**) (coronal slice acquisition) as well as on individual sections (**e**) both the patent lumen and the mural thrombus containing methemoglobin appear with a bright signal

tic follow-up. In acute dissection the false channel may be thrombosed when no reentry has developed. Furthermore, an enlarged diameter of the false lumen can significantly compress the true channel. In these cases acute occlusion of the distal abdominal aorta, visceral arteries, renal arteries, or iliac arteries can occur, which demands immediate surgical intervention. Preoperative planning requires diagnostic d 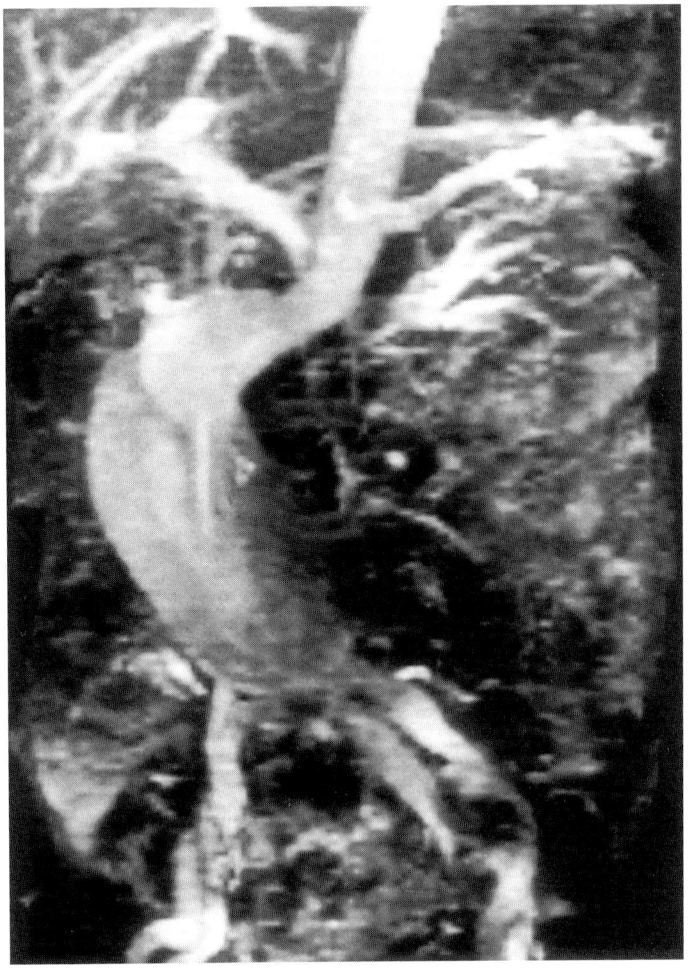 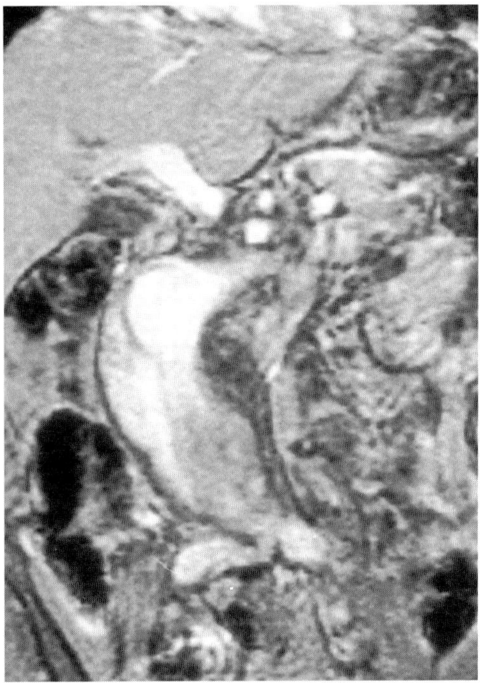 e

Fig. 15.7d,e

information on the origin and extent of the dissection, the site of entry and reentries, and the status of visceral, renal, and iliac arteries.

Ultrasonography usually can detect intimal flaps in the abdominal aorta; this is particularly true when color Doppler US is used, by which means flow conditions can be reliably determined. However, evaluation of involved aortic branches is often limited by extensive bowel gas, obesity, and technical problems. Thus, contrast-enhanced CT is accepted today as the diagnostic method of choice in the evaluation of AAD, and displays a high accuracy in detecting intimal flaps and in determining flow conditions in the true and false lumen. The development of helical CT technology has reduced the pitfalls associated with mistimed bolus injection, partial volume effects, or motion artifacts, which had sometimes led to the intimal flaps being obscured. The main advantage of conventional angiography over US and CT is the accurate detection and localization of reentries in the abdominal aorta, and the reliable determination of renal and visceral blood supply from the true or false lumen.

In evaluating AAD with MRI techniques, bright blood GRE MRA has been shown to be superior to SE MRI. Thrombus which cannot be safely differentiated from slow or inverse blood flow in SE sequences appears with low signal in flow-sensitive GRE sequences, whereas flowing blood has a bright signal in both the true and the false channel, clearly delineating the intimal flap (Fig. 15.10a,b) (Arlart et al. 1992). Because flow velocity is usually faster in the true lumen, it often appears brighter than in the false channel on MRA images. In cardiac gated GRE cine studies, the true lumen appears brightest shortly after peak systole, whereas the false lumen may have a delay in peak flow and can show reversed flow on velocity-encoded images. When using a sequential breath-hold 2D GRE FLASH sequence, and particularly on transversely acquired individual sections,

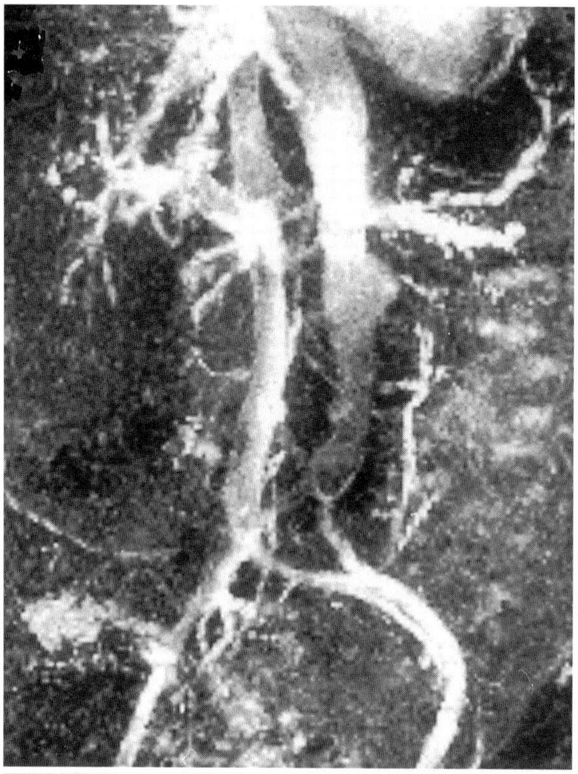

a

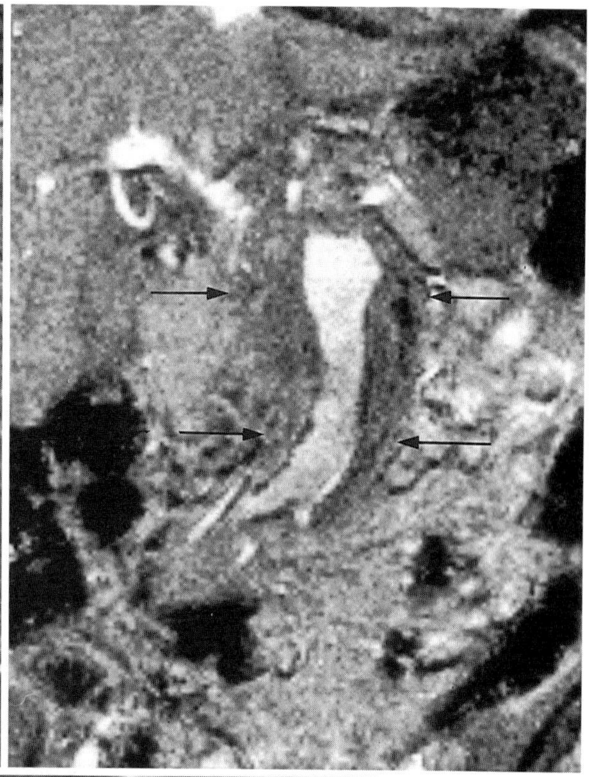

b

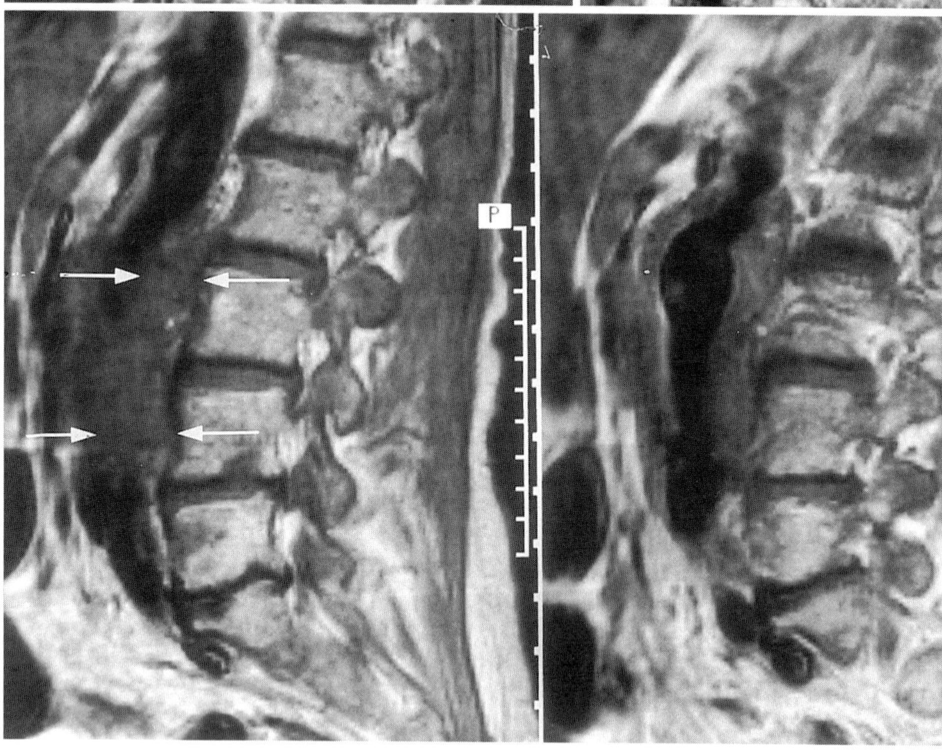

c

Fig. 15.8a–e. Inflammatory abdominal aortic aneurysm. On a 2D TOF MR-projection angiogram (**a**) as well as on a coronal (**b**) and transverse (**e**) single slice, the patent aortic lumen is separated from the IVC by a mass of intermediate signal intensity caused by perivascular inflammation of the aortic wall (*arrows*). On T1 SE images (1.5 T, body coil, TR/TE = 400/25 ms, slice thickness = 5 mm, number of acquisitions = 6, field of view = 400 mm, matrix = 192 × 256) this inflammatory mass appears with low signal (**c**). Diagnostic information, including patent lumen and mural thickening, is similar on a contrast-enhanced CT slice (**d**) and the individual transverse MR section (**e**) except for calcifications, which cannot be identified by MRI.

Fig. 15.8d,e

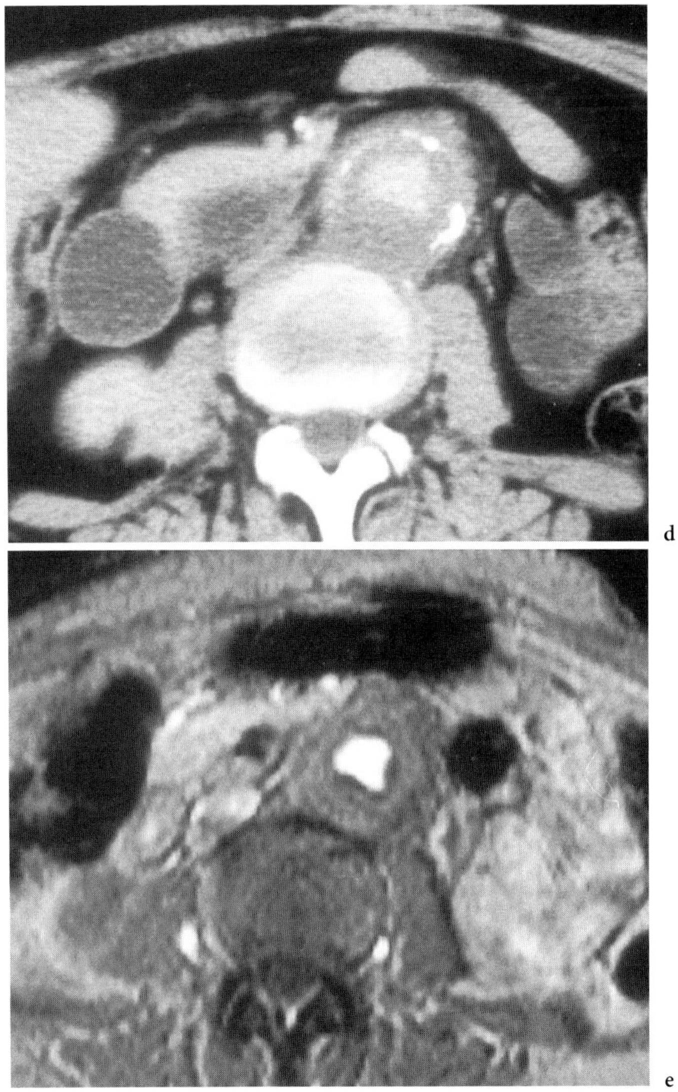

the dissecting membrane can be delineated safely and the relationship to renal arteries or the SMA can be evaluated (Fig. 15.10c–e). However, smaller reentries frequently found with conventional selective catheter angiography at the level of visceral and renal arteries may be absent on MR angiograms.

15.3.2.4 Renal Arteries

Indications for renal artery imaging include renovascular hypertension (RVH), renal trauma, preoperative vascular mapping in malignant tumors of the kidney, and assessment of potential kidney donors. The renal artery itself can be involved by a variety of lesions such as aneurysms, arteriovenous fistulas or malformations, hemangiomas, dissections, thrombembolic occlusion, occlusion due to progressive atherosclerosis, and stenotic disorders (Kadir 1986). Arterial hypertension is a clinical symptom which frequently is associated with renal artery disease. The prevalence of RVH is estimated at 1%–5% among the entire hypertensive population. Despite the low incidence, screening for RVH is useful because enhanced arterial blood pressure can be decreased or normalized in the majority of patients when renal artery stenosis (RAS) is successfully managed by percutaneous transluminal angioplasty. In addition, the development of chronic renal failure can be prevented by this therapeutic procedure.

Approximately 60% of cases of RVH (mainly in the older male population) are caused by atherosclerotic disease, and about 35% (mainly in the

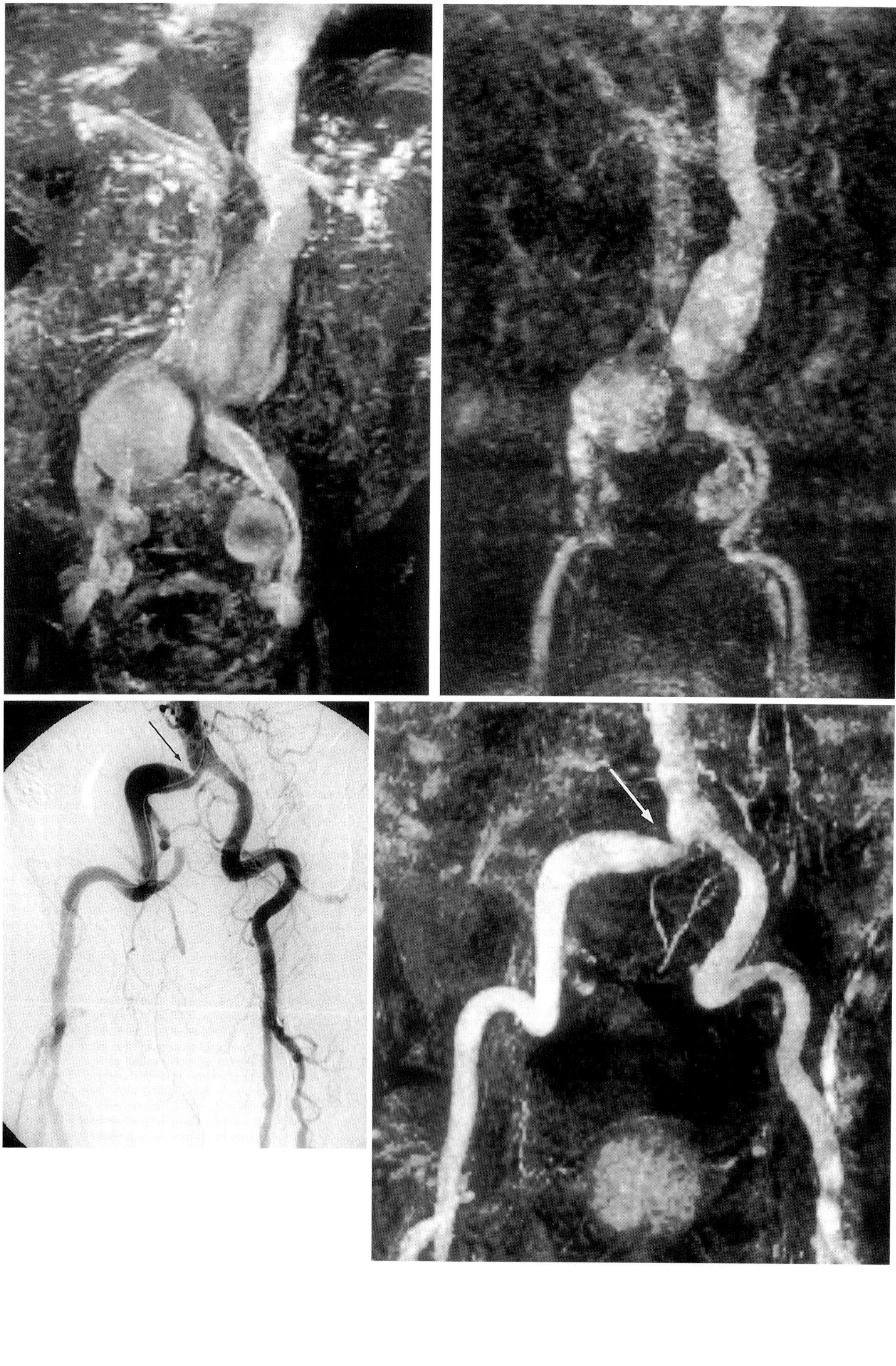

a b c d

younger female population) are due to fibromuscular dysplasia (FMD). However, stenosing atherosclerotic plaques in the renal artery are also found in a significant number of the normotensive population. Other etiologies of RVH are rare and include congenital stenosis, neurofibromatosis, arteritis, intimal dissection, and aneurysm. Atherosclerotic RAS can be observed bilaterally in about 30% of cases; in more than 90% the main RA is involved near its origin. The morphological appearance of atherosclerotic RAS is commonly eccentric and irregular, with a poststenotic dilatation. FMD is found bilaterally in about two-thirds of cases; the main renal artery is involved in 79%, segmental arteries additionally in 17%, and peripheral arteries in only 4%. The morphology of FMD depends on the type of lesion (HARRISON and MCCORMACK 1971). Mostly the lesion appears centric with a smooth border, and varies from short segmental narrowing to alternating areas of aneurysmal formation and narrowing ("string of beads" appearance).

In patients in whom RVH is suspected, diagnostic vascular imaging requires reliable determination of the number of renal arteries, including high-quality imaging of pole arteries, accessory arteries, and segmental branches, as well as reliable detection of RAS and accurate grading of the stenosis. To determine the hemodynamic significance of RAS in unclear cases, additional procedures are recommended, such as renal vein renin sampling, determination of converting enzyme inhibitor levels, or dynamic radionuclide scanning during converting enzyme inhibition.

Noninvasive Doppler US has advanced in recent years to become an important diagnostic tool in the detection of RVH. However, in the literature sensitivity and specificity vary widely, from 0% to 82% and 37% to 100%, respectively, due to the use of different parameters for analysis of flow phenomena and different equipment. Adequate or complete examination of the main renal arteries is not possible in approximately 30%–40% of cases. For RAS grading on the basis of Doppler US evaluation, indirect parameters such as renal resistive index have been shown to have a sensitivity of 82% and a specificity of 92% when RAS exceeds 50%, whereas sensitivity and specificity have been reported to be 100% and 94%, respectively, when RAS exceeds 60% (SCHWERK et al. 1994). However, US may not reliably detect stenosed accessory renal arteries, and results depend on extensive examination experience. Minimal invasive intravenous DSA has been considered a useful method, with a reported sensitivity of 88% and a specificity of 89% (HAVEY et al. 1985). As a major limitation, large contrast volumes are needed in most cases and vascular overlap may occur as well as significantly reduced image quality. Thus, conventional intra-arterial arteriography is accepted as the gold standard for detecting and evaluating RAS. In spite of its invasiveness, the risk to the patient is minimal when modern catheter equipment is used by experienced personnel. However, its invasiveness and technical and personnel costs limit renal arteriography as a general screening modality in the hypertensive population, particularly in view of the low prevalence of the disease. Today, arteriography is widely used in preselected patients in whom RVH is suspected due to specific clinical symptoms and data from the history. A major advantage of catheter angiography is the fact that accessory renal arteries as well as RAS can be detected safely in different localizations (mentioned above) and may be graded reliably. Moreover, percutaneous catheter management (balloon angioplasty, stent placement) of a proven RAS can be performed immediately after diagnosis within the same procedure.

As an alternative to angiography, the recently developed contrast-enhanced helical CT angiography (CTA) is able to provide excellent images of renal arteries. In 3D reconstructions of shaded surface display (SSD) and maximum-intensity projection (MIP), main renal arteries including the segmental bifurcation and accessory renal arteries can be evaluated, vascular calcifications can be visualized, and stenotic lesions can be graded. Promising results of CTA in detecting RAS >50% have shown that the MIP technique is superior to the SSD technique and has a sensitivity and specificity of more than 90% and 80%, respectively (GALANSKI et al. 1992; RUBIN et

◄ **Fig. 15.9a–d.** Contrast-enhanced MRA of the abdominal aorta and iliac arteries in aneurysmal disease of the abdominal aorta, the right common iliac artery, and the left internal iliac artery (**b**) and in iliac occlusive disease (**d**). Compared with imaging by conventional 2D TOF MRA (coronal slice acquisition, MIP angiogram) (**a**), aneurysmal disease of the aorta and iliac arteries can be evaluated more accurately on a contrast-enhanced 3D TOF MIP angiogram (**b**) (GRE FISP, coronal slice acquisition, i.v. infusion of 30 ml of Gd-DTPA = 0.15 mg/kg, 1.5 T, TR/TE = 30/6 ms, flip angle = 35°, acquisition time of 4 min for a 48-mm slab with 32 partitions, slice thickness = 1.5 mm, number of acquisitions = 1, field of view = 350 mm, matrix = 256 × 256). Additionally, iliac stenotic disease (*arrow*) and kinked vessels can be evaluated with the same reliability on the DSA image (**c**) and gadolinium-enhanced MR-angiogram (**d**)

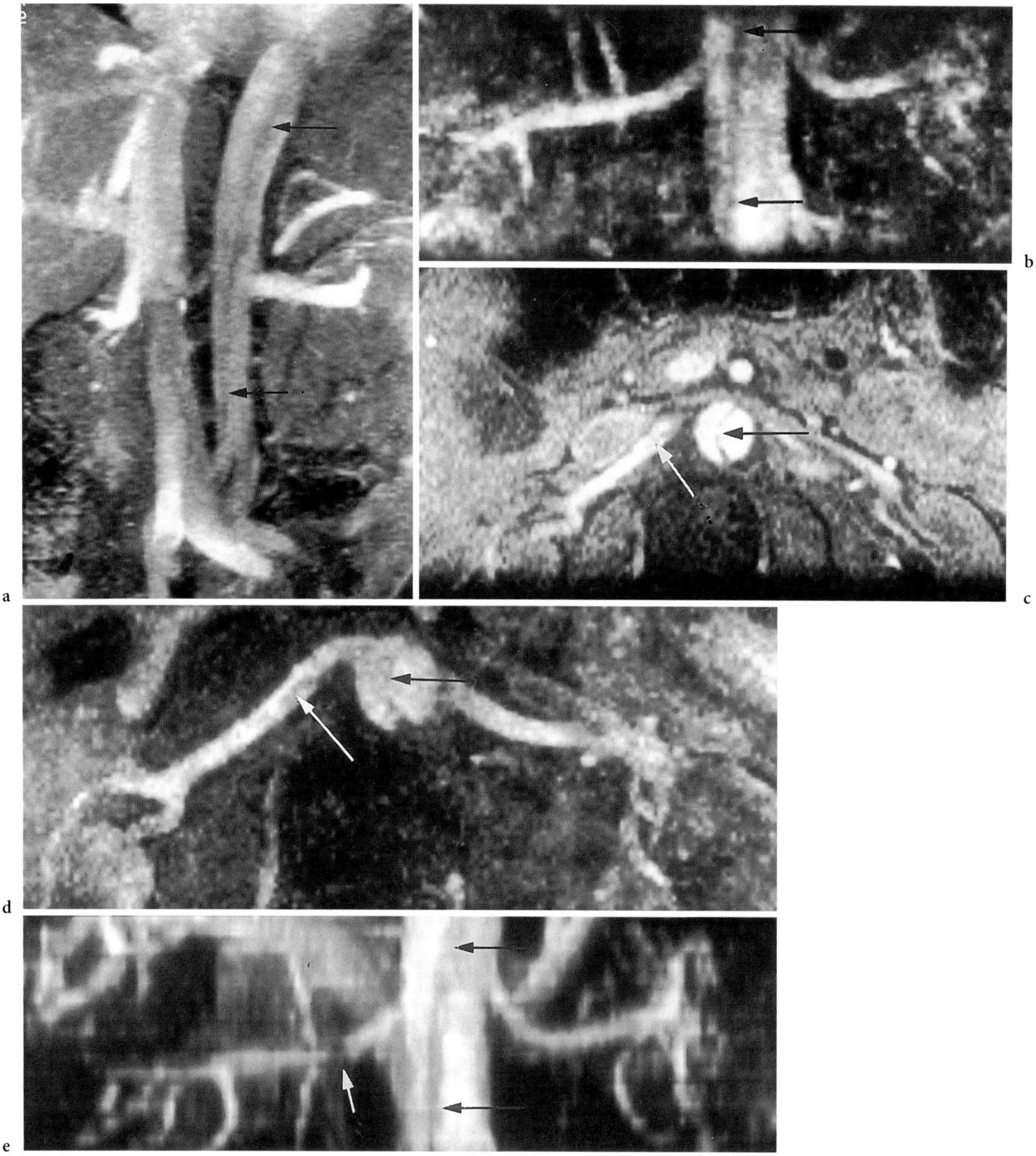

Fig. 15.10a–e. Aortic dissection of type III (DeBakey classification) involving the abdominal aorta and the right renal artery (angiographically proven). On the coronal 2D TOF MR-projection angiogram (**a**) an intimal flap with low signal can be clearly distinguished from bright patent true and false lumens (*arrow*), resulting in a typical "double-barrel" aorta. By means of 3D TOF MIP MRA (GRE FISP, TONE) (**b**) the intimal flap can be delineated in the aorta whereas normal flow signal is seen in both renal arteries. In contrast, on transverse 2D single sections (**c**) and an axially reconstructed MIP angiogram (**d**) the intimal flap can be delineated in the aortic lumen (*black arrows*) as well as in the right renal artery (*white arrow*). In a coronal MIP angiogram reconstructed from transversely acquired slices (**e**) the intimal flap within the renal artery mimics stenosis (*white arrow*)

al. 1994). Accessory renal arteries detectable by arteriography have been reported also to be detectable by CTA.

Compared with noninvasive MRA, CTA necessitates intravenous injection of contrast agents. Nevertheless, both methods are currently in competition for renal artery imaging. MRA includes two-dimensional and three-dimensional TOF and PC techniques that are also considered competitive methods (Fig. 15.11a–e). For evaluating the number of renal arteries a coronal 2D TOF GRE sequence has been shown to be most useful (Kim et al. 1990) (Figs. 15.1c, 15.12) whereas for detecting RAS transverse 2D slice acquisition or 3D MRA has proved superior in the majority of cases (Figs. 15.13–15.15). In RAS >50%, sensitivity and specificity of 2D TOF MRA and 3D TOF MRA are reported ranging from 78% to 100% and from 80% to 94%, respectively (Arlart et al. 1992b; Kent et al. 1991; Kim et al. 1990; Loubeyre et al. 1994). It has been demonsterated that both sensitivity and specificity of 3D TOF MRA can be improved by the TONE technique (Strotzer et al. 1993), and may be positively influenced by fat saturation and systolic peak flow acquisition (Li et al. 1994). However, visualization of accessory renal arteries remains a problem and detection is reported only in approximately 50% of cases. 2D and 3D renal PC MRA seems to be superior to TOF MRA when data from the literature are considered (Richter et al. 1993), including a comparative study by Debatin et al. (1989). Furthermore, breath-hold PC MRI allows quantification of renal artery blood flow which may be utilized in evaluating the hemodynamic significance of RAS (Debatin et al. 1994). In order to optimize the accuracy of MRA in detecting RAS, a combination of 2D and 3D techniques in different acquisition planes is recommended by most authors, using TOF as well as PC techniques (Anderson et al. 1993; Arlart et al. 1992b; Debatin et al. 1991; Yucel et al. 1993a).

Although correct grading of RAS on MR angiograms is reported in the literature to range from 67% to 91%, in our opinion classification of stenosis remains problematic with both TOF and PC MRA. As a consequence, overestimation has been reported in 7%–33% of cases (Arlart et al. 1992b; Kent et al. 1991; Kim et al. 1990; Richter et al. 1993). In our experience a RAS can be considered hemodynamically significant when the MIP angiogram shows a complete signal loss in the involved area that can be confirmed in the individual slice (Figs. 15.13b, 15.14d, 15.15c). A moderate signal void in the area of suspected RAS, however, frequently may be caused by nonstenosing plaque material (Fig. 15.16b). It remains questionable whether RAS due to FMD can be detected safely by MRA. Diverse technical features such as selective venous presaturation, TONE, MOTSA, fat saturation, acquisition during peak systole, k-space segmentation, or the recently reported STAR MRA with echoplanar imaging presumably can improve the image quality and accuracy of MRA. However, in spite of the availability of these different acquisition modalities, the main renal arteries cannot be visualized satisfactorily distal to the segmental bifurcation (Figs. 15.11, 5.13–15.16), nor can smaller accessory renal arteries be detected reliably at present (Fig. 15.17).

The use of gadolinium-enhanced 3D TOF MRA, recently introduced for detecting RAS and accessory renal arteries, does not depend on blood inflow or blood motion and thereby circumvents many of the problems of conventional TOF MRA. Thus, it may reduce some of the drawbacks mentioned above and eventually may become the most successful MR technique in imaging the renal arteries when spatial resolution can be further improved and acquisition time shortened. Preliminary results published by Prince (1994) are promising and include the detection of RAS with 85% sensitivity and 93% specificity. However, in our experience vascular overlap of renal vessels, which will reduce diagnostic accuracy, can not be avoided completely (Figs. 15.11i, 15.12c) in spite of a short gadolinium infusion time and rapid image acquisition.

15.4 Conclusion

Compared with the excellent image quality of MRA in the head and neck region which causes this noninvasive technique to be indicated in a number of cerebral disorders, MRA of abdominal arteries still has a variety of limitations. Acquisition and imaging problems have not yet been adequately solved and include susceptibility artifacts, reduced spatial resolution, and in-plane saturation effects. Moreover, reduced signal intensity occurs in tortuous vessels and arteries of small diameter, and under slow or turbulent flow conditions. In the abdomen further imaging problems are caused by respiratory motion artifacts and artifacts due to bowel movement or vasular pulsation. Although different imaging strategies have been developed, intraparenchymal arterial vasculature usually cannot be evaluated satisfactorily at the moment. However, in aortoiliac occlusive disease as well as in aortoiliac aneurysm MRA results

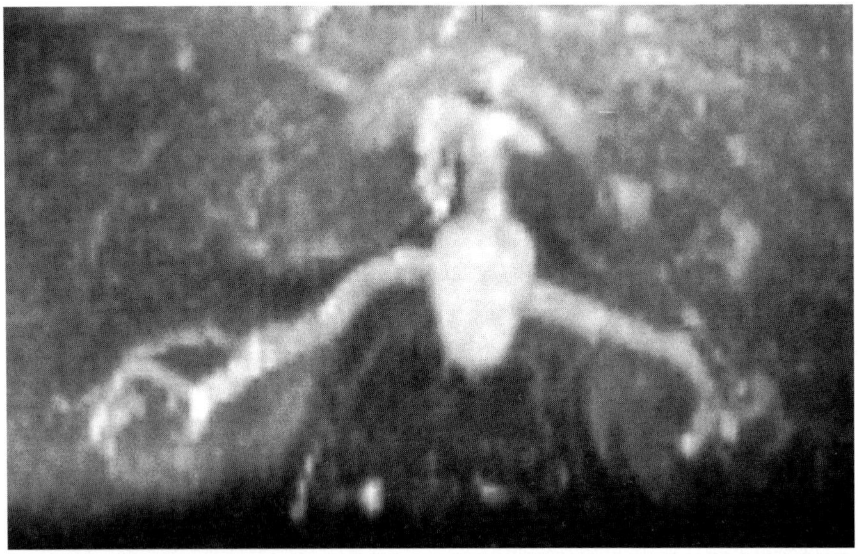

a

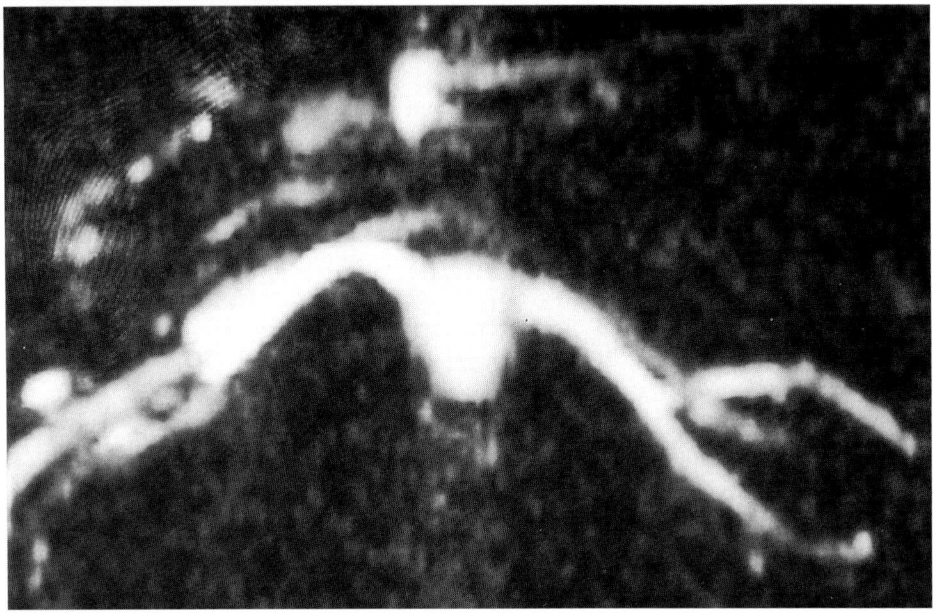

b

Fig. 15.11a–i. Renal arteries in normal individuals. **a** 2D TOF GRE FLASH MRA (flow compensation, transverse single slice acquisition, 1.5 T, body coil, TR/TE = 26/8 ms, flip angle = 35°, slice thickness = 3–5 mm with 1 mm of overlap, field of view = 340 mm, matrix = 192 × 256, presaturation pulse through both kidneys and the IVC eliminates venous signal). **b, c** 2D PC MRA [transverse (**b**) and coronal (**c**) MIP angiogram, 1.5 T, body coil, TR/TE = 40/8.4 ms, flip angle = 60°, slice thickness = 7 mm with 2 mm of overlap, matrix = 128 × 256]. (Courtesy of Dr. Jörg F. Debatin, MRI Center of Zurich University Hospital). **d–g** 3D TOF GRE FISP MRA (1.5 T, body coil, transverse slice acquisition, TR/TE = 35/7 ms, TONE, 75-mm slab with 64 partitions, slice thickness = 1.2 mm, number of acquisitions = 2, field of view = 300 mm, matrix = 192 × 256). Images show bilaterally two renal arteries (**d**), proximal visceral arteries (**e**) (*CT*, celiac trunk; *SMA*, superior mesenteric artery; *LRA*, left renal artery), left lower pole artery (**f**) and transplant renal artery in a double-slab technique (**g**) (*CIA*, common iliac artery; *IIA*, internal iliac artery; *EIA*, external iliac artery). **h, i** Intra-arterial DSA (**h**) and gadolinium-enhanced 3D TOF GRE FISP MRA (**i**) (1.5 T, coronal slice acquisition, TR/TE = 30/6 ms, 48-mm slab with 32 partitions, slice thickness = 1.5 mm, number of acquisitions = 1, field of view = 350 mm, matrix = 256 × 256, i.v. infusion of 30 ml of Gd-DTPA over the acquisition time of 4 min). On the MIP angiogram the ostium of the left renal artery cannot be evaluated due to arteriovenous overlap)

are promising when employing TOF techniques, in particular when gadolinium contrast is used. For complete aneurysmal staging including evaluation of renal and iliac arteries, a combination of coronal and transverse acquisition techniques as well as selective presaturation of venous flow has been recommended. MRA may be indicated in these cases particularly when other noninvasive vascular exami-

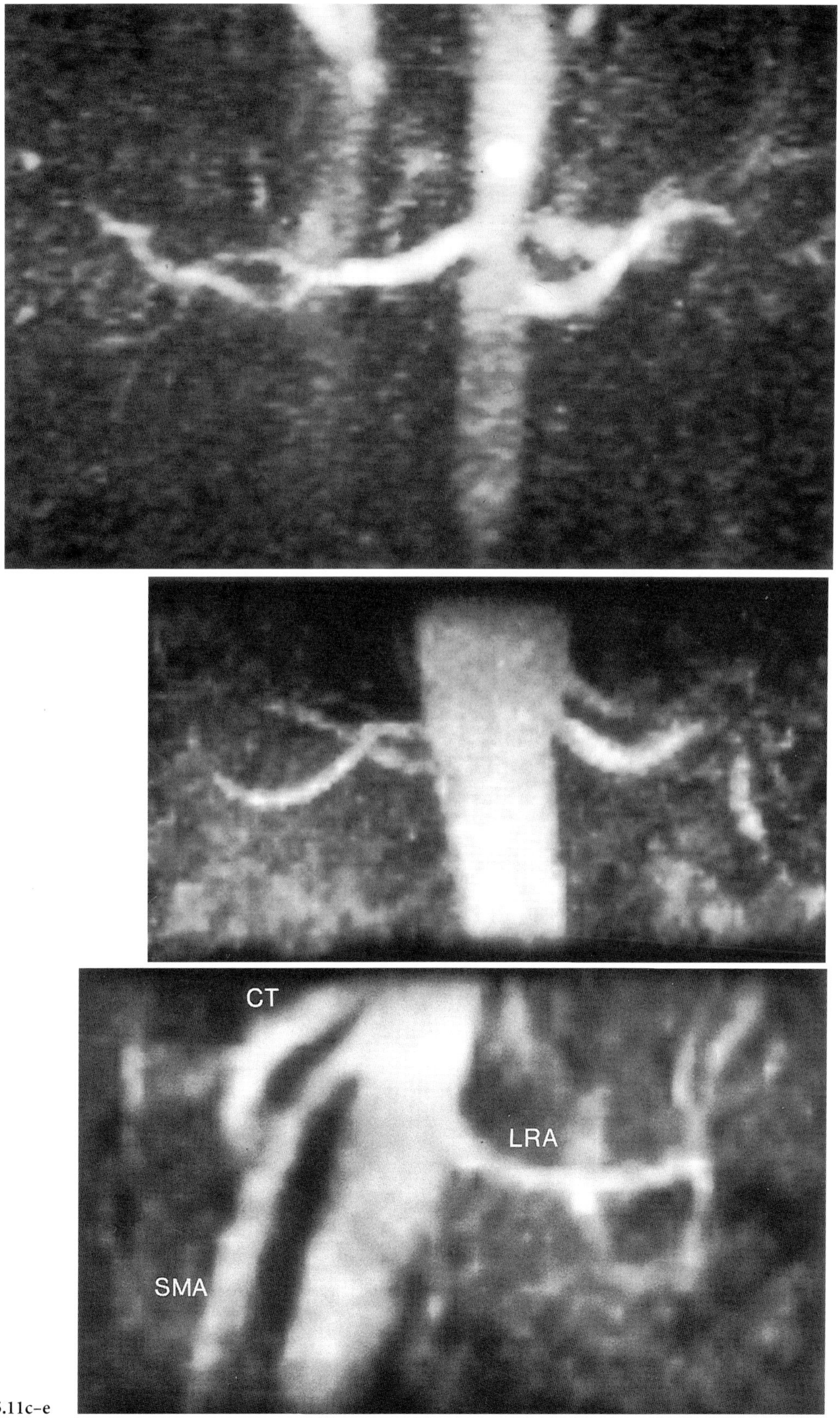

Fig. 15.11c–e

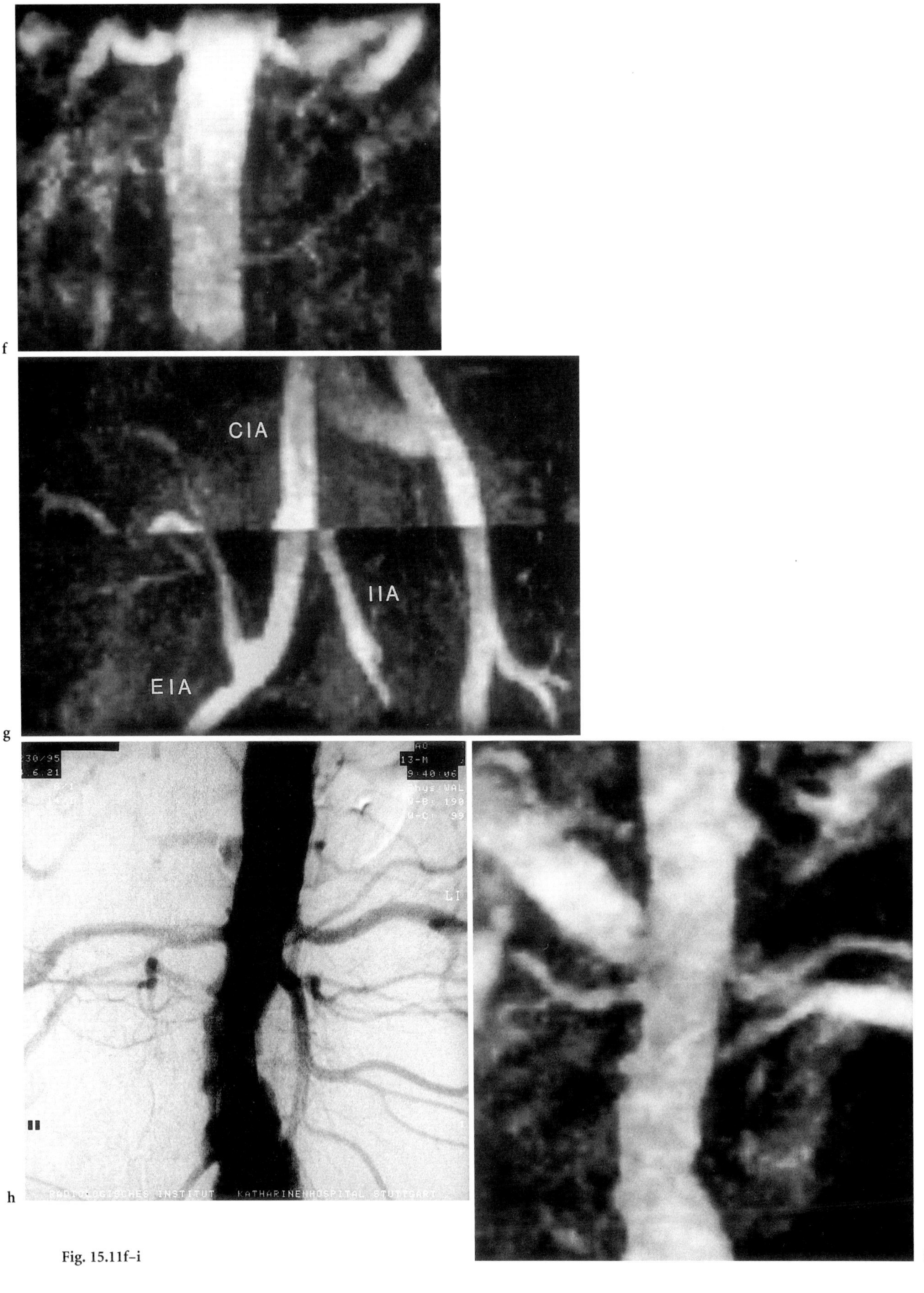

Fig. 15.11f–i

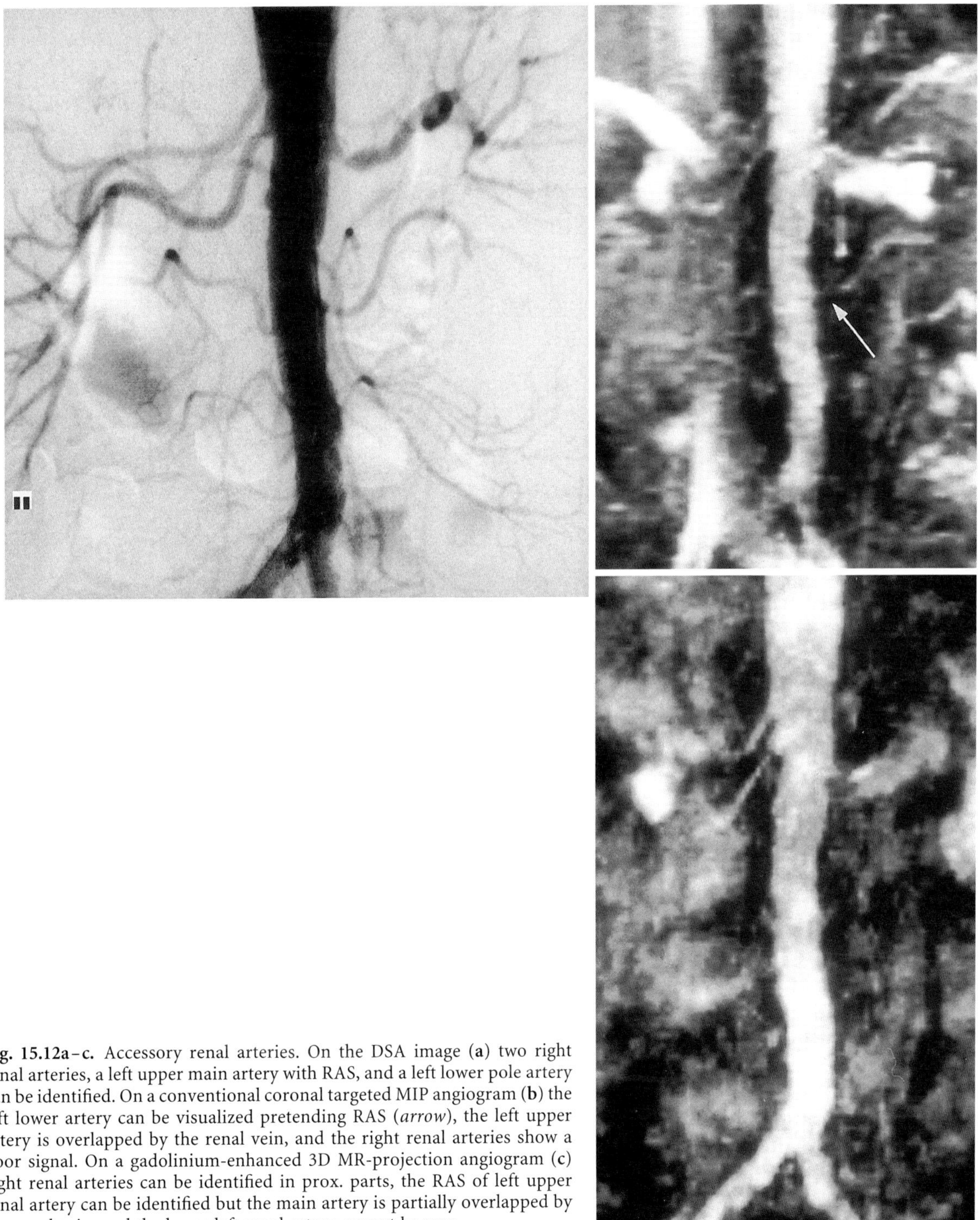

a b c

Fig. 15.12a–c. Accessory renal arteries. On the DSA image (**a**) two right renal arteries, a left upper main artery with RAS, and a left lower pole artery can be identified. On a conventional coronal targeted MIP angiogram (**b**) the left lower artery can be visualized pretending RAS (*arrow*), the left upper artery is overlapped by the renal vein, and the right renal arteries show a poor signal. On a gadolinium-enhanced 3D MR-projection angiogram (**c**) right renal arteries can be identified in prox. parts, the RAS of left upper renal artery can be identified but the main artery is partially overlapped by the renal vein, and the lower left renal artery cannot be seen

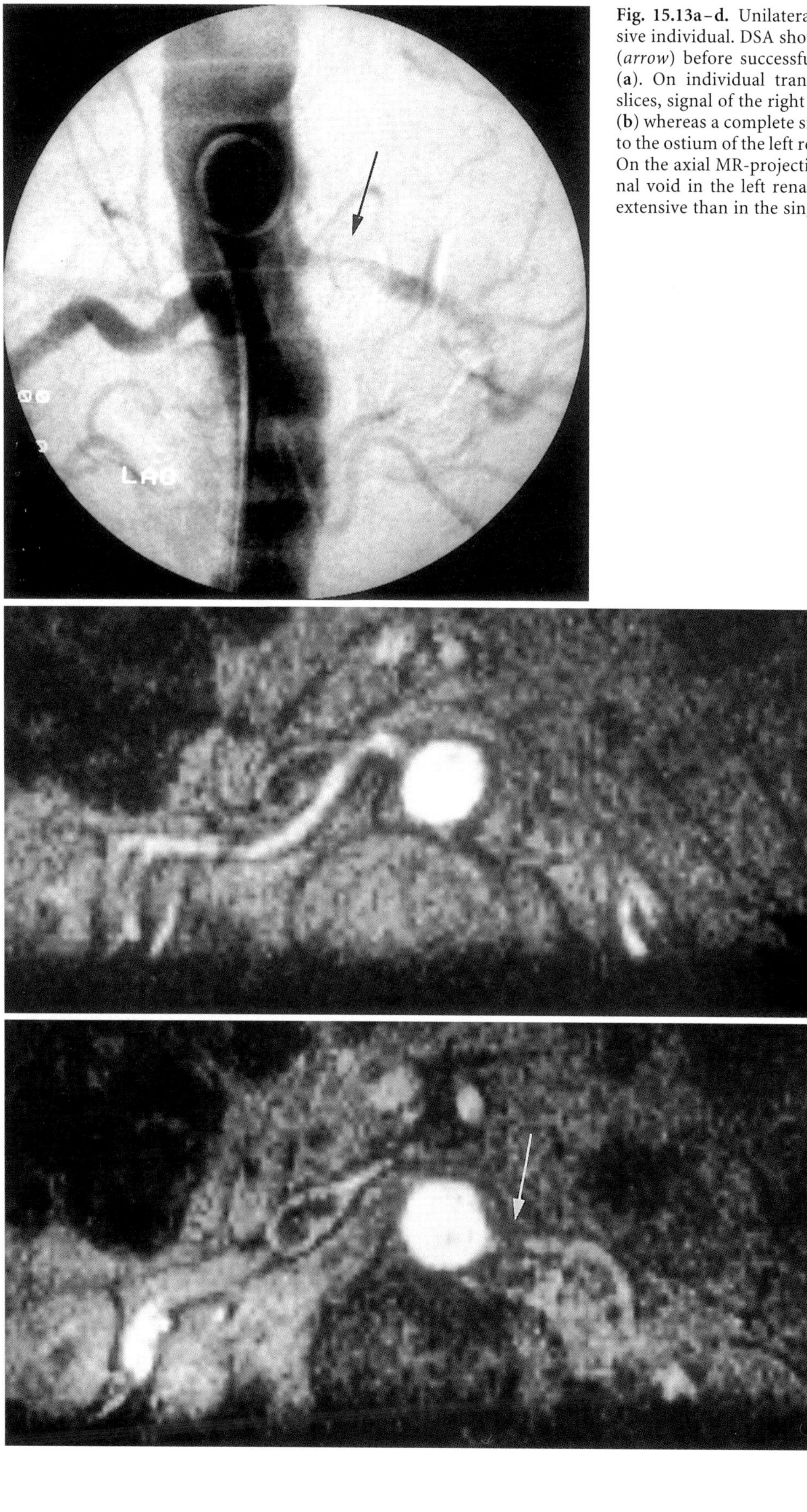

a

b

c

Fig. 15.13a–d. Unilateral RAS in a hypertensive individual. DSA shows significant left RAS (*arrow*) before successful balloon angioplasty (**a**). On individual transverse 2D TOF MRA slices, signal of the right renal artery is normal (**b**) whereas a complete signal void is seen close to the ostium of the left renal artery (**c**) (*arrow*). On the axial MR-projection angiogram (**d**) signal void in the left renal artery appears more extensive than in the single slice (*arrows*)

Fig. 15.13d

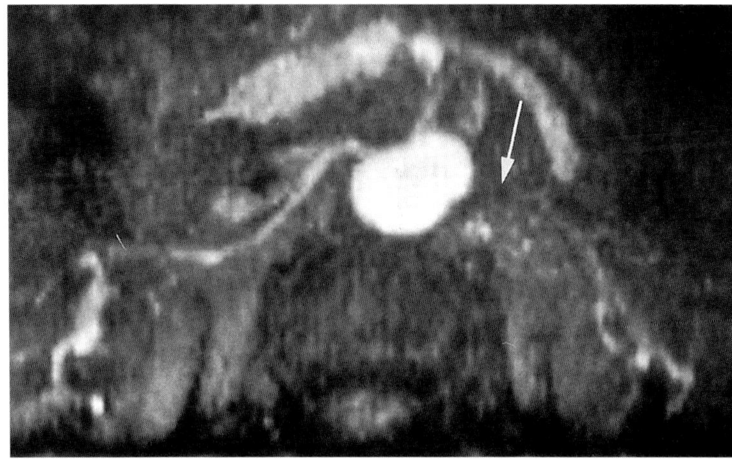

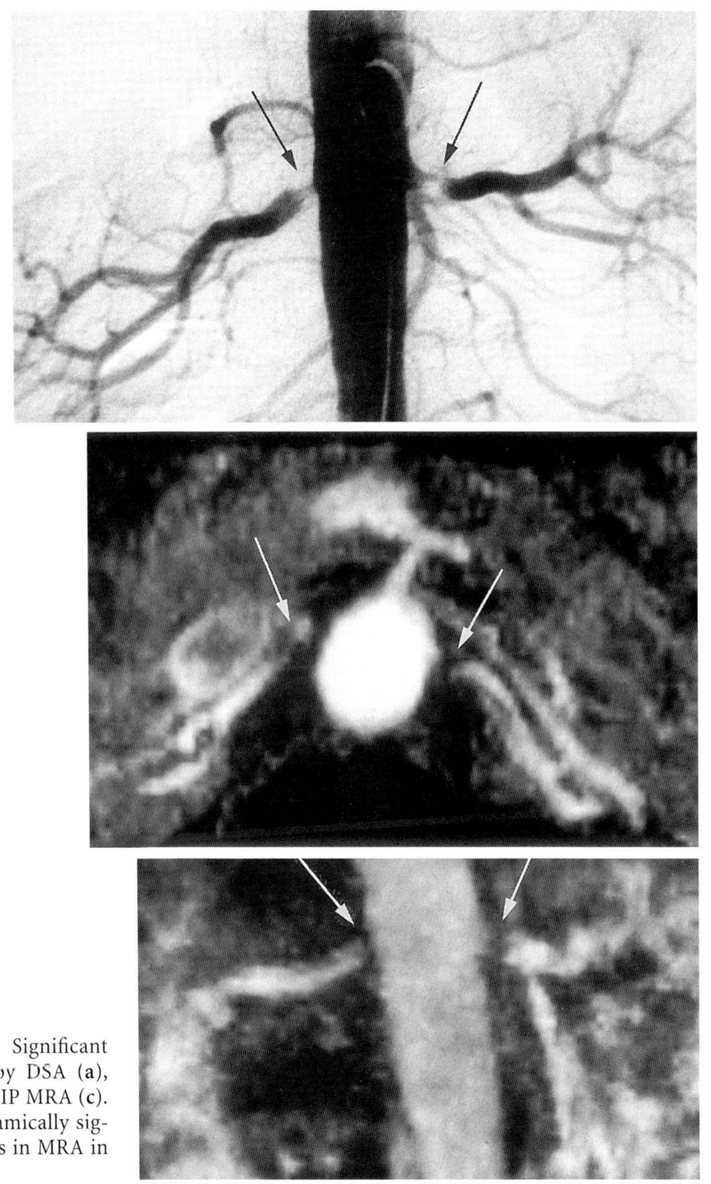

Fig. 15.14a–c. Bilateral atherosclerotic RAS. Significant stenoses (*arrows*) can be accurately depicted by DSA (**a**), transverse 2D TOF MIP MRA (**b**), and 3D TOF MIP MRA (**c**). Note that RAS should be evaluated as hemodynamically significant only when there is a complete signal loss in MRA in the area of stenosis

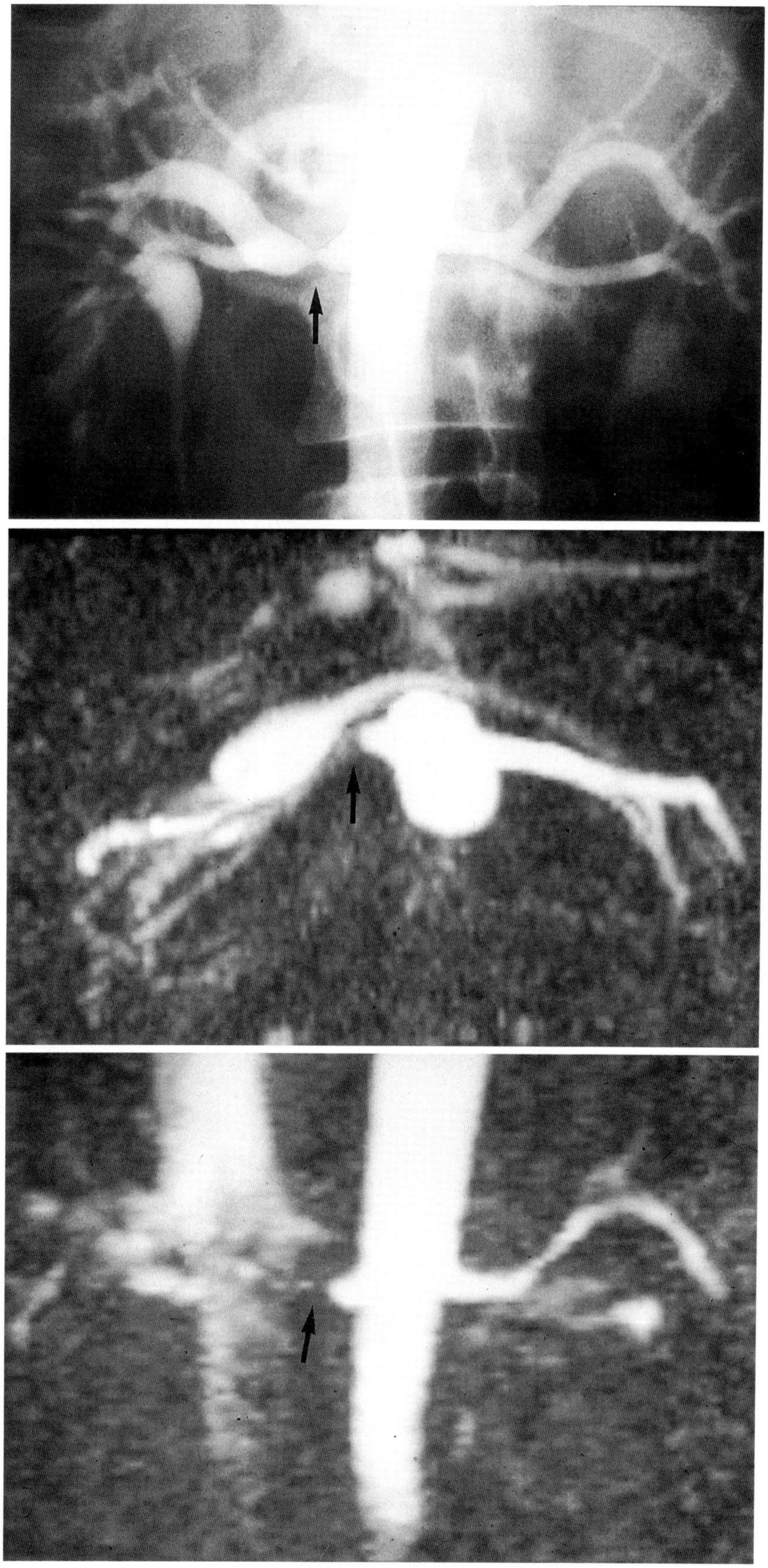

Fig. 15.15a–c. Atherosclerotic RAS of the right kidney. Stenosis and poststenotic dilatation in a conventional angiogram (**a**) can be detected accurately on transverse (**b**) and coronal (**c**) images using a 2D PC MRA technique (*arrows*). A complete signal void is documented in the area of stenosis. (Courtesy of Dr. Jörg F. Debatin, MRI Center of Zurich University Hospital) (Andersen et al., Clin. Magnet. Res. Angiogr., Raven Press 1993, Chap. 18, pp. 432–433)

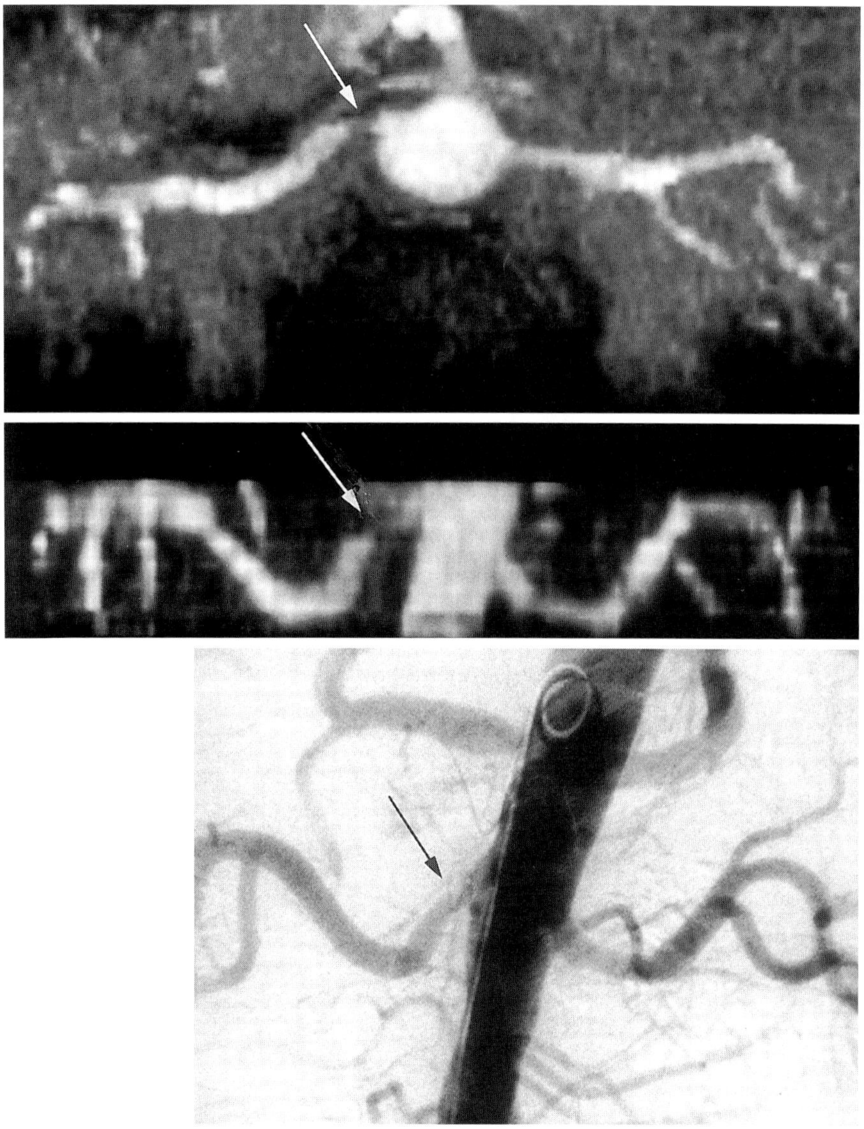

a

b

c

Fig. 15.16a–c. Atheromatous plaque at the ostium of the right renal artery. In a 72-year-old male patient with arterial hypertension, renal arteries were examined initially by MRA in order to exclude a renovascular etiology. On both transverse (**a**) and the coronal (**b**) 2D TOF MR-projection angiograms (MIP reconstruction from transverse slices) a significant signal reduction could be observed close to the origin of the right renal artery (*arrow*), suggesting the diagnosis of RAS. However, when i.a. DSA was performed in this region only a small atherosclerotic plaque was found (**c**), indicating that the signal void was due to turbulent flow

nation techniques such as US show limitations, and when iodinated contrast cannot be used due to the need to prevent renal failure in patients with preexisting renal insufficiency, which frequently is associated with diabetes and hypertension. In detecting RAS, PC MRA techniques which allow additional flow measurement seem superior to conventional TOF techniques. Nevertheless, a combination of 2D and 3D acquisition in coronal and transverse planes should be recommended in order to improve the accuracy of MRA.

Despite the foregoing considerations, renal MRA results have not been found acceptable by most of the users – hence the efforts to improve image quality and accuracy. In spite of the introduction of a variety of different modalities on the basis of TOF MRA, including breath-holding, presaturation techniques, fat saturation, peak flow acquisition, k-space segmentation, TONE, MOTSA, STAR, and echo planar imaging, detection of accessory renal arteries and RAS is not satisfactory and RAS grading is currently unreliable as compared with conventional

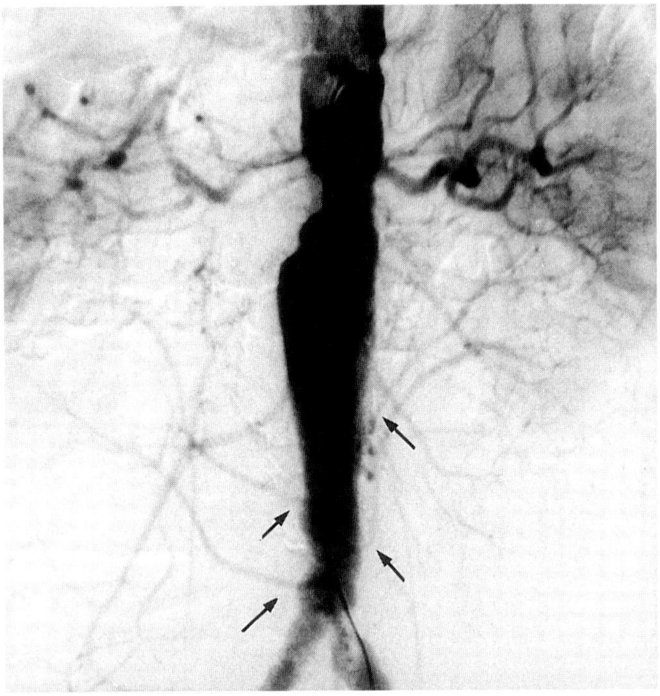

a

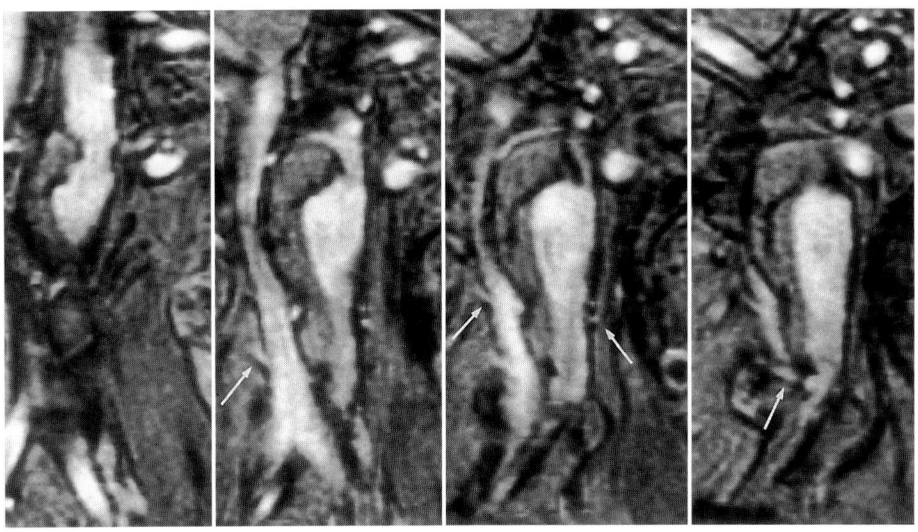

b

Fig. 15.17a–c. Multiple accessory renal arteries in atherosclerotic abdominal aortic aneurysmal disease. On the DSA image (**a**) both kidneys are supplied by an upper main renal artery and two lower accessory arteries (*arrows*) originating from the aneurysmal sac. These accessory renal arteries (*arrows*) cannot be reliably identified without knowledge of the DSA findings either on individual 2D coronal MRA slices (**b**) or on a coronal 2D MIP angiogram (**c**) reconstructed from transversely acquired slices

angiography. Moreover, current data from helical CTA demonstrate that this technique seems more reliable than MRA. More clinical experience is necessary in the field of both CTA and newer MRA techniques, in particular gadolinium contrast MRA. As a screening modality for renovascular hypertension MRA has to be faster, more comfortable, more accurate, and less expensive.

Acknowledgment. The authors thank Mrs. E. Griepentrog and Mrs. I. Meurer for assistance with data acquisition and preparation of image material.

References

Amanuma M, Mohiaddin RH, Hasegawa M, Heshiki A, Longmore DB (1992) Abdominal aorta: characterization of blood flow and measurement of its regional distribution by cine magnetic resonance phase-shift velocity mapping. Eur Radiol 2:559–564

Amparo EG, Hoddick WK, Hricak H, Sollitto R, Justich E, Filly RA, Higgins CB (1985) Comparison of magnetic resonance imaging and ultrasonography in the evaluation of abdominal aortic aneurysm. Radiology 154:451–456

Anderson CM, Edelman RR, Turski PA (1993) Clinical magnetic resonance angiography. Raven Press, New York

Fig. 15.17c

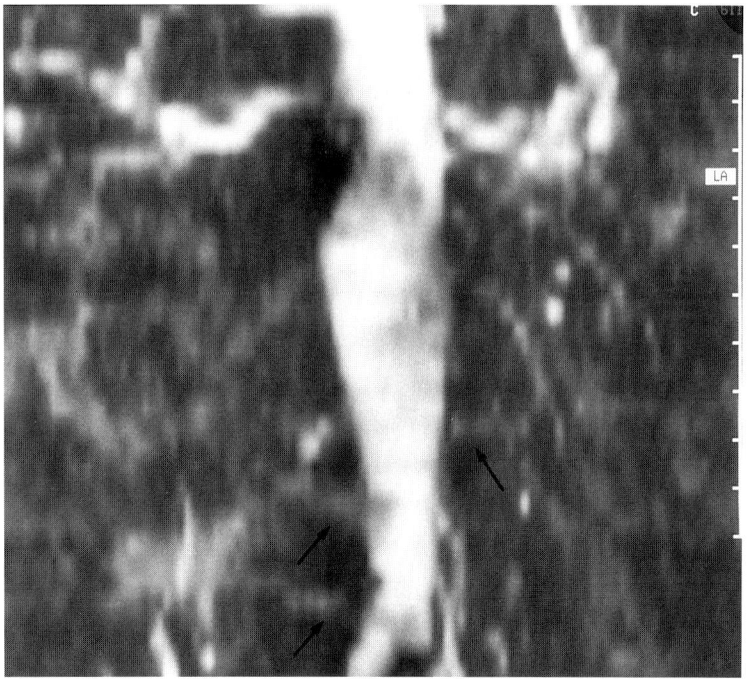

Arlart IP, Guhl L, Edelman RR (1992a) Magnetic resonance angiography of the abdominal aorta. Cardiovasc Intervent Rodiol 15:43-50

Arlart IP, Guhl L, Hausmann R (1992b) Evaluation of 2D- and 3D-time-of-flight MRA in renal artery stenosis. ROFO 157:59-64

Cullenward MJ, Scanlan KA, Pozniak MA, Acher CA (1986) Inflammatory aortic aneurysm (periaortic fibrosis): radiologic imaging. Radiology 159:75-82

Darling RC, Messina CR, Brewster DC, Ottinger LW (1977) Autopsy study of unoperated abdominal aortic aneurysms: the case for early resection. Circulation 56 (suppl II): 161-164

Debatin JF, Spritzer CE, Grist TM, Beam C, Svetkey LP, Newman GE, Sostman HD (1991) Imaging of the renal arteries: value of MR angiography. AJR 157:981-990

Debatin JF, Ting RH, Wegmüller H et al. (1994) Renal artery blood flow: quantification with phase contrast MR imaging with and without breath holding. Radiology 190:371-378

Dumoulin CL, Yucel EK, Vock P, Souza SP, Terrier F, Steinberg FL, Wegmuller H (1990) Two- and three-dimensional phase contrast MR angiography of the abdomen. Comput Assist Tomogr 14:779-784

Ecklund K, Hartnell GG, Hughes LA, Stokes KR, Finn JP (1994) MR angiography as the sole method in evaluating abdominal aortic aneurysms: correlation with conventional techniques and surgery. Radiology 192:345-350

Edelman RR, Wentz KU, Mattle H, Zhao B, Liu C, Kim D, Laub G (1989) Projection arteriography and venography: initial clinical results with MR. Radiology 172:351-357

Evancho M, Osbakken M, Weidner W (1985) Comparison of NMR imaging and aortography for preoperative evaluation of abdominal aortic aneurysm. Magn Reson Med 2:41-45

Galanski M, Prokop M, Chavan A, Schaefer CM, Nischelsky JE (1992) Renal artery stenoses: spiral CT angiography. Radiology 189:185-192

Gross-Fengels W, Friedmann G, Lemaitre F, Schmidt R, Lanfermann H, Erasmi H (1991) Conventional and cine-MRI in patients with aneurysms of the abdominal aorta. ROFO 155:337-343

Harrison EG Jr, McCormack LJ (1971) Pathologic classification of renal arterial disease in renovascular hypertension. Mayo Clin Proc 46:161-167

Havey RJ, Krumloysky JE, DelGreco F, Gartner HM (1985) Screening for renovascular hypertension. Is renal digital subtraction angiography the preferred non-invasive test? JAMA 254:388-393

Kadir S (1986) Diagnostic angiography. Saunders, Philadelphia

Kandarpa K, Piwnica-Worms D, Chopra PS et al. (1992) Prospective double-blinded comparison of MR imaging and aortography in the preoperative evaluation of abdominal aortic aneurysms. JVIR 3:83-89

Kent KG, Edelman RR, Kim D, Steinman TI, Porter DH, Skillman JJ (1991) Magnetic resonance imaging: a reliable test for the evaluation of proximal atherosclerotic renal arterial stenosis. J Vasc Surg 13:311-318

Kim D, Edelman RR, Kent KG, Porter DH, Skillman JJ (1990) Abdominal aorta and renal artery stenosis: evaluation with MR angiography. Radiology 174:727-731

Koslin DB, Kenney PJ, Keller FS, Koehler RE, Gross GM (1988) Preoperative evaluation of abdominal aortic aneurysm by MR imaging with aortography correlation. JCVIR 11:329-335

Li D, Haacke EM, Mugler JP, Berr S, Brookeman JR, Hutton MC (1994) Three-dimensional time-of-flight MR angiography using selective inversion recovery RAGE with fat saturation and ECG-triggering: application to renal arteries. Magn Reson Med 31:414-422

Loubeyre P, Revel D, Garcia P et al. (1994) Screening patients for renal artery stenosis: value of three-dimensional time-of-flight MR angiography. AJR 162:847–852
Owen RS, Baum RA, Carpenter JP, Holland GA, Cope C (1993) Symptomatic peripheral vascular disease: selection of imaging parameters and clinical evaluation with MR angiography. Radiology 187:627–635
Parker DL, Yuan C, Blatter DD (1991) MR angiography by multiple thin slab 3D acquisition. Magn Reson Med 17:434–451
Prince MR (1994) Gadolinium-enhanced MR aortography. Radiology 191:155–164
Prokop M, Schaefer C, Kalender WA, Polacin A, Galanski M (1993) Vascular imaging with spiral CT. Radiologe 33:694–704
Purdy D, Cadena G, Laub G (1992) The design of variable flip angle slab selection (TONE) pulses for improved 3D MR angiography. In: Book of abstracts, vol 1. Eleventh annual meeting, Soc Magn Res Med, Berkeley, Calif., USA, p 882
Richter CS, Krestin GP, Eichenberger AC, Schöpke W, Fuchs WA (1993) Assessment of renal artery stenosis by phase-contrast magnetic resonance angiography. Eur Radiol 3:493–498
Rubin GD, Dake MD, Napel S et al. (1994) Spiral CT of renal artery stenosis: comparison of three-dimensional rendering techniques. Radiology 190:181–190
Ruff SJ, Watson MR (1988) Magnetic resonance imaging versus angiography in the preoperative assessment of abdominal aortic aneurysm. Am J Surg 155:651–654
Schwerk WB, Restrepo IK, Stellwaag M, Klose KJ, Schade-Brittinger C (1994) Renal artery stenosis: grading with image-directed Doppler US evaluation of renal resistive index. Radiology 190:785–790
Siegel CL, Cohan RH (1994) CT of abdominal aortic aneurysms. AJR 163:17–29
Strotzer M, Pruell CM, Geissler A, Kohler SM, Kraemer BK, Gmeinwieser J (1993) Noninvasive assessment of renal artery stenosis: a comparison of MRA, angiography, color Doppler sonography, and intraarterial angiography. Radiology 189 (suppl): 189
Vollmar J (1985) Das Bauchartenaneurysma: Wandel in der Diagnostik und chirurgischen Therapie. Chirurg 56:238–242
Yucel EK, Kaufman JA, Prince M, Bazari H, Fang LST, Waltman AC (1993a) Time of flight renal MR angiography: utility in patients with renal insufficiency. Magn Reson Imaging 11:925–930
Yucel EK, Kaufman JA, Geller SC, Waltman AC (1993b) Atherosclerotic occlusive disease of the lower extremity: prospective evalation with two-dimensional time-of-flight MR angiography. Radiology 187:637–641

16 Arteries of the Lower Extremities

R. VOSSHENRICH

CONTENTS

16.1 Introduction

The widespread prevalence of atherosclerotic vascular disease has given rise to the need for a simple imaging examination. The angiographic capabilities of magnetic resonance imaging (MRI) have recently become applicable in the clinical setting. MR angiography (MRA) has proved useful in the evaluation of intracranial, carotid, and abdominal vessels (ALFIDI et al. 1987; EDELMANN 1993). MRA of the lower extremities can provide noninvasive arterial images. Recent studies on phantoms and examinations in healthy subjects or in patients with peripheral occlusive disease show encouraging results (BORELLO 1993; CAPUTO et al. 1992; MULLIGAN et al. 1991; YUCEL et al. 1992). Examinations of the lower limbs, however, have been hindered by the need to depict long segments of vessels and the requirements of a high signal-to-noise ratio, high spatial and contrast resolution, and acceptably short imaging times.

Peripheral vascular disease has been evaluated using conventional imaging modalities such as contrast medium-enhanced x-ray angiography to depict the morphology, and duplex ultrasonography to evaluate the hemodynamic significance of a stenosis.

R. VOSSHENRICH, MD, Georg-August-Universität, Zentrum Radiologie, Abt. Röntgendiagnostik I, Robert-Koch-Straße 40, 37075 Göttingen, FRG

Since its introduction in the 1950s, conventional x-ray arteriography has provided the anatomical information necessary to plan vascular intervention therapy. Although modern conventional or digital angiography techniques are generally safe, they are not without risk and discomfort. Puncture site complications include bleeding, thrombosis, infection, dissection, pseudoaneurysm, and arteriovenous fistula. Most catheter-induced hematomas are benign and self-limited. In some patients, however, puncture-site complications can require surgical repair and prolonged hospital stay. With the exception of contrast medium-induced renal failure, major systemic reactions due to conventional arteriography are rare but include anaphylaxis, angina pectoris, and congestive heart failure (WAUGH and SACHARIAS 1992).

Noninvasive diagnostic methods for peripheral vascular disease include a series of Doppler-assisted pressure measurements, analysis of the femoral and popliteal artery waveforms by continuous wave Doppler, and more recently the use of duplex Doppler to evaluate focal and diffuse arterial disease. The four most commonly used noninvasive diagnostic tests are the ankle-to-brachial index, segmental arterial pressures, toe pressures, and the femoral artery pulsatility index. These may be useful in localizing segmental arterial disease but do not determine the extent or severity of disease. Color Doppler flow imaging has evolved from duplex Doppler ultrasonography. By providing an anatomical arterial map and defining the hemodynamic significance of a stenosis, it can identify lesions amenable to interventional vascular procedures. Color-coded ultrasonography can be used as a noninvasive screening test (BAUMGARTNER et al. 1993; FOLEY 1991).

16.2 MRA Techniques

One of the major difficulties encountered in MRA is the suppression of the large static sample signal.

Several techniques have been developed to remedy this problem, including the time-of-flight or inflow method and the phase-contrast method. The former uses bipolar gradient pulses for the flow encoding while the latter uses the saturation effect of the static spins.

16.2.1 Time-of-flight Technique

Since MRA of the lower limbs requires imaging of a large area of the body, the long scanning time makes standard 3D techniques impractical.

Transverse 2D TOF MRA is widely used for evaluation of the extremities. The signal from flowing blood is greatest when the plane is as perpendicular as possible to the longitudinal axis of the blood vessels. In the lower extremities this is achieved by imaging in the axial plane. Inflow enhancement occurs for spins entering the slice from either direction. This ordinarily leads to an image depiction of both arteries and veins. A simple way to reduce the inflow enhancement of the venous blood is to use presaturation pulses placed distal to the imaging slice. The purpose of the saturation band is to reduce the longitudinal magnetization of the blood by the application of 90° pulses in combination with a spoiler gradient. A sufficiently thick band should be used so that high-velocity unsaturated spins cannot pass through this region without experiencing a presaturation pulse. For best results, this saturation band needs to be placed adjacent to the imaging slice. An excessive gap between these regions will permit the longitudinal magnetization of the spins before they enter the imaging slice. The loss of presaturation is directly related to the time that the spins have to recover. The longer the distance between the presaturation region and the imaging slice, and the slower the velocity, the less efficient the presaturation becomes. To obtain best presaturation, the saturation band should remain adjacent to the slice as the slice position changes. This is referred to as a "travelling" or "tracking" saturation pulse.

Contrast between rapidly flowing blood and stationary tissue depends on the repetition time (TR) and the flip angle (FA). Generally TR should be as short as possible to minimize the acquisition time. A TR of 27–50 ms was found to provide high vascular signal intensity in reconstructed vascular segments and small collateral vessels. In addition it enables the shortest imaging time possible and provides excellent soft tissue suppresion. The choice of the flip angle also has a profound effect on the contrast between vascular and soft tissue signal intensity and the signal of the flowing spins. The vascular signal intensity seems to be relatively independent of an FA greater than 40°. However, when such flip angles are used, increased ghost intensity drops occur. This demonstrates that retrograde flow in the artery is a significant problem. Therefore an FA of 40–60° is recommended as a good compromise.

Furthermore, the technique should cover the entire vascular tree from the distal aorta to the ankles (approximately 120 cm) within a comfortable imaging time. Since spatial resolution into z direction does not need to be as high as in the carotid arteries, a 5-mm slice thickness in the pelvis and the thighs seems reasonable. It allows complete imaging from the aortic bifurcation to the popliteal arteries in approximately 20 min while retaining adequate spatial resolution. In the calf, the anterior tibial, the posterior tibial, and the peroneal arteries may be 2 or 3 mm in diameter. Therefore, 3-mm slices combined with a higher resolution (192 × 256) than in the upper vascular tree (160 × 256) will yield images of better quality. In addition, the rather horizontal course at the origin of the anterior tibial artery causes partial volume effects which hamper visualization of this segment when using 5-mm-thick slices. With this technique the average examination time for body coil imaging of the lower extremities is less than 30 min.

Powerful computer performance is advisable when reconstructing the large number of images into projection angiograms. Using the maximum-intensity projection (MIP) algorithm, the lower extremity arterial tree can be imaged on a single coronal projection, supplemented by 45° oblique views of the iliac and femoral bifurcation.

16.2.2 Phase-Contrast Technique

With the PC technique, flow is detected and imaged on the basis of the shift in the phase of magnetization that occurs when spins move in the presence of a magnetic field gradient (Dumoulin and Hart 1986; Dumoulin et al. 1988). The PC technique can also be implemented in 2D and 3D slice acquisitions comparable to standard gradient-echo acquisition modes. Since it does not depend on inflow, a unique feature of PC MRA is the possibility of obtaining direct projection similar to conventional angiograms, without the need for any slice selection.

Flow contrast in a standard PC study depends on the ratio of the velocity-encoding gradient (VENC) to the velocity. The best contrast is achieved when the VENC is set just higher than the maximal velocity since this produces the largest phase shift without aliasing. If one combines the images from a standard PC study using a constant VENC, then good flow contrast is obtained for peak systolic flow. There is little signal contribution other than noise from slower diastolic flow. To overcome this problem, the VENC must be set to a different value for each phase of the cardiac cycle so that the flow contrast is maximized throughout the R-R interval.

The advantage of PC is that it is inherently quantitative. Since the amplitude and duration of the flow-encoding pulses are known, the velocity of blood flow can readily be calculated from the observed phase shift.

16.3 Applications of MRA

16.3.1 Normal Anatomy

Before starting MR examinations of the lower limbs, some anatomical characteristics of the peripheral arteries in the lower extremities must be taken into account. To provide an adequate evaluation of peripheral arterial disease, it is necessary to image a large area from the distal aorta to the calf (Fig. 16.1).

The aorta enters the abdominal cavity thought the aortic hiatus of the diaphragm anterior to the 12th vertebral body. It generally lies anterior to the lumbar vertebral bodies, to the left of the midline, and follows a relatively straight course to its bifurcation near the L4 vertebral body. In older individuals and those with atherosclerotic disease, it can pursue a tortuous course. The bifurcation gives rise to the common iliac arteries that run inferolaterally along the medial aspect of the psoas muscle. The bifurcation into internal and external iliac arteries usually occurs near the midportion of the sacroiliac joint. The internal iliac arteries supply the pelvic viscera and have variable anatomy. They are important as collateral pathways for blood flow in occlusive disease of the common and external iliac arteries. The external iliac arteries have only two branches, the deep circumflex iliac artery and the inferior epigastric artery, that also function as collaterals in occlusive disease. The external iliac arteries act mainly as conduits from the pelvis to the thigh. Below the inguinal ligament, they become the common femoral arteries.

The common femoral arteries are readily palpable near the inguinal crease. They are very superficial here, lying anterior to the femoral head and iliopsoas tendon. At the common femoral bifurcation, the deep femoral artery runs posterolaterally or directly posteriorly, while the superficial femoral artery pursues a caudal and somewhat medially directed course. The deep femoral artery is the most important vessel in the thigh by virtue of the collateral network that can develop from its circumflex femoral branches. The superficial femoral artery acts as a conduit between the thigh and the calf, having few branches. One branch, the descending genicular artery, a potential collateral about the knee, is the angiographic marker of the distal end of the superficial femoral artery and the beginning of the popliteal artery.

The popliteal artery is palpable lateral to the midline in the popliteal fossa. The four genicular arteries, which arise from the popliteal artery, create an important collateral network about the knee joint.

At the inferior border of the popliteal muscle, the popliteal artery gives rise to the anterior tibial artery and the tibioperoneal trunk. The anterior tibial artery pierces the intraosseous membrane to enter the anterior muscle compartment of the calf. It becomes the dorsalis pedis artery in front of the ankle and enters the foot. The tibioperoneal trunk is of variable length and bifurcates to give rise to the posterior tibial and peroneal arteries. Both run through the posterior muscle compartment of the calf. The posterior tibial artery enters the foot just posterior to the medial malleolus.

Furthermore, specific flow phenomena must be noted, especially in the popliteal and tibioperoneal arteries. While flow in the carotid arteries occurs throughout the cardiac cycle, resulting in a relatively constant flow pattern, the flow pattern found in the lower extremities exhibits a triphasic wave form. Forward flow occurs during systole, followed by a brief period of retrograde flow in the early diastole, and a low foward flow in mid and late diastole. These physiological findings have an important effect on the images obtained with different MRA techniques.

16.3.2 Pathology

16.3.2.1 Peripheral Arterial Occlusive Disease

Peripheral arterial disease reflects a systemic process which often occurs with concomitant involvement of

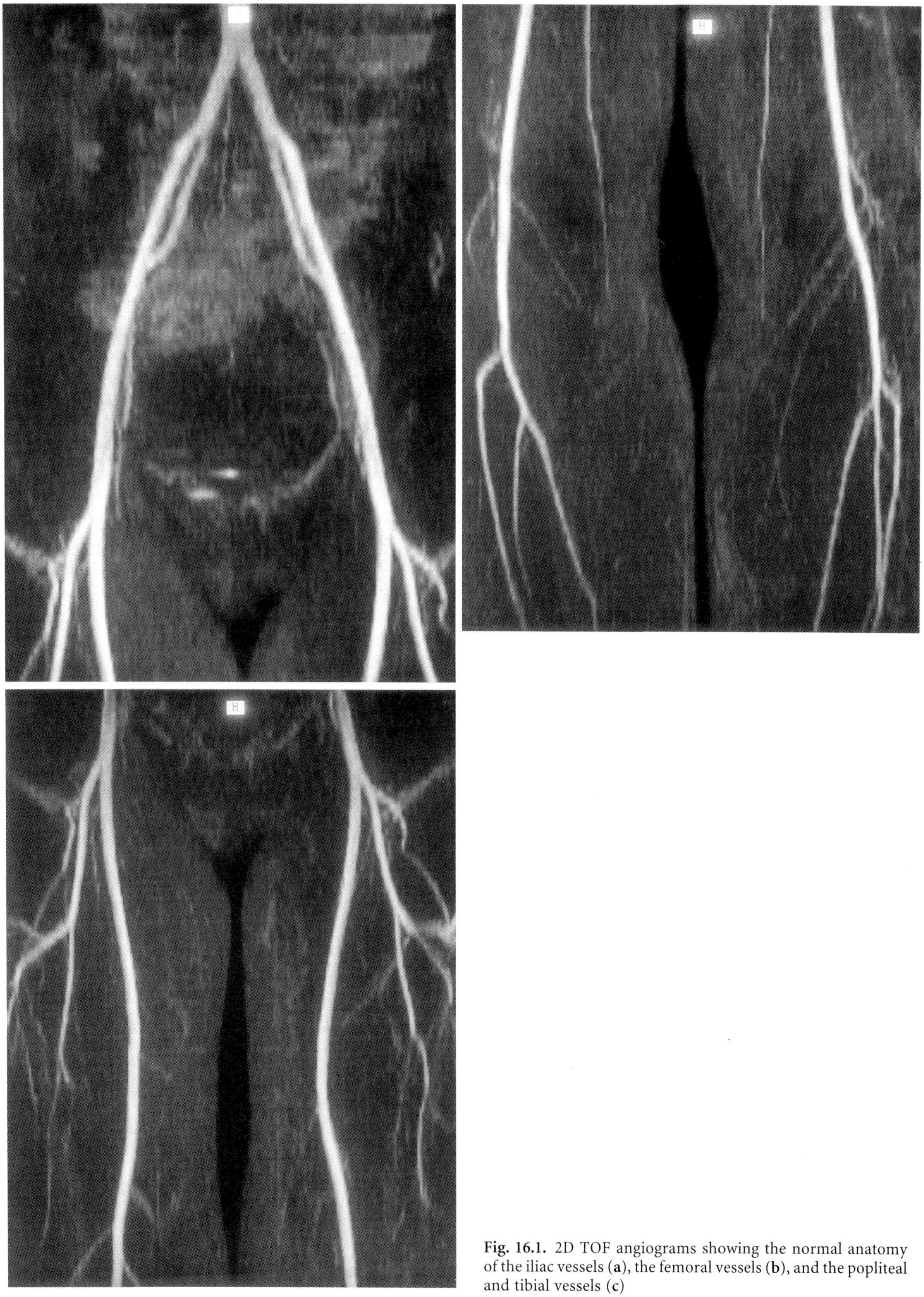

a b c

Fig. 16.1. 2D TOF angiograms showing the normal anatomy of the iliac vessels (**a**), the femoral vessels (**b**), and the popliteal and tibial vessels (**c**) the cerebral, coronary, and renal vessels (DUPREZ 1992; HERTZER 1991). Damage to arteries promotes the development of intimal hyperplasia and atherosclerotic plaque, which results in luminal narrowing and possible ischemia. Patients with peripheral arterial occlusive disease may present with claudication or an acute limb-threatening event. Atherosclerotic lesions develop in regions of mechanical stress or vascular branching. The most common peripheral site of involvement is the distal superficial femoral artery in the region of the adductor hiatus. Additional sites, in order of decreasing frequency, are the iliac vessels, calf vessels, and popliteal and common femoral arteries (RUDOLFSKY 1988). Isolated aortoiliac lesions tend to occur in young patients (40–60 years old) with a history of smoking. Femoropopliteal disease is seen in older patients (60–80 years old). Distal disease is often associated with diabetes.

In the past, treatment options were limited to conservative management of symptomatic disease or amputation of threatened limbs. In addition to surgical bypass or thromboendarterectomy, recently introduced percutaneous interventional treatments (angioplasty, arterectomy, stenting, and thrombolytic therapy) have expanded the spectrum of treatable lesions.

In patients with symptoms or signs of peripheral arterial occlusive disease, the vascular surgeon must determine whether surgery is indicated or not. Contrast arteriography is considered when the workup suggests that surgery should be done. Arteriography provides the surgeon with information about the anatomical location of disease and the length of the diseased segment. The length of the diseased segment is important because short-segment disease, whether stenotic or occlusive, is amenable to angioplasty. Thus, arteriography is regularly used as a preoperative road map. This makes MRA an attractive option in the evaluation of the lower limb.

Several MRA studies in patients with peripheral arterial occlusive disease produced excellent results. Using x-ray angiography as a standard of comparison, a sensitivity of 79%–100% and a specificity of 90%–100% in stenoses of 50% or greater was demonstrated (OWEN et al. 1993; SCHNALL et al. 1993; YUCEL et al. 1993). Other studies show different results (MCCAULEY et al. 1994; QUINN et al. 1993; VOSSHENRICH et al. 1993). The major limitation of all MRA techniques is their susceptibility to signal loss from intravoxel phase dispersion. This leads to an overestimation of the degree of stenosis for higher grades of obstruction to the point that, in severe cases, a complete signal void may be produced at the site of stenosis (Fig. 16.2). Several methods are or will be available to decrease intravoxel phase dispersion and improve the imaging of stenotic lesions. Most important is the reduction of echo time (TE). If TE is reduced to 6 ms or less, a significant improvement occurs in the visualization of stenoses. Intravoxel phase dispersion can also be reduced by decreasing voxel size or by decreasing the slice thickness. In MRA of the lower limb both methods have limited application because they prolong the examination time and reduce the signal-to-noise ratio of the images.

Another difficulty to be aware of is the typical appearance of vessels with horizontal flow. These vessels tend to have a signal of lower intensity and show artifactual stenosis or complete occlusion. The iliac arteries are a typical example. Because of the more oblique course of these segments, and abdominal and bowel motion artifacts, a substantial amount of in-plane flow may result. This in-plane flow causes partial saturation of the flowing spins and therefore diminishes signal intensity from flowing blood. The loss of signal intensity is more marked on the maximum-intensity-projection images (Fig. 16.3). To avoid misdiagnosis, it is necessary to examine the actual axial sections. In clinical pratice, correlation of suspected iliac stenosis with hemodynamic measurements is mandatory. PC MRA may prove superior to TOF MRA in tortuous iliac segments because of its ability to depict blood flow in any desired plane. Further studies are required, however, to verify this potential.

The evaluation of horizontal vessel segments in the tibioperoneal trunk is also difficult using 2D TOF MRA. As described for the iliac arteries, in-plane flow can cause partial saturation of the flowing spins, resulting in diminished flow-related enhancement. In this situation, the additional use of the 3D method and reduction of the slice thickness provide an improvement in visualization (Fig. 16.4). The time demands of 3D acquisitions resulting from the tendency of flowing blood to become saturated within the imaged volume make this approach somewhat impractical for use in the examination of the whole lower limb.

Two approaches may reduce this problem. One is gadolinium enhancement, which prolongs the distance over which the flow may be imaged using 3D methods (LOSSEF et al. 1992). Its disadvantage is that the accompanying veins also show a marked enhancement, which makes the evaluation of the adja-

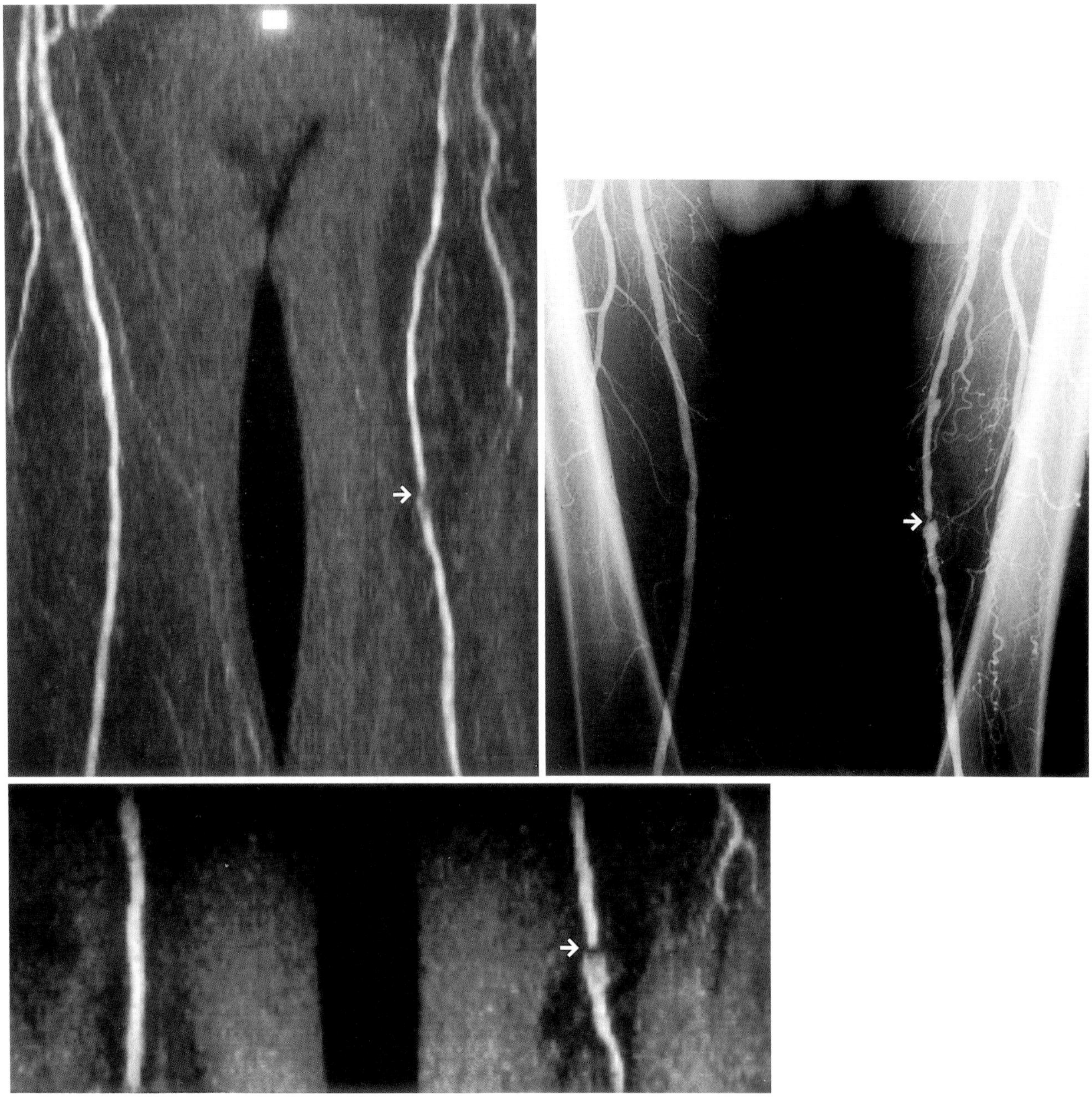

Fig. 16.2. 2D TOF angiogram of the femoral arteries (**a**) demonstrating a short loss of signal in the left femoral artery (*arrow*). The corresponding conventional arteriogram (**b**) shows a severe stenosis and a poststenotic dilatation. An additional 3D TOF angiogram (**c**) confirms this finding

cent arteries more difficult. The other recently developed method is the multislab 3D technique, which combines aspects of both 2D and 3D imaging modes. In this technique, which has been applied to the carotid and cerebral arteries, multiple small 3D slabs are acquired and then concatenated before reprojection. Increased scanning time is still required compared with the 2D TOF techniques because of the need to overlap the slabs.

An additional artifact resulting from pulsatile flow is the presence of horizontal bands of the projected images. This results from each section being acquired during a different point in the cardiac cycle, and thus showing various amounts of flow-related enhancement. This causes the vessel to show different signal intensities in each section, which is appreciated as a horizontally oriented banding artifact. Furthermore, vessels with triphasic flow may have

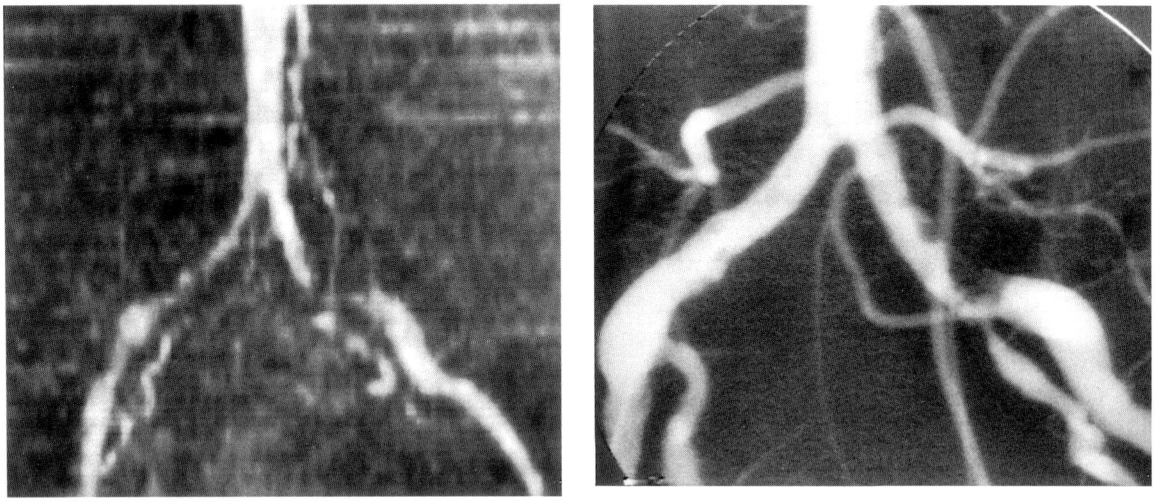

a b

Fig. 16.3. 2D TOF angiogram of the iliac arteries (**a**) demonstrating a short loss of signal in the left iliac artery and a severe stenosis in the right iliac artery. The corresponding digital arteriogram (**b**) shows a severe stenosis in the left iliac artery and no pathology in the right iliac artery

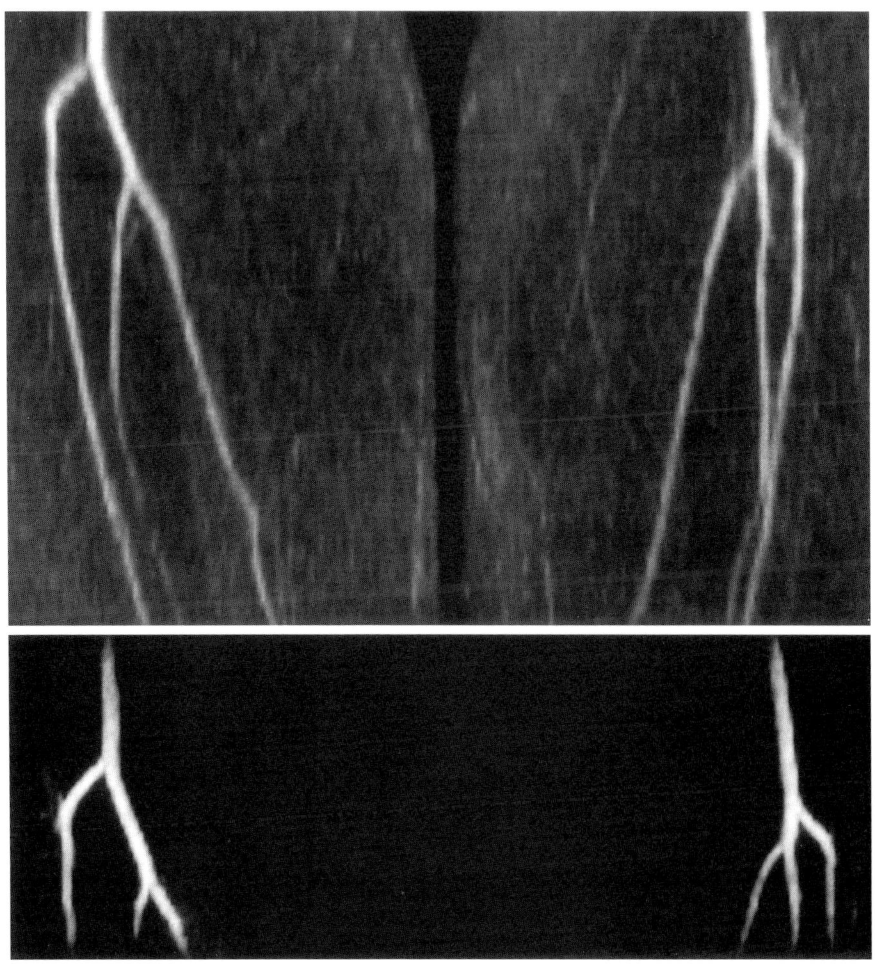

a b

Fig. 16.4. 2D TOF angiogram of the tibioperoneal trunk (**a**) demonstrating the runoff anatomy of the tibial arteries. 3D TOF angiography (**b**) better visualizes the horizontal parts of the anterior tibial arteries more marked banding artifacts, since retrograde flowing blood may transiently enter the region of the tracking saturation pulse. This saturated blood then reverses its flow direction and enters the imaging plane, producing horizontally oriented hypointense bands. This artifact is not typically seen in patients with peripheral arterial occlusive disease. Its occurrence can be diminished by increasing the spacing between the saturation section and the imaging plane. Cardiac gating is another remedy, but it may not be currently available on all commercial 2D TOF software. Beyond this limitation, gating is often difficult in patients with chronic ischemia of the lower limb. Especially in older patients, the application of ECG-triggered sequences is sometimes impractical because of myocardial insufficiency and arrhythmia.

Motion artifacts can also simulate abnormalities in the vasculature. Motion of the patient from section to section during the acquisition of a 2D TOF angiogram results in the apparent offset of the vessel, which can simulate a vascular lesion. The presence of motion can be verified by evaluating the outer contour of the skin and each of the individual sections.

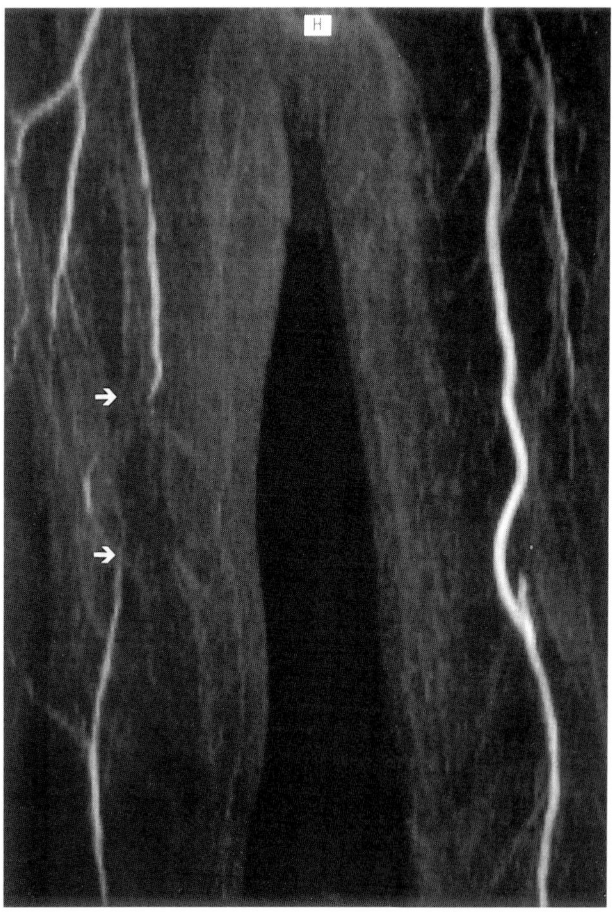

Fig. 16.5. 2D TOF angiogram of the thighs demonstrating a normal femoropopliteal bypass on the patient's left side and a verified occlusion of the right distal femoral artery (*arrows*)

After surgical treatment, it is necessary to document the results (Vosshenrich et al. 1992). The visualization of normal bypass anatomy is no problem (Fig. 16.5). The presence of metallic clips or implants can, however, result in additional artifacts. These structures cause substantial distortion of the static magnetic field, causing signal loss in their vicinity. This is a particular problem in MRA because the used gradient-echo sequences are very susceptible to this effect. The result is an apparent vessel segment occlusion or stenosis. The axial images help confirm the artifactual nature of the finding by demonstrating the signal dropout caused by the metallic body.

It is also possible to use MRA to image the lower limb after percutaneous interventional treatments like angioplasty or stenting, but there is little clinical experience with this application to date. To document the technical result after angioplasty, 2D TOF MRA can be used (Fig. 16.6). Approximately 6 weeks after stenting, these devices become firmly incorporated into the vessel wall and it is unlikely that they would become dislodged by attraction to the magnetic field. However, to image intravascular stents using MRA, adequate sequence parameters will have to be evaluated in the future.

16.3.2.2 Aneurysm of the Lower Limb

The majority of peripheral artery aneurysms are atherosclerotic. Therefore, the detection of an aneurysm in the lower limb indicates the presence of aneurysmal disease at other sites (Fig. 16.7). Popliteal artery aneurysms account for two-thirds of all extremity aneurysms. In one-third of cases there is a concurrent aneurysm of the abdominal aorta. In more than 50% bilateral popliteal aneurysms can be found; the majority are fusiform. Mycotic or traumatic aneurysms are typically saccular (Fig. 16.8).

Fifty percent of the aneurysms are asymptomatic. Especially in popliteal artery aneurysm, the symptoms, when present, usually reflect peripheral ischemia. A local mass effect results in nerve compression and occlusion of the popliteal vein. The diagnosis is facilitated by ultrasonography. Doppler

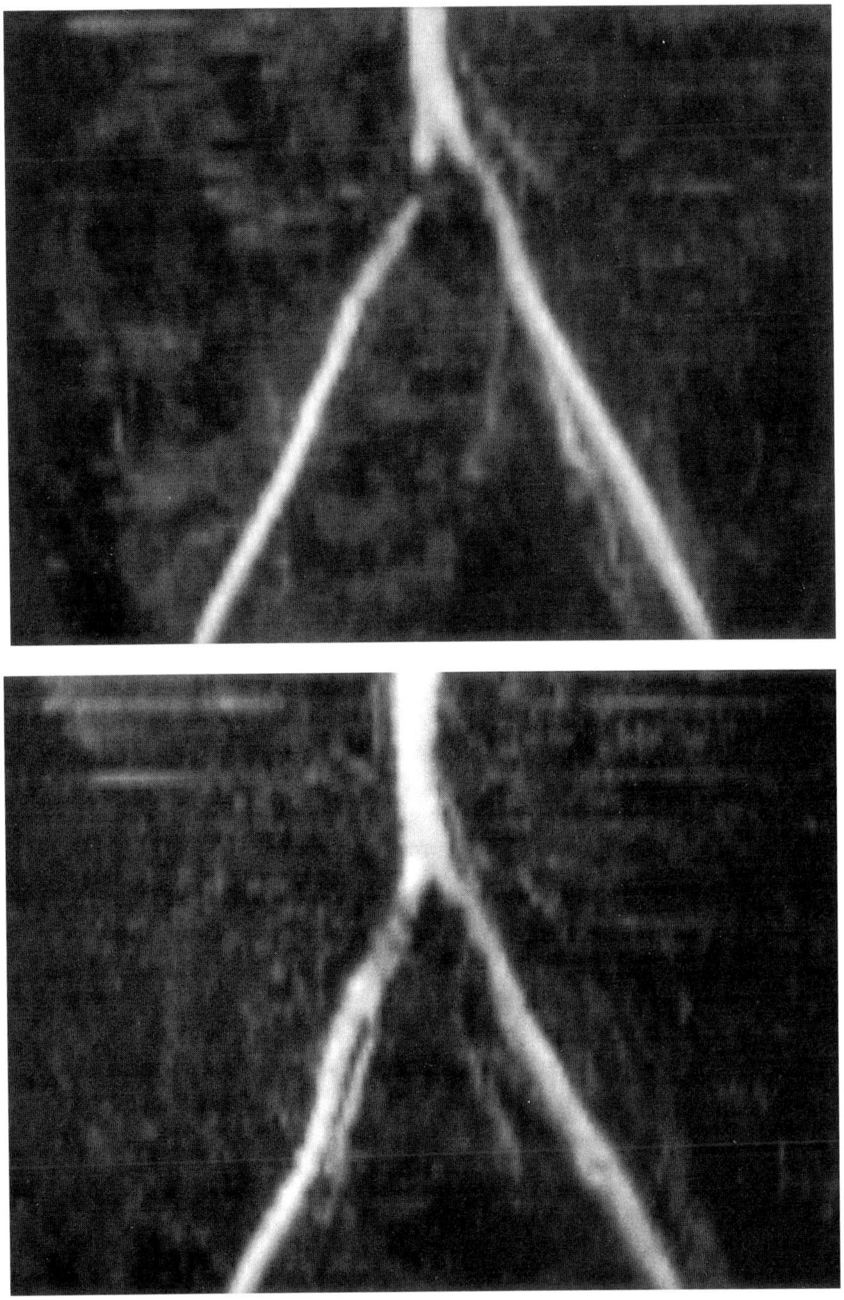

Fig. 16.6. 2D TOF angiogram of the iliac arteries showing a severe stenosis in the right iliac artery before angioplasty (**a**). After dilatation an inhomogeneous signal intensity remains in the right iliac artery (**b**)

ultrasonography helps to distinguish the artery from a Baker's cyst. 2D TOF MRA can visualize the aneurysm and determine its length.

16.4 Conclusions

Although MRA of the lower extremites is in its infancy, it appears to have great potential. Many questions must be satisfactorily answered before MRA can routinely be used as a screening technique or as the definitive diagnostic procedure in these arteries. MRA is able to detect arterial stenoses and determine their length. The degree of the stenosis, however, is often overestimated. Although the spatial resolution of MRA now allows the depiction of the tibial vessels, evaluation of their major branches must be possible before it can be used routinely as a preoperative dia

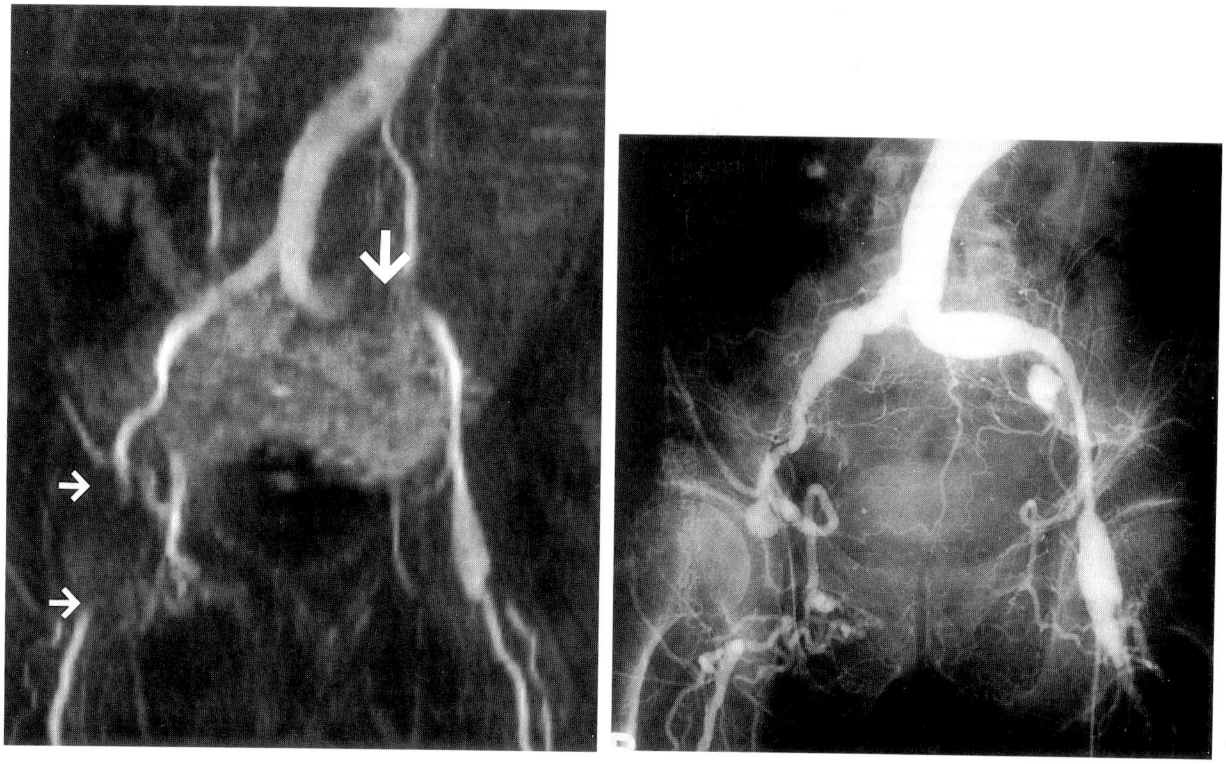

b

Fig. 16.7. 2D TOF angiogram of the iliac arteries (**a**) demonstrating aneurysms and an occlusion of the right common femoral artery (*small arrows*). It suggests an occlusion of the left common iliac artery (*large arrow*). The corresponding arteriogram (**b**) confirms the findings on the patient's right side and shows an aneurysm of the left common iliac artery

a,b

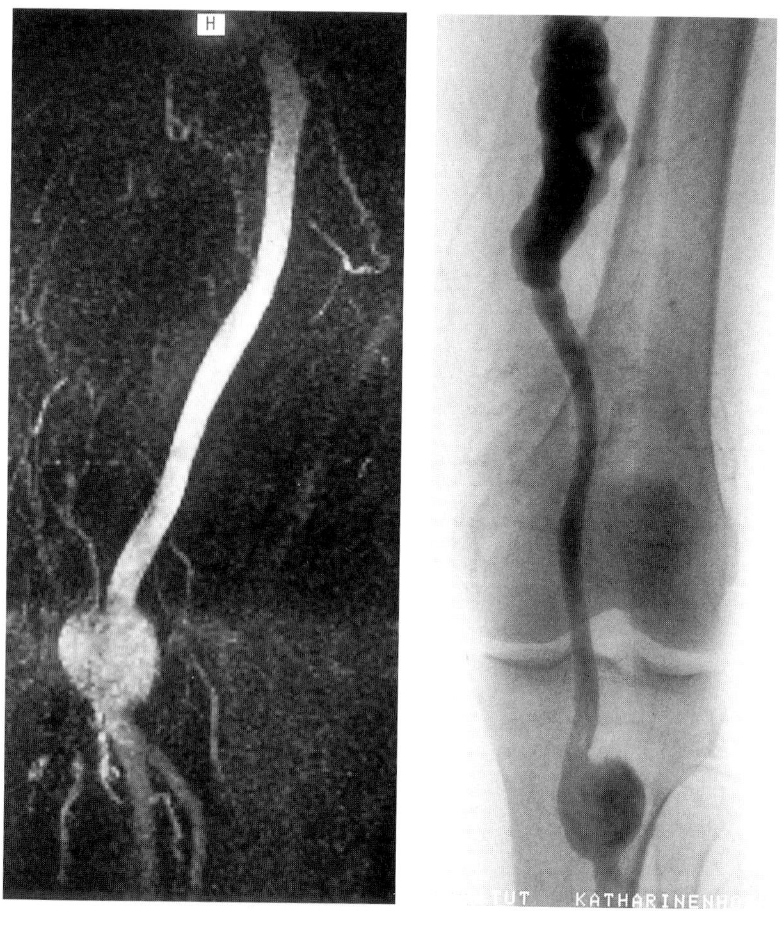

Fig. 16.8. 2D TOF angiogram of the left popliteal artery (**a**) showing a saccular aneurysm. The corresponding arteriogram (**b**) confirms this finding. (Courtesy of I.P. Arlart, Stuttgart, FRG)

agnostic procedure. The postoperative examination of patients with bypasses seems to be a practical application of MRA as an alternative to conventional or digital arteriography.

At present, MRA is cost-effective, but not as readily available as conventional angiography. MRA technology continues to advance rapidly. With improved technology, its judicious use, and appropriate patient selection. MRA should assume a role in the evaluation of peripheral arterial disease in the future.

References

Alfidi RJ, Masaryk TJ, Haacke EM et al. (1987) MR angiography of peripheral, carotid and coronary arteries. Am J Roentgenol 149:1097–1109

Baumgartner I, Maier SE, Koch M, Schneider E, von Schulthess GK, Bollinger A (1993) Magentresonanzarteriographie, Duplexsonographie und konventionelle Arteriographie zur Beurteilung der peripheren arteriellen Verschlußkrankheit. Fortschr Rontgenstr 159:167–173

Borello JA (1993) MR angiography versus conventional X-ray angiography in the lower extremites: everyone wins. Radiology 187:615–617

Caputo GR, Masui T, Gooding GAW, Chang JM, Higgins CB (1992) Popliteal and tibioperoneal arteries: feasibility of two-dimensional time-of-flight MR angiography and phase velocity mapping. Radiology 182:387–392

Dumoulin CL, Hart HR (1986) Magnetic resonance angiography. Radiology 161:717–720

Dumoulin CL, Souza SP, Walker MF, Yoshitome E (1988) Time-resolved magnetic resonance angiography. Magn Reson Med 6:275–286

Duprez D (1992) Natural history and evolution of peripheral obstructive arterial disease. Int Angiol 11:165–168

Edelmann RR (1993) MR angiography: present and future. Am J Roentgenol 161:1–11

Foley WD (1991) Color Doppler flow imaging. Andover, Boston

Hertzer NR (1991) The natural history of peripheral vascular disease. Circulation 83 [suppl I] 12–19

Lossef St V, Rajan SS, Patt RH et al. (1992) Gadolinium-enhanced magnitude contrast MR angiography of popliteal and tibial arteries. Radiology 184:349–355

McCauley TR, Monib A, Dickey KW et al. (1994) Peripheral vascular occlusive disease: accuracy and reliability of time-of-flight MR angiography. Radiology 192:351–357

Mulligan SA, Matsuda T, Lanzer P et al. (1991) Peripheral arterial occlusive disease: prospective comparison of MR angiography and color duplex US with conventional angiography. Radiology 178:695–700

Owen RS, Baum RA, Carpenter JP, Holland GA, Cope C (1993) Symptomatic peripheral vascular disease: selection of imaging parameters and clinical evaluation with MR angiography. Radiology 187:627–635

Quinn St F, Demlow TA, Hallin RW, Eidemiller LR, Szumowsli J (1993) Femoral MR angiography versus conventional angiography: preliminary results. Radiology 189:181–184

Rudolfsky G (1988) Kompaktwissen Angiologie. Perimed, Erlangen

Schnall MD, Holland GA, Baum RA, Cope C, Schiebler ML, Carpenter JP (1993) MR angiography of the peripheral vasculature. Radiographics 13:920–930

Vosshenrich R, Muller E, Hausmann R, Fischer U, Vettelschoss M, Grabbe E (1992) 2D time-of-flight MR angiography and phase velocity mapping in patients with femoropopliteal bypasses. Radiology 185 (P): 133

Vosshenrich R, Fischer U, Grabbe E (1993) Erste Erfahrungen mit der MR magnitude contrast angiography an den unteren Extremitäten. Fortschr Rontgenstr 159:393–397

Waugh JR, Sacharias N (1992) Arteriographic complications in the DSA era. Radiology 182:243–246

Yucel EK, Dumoulin CL, Waltman AC (1992) MR angiography of lower-extremity arterial disease: preliminary experience. J Magn Reson Imaging 2:303–309

Yucel EK, Kaufman JA, Geller St C, Waltman AC (1993) Atherosclerotic occlusive disease of the lower extremity: prospective evaluation with two-dimensional time-of-flight MR angiography. Radiology 187:637–641

17 Veins of the Mediastinum and Retroperitoneum

I.P. Arlart and L. Guhl

CONTENTS

17.1 Introduction

Non-invasive ultrasonography (US) is widely used with great success in the upper cervicothoracic region as well as in the retroperioneal space and for evaluating the splenoportal system. However, US plays only a limited role in the diagnostic approach to intrathoracic veins due to technical limitations. The detection of vascular abnormalities suitable for US investigation is facilitated by additional use of color Doppler. Although this technique can significantly improve the accuracy of US due to the ability to perform morphological examination and functional measurements, problems of inaccuracy may occur in quantitative flow determination.

Contrast-enhanced computed tomography (CT) is accepted today for diagnostic body imaging of intrathoracic, mediastinal, and retroperitoneal venous structures, particularly when extravascular tissues have to be evaluated that may involve the venous system.

Indications for conventional x-ray angiography have changed significantly since the introduction of noninvasive techniques that permit visualization of both intraluminal and perivascular disorders that may cause veno-occlusive disease. Use of venous angiography has been reduced to disorders in which detailed intraluminal changes have to be identified or in which the evaluation of pathological collateral or malformed venous structures is required with high-resolution imaging. Furthermore, indications for conventional angiography exist in cases in which noninvasive techniques have yielded equivocal findings or shown limitations due to technical problems, and when interventionl procedures are planned, such as local fibrinolysis in acute occlusive disease, placement of cava filters in recurrent pulmonary embolism, or implantation of endoluminal stents.

Since the introduction of noninvasive MRA into the diagnostic armamentarium for venous disorders, it has been demonstrated that this technique is able to provide information on not only perivascular structures but also intraluminal venous blood flow; this is achieved by means of reconstructed projection angiograms that are similar to conventional angiograms due to use of a two-dimensional sequential slice acquisition mode employing flow-sensitive gradient-echo (GRE) pulse sequences. In addition, successful functional qualitative and quantitative determination of flow velocity by means of MR techniques has been described. In order to validate the usefulness and diagnostic accuracy of MRA it has to be compared with Doppler US and, in particular, with contrast-enhanced CT and conventional angiography, which are accepted as the diagnostic gold standards.

17.2 Technical Considerations

When spin-echo (SE) sequences were used for imaging of mediastinal or abdominal veins, flow effects could be observed which were unpredictable and potentially confusing, and from flow-void images no information was gained about flow direction or velocity. These drawbacks were successfully overcome by the introduction of GRE MRA techniques

I.P. Arlart, Professor, Dr., Ärztlicher Direktor, Radiologisches Institut, Katharinenhospital, Kriegsbergstraße 60, 70174 Stuttgart, FRG
L. Guhl, Dr., Radiologisches Institut, Katharinenhospital, Kriegsbergstraße 60, 70174 Stuttgart, FRG

for venous flow imaging (EDELMAN 1993). Today, SE imaging nevertheless still plays a role as an additional procedure which is recommended when MRA findings are unclear or equivocal (ARRIVE et al. 1991; EDELMAN 1993). Particularly when a 2D time-of-flight (TOF) MRA technique is selected, images of excellent quality can be obtained (ARLART et al. 1991; EDELMAN et al. 1989a; FINN et al. 1993), whereas 3D TOF techniques and phase-contrast (PC) techniques have shown limitations in the assessment of venous structures in the thoracic and abdominal region.

17.2.1 2D TOF MRA

In order to obtain images of thoracic or abdominal vessels without blurring due to motion artifacts it is mandatory to acquire the sections within a breath-hold. Because venous flow is sufficiently steady during breath-holding, cardiac gating is unnecessary, and the presence of ghosting artifacts is not a problem. When a sequential standard 2D TOF GRE FLASH sequence is used, normally both the arterial and the venous flow signal appears with a bright signal. This sequence has been applied most successfully in our institution for imaging of the abdominal vasculature (see Chap. 15). However, when venous structures have to be evaluated, arteriovenous overlap may significantly limit diagnostic accuracy. Thus, arterial flow signal and arterial phase-encoding ghost artifacts should be removed by application of a presaturation pulse (EDELMAN et al. 1989a) (Fig. 17.1a–c). Following a localizer slice through the area of interest, a 5- to 8-cm transverse presaturation band should be localized through the supravalvular ascending aorta to eliminate arterial flow signal in the mediastinum as well as in the abdomen. Using a multislice technique two separate slices can be acquired within a breath-hold (typical sequence parameters used in our institution are cited in the legend to Fig. 17.2). For optimization of flow signal in venous examinations it has proved useful to select a slightly larger flip angle than for arterial examinations.

In coronal slice acquisitions the number of sections required to image the mediastinal or retroperitoneal veins usually ranges from 10 to 15 when a 5 mm slice thickness in chosen. However, for imaging the entire abdomen up to 40 slices may be necessary in order to visualize deep lumbar veins as well as superficial epigastric collateral pathways. The major advantage of coronal imaging is the visualization of the vascular anatomy over a large field of view (up to 45 cm) that covers the complete thorax or abdomen. In-plane saturation can be limited in most cases by a slightly oblique course of the vessels through the sections. However, when focal signal void occurs, which may mimic a vascular lesion, imaging at right angles to the vessel axis is recommended (transverse and sometimes sagittal slice orientation) in order to minimize in-plane saturation of caudocranial venous flow [superior vena cava (SVC), azygos/hemiazygos vein, inferior vena cava (IVC), hepatic veins, mesenteric vein, spermatic/ovarian vein] or horizontal venous flow (brachiocephalic veins, splenoportal vein, renal veins) due to high inflow contrast resulting from the flow direction perpendicular to the slice (Fig. 17.2a–c). When transverse sections are acquired, the sequence design can be used as described above, including the supradiaphragmatic presaturation of arterial flow. In order to cover the hepatic vasculature a mean number of 36 (2 × 18) slices are required. However, when the entire IVC has to be examined the mean number of slices may increase to as many as 70 (2 × 35), which will prolong the examination time significantly. A drawback of supradiaphragmatic presaturation is that in cases with slow arterial flow a desaturation phenomenon may occur in the distal part of the abdominal aorta as well as in the supra-aortic branches. Thus, to eliminate arterial flow signal more effectively a transverse sequential 2D acquisition technique is recommended in which a migrating parallel saturation pulse is integrated (typical sequence parameters are cited in the legend to Fig. 17.2.). Examination time can be reduced without relevant void of resolution by a specific sequence design which includes shortening of TR that enables the acquisition of two slices within a breath-hold of approximately 15 s, the use of a rectangular field of view, and a reduction of phase-encoding steps with a matrix of 128 × 256.

For accurate identification of a vascular lesion and its etiology, analysis of both the individual slices and the MIP angiogram established from single slices has been shown to be mandatory. In addition, targeted MIP angiograms established from selected single sections of interest can reduce disturbing vascular overlap. Intravascular thrombus structures or perivascular lesions can be detected more reliably on individual slices, whereas projection angiograms reconstructed in the transverse or coronal view are predominantly useful for demonstration of venous anatomy. Motion artifacts due to incorrect breath-holding are more relevant in transverse than in

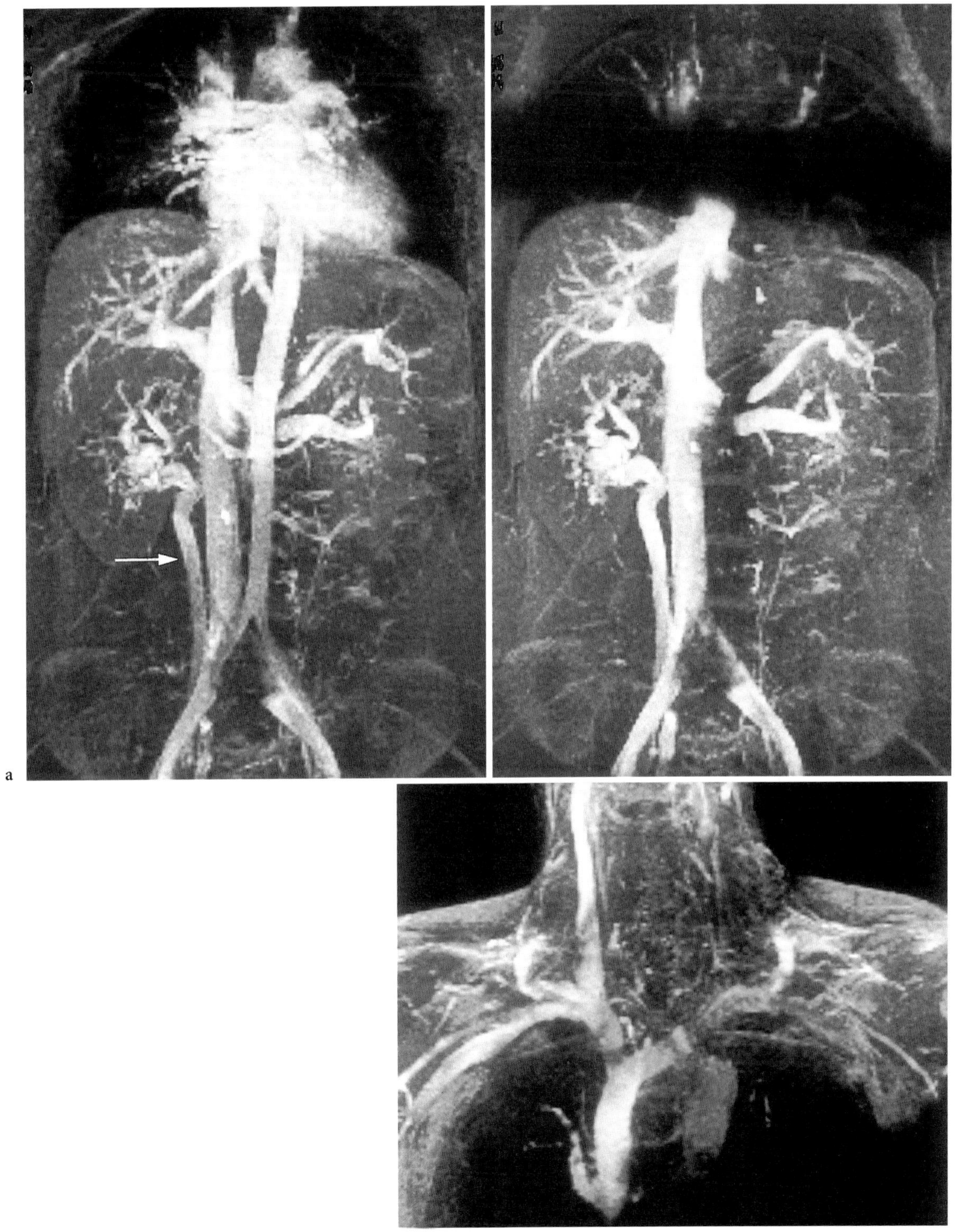

Fig. 17.1a–c. Normal individual. MR-projection angiogram on the basis of a coronal 2D TOF GRE FLASH sequence of the chest and abdomen without (**a**) and with (**b**) presaturation in order to obtain pure MR venograms. By localization of a presaturation band through the aortic root, arterial flow signal can be eliminated in the abdomen (**b**) and chest (**c**). Note the atypical course of the right renal vein (*arrow*)

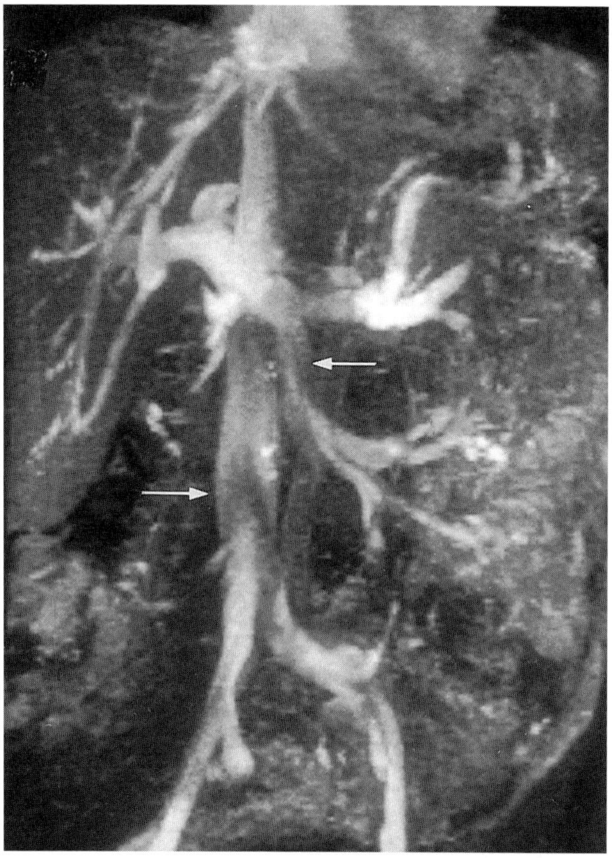

a

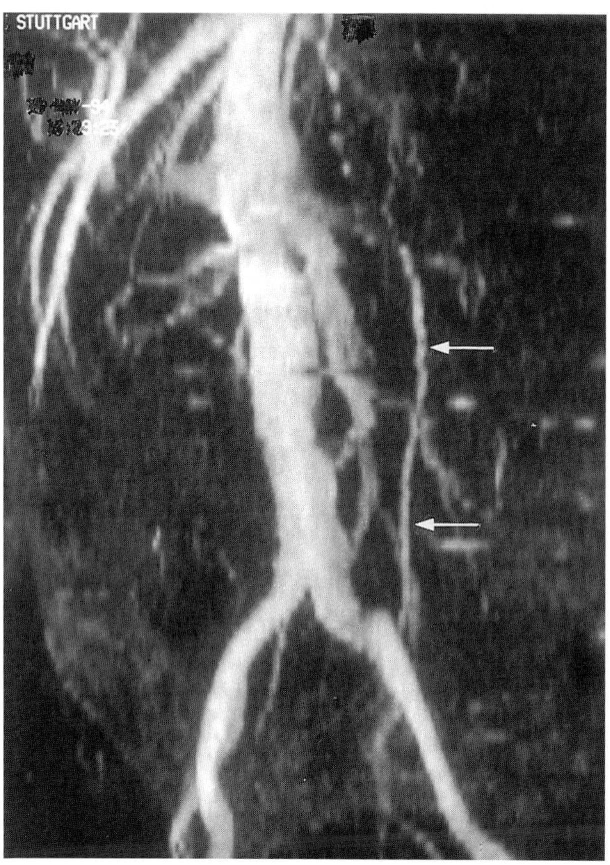

b

Fig. 17.2a–c. Normal individual. Demonstration of signal void due to in-plane saturation on a 2D TOF MR-projection angiogram reconstructed from slices acquired in different planes. In the coronal acquisition plane (**a**) (1.5 T, body coil, two slices within a breath-hold of 15 s, TR/TE = 60/8 ms, flip angle = 30°, slice thickness = 5 mm with 1 mm of overlap, FOV = 450 mm, matrix = 192 × 256, supradiaphragmatic presaturation pulse) signal void occurs in the lower IVC and the middle part of the mesenteric vein (*arrows*); the left spermatic vein cannot be identified. In the transverse acquisition plane (**b**) (1.5 T, body coil, two slices within a breath-hold of 16 s, TR/TE =35/9 ms, flip angle = 40°, slice thickness = 5 mm with 1 mm of overlap, matrix = 180 × 256, parallel presaturation pulse) these particular vascular areas, including the left spermatic vein (*arrows*), show a high signal due to perpendicular inflow. However, signal of both renal veins and the splenoportal system is in-plane saturated. In the sagittal acquisition plane (**c**) (1.5 T, two slices within a breath-hold, TR/TE = 60/8 ms, flip angle = 30°, slice thickness = 5 mm with 1 mm of overlap, matrix = 192 × 256, supradiaphragmatic presaturation pulse) signal of inflowing renal veins and the splenoportal vein is high (*arrows*)

Fig. 17.2c

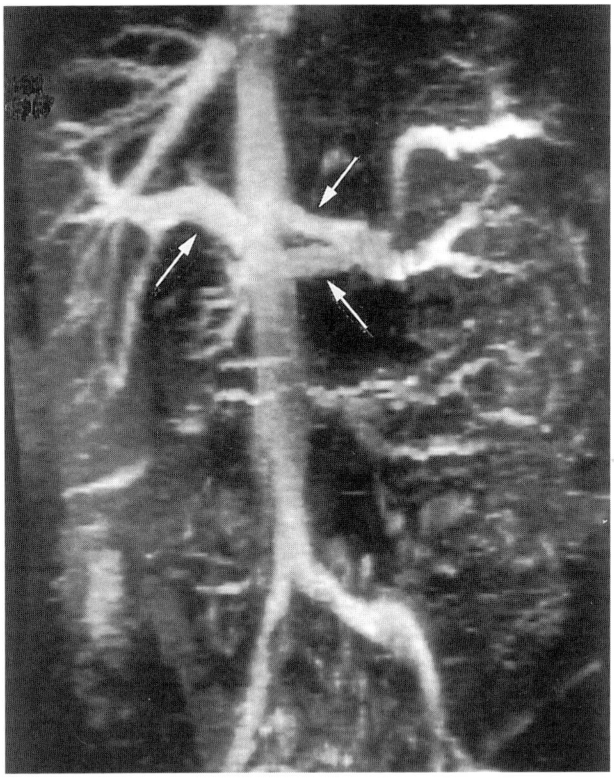

coronal slices. Because functional pseudo-obstruction of the IVC and SVC can be caused by a Valsalva maneuver, patients have to be informed that breath-holding during slice acquisition should be done without this maneuver in order to avoid false-positive signal loss. The presence of signal-intense perivascular fat may degrade flow contrast as well as ascites or perivascular hematoma that contains methemoglobin. For equivalent pulse sequences and imaging parameters, signal-to-noise ratio and vascular contrast-to-noise ratio are lower at 1.0 T than at 1.5 T. This can be a limiting factor when short acquisition times are desirable, as with breath-hold techniques (FINN et al. 1992).

17.2.2 PC MRA

PC MRA generally is well suited for imaging a wide range of velocities, including relatively slow blood flow. 2D and 3D PC MRA techniques have been demonstrated to be applicable in imaging of the venous system of the body (DUMOULIN et al. 1990). Flow-sensitive images are generated by a 2D PC technique which can be extended to three dimensions for volume studies. Three flow-sensitive directions have to be measured, and are combined into the resultant image. In 2D PC MRA breath-holding has proved useful during slice acquisition, whereas with the 3D technique patients are allowed to breath normally. Artifacts due to cardiac motion and pulsatile blood flow may occur and can be minimized by selecting the phase-encoding direction so that it is coincident with the patient's right-left axis, and by an 8-cm transverse presaturation band at the level of the heart. The right-left flow-sensitive image gives the best depiction of the splenic vein, the main portal vein, the mesenteric vein, and hepatic veins. An additional inferosuperior flow direction image completes venous flow signal in the abdomen (IVC, confluence of mesenteric vein), whereas anteroposterior flow-sensitive imaging sometimes provides little additional information. 2D PC MRA is preferable to 3D PC MRA of the abdominal veins because saturation effects are less pronounced.

17.2.3 Quantification of Venous Blood Flow

For blood flow quantification different techniques have been described in the literature that can be applied in the arterial and venous vasculature, such as a combination of flow presaturation (bolus tracking) and low flip angle GRE cine angiography (EDELMAN et al. 1989b), cine MR velocity mapping on the basis of a field-even-echo-rephasing (FEER) sequence (Mohiaddin et al. 1990), flow imaging on the basis of 2D PC angiography (Walker et al. 1988), and a GRE-RACE sequence (Helzel and Mueller 1993). These flow measurements may have important potential applications for the investigation of disorders of central veins and the right side of the heart (see also Chap. 7).

17.3 Application of MRA

The venous system of the trunk mainly comprises the mediastinal veins, the retroperitoneal veins, and the splenoportal vein. This chapter deals with mediastinal and retroperitoneal veins, whereas the splenoportal vein and the pelvic veins are discussed in Chaps. 18 and 19.

17.3.1 Venous System of the Chest

17.3.1.1 Normal Anatomy and Variants

As the main mediastinal vein, the SVC is formed by venous drainage of the head and neck, and the upper extremities. The axillary veins continue as subclavian veins at the lateral border of the first rib. At the medial border of the anterior scalene muscle the subclavian veins continue as brachiocephalic veins after confluence of the external and internal jugular veins. The SVC and brachiocephalic veins have no valves.

The second important venous system in the mediastinum includes the azygos and hemiazygos vein. The azygos vein originates at the level of the right renal vein, and the hemiazygos vein at the level of the left renal vein in continuation of the ascending lumbar vein. At the level of T4 the azygos vein joins the SVC, and at the level of T8 the hemiazygos vein normally joins the azygos vein.

The most important abnormalities of the SVC include persistence of the left anterior cardinal vein forming a left SVC, and duplication of the SVC.

17.3.1.2 Pathology

The dominant disorders of chest veins are the thoracic inlet syndrome (TIS) and the superior vena cava syndrome (SVCS). In patients with thoracic outlet syndrome and TIS, nerve compression is responsible for symptoms in 95% and vascular compression in the remaining 5%. The upper extremity neurovascular bundle traverses three narrow spaces, the interscalene triangle, the costoclavicular triangle, and the subcoracoid space. In TIS the subclavian vein is most commonly compressed at the costoclavicular triangle; such compression may be provoked by hyperabduction of the arm (Ehringer et al. 1979). Acute thrombosis of the subclavian vein may result from mechanical involvement on the basis of chronic microtrauma of the vascular intima. Clinical symptoms of TIS are limited to the involved extremity and are far less severe than those of SVCS.

SVCS produces a characteristic clinical picture of marked cyanosis, swelling of the head, neck, and bilateral upper extremities, and a superficial collateral circulation. The hemodynamic significance of venous obstruction depends on its site and degree, and the development of collateral pathways. When acute SVCS occurs, it demands a rapid diagnostic evaluation in order to institute suitable management. Before treatment is planned, precise determination of the underlying pathology is required. In the etiology of the SVCS, intrinsic lesions, i.e., thrombus due to an idiopathic disorder or an iatrogenic cause such as prolonged use of indwelling venous catheters or pacemaker electrodes, may play an important role. If diagnosed and treated promptly by fibrinolysis, this process is reversible, whereas the development of organized thrombus is irreversible. Furthermore, space-occupying processes adjacent to the SVC may compress, obstruct, or infiltrate the vessel; in up to 70% of cases these effects are due to metastatic or bronchogenic carcinomas, in approximately 20% to malignant lymphomas, and in the remaining cases to other lesions (granulomatous or fibrosing mediastinitis, aortic aneurysm, pericardial disease, dermoid or bronchogenic cysts, substernal goiter, hematoma) (Hirsh et al. 1986). Benign lesions can be managed successfully by surgery, whereas malignant disease involving the SVC is associated with a poor prognosis. However, palliative treatment with radiation therapy alone or in combination with chemotherapy may provide symptomatic relief, as may the application of self-expanding endovascular stents (Davenport et al. 1978; Dombernowsky and Hansen 1978; Roesch et al. 1992).

Radiologic diagnostic work-up includes US, venography, CT, and MRI in order to clarify the site and severity of venous obstruction and to detect vessels accessible to cannulation. Guided fine-needle aspiration biopsy offers the advantage of a percutaneous technique with low morbidity and high accuracy (ROSENBERGER and ADLER 1980). The sensitivity and specificity of color Doppler US in detecting axillary and subclavian vein thrombosis is reported to range from 92% to 100% (BAXTER et al. 1991; LONGLEY et al. 1992). However, duplex US is technically limited in the evaluation of central thoracic veins (HAIRE et al. 1991). Cavography via simultaneous bolus injection of contrast material via both cubital veins is the diagnostic procedure of choice to evaluate the site and extent of the occlusion (BENENATI et al. 1986).

Noninvasive MRI of the upper mediastinum using a conventional SE sequence is able to depict extravascular disorders which obstruct the SVC as accurately as CT (WEINREB et al. 1986) (Fig. 17.3a), whereas collateral pathways can be better visualized by contrast-enhanced CT (ENGEL et al. 1983; MCMURDO et al. 1986). MRA with sequential acquisition of slices within a breath-hold using a flow-compensated 2D TOF GRE FLASH sequence can demonstrate the vascular lesion without contrast medium with an efficacy similar to 3D reconstructions on the basis of contrast-enhanced spiral-CT scanning (TELLO et al. 1993). Compared with SE images, GRE imaging has been shown to be more sensitive and specific in detecting venous thrombosis and in visualizing collaterals (HANSEN et al. 1990) (Fig. 17.3b). A major advantage of MRA is that the evaluation of the SVC is not limited by overlapping arterial vessels due to presaturation of the arterial signal. Findings with MRA are comparable to those of venography for the diagnosis of venous occlusion and nonocclusive thrombosis (Fig. 17.4). Moreover, MRA is able to show altered blood flow readily with selective presaturation or bolus tracking, and provides more comprehensive information than catheter venography on central venous anatomy, blood flow, and collateral pathways (FINN et al. 1993). Thus, MRA has become an accurate and graphic technique for the evaluation of thoracic veins and their tributaries. A limitation of MRA is its frequent inability to satisfactorily depict collaterals in the shoulder and neck area, and these are better defined by means of DSA (Fig. 17.5).

17.3.2 Abdominal Venous System

17.3.2.1 Normal Anatomy and Variants

Blood of the lower extremities is drained by the common femoral veins and the external iliac veins. The external iliac vein is joined by the internal iliac vein, forming the common iliac vein. The IVC originates at the confluence of the common iliac veins at the level of L5. The main branches of the IVC are the renal and lumbar veins, the right gonadal and right adrenal vein, the inferior phrenic vein, and the hepatic veins. Left gonadal veins join the left renal vein. The right renal vein is short and joins the IVC at the level of L1 whereas the left renal vein follows a relatively horizontal course and usually joins the IVC at a somwhat higher level. Occasionally, a single hilar vein bifurcates before entering the IVC. Circumaortic and retroaortic veins occur in 17% and 3%, respectively, and multiple renal veins are seen in approximately 28% of individuals (KADIR 1986).

A duplication of the IVC with confluence at the level of the renal veins is reported in approximately 2% of individuals whereas a solitary left IVC occurs in about 0.2%. Interruption of the hepatic segment of the IVC is observed in 0.6% of patients with congenital heart disease, inducing venous drainage to the heart via the azygos/hemiazygos system. In the presence of occlusion of the IVC, numerous deep and superficial collateral pathways are available for venous return to the right heart, such as the paravertebral plexus, ascending lumbar veins, azygos/hemiazygos vein, left gonadal vein, superficial abdominal wall communications via superficial epigastric veins, and the portal system via the superior hemorrhoidal-inferior mesenteric vein route.

17.3.2.2 Pathology

Because of its high technical status, abdominal B-scan ultrasonography (US) combined with the color Doppler technique is able to provide excellent image quality and to detect most of the venous abnormalities in this area with high accuracy when the diagnostic procedure is performed by experienced users. Thus, the most important applications of color Doppler imaging in the abdomen are for differentiation of vascular from nonvascular structures, identification of vessels, mapping of vessels for duplex sampling, determination of the direction of blood flow, detection of vessel narrowing or occlusion, identification of collateral vessels, detection of arteriovenous malformations, diagnosis of aneurysms and pseudoaneurysms, characterization of flow to organs, evaluation of transplant vascular sup-

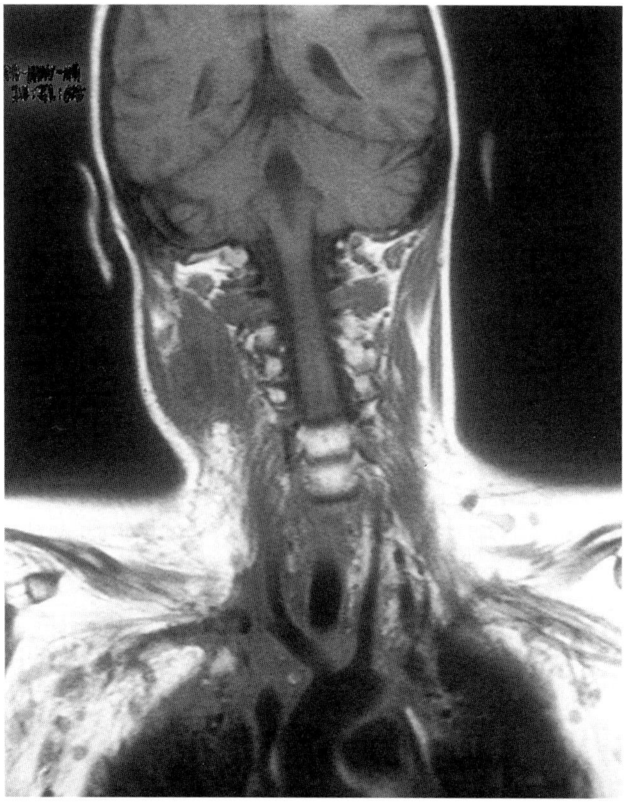

a

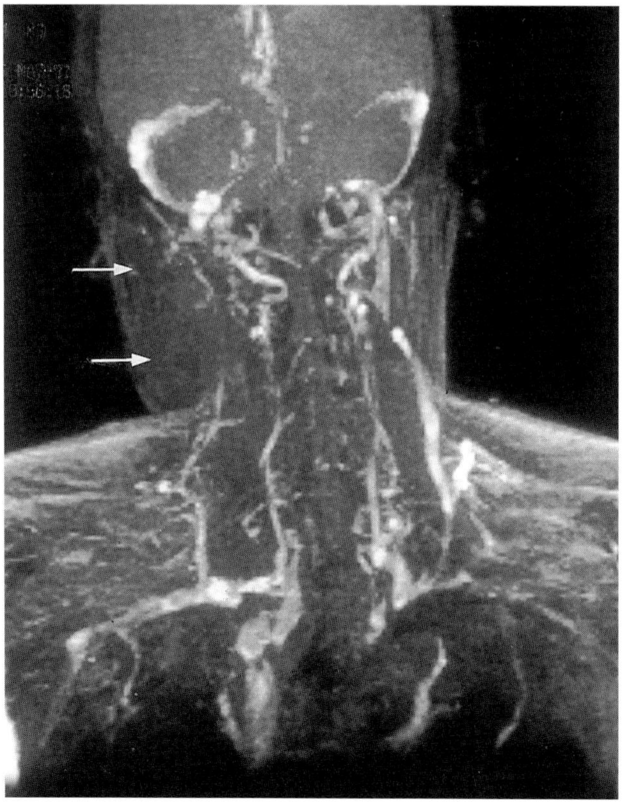

b

Fig. 17.3a,b. Extrinsic occlusion of the right internal jugular vein and brachiocephalic veins due to malignant lymphoma. Coronal T1 SE images (1.5 T, body coil, TR/TE = 600/25 ms, slice thickness = 5 mm, matrix = 192 × 256) show right cervical soft tissue infiltration and perivascular infiltration in the upper mediastinum (intermediate signal) (**a**). On a coronal 2D TOF MR-projection angiogram (**b**) signal void can be observed in the right internal jugular vein and the right and left brachiocephalic vein (*arrows*)

Fig. 17.4. Thoracic inlet syndrome. On a coronal 2D TOF MR-projection angiogram signal void is seen in the right axillary and subclavian veins due to acute thrombosis (*arrow*). Smaller brachial veins, including the cephalic vein and suprascapular collaterals, can be identified with high flow signal

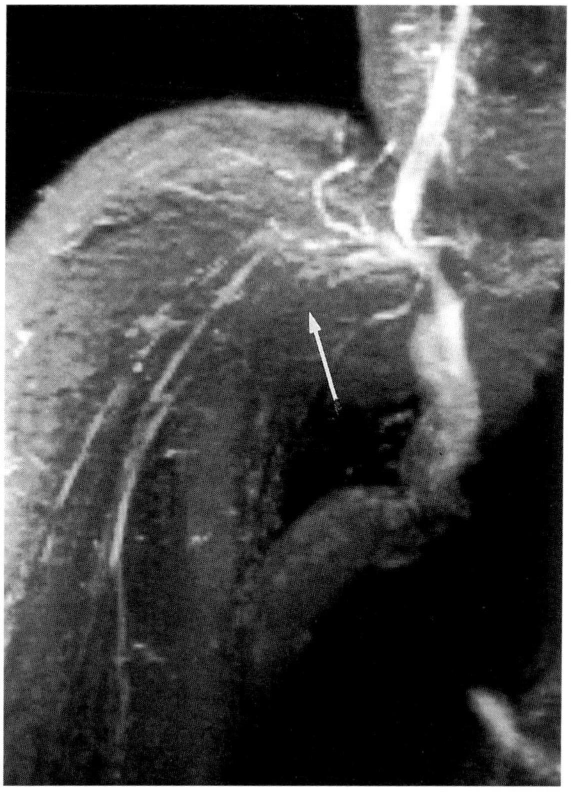

Fig. 17.5a,b. Acute thrombosis of the left subclavian and brachiocephalic vein due to longstanding central venous catheter. On both DSA (**a**) and 2D TOF MR-projection angiography (**b**) venous occlusion can be identified identically (*arrow*). Note that collateral pathways not extensively developed can be seen significantly better with DSA than with MRA. In many cases transverse slice acquisition has proved more useful in imaging axillary, subclavian, and brachiocephalic veins due to higher flow signal in this plane.

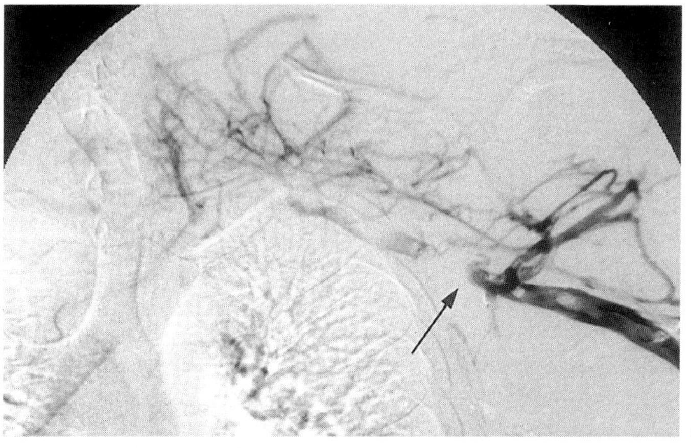

a

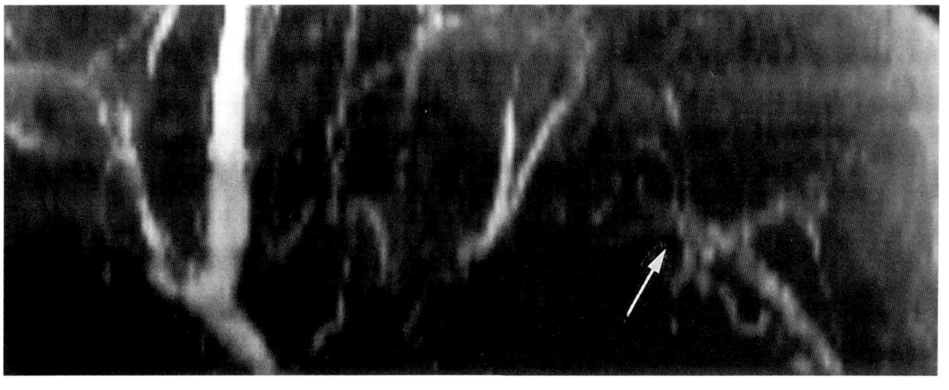

b

ply, and characterization of malignant tumors (Merritt 1992). Vascular abnormalities visible on US may result from extrinsic causes producing vessel compression, displacement, encasement, or changes in relative wall echogenicity. US may also demonstrate intrinsic abnormalities, including tumor, thrombus, aneurysm, or dilatation resulting from a wide range of causes such as primary and secondary malignant disease, diffuse infiltration, and inflammatory disease.

Conventional x-rax angiography of the IVC and its branches, the gold-standard for many years, can be restricted to subjects in whom the value of US is limited by bowel gas, obesity, or deep localization of vascular structures. MRA, which is able to provide both vascular imaging and imaging of perivascular tissue structures, can resolve the issues concerning vascular disorders more effectively than does Doppler US. Furthermore, due to its large unimpeded field of view, MRA has been shown to be better than duplex US in determining the patency of large veins and in depicting collateral vessels when occlusive disease is present.

17.3.2.2.1 Inferior Vena Cava. Indications for specific vascular imaging of the IVC include the evaluation of anatomical variants or congenital disorders which are usually detected coincidentally by US or CT (Figs. 17.1b, 17.6) and obstruction or displacement. Intrinsic obstruction and thrombosis of the IVC are more frequently caused by malignancies than by non-neoplastic disorders. A caval thrombus represents a potential source of life-threatening pulmonary embolus. Extrinsic compression or displacement of the IVC may occur at any level. The mid-IVC is most frequently involved by lymphadenopathy (Fig. 17.7), renal tumor (Fig. 17.8), adrenal tumor, hepatic or pancreatic tumor, and mesenchymal retroperitoneal or pelvic masses (Fig. 17.9). Primary tumors of the IVC (leiomyosarcoma) are rare (Fig. 17.10).

On MRA, obstruction of the IVC shows a complete signal void, and the extension is frequently overestimated in projection angiograms. Thus, for reliable identification of the true extension of the lesion individual slices have to be analyzed (Fig. 17.9c). Collateral pathways can be well documented by MRA in projection angiograms. Imaging of the lumbar/paravertebral venous plexus has been shown to be most successful when transverse sections are acquired as this optimizes inflow enhancement (Fig. 17.11). Moreover, this plane should be selected in cases with slow venous flow in a collapsed IVC, as sometimes may occur in the supine patient. In these individuals, signal in the IVC may be reduced significantly in coronal or sagittal planes in spite of a normally patent vessel (Arlart et al. 1991). The protective insertion of a vena cava filter in patients with recurrent pulmonary embolism due to acute or subacute femoroiliac deep vein thrombosis does not preclude the use of MRA. Dislodging of the filter during MR examination is unlikely when it has been in place for more than 6 weeks. However, severe susceptibility artifacts may occur when the filter material is ferromagnetic (Fig. 17.12), obscuring clot material near the filter.

17.3.2.2.2 Hepatic Veins. Obstruction of the hepatic veins can arise due to a space-occupying intrahepatic mass, and at the level of the IVC due to thrombus or tumor invasion. This may cause the Budd-Chiari syndrome (BCS), which comprises several features including ascites, hepatomegaly, and jaundice. US has been reported as the method of choice for initial diagnosis of BCS (Menu et al. 1985). In addition, the results of pulsed Doppler US have been found to correlate well with angiographic findings in cases of hepatic venous outflow obstruction, demonstrating characteristic blood flow signal from the IVC and/or hepatic veins that changes from phasic to absent, reversed, turbulent, or continuous (Hosoki et al. 1989). A characteristic appearance of BCS has also been reported by Stark et al. (1986) when employing MRI with SE sequences. However, typical patterns of hepatic vein occlusion and IVC involvement can be identified more reliably by use of GRE MRA (Fig. 17.13). Although characteristic signs can be found, such as luminal narrowing and partial or complete signal void in the hepatic veins, very similar findings may also occur in different forms of cirrhotic liver disease (Anderson et al. 1993). MRA has been shown to be extremely useful for mapping the intrahepatic venous anatomy in tumorous disease before surgical intervention (Fig. 17.14) or when TIPS procedures are planned. In these cases both coronal and transverse acquisition planes should be selected for MRA inorder to optimize flow signal and to obtain multidimensional information on the intrahepatic vasculature.

17.3.2.2.3 Renal Veins. Most disorders of the renal veins are due to thrombotic disease. The latter may occur spontaneously in renal disease, hypercoagulopathy, dehydration, paraneoblastic states, and following trauma or surgical intervention.

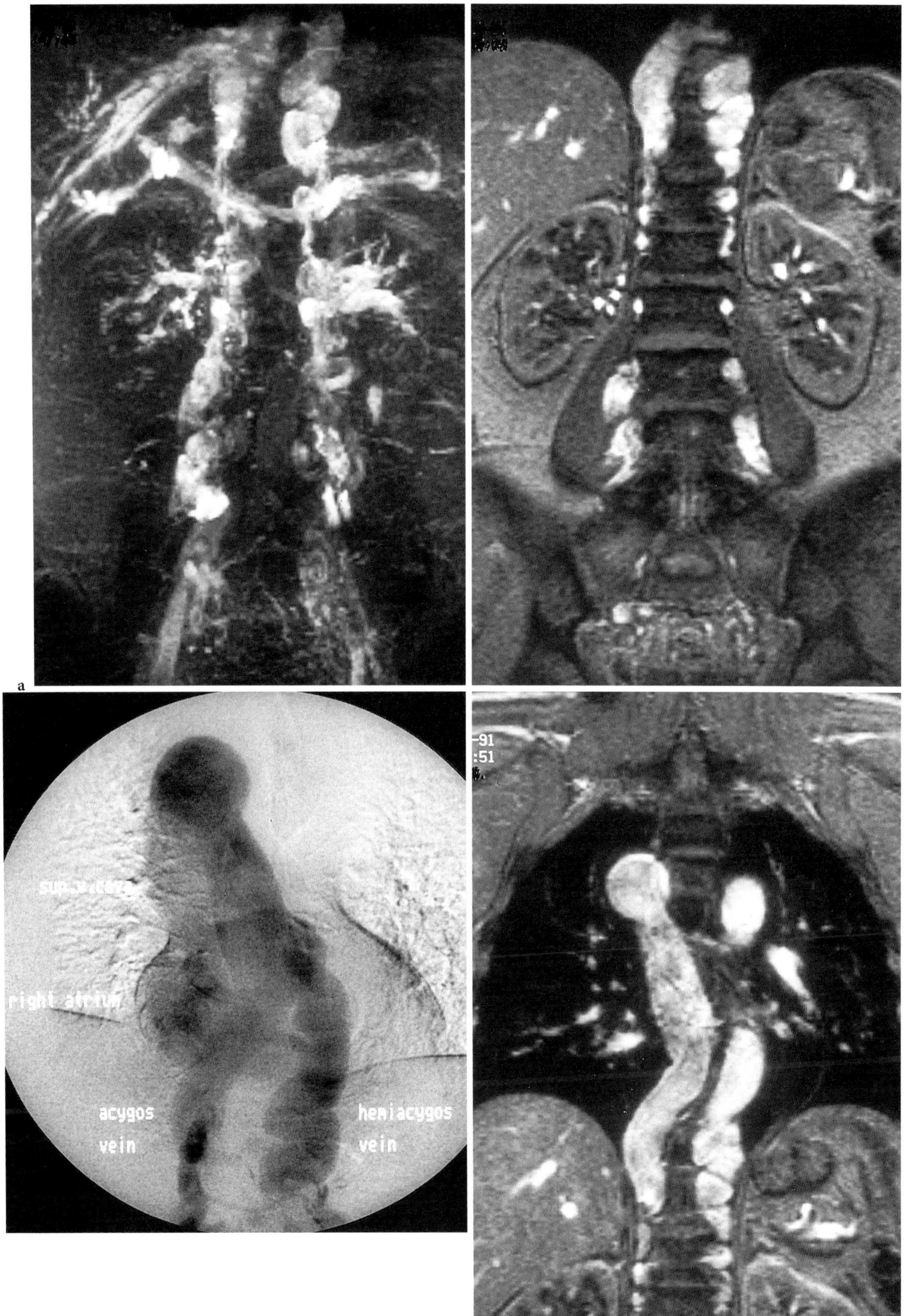

Fig. 17.6a–d. Atresia of the IVC in a 52-year-old male patient with chronic lower extremity swelling. In the 2D TOF MR-projection angiogram (**a**) and individual coronal section (**b**) venous flow can be identified in both the ascending lumbar veins. Renal veins are drained via the azygos vein and hemiazygos vein. Enlarged mediastinal azygos and hemiazygos veins appear similar on DSA (**c**) and MRA (**d**)

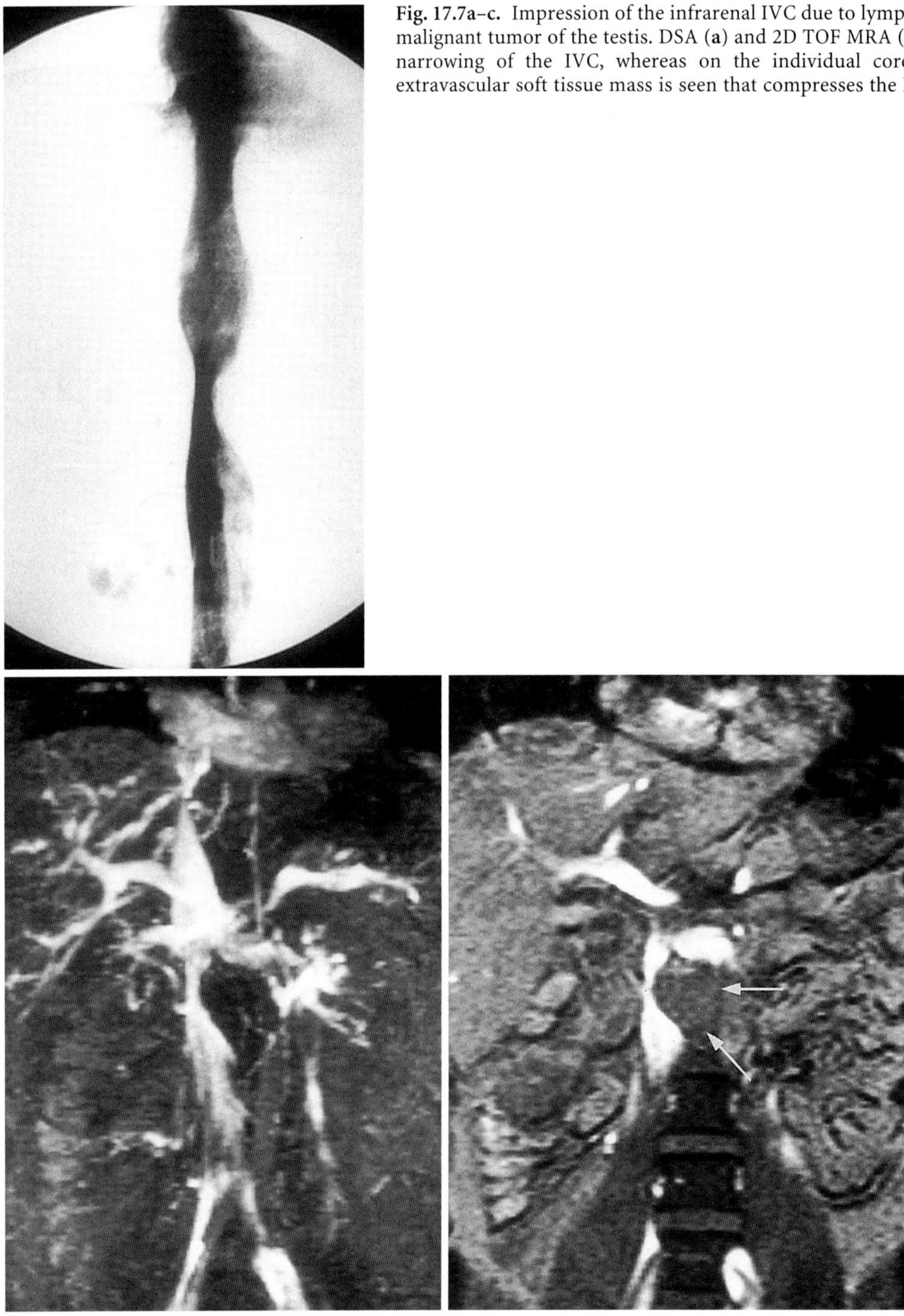

a b c

Fig. 17.7a–c. Impression of the infrarenal IVC due to lymphoma in a patient with malignant tumor of the testis. DSA (**a**) and 2D TOF MRA (**b**) demonstrate a focal narrowing of the IVC, whereas on the individual coronal MR slice (**c**) an extravascular soft tissue mass is seen that compresses the IVC (*arrows*)

However, more often secondary involvement of the renal veins can be observed as a result of tumor- and lymphoma-induced vascular compression or intravascular extension of a tumor thrombus in patients with renal cell carcinoma.

Cinical manifestations of renal vein thrombosis (RVT) may develop gradually and depend on the degree of occlusion and the presence of collaterals. They include flank pain, nephrotic syndrome, proteinuria, hematuria, azotemia, and occasionally arterial hypertension. In patients with renal cell carcinoma, intravascular growth involving the renal a 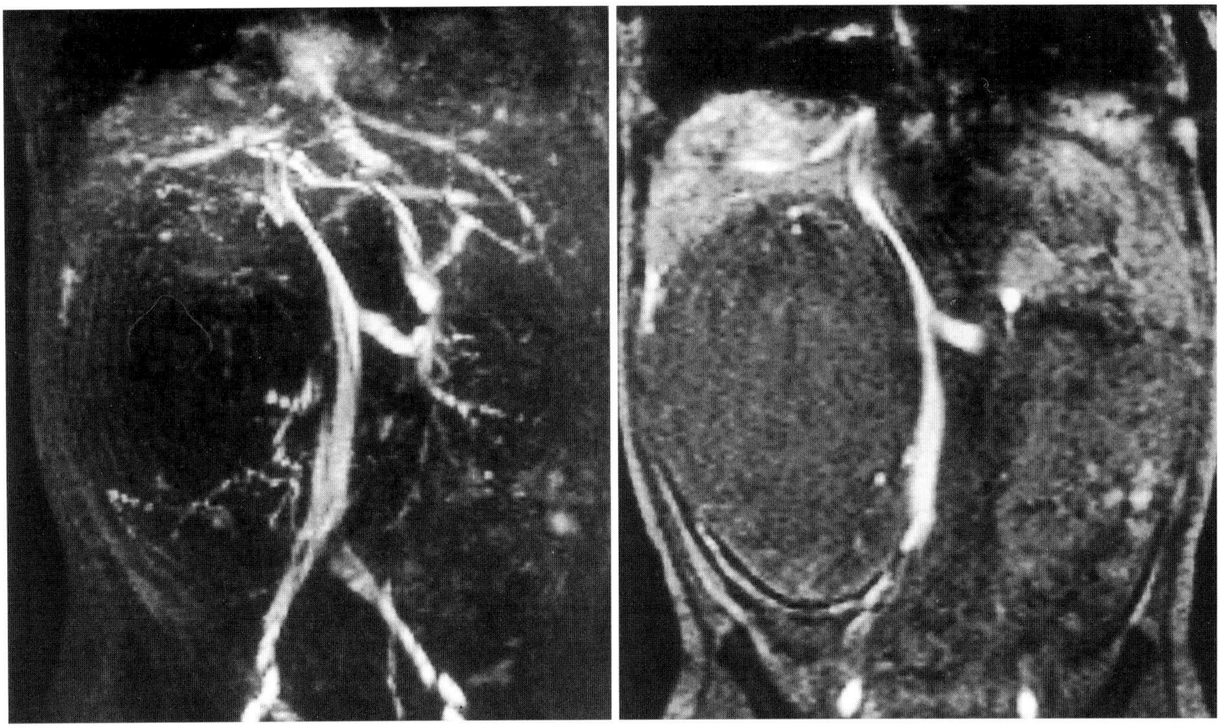 b

Fig. 17.8a,b. Displacement of the IVC by a large nephroblastoma in a 5-year-old child (**a** 2D TOF MR-projection angiogram, **b** coronal individual MR section). In this particular case MRA was performed without breath-holding

vein and the IVC can be observed in 21%–35% and 5%–10% of cases, respectively. In localized renal cell carcinoma the existence of tumor thrombus and the level of thrombus extension into the IVC affect the survival rate, which is far less than with tumor invasion of the capsule or lymphadenopathy (CHERRIE et al. 1982; SCHEFFT et al. 1978). Nevertheless, surgical intervention is indicated in patients with a large tumor thrombus which may induce vascular complications or pulmonary embolism. Successful removal of intraluminal tumor requires accurate preoperative delineation of the superior tumor extent, as this determines the operative approach.

Angiography and dynamic CT have been established as the mainstays of vascular evaluation in renal cell carcinoma, while noninvasive US can provide important additional information. Although 100% sensitivity and specificity of venacavography in detecting tumor thrombus in the IVC have been reported (HIETALA et al. 1988; KALLMAN et al. 1992; LANG 1984), the extent of thrombus may be estimated incorrectly due to early filling of collaterals and a lack of opacification of the vessel behind the occlusion. Furthermore, when examining the renal veins selectively, thrombolysis can be achieved by catheter procedures. In US the upper border of tumor occlusion in the IVC can be delineated in most cases. However, echogenic thrombus material can be detected more reliably in the right than in the left renal vein. The accuracy of US may be significantly impaired by bowel gas, obesity, and deep location of venous structures. The literature contains many reports on the accuracy of US and CT in the detection of intravascular tumor thrombus in renal cell carcinoma. The reported sensitivity and specificity of US range from 33% to 100% and from 92% to 100%, respectively, when the thrombus is located in the IVC, and from 18% to 95% and 76% to 93%, respectively, when it is located in the renal vein. Dynamic contrast-enhanced CT is able to delineate tumor thrombus in the renal vein and IVC more reliably, with a sensitivity and specificity of 66%–100% and 76%–100%, respectively (GIULIANI et al. 1987; HIETALA et al. 1988; KALLMAN et al. 1992; LANG 1984; LONDON et al. 1989; SCHWERK et al. 1985). Reported pitfalls of CT include difficulty in delineating the superior extent of the tumor when it is near the right atrium or the hepatic venous confluence, and the risk of mistaking unopacified blood or streaming of contrast material in laminar blood flow for clot-like filling defects.

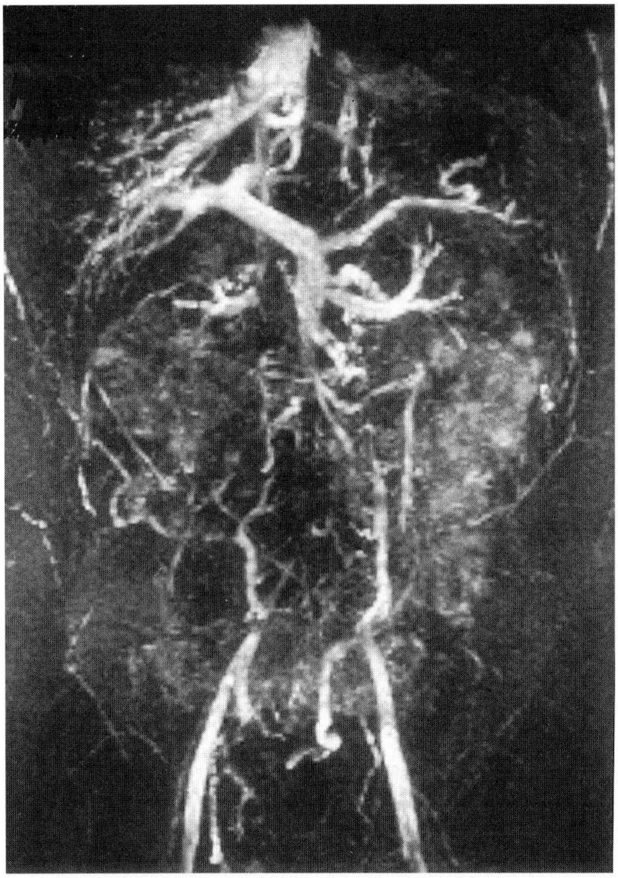

a

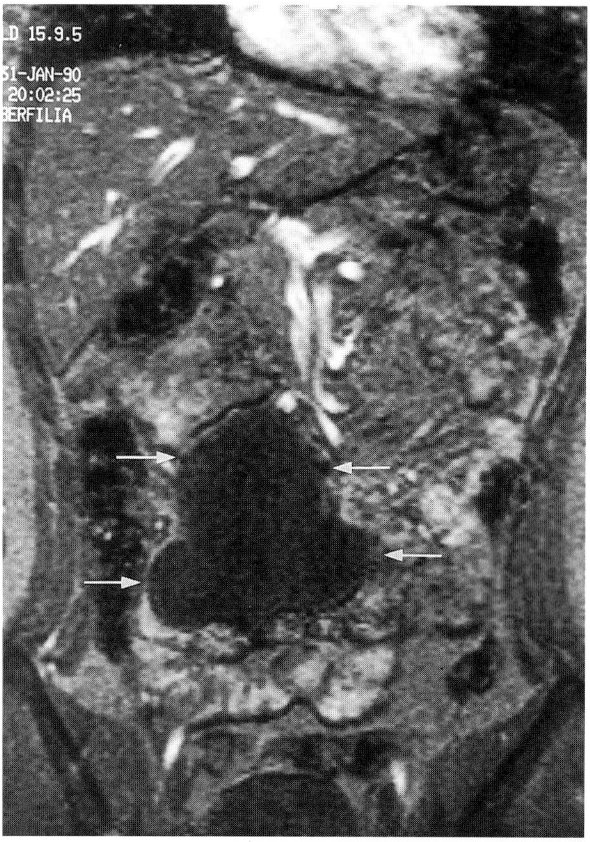

c

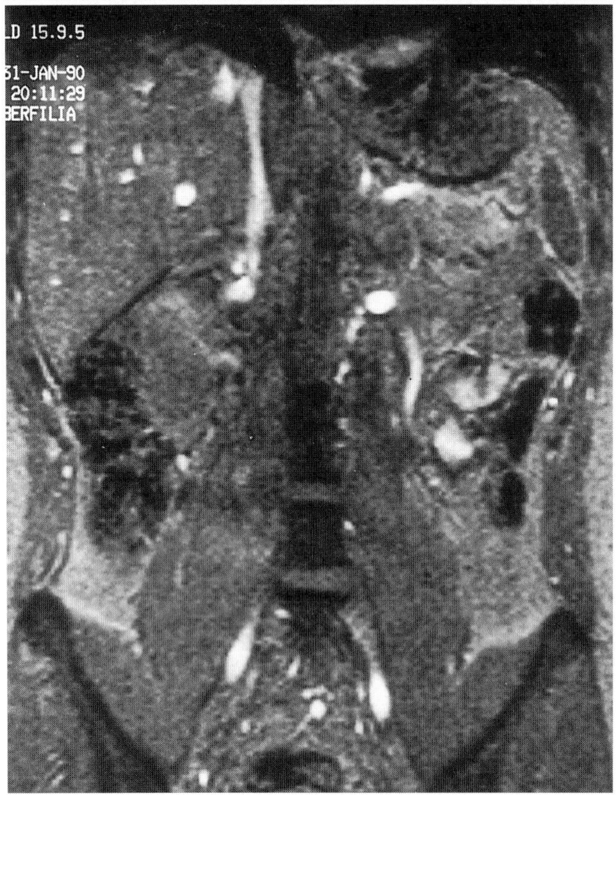

b

Fig. 17.9a–c. Complete occlusion of the infrarenal IVC due to a pelvic malignancy. On a coronal 2D TOF MR-projection angiogram (**a**) signal void is seen in the entire IVC. Collateral pathways via ascending lumbar veins and superficial epigastric veins are visualized. Drainage of the renal veins into a patent suprarenal IVC and site of the pelvic mass (*arrows*) can be evaluated only in individual coronal MR sections (**b**,**c**) (Andersen et al., Clin. Magnet. Res. Angiogr., Raven Press 1993, Chap. 17, p. 390)

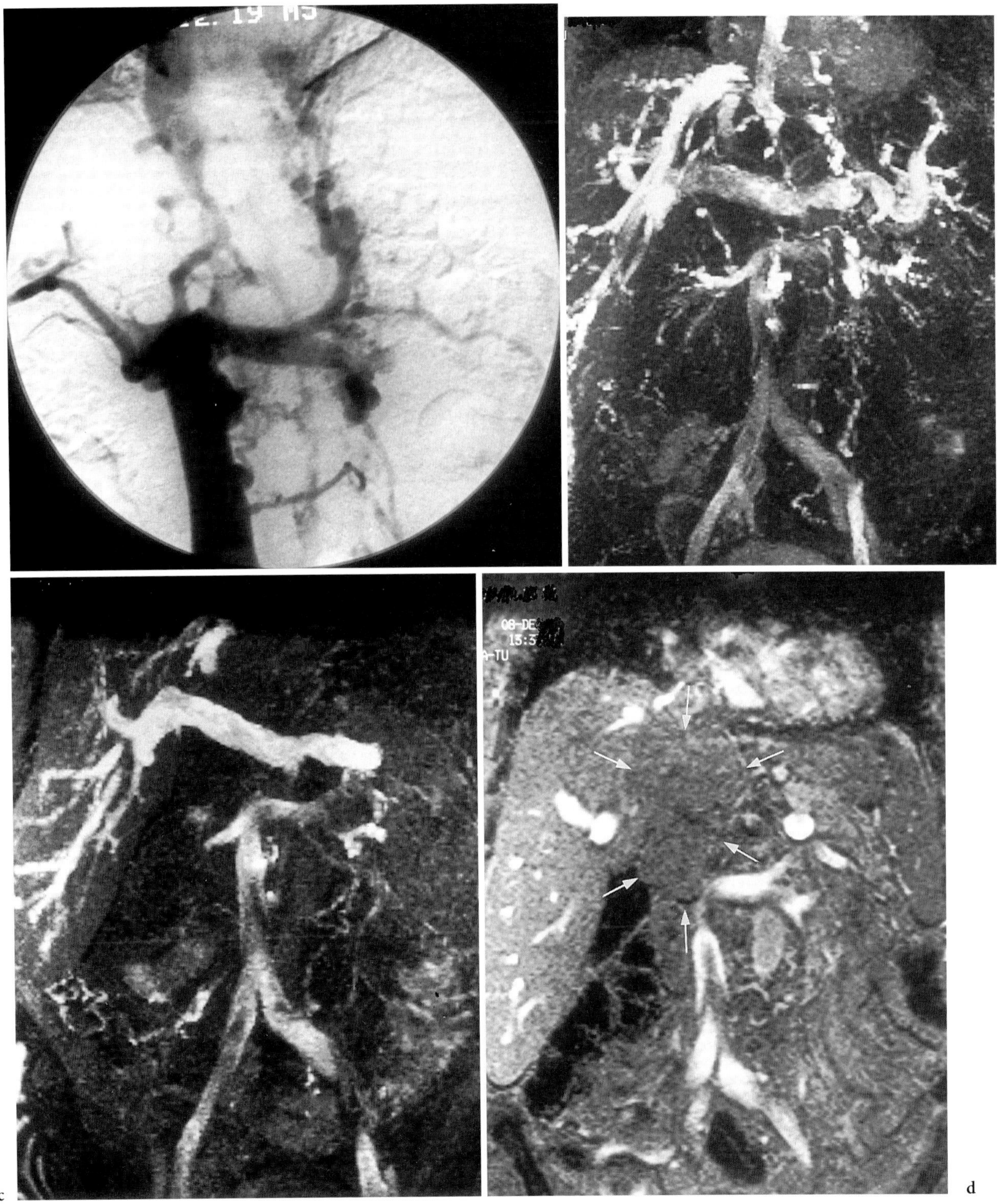

Fig. 17.10a–e. Occlusion of the suprarenal IVC due to a leiomyosarcoma of the IVC (histologically proven). DSA (**a**) and coronal 2D TOF MIP MRA (**b**) show the occluded area similarly, including collateral pathways. The MIP angiogram reconstructed from a selected slice collection (targeted MIP) demonstrates the occlusion without vascular overlap (**c**), while the tumor mass can be evaluated on a 2D single section (**d**) (*arrows*) as well as on a T2-weighted SE image (TR/TE = 2700/90 ms, 160 × 256) (**e**)

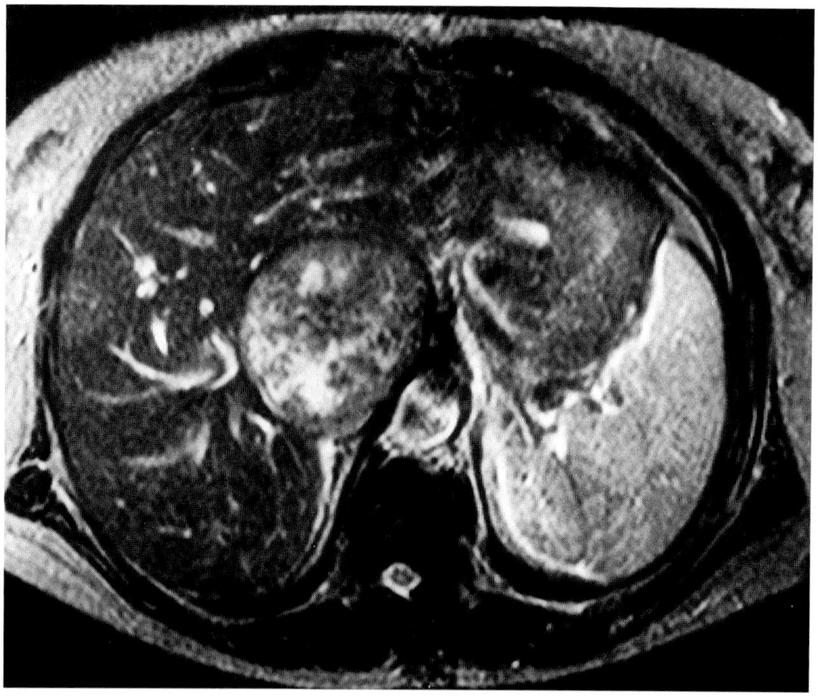

Fig. 17.10e

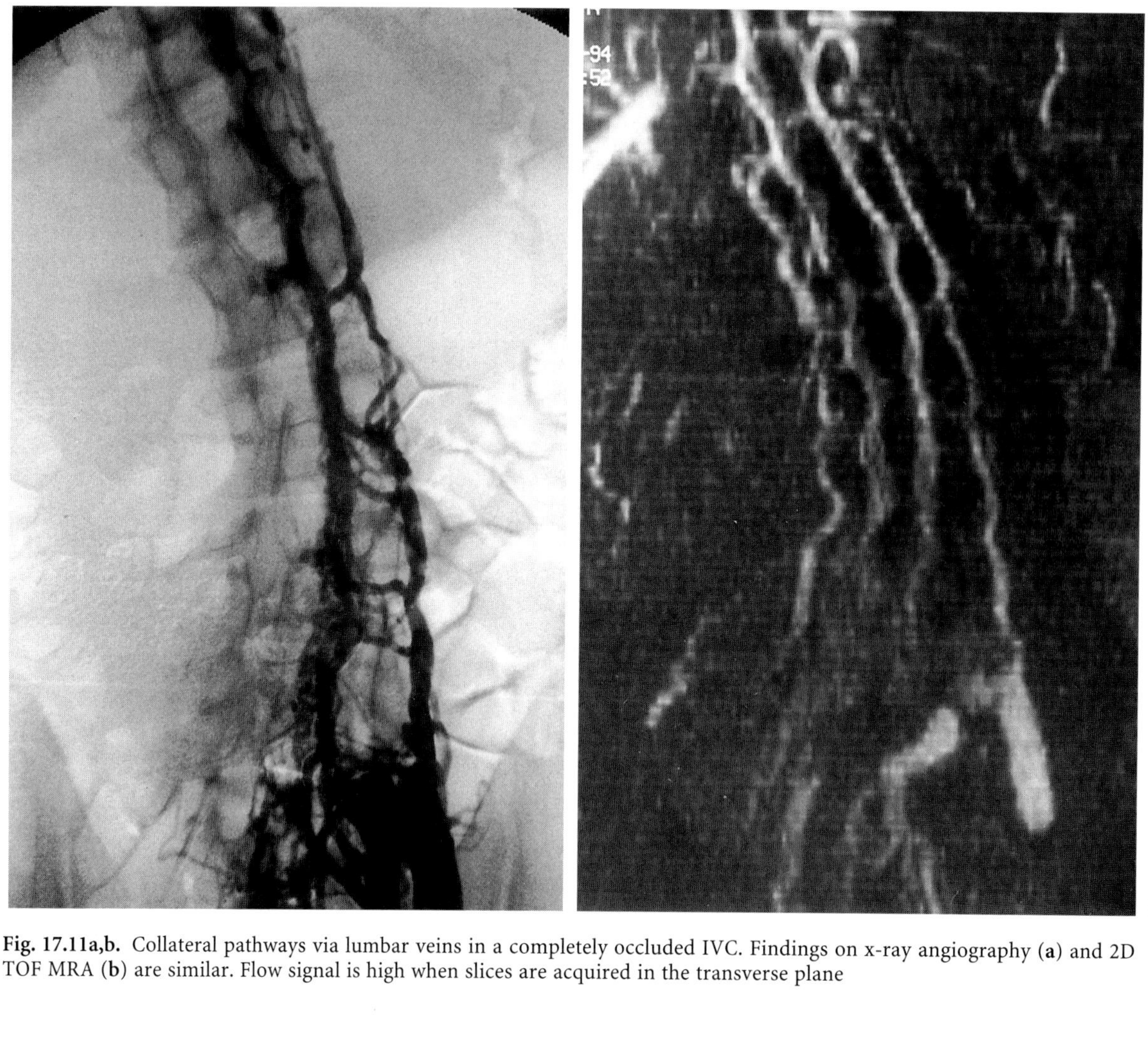

a

b

Fig. 17.11a,b. Collateral pathways via lumbar veins in a completely occluded IVC. Findings on x-ray angiography (**a**) and 2D TOF MRA (**b**) are similar. Flow signal is high when slices are acquired in the transverse plane

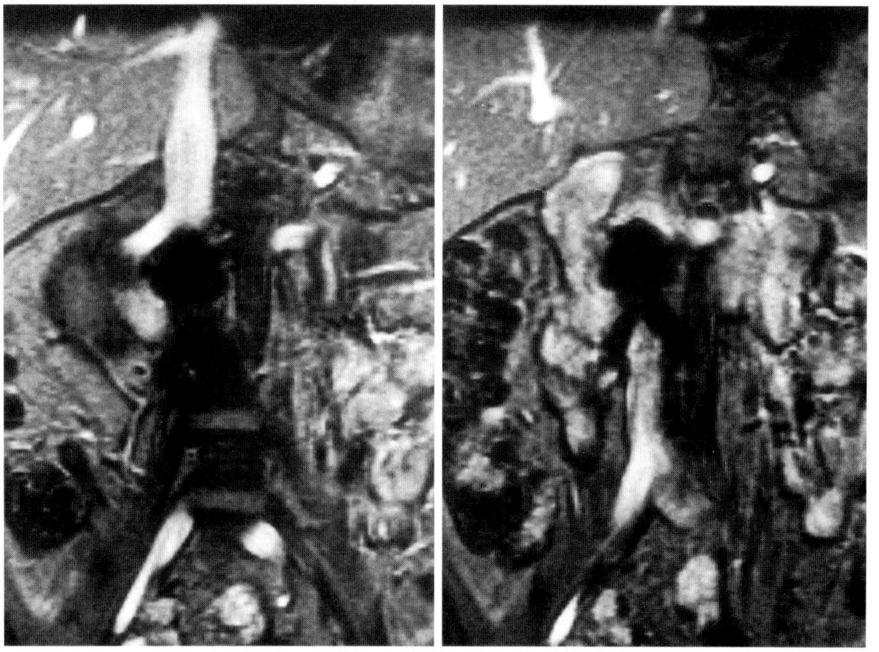

a,b

Fig. 17.12a,b. Placement of a vena cava filter in a typical localization. On single coronal TOF MR slices (2D GRE FLASH) focal signal loss can be observed in the IVC due to susceptibility artifacts of ferromagnetic filter material

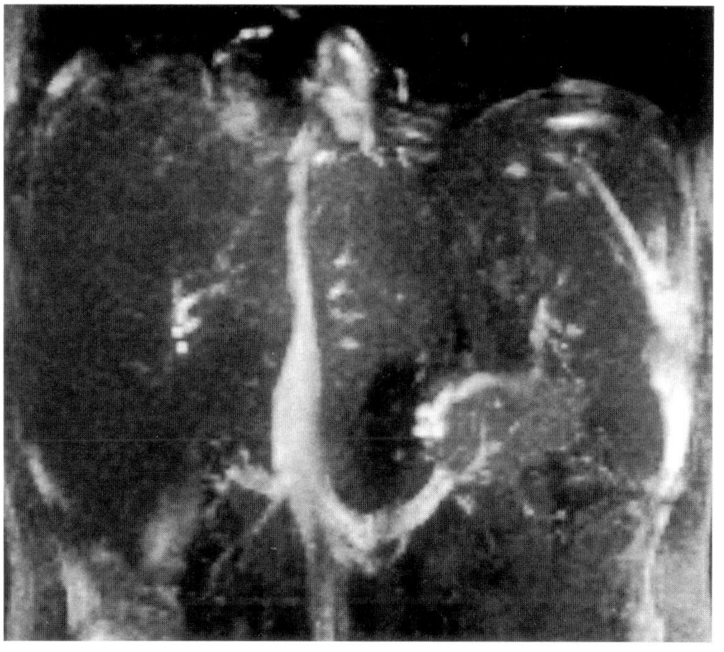

a

Fig. 17.13a–c. Budd-Chiari syndrome in a 28-year-old female patient with coagulation disorder of unknown origin. Coronal and transverse 2D TOF MR-projection angiograms (**a**,**b**) show a patent splenoportal vein, small intrahepatic portal branches, and a reduced signal in the hepatic veins. In a transversely acquired targeted MIP angiogram (**c**) the number of hepatic veins is significantly reduced and a narrowing of central veins is demonstrated. Signal-intense areas around the liver and spleen are caused by ascites

Renal veins and the IVC can both be demonstrated by MRI without contrast material in most cases. Using a conventional T1-weighted SE technique, HORAN et al. (1989) reported a sensitivity and specificity of 82% and 97%, respectively, in evaluating tumor extension into the IVC, and of 65% and 81%, respectively, in detecting renal venous tumor thrombus. Other reports have confirmed a sensitivity of up to 100% (GOLDFARB et al. 1990), particularly when additional GRE MRA has been used (KALLMAN et al. 1992; ROUBIDOUX et al. 1992) in order to differentiate accurately slow flow from thrombus material. 2D TOF GRE MRA has proved the most useful technique for providing diagnostic information on vascular tumor extension, its accuracy apparently being similar to that of contrast-enhanced CT and angiography (Figs. 17.15, 17.16). In a recently published study by our group (ARLART et al. 1992) it could be demonstrated that MRA in a coronal acquisition plane had a sensitivity of 100% in evaluating

Fig. 17.13b,c

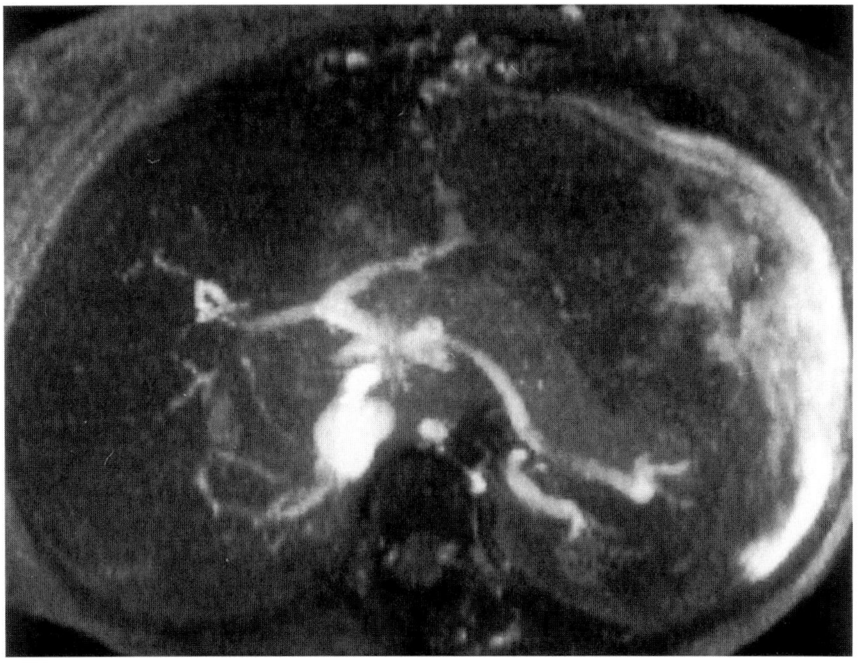

b

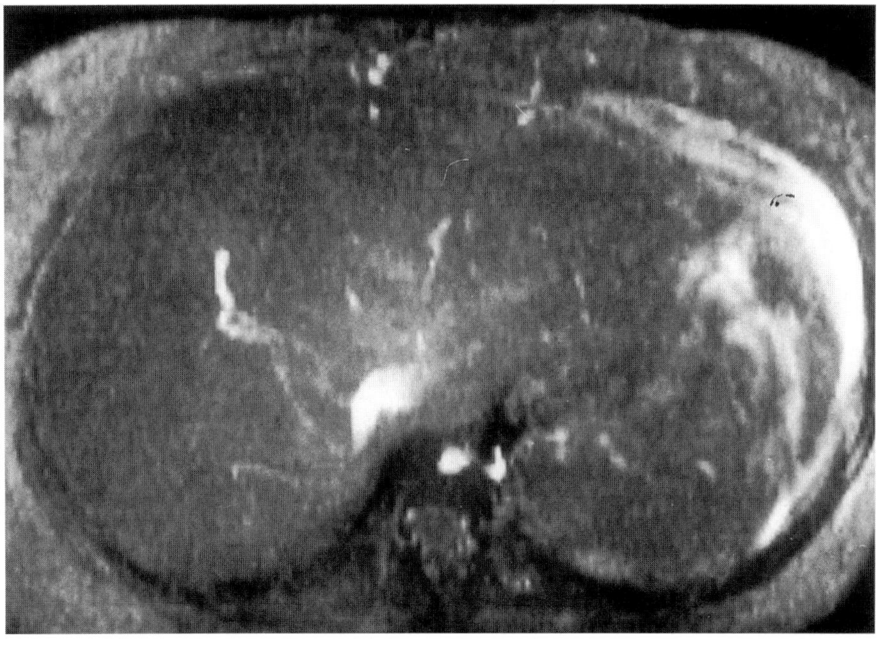

c

tumor extension into the IVC, and 84% in detecting tumor thrombus in the renal veins. When MRA sections were acquired in both coronal and transverse planes, tumor thrombus in the renal veins could be identified in all examined cases, and identification was facilitated by analysis of individual sections in addition to the MIP MR angiogram (Fig. 17.17). The possibility of misdiagnosis exists when signal void occurs due to in-plane saturation which mimics thrombus material, or when thrombus material includes methemoglobin, which appears bright on TOF acquisitions and may mimic a patent lumen. In these cases, additional GRE MRA studies are recommended using a different slice acquisition mode as well as additional T1 SE MR examinations. Although CT is the preferred technique of staging renal cell carcinoma, MRA can better show the extension of tumor thrombus into the renal veins and IVC, and,

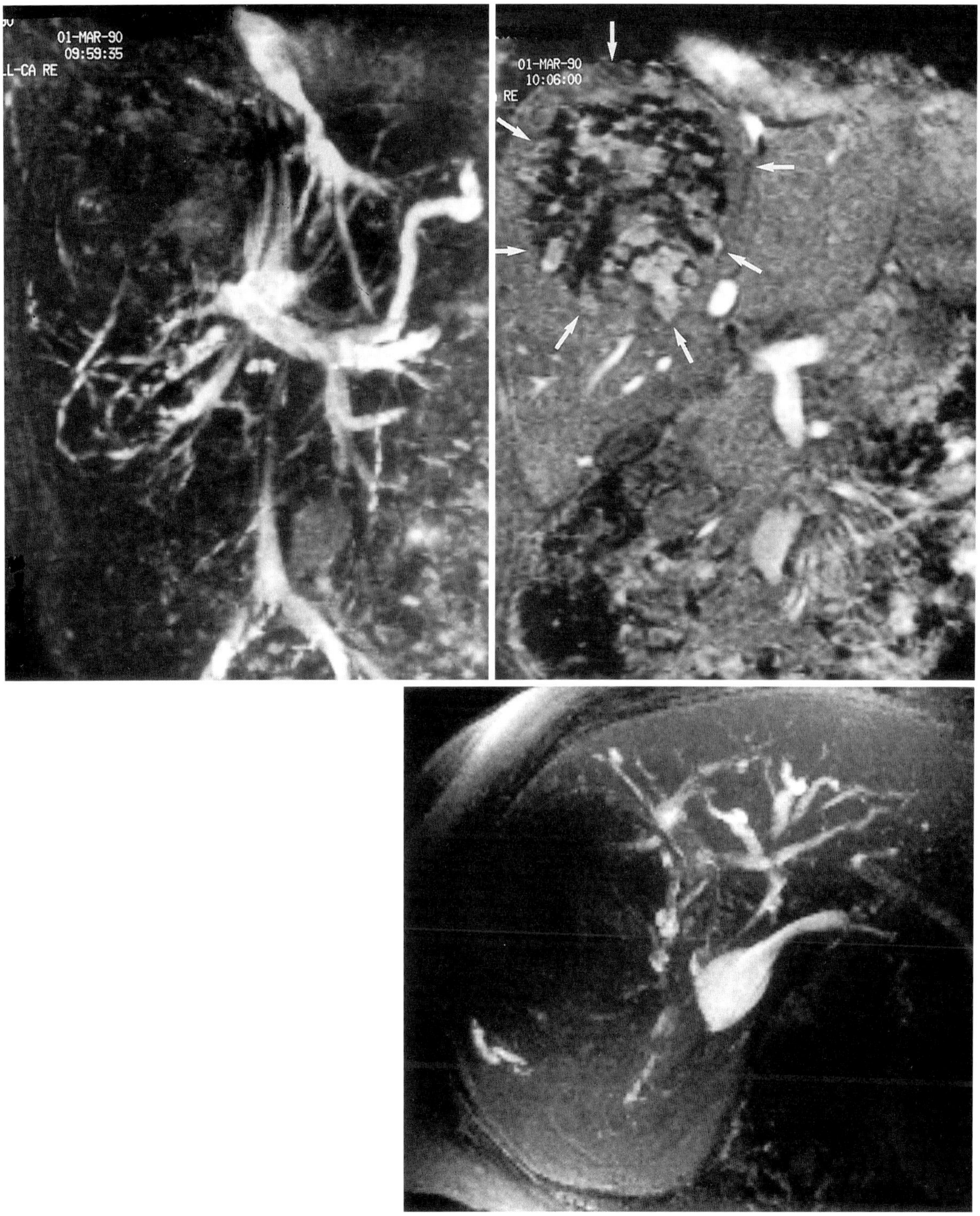

a c b

Fig. 17.14a–c. Malignant hepatoma in the right liver lobe. Displacement of intrahepatic vessels and the IVC is seen on coronal (**a**) and transverse (**b**) 2D TOF MIP MRA. On an individual coronal 2D section (**c**) the tumor mass can be evaluated clearly (*arrows*). By rotation of projection-angiograms along the long or cross axis of the body, optimized multidimensional information is obtained (Andersen et al., Clin. Magnet. Res. Angiogr., Raven Press 1993, Chap. 17, p. 388) most importantly, the involvement of hepatic veins or extension through the diaphragm; thereby it may influence the therapeutic approach.

17.4 Conclusion

Disorders of the large veins are mainly the result of extrinsic and intrinsic causes which may induce displacement, stenosis, or occlusion. In addition, the numerous anatomical variants of venous structures may be confusing when employing a routine diagnostic approach. In these cases bright blood MRA has proved to be an excellent imaging modality with high accuracy in identifying primary vascular lesions as well as veno-occlusive disease secondary to perivascular disorders. As the most useful MRA technique a 2D TOF GRE FLASH sequence has been introduced into clinical practice. MR venography is able to combine all the advantages of Doppler US, contrast-enhanced CT, and conventional venography, without any invasiveness or the necessity of administering contrast material. Functional quantitative measurement of flow velocity is possible by different MR techniques. However, more scientific work has to be done to evaluate the value of these procedures in routine clinical practice. While the diagnostic accuracy of MRA in patients with arterial lesions is currently limited by a variety of problems, MR venography, if available, can be recommended as the diagnostic method of choice in the majority of lesions of the large central veins. Noninvasive MR venography is especially appropriate in patients in whom administration of larger volumes of iodinated contrast material is contraindicated by risk factors such as cardiovascular disease, diabetes, and renal insufficiency, or in whom x-ray exposure has to be avoided.

Acknowledgment. The authors thank Mrs. E. Griepentrog and Mrs. I. Meurer for assistance with data acguisition and preparation of image material.

References

Anderson CM, Edelman RR, Turski PA (1993) Clinical magnetic resonance angiography. Raven Press, New York

Arlart IP, Guhl L, Fauser L, Edelman RR, Kim D, Laub G (1991) Magnetic resonance angiography (MRA) of the abdominal veins. Radiologe 31:192–201

Arlart IP, Guhl L, Edelman RR (1992) MRA of renal veins and inferior vena cava for staging of renal cell carcinoma. ROFO 157:584–590

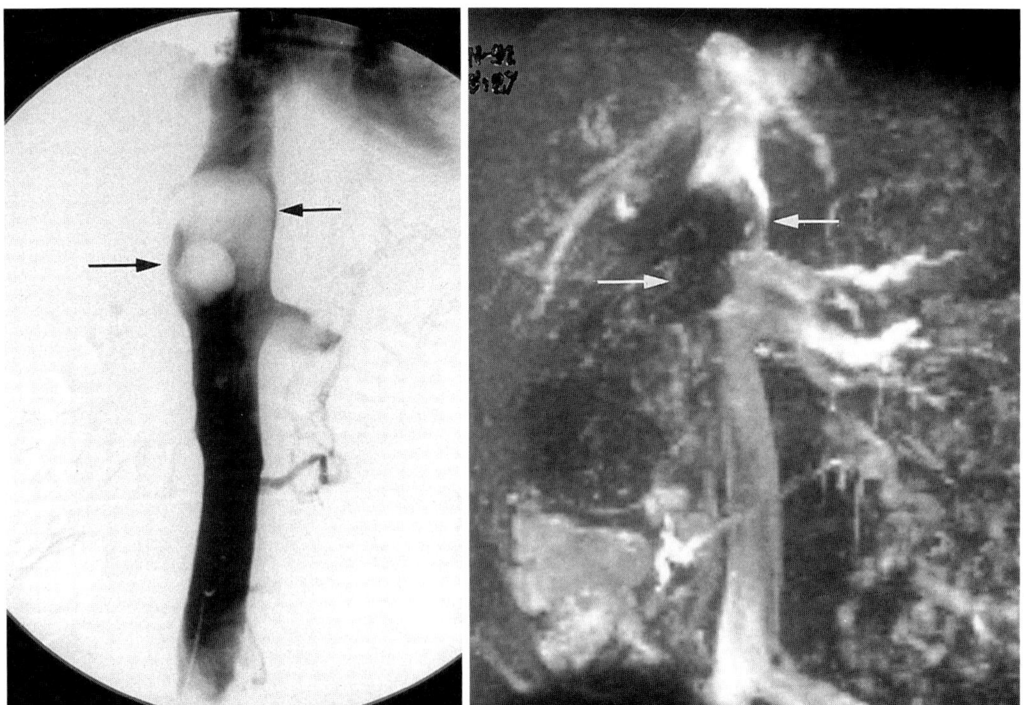

a b

Fig. 17.15a,b. Right-sided renal cell carcinoma with venous tumor thrombus. On both the x-ray angiogram (**a**) and the coronal 2D TOF MR-projection angiogram (**b**) a filling defect and signal void (*arrows*) can be seen in the IVC at the level of the right renal vein due to tumor thrombus

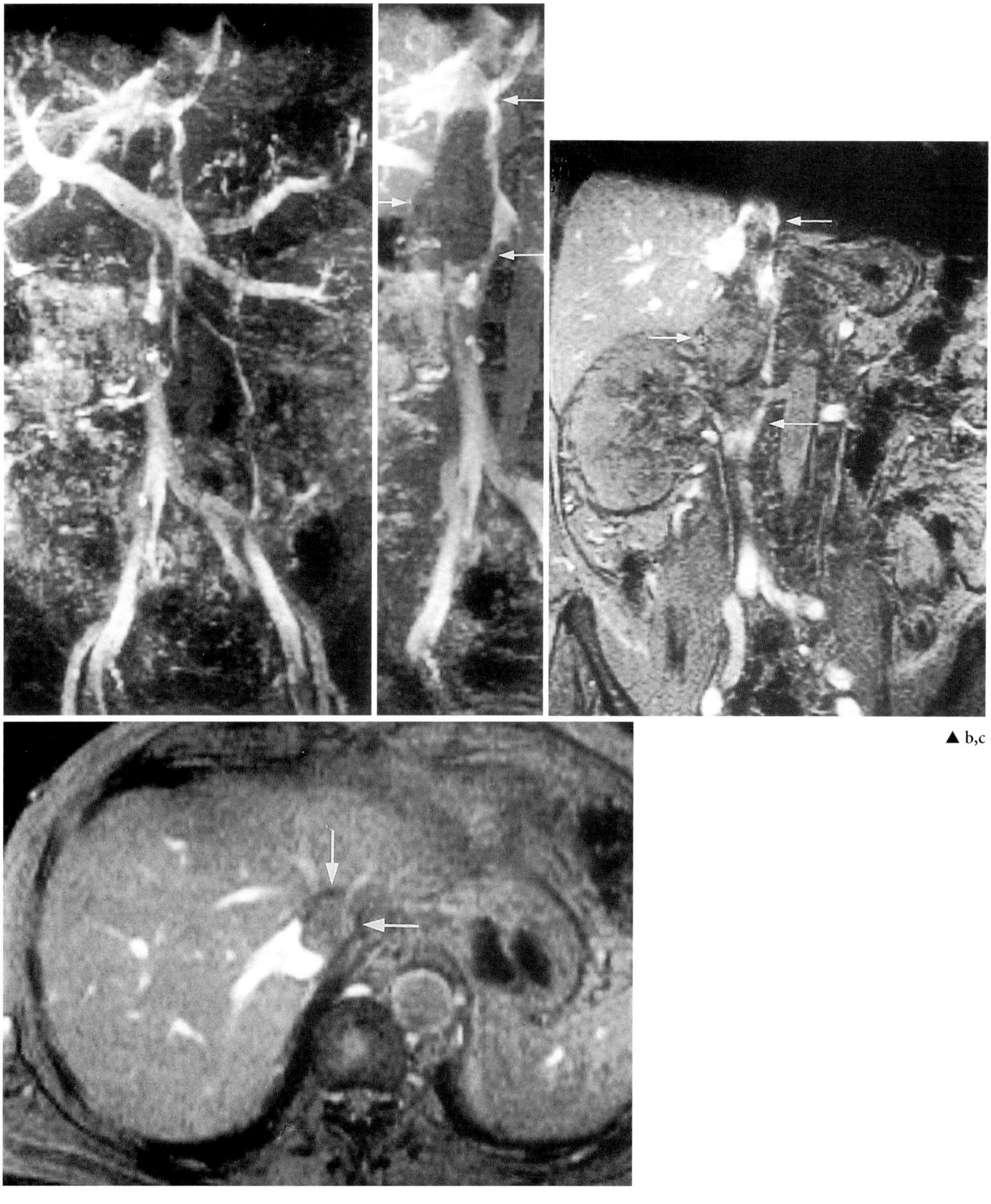

a ▲ b,c d

Fig. 17.16a–d. Large tumor thrombus of right-sided renal cell carcinoma. Occlusion of the IVC is demonstrated on a 2D TOF MR-projection angiogram (**a**,**b**) and on individual single slices (**c**,**d**). On a targeted MIP-angiogram (**b**) and on single sections the tumor thrombus can be delineated accurately up to the right atrium (*arrows*)

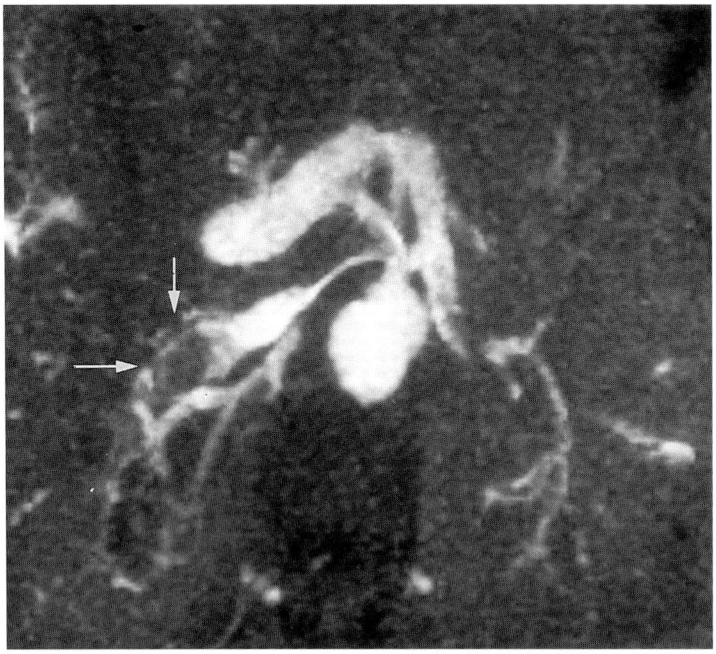

a

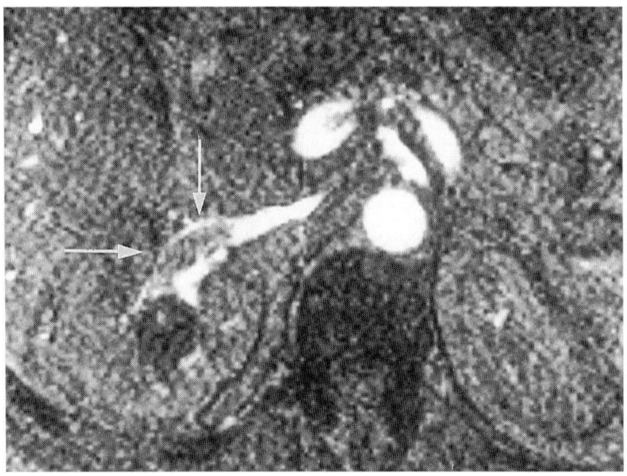

b

Fig. 17.17a,b. Renal cell carcinoma with small tumor thrombus in the right renal vein (*arrows*). For detection of smaller intravascular tumors, 2D TOF MRA in the transverse acquisition plane (**a** MIP angiogram, **b** single slice) is superior to 2D TOF MRA in the coronal plane in most cases

Arrive L, Menu Y, Dessart I et al. (1991) Diagnosis of abdominal venous thrombosis by means of spin-echo and gradient-echo MR imaging: analysis with receiver operating characteristic curves. Radiology 181:661–668

Baxter GM, Kincaid W, Jeffrey RF, Millar GM, Porteous C, Morely P (1991) Comparison of colour Doppler ultrasound with venography in the diagnosis of axillary and subclavian vein thrombosis. Br J Radiol 64:777–781

Benenati JF, Becker GJ, Mail JT, Holden RW (1986) Digital subtraction venography in central venous obstruction. AJR 147:685–688

Cherrie RJ, Goldman DG, Lindner A, de Kernion JB (1982) Prognostic implications of vena caval extension of renal cell carcinoma. J Urol 128:910–912

Davenport D, Ferree C, Blake D, Raben M (1978) Radiation therapy in the treatment of malignant superior vena caval obstruction. Cancer 42:2600–2603

Dombernowsky P, Hansen HH (1978) Combination chemotherapy in the management of superior vena caval obstruction in small cell anaplastic carcinoma of the lung. Acta Med Scand 204:513–516

Dumoulin CL, Yucel EK, Vock P, Souza SP, Terroer F, Steinberg FL, Wegmueller H (1990) Two- and three-dimensional phase contrast MR angiography of the abdomen. Comput Assist Tomogr 14:779–784

Edelman RR (1993) MR angiography: present and future. AJR 161:1–11

Edelman RR, Wentz KU, Mattle H, Zhao B, Liu C, Kim D, Laub G (1989a) Projection arteriography and venography: initial clinical results with MR. Radiology 172:351–357

Edelman RR, Mattle HP, Kleefield J, Silver MS (1989b) Quantification of blood flow with dynamic MR imaging and presaturation bolus tracking. Radiology 171:551–556

Ehringer H, Fischer H, Netzer CO, Schmutzler R, Zeitler E (1979) Venöse Abflußstörungen. Enke, Stuttgart, pp 407

Engel IA, Auh YH, Rubenstein WA, Sniderman K, Whalen JP, Kazam E (1983) CT diagnosis of mediastinal and thoracic inlet venous obstruction. AJR 141:521–526

Finn JP, Goldmann A, Edelman RR (1992) Magnetic resonance angiography in the body. Magn Reson Q 1:1–22

Finn JP, Zisk JHS, Edelman RR, Wallner BK, Hartnell GG, Stokes KR, Longmaid HE (1993) Central venous occlusion: MR angiography. Radiology 187:245–251

Giuliani L, Giberti C, Martorana G, Iserta A, Neumaier CE (1987) Value of computerized tomography and ultrasonography in the preoperative diagnosis of renal cell carcinoma extending into the inferior vena cava. Eur J Radiol 13:26–30

Goldfarb DA, Novick AC, Lorig R et al. (1990) Magnetic resonance imaging for assessment of vena caval tumor thrombi: a comparative study with vena cavography and computerized tomography scanning. J Urol 144:1100–1104

Haire WD, Lynch TG, Lund GB, Liebermann RP, Edney JA (1991) Limitations of magnetic resonance imaging and ultrasound directed (duplex) scanning in the diagnosis of subclavian vein thrombosis. J Vasc Surg 13:391–397

Hansen ME, Spritzer CE, Sostman HD (1990) Assessing the patency of mediastinal and thoracic inlet veins: value of MR imaging. AJR 155:1177–1182

Helzel MV, Mueller E (1993) Real-time flow-determination of vena cava inferior on two different levels via RACE pulse sequence in MR. ROFO 158:230–237

Hietala SO, Ekelund L, Ljungberg B (1988) Venous invasion in renal cell carcinoma: a correlative clinical and radiological study. Urol Radiol 9:210–216

Hirsh J, Hull RD, Raskob GE (1986) Epidemiology and pathogenesis of venous thrombosis. J Am Coll Cardiol 8:104–113

Horan JJ, Robertson CN, Choyke PL, Frank JA, Miller DL, Pass HI, Linehan WM (1989) The detection of renal cell carcinoma extension into the renal vein and inferior vena cava: a prospective comparison of venography and magnetic resonance imaging. J Urol 142:943–948

Hosoki T, Kuroda C, Tokunaga K, Marukawa T, Masuike M, Kozuka T (1989) Hepatic venous outflow obstruction: evaluation with pulsed duplex sonography. Radiology 170:733–737

Kadir S (1986) Diagnostic angiography. Saunders, Philadelphia

Kallman DA, King BF, Hattery RR, Charboneau JW, Ehman RL, Guthman DA, Blute ML (1992) Renal vein and inferior vena cava tumor thrombus in renal cell carcinoma. J Comput Assist Tomogt 16:240–247

Lang EK (1984) Comparison of dynamic and conventional computed tomography, angiography, and ultrasonography in the staging of renal cell carcinoma. Cancer 54:2205–2214

London NJM, Messios N, Kinder RB, Smart JG, Osborn DE, Watkin EM, Flynn JT (1989) A prospect study of the value of conventional CT, dynamic CT, ultrasonography, and arteriography for staging renal cell carcinoma. Br J Urol 64:209–217

Longley DG, Yedlicka JW, Molina EJ, Schwabacher S, Hunter DW, Letourneau JG (1992) Thoracic outlet syndrome: evaluation of the subclavian vessels by color duplex sonography. AJR 158:623–630

McMurdo KK, DeGeer G, Webb WR, Gamsu G (1986) Normal and occluded mediastinal veins: MR imaging. Radiology 159:33–38

Menu Y, Alison D, Lorphelin JM, Valla D, Belghiti J, Nahum H (1985) Budd-Chiari syndrome: US evaluation. Radiology 157:761–764

Merritt CRB (1992) Doppler color imaging: abdomen. Churchill Livingstone, New York

Mohiaddin RH, Wann SL, Underwood R, Firmin DN, Rees S, Longmore DB (1990) Vena caval flow: assessment with cine MR velocity mapping. Radiology 177:537–541

Roesch J, Uchida BT, Hall LD et al. (1992) Gianturco-Roesch expandable Z-stents in the treatment of superior vena cava syndrome. Cardiovasc Intervent Rodiol 15:319–327

Rosenberger A, Adler O (1980) Superior vena cava syndrome: a new radiologic approach to diagnosis. Cardiovasc Intervent Radiol 3:127–130

Roubidoux MA, Dunnick NR, Sostman HD, Leder RA (1992) Renal carcinoma: detection of venous extension with gradient-echo MR imaging. Radiology 182:269–272

Schefft P, Novick AC, Straffon RA, Stewart BH (1978) Surgery for renal cell carcinoma extending into the inferior vena cava. J Urol 120:28–31

Schwerk WB, Schwerk WN, Rodeck G (1985) Venous renal tumor extension: prospective US evaluation. Radiology 156:491–495

Siminovitch JMP, Moutie J, Straffin RA (1982) Inferior venacavography in the preoperative assessment of renal adenocarcinoma. J Urol 128:908–909

Stark DD, Hahn PF, Trey C, Clouse MR, Ferrucci JT (1986) MRI of the Budd-Chiari syndrome. AJR 146:1141–1148

Tello R, Scholz E, Finn JP, Costello P (1993) Subclavian vein thrombosis detected with spiral CT and three-dimensional reconstruction. AJR 160:33–34

Walker MF, Souza SP, Dumoulin CL (1988) Quantitative flow measurement in phase contrast MR angiography. J Comput Assist Tomogr 12:304–313

Weinreb JC, Mootz A, Cohen JM (1986) MRI evaluation of mediastinal and thoracic inlet venous obstruction. AJR 146:679–684

18 The Splenoportal Venous System

L. Guhl and I.P. Arlart

CONTENTS

18.1 Introduction

Portal hypertension has a high prevalence even in well-developed countries. In the Federal Republic of Germany approximately 3% of the population suffer from portal hypertension (Kraas et al. 1984). About 40% of patients with portal hypertensive disease have esophageal varices (Berchtold 1976), and about one-half of these undergo one or more episodes of bleeding. Strong esophageal hemorrhage has a lethality of approximately 50%. Therapeutic possibilities in patients with symptomatic portal hypertension are limited. Sclerotherapy is the most important method for management of esophageal or gastric hemorrhage. Portal venous pressure may be diminished by means of shunt surgery. However, problems of surgical intervention include the development of hepatic encephalopathy and shunt occlusion. As an alternative therapeutic procedure, transjugular portosystemic shunt (TIPS), performed by placement of a stent into an intrahepatic connection between a hepatic vein and a branch of the portal vein, has been introduced successfully.

Conventional angiography has been accepted as the method of choice for examinations of the portal venous system during the last decades. Two methods are in use. The more common is indirect visualization via selective arteriography of the celiac axis, the splenic artery, or the superior mesenteric artery. Direct splenoportography via percutaneous transhepatic catheterization delivers better results but is more invasive. The transsplenic route for direct phlebography of the portal venous system has been abandoned because of its significantly higher complication rate.

Despite its ability to display the portal venous structures, conventional CT has never achieved great importance for the assessment of this vessel system. This may change with the modern volume scan techniques.

In recent years gray-scale and particularly color duplex ultrasonography (US) has attained a more and more important role in the evaluation of the splenoportal veins, including the depiction of collateral pathways and the determination of flow direction and flow rates. The main reasons for this are the low cost, wide availability, and noninvasiveness of US. Nevertheless, US shows some limitations, making a more reliable method desirable. MR phlebography is a promising method that may be able to solve this problem.

18.2 Technical Considerations

Black blood [spin-echo (SE)] imaging was the first MR technique available for vascular imaging in the upper abdomen. Despite its disadvantages, such as a signal character that is very variable and confusing (Williams et al. 1985), this method has retained some importance for the examination of the splenoportal veins. Occasionally T1-weighted SE images are superior to gradient-echo (GRE) images

L. Guhl, Dr., Radiologisches Institut, Katharinenhospital, Kriegsbergstraβe 60, 70174 Stuttgart, FRG
I.P. Arlart, Professor, Dr., Ärztlicher Direktor, Radiologisches Institut, Katharinenhospital, Kriegsbergstraβe 60, 70174 Stuttgart, FRG

in the discrimination between thrombus and flowing blood and can help to avoid mistakes in other situations, where susceptibility or T1 effects may simulate flow in GRE images (Fig. 18.1).

Using bright blood MRA, the most important technique for examinations of the splenoportal veins has proved to be 2D time-of-flight (TOF) MRA. Phase-contrast (PC) and 3D methods are also in use but have shown some limitations.

18.2.1 2D TOF MRA

Short repetition times (TR), the precondition for inflow effect imaging, can only be achieved with GRE techniques if sufficient signal-to-noise ratios are attained. These TOF techniques have less confusing signal behavior than SE images. Washout effects are negligible (Edelman et al. 1990). Gradient motion refocusing (Laub and Kaiser 1988) leads not only to a decrease in artifacts due to peristaltic motion, but also to a further improvement of signal intensity. Projection images are easily obtained by means of the maximum-intensity projection (MIP) algorithm (Edelman et al. 1989b; Laub and Kaiser 1988) and are less confusing than corresponding black blood projection images. Saturation techniques (Edelman et al. 1988; Arlart et al. 1991) that allow the discrimination of arteries and veins are possible. Short TR permits the acquisition of images with sufficient matrix dimensions using the breath-hold technique.

Flow in the splenoportal system shows low pulsatility. Thus ghosting artifacts play only a minor role. Inconstant breathing may impair image quality but has been proved to be a minor problem in clinical practice, especially when the coronal slice direction has been chosen.

Arterial flow signal can be suppressed effectively by a presaturation band (Edelman et al. 1988) in the transverse orientation, placed slightly cephalad to the diaphragm. Typical sequence parameters are: spoiled GRE multislice (two slices) sequence with motion compensation in the readout and slice selection direction, TR = 60 ms, TE = 8 ms, flip angle = 30°, slice thickness = 5 mm with 1 mm overlap, FOV = 450 mm, matrix = 192 × 256. These parameters give rise to a data acquisition time of 15 s. Imaging in two orthogonal slice directions (e.g., coronal and transverse) is recommended to minimize diagnostic problems due to in-plane flow. Migrating saturation techniques may be useful for examination in the transverse slice orientation. However, this technique offers no substantial benefit, because suppression of the aortic flow signal in the upper abdomen using a stationary supradiaphragmatic presaturation pulse is sufficient in most cases.

Depending on the imaging system under use, image acquisition advancing step by step through the abdomen may be ponderous. This problem may be solved by automatic step procedures on the basis of dedicated stand-alone programs or procedures. Image quality is dependent on the degree of cooperation from the patient. Before the examination is commenced, patients have to be informed how to breath. Constancy is more important than the degree of expiration. Peristaltic motion should be reduced by administration of glucagon or scopolamine-*N*-butylbromide.

Signal-to-noise ratio is a problem when short TRs are used. Even at 1.5 T smaller fields of view, which improve spatial resolution, are only possible when special hardware is used, e.g., surface coils of the Helmholtz or wrap-around type. With such equipment even small veins can be displayed (Fig. 18.2), especially if slice orientation is perpendicular to the course of the vessel.

In order to achieve a high diagnostic accuracy it has been shown to be mandatory to analyze not only the projection angiograms but also the individual slices, particularly in cases with small intra- and extravascular lesions. Smaller vessels may be displayed in a projection angiogram by use of a "targeted MIP" technique, selecting only representative slices which contain the specific information (Anderson et al. 1990). In addition, targeted MIP may avoid disturbing vascular overlaps.

By sophisticated use of the presaturation technique portal and hepatic veins can be displayed selectively. The portal venous flow signal can be suppressed by an oblique stationary presaturation band near the hepatic hilum (Otake et al. 1990), which is orientated orthogonal to the direction of the vena portae. The intrahepatic branches of the portal vein may be visualized selectively by using a migrating saturation area close to the right side of the slice and a sagittal imaging plane (Nabeshima et al. 1993). However, this technique only works in the right hepatic lobe.

18.2.2 3D and PC Techniques

In imaging of venous vascular structures 3D TOF techniques have not proved useful owing to strong suppression of venous flow signal. Therefore 3D imaging of the portal venous system is only possible

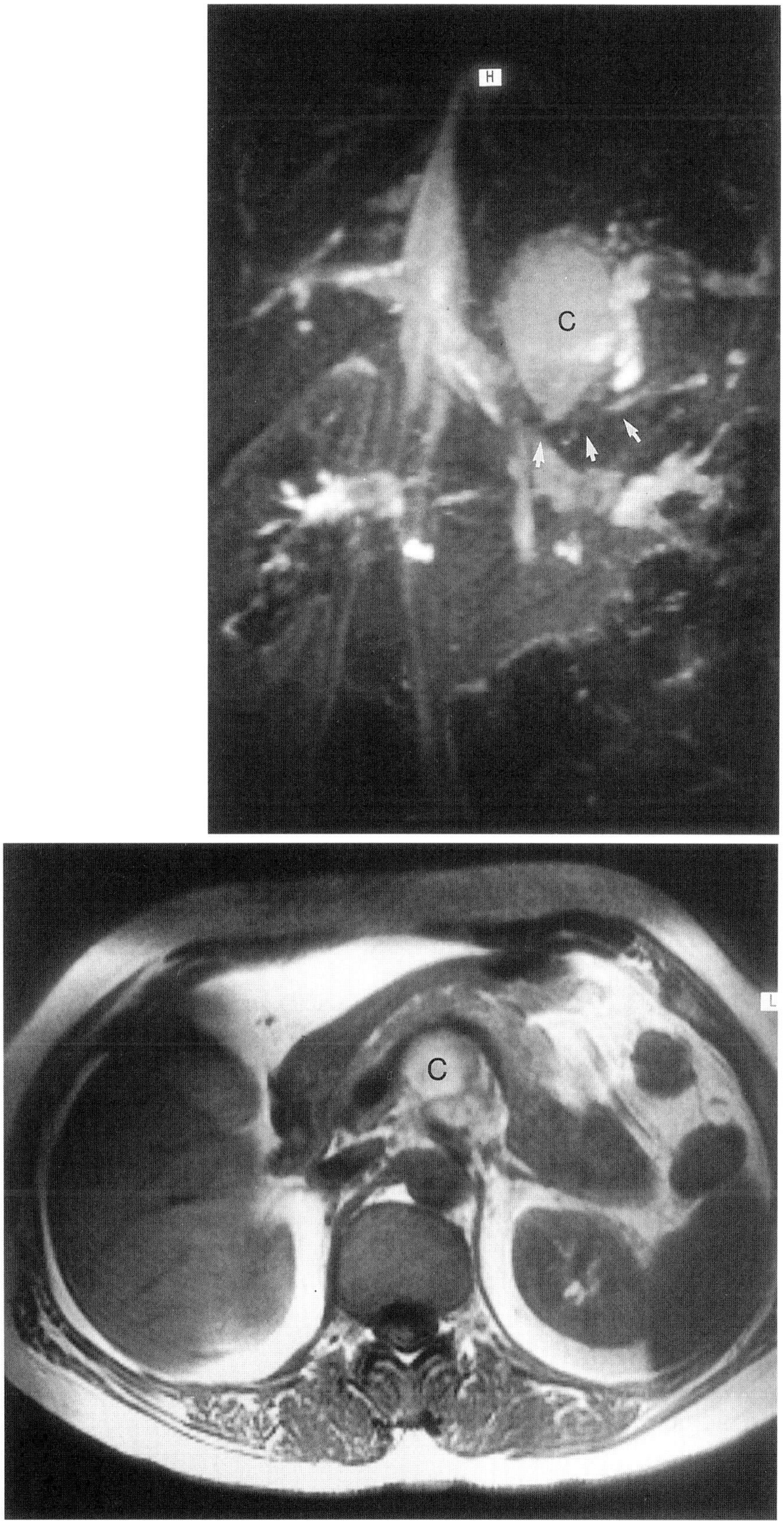

a

b

Fig. 18.1a,b. Pseudo-obstruction of the splenic vein. **a** On the 2D TOF maximum intensity projection image, the splenic vein seems to be obstructed (*arrows*). Near the vein a roundish structure with high signal (*C*) can be seen. **b** T1-weighted SE image showing a patent splenic vein and high signal in a pancreatic pseudocyst (*C*) with older blood. The MIP algorithm cannot discriminate flow and high signal caused by a short T1 time. Consequently in the environment of old blood, vessel structures may disappear

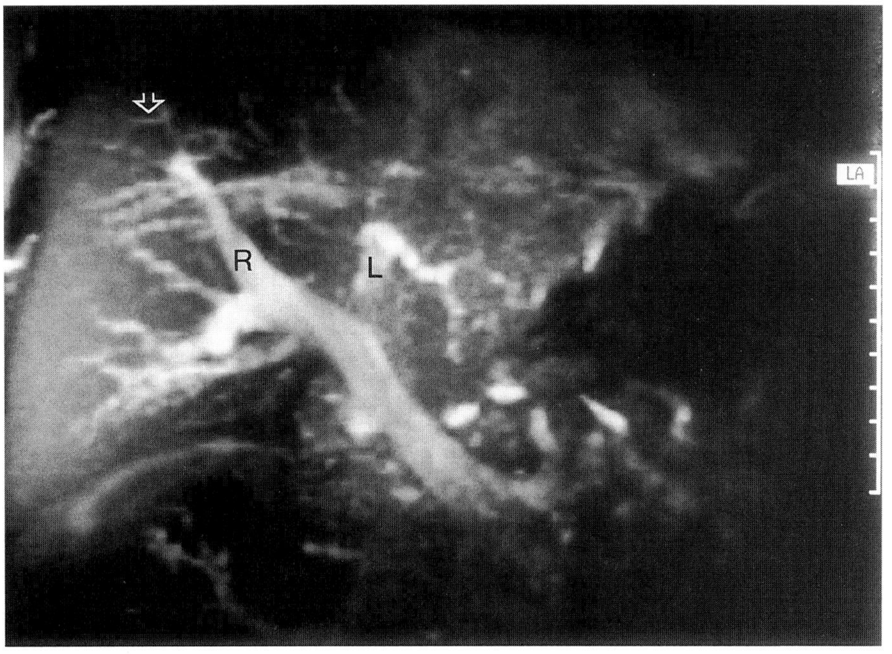

Fig. 18.2. 2D TOF examination of the venous system of the liver (projection image). This examination of a healthy volunteer was performed with a dedicated "body" coil of the Helmholtz type. With this technique fifth order branches of the portal vein are visible (*arrow*). *R*, right branch, *L*, left intrahepatic branch of the portal vein

when PC techniques are used, yielding superior signal-to-background ratios. However, the main disadvantage of PC is its high sensitivity to motion artifacts and the possibility of information loss due to improper velocity settings. Nevertheless, 2D PC achieves considerable image quality even in the upper abdomen (NGHIEM et al. 1994a; SPRITZER et al. 1990), especially if the phase-encoding direction is selected in the patient's right-left orientation (DUMOULIN et al. 1990). Typical sequence parameters that are recommended for PC velocity imaging (at 1.5 T) are GRE technique, TR = 30 ms, TE = 8.5 ms, FA = 30°, slice thickness = 5–7 mm with 2 mm overlap, FOV = 340–400 mm, matrix = 128 × 256, velocity encoding of 20 cm/s in all axes (NGHIEM et al. 1994a). PC MRA may be combined with gadolinium administration. With this combination a good correlation with conventional angiography and US may be achieved (NGHIEM et al. 1994b). However, after the administration of gadolinium suppression of the arterial flow signal is not possible, resulting in undesirable vascular overlaps.

18.2.3 Quantification of Venous Blood Flow

Currently, Doppler US is widely used for the assessment of portal venous blood flow (BURNS et al. 1993; D'ALIMONTE et al. 1993). However, this technique may be limited by the patient's habitus and bowel gas. Its accuracy may be compromised if the available acoustic window does not allow an adequate angle between the transducer and the vessel.

Flow velocity in the splenoportal system can be determined by different phase-encoding methods, including PC, cine PC (BURKART et al. 1992, 1993), and RACE (MUELLER et al. 1988, 1989). All methods work with motion-induced phase shifts that arise under the influence of a magnetic gradient field. Cine PC has been shown to be highly accurate (BURKART et al. 1992) and practicable for the evaluation of flow rates in the portal venous system. However, as a disadvantage, it is relatively time consuming, because phase encoding must be done in all three gradient directions to obtain accurate results.

Bolus tracking (EDELMAN et al. 1989a,c, 1990; FINN et al. 1991a; TAMADA et al. 1989) is a TOF method that allows flow quantification within a breath-hold. A bolus of blood is "tagged" by a thin saturation band orientated perpendicular to the vessel of interest and is displayed on a series of images acquired shortly after the application of the presaturation pulse (Fig. 18.3). The velocity of the bolus can be calculated as the ratio of bolus displacement and time delay between two corresponding images. The main advantage of the bolus tracking

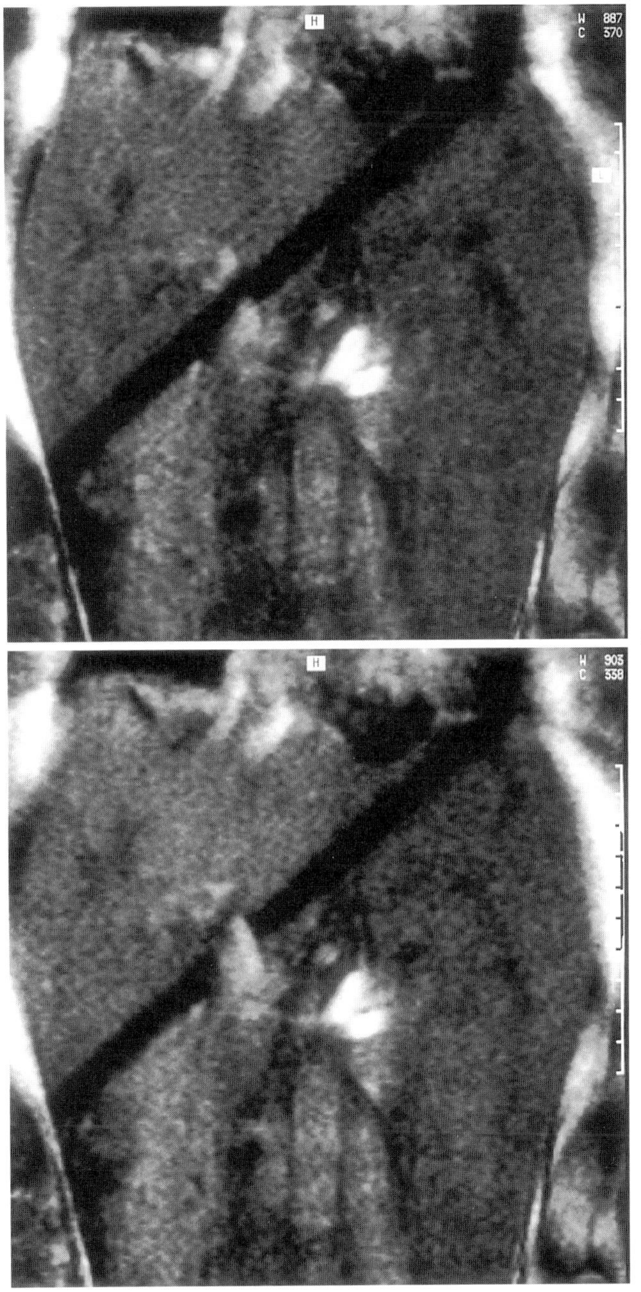

a

b

Fig. 18.3a,b. Bolus tracking. First (**a**) and last (**b**) image of a bolus tracking series, displaying motion of "tagged" blood in a portal vein. In this normal volunteer not only the displacement but also the direction of the flow-related parabolic distorsion of the presaturation area indicates hepatopetal flow

technique is the visual display of flow motion, while phase methods are very complex. This technique shows a strong correlation with duplex US. As flow in the portal system normally shows no pulsatility, this method is accurate in the majority of cases. When pulsatility occurs, cardiac gating will be necessary (Edelman et al. 1989a,c; Tamada et al. 1989). Angulation between the course of the vessels and the imaging planes may cause misleading results, which can be corrected by means of a simple algorithm (Edelman et al. 1989c).

18.3 Applications of MRA to the Portal Venous System

18.3.1 Normal Anatomy and Variants

The splenoportal veins form the third part of the vessel system of the body. This vessel group is characterized by the absence of a direct connection to the heart. It includes veins draining the intestine, the spleen, the gallbladder, and the pancreas. The central part of this vessel system is the portal vein, through which nearly all blood from these viscera passes on its way to the liver. The vena portae, which shows little variability, contains no valves. It arises at the confluence of the superior mesenteric vein (SMV) and the splenic vein and has a length of about 6 (3–9) cm. Other feeders are the cystic and the right gastric veins.

At the liver hilum the portal vein divides into a right and a left main branch, and sometimes into three branches (Okudira 1991). After entering the left lobe, the left main branch divides into branches running to the caudate and quadrate lobes and the hepatic segments 2 and 3. At the level of the falciform ligament it is joined by the paraumbilical vein, a remnant of the obliterated left umbilical vein. The right main branch divides in most cases into two rami (r. anterior and r. posterior).

In the normal liver intrahepatic portal venous branches show few anastomoses between themselves. They attain a diameter of 0.7–0.8 mm after the seventh branching. The terminal portal veins are formed by the 12th order branches (Okudira 1991).

The SMV drains the transverse and ascending colon, the cecal region, the jejunal and ileal part of the small bowel, the greater curvature of the stomach, and the pancreas. The veins draining the small bowel empty into the left side of the SMV, often forming a main branch, which may be as big as the main stem of the SMV, giving the impression of a double superior mesenteric vein (Reichardt et al. 1979).

Besides the main vessels draining the spleen, other feeders of the splenic vein are the inferior mesenteric vein (IMV) in 40% of patients (Reichardt et al. 1979) and the left gastroepiploic, the short gastric, and the left gastric vein, each of which is present in most cases. The IMV drains the descending colon, the sigmoid colon, and the rectum. In 30% of cases it enters the confluence of the splenic and superior mesenteric veins and in another 30%, the upper part of the SMV. The right and left gastric veins form the coronary vein, running along the lesser curvature of the stomach.

The 2D TOF GRE technique has proved to be a very stable method for imaging of the larger vessels of the upper abdomen. In our own study (Guhl and Arlart 1992), image quality allowed the assessment of the portal and splenic veins in 99% and of the intrahepatic branches of the portal vein in 81% of patients. With sophisticated equipment, for example body coils of the Helmholtz type, fifth order intrahepatic branches of the portal vein (Fig. 18.2) may be seen (Guhl et al. 1993). However, imaging of veins with a smaller diameter, such as the IMV or the gastric- and gastroepiploic veins, may be a problem.

18.3.2 Stenoses and Occlusions

The most frequent causes of stenoses and occlusions of the portal and splenic veins are tumors and sequelae of thrombosis. Furthermore, idiopathic stenosis and phlebosclerosis may occur.

Because US has been shown to be a very sensitive method in the detection of occlusions (Finn et al. 1991a), both conventional angiography and MRA may play a role only in cases in which bowel gas or obesity does not allow sufficient sonographic evaluation. Contrast-enhanced computed tomography (CT) has proved useful in the accurate evaluation of stenoses or occlusions of the splenic or main portal vein in cases of pancreatic cancer (Savader et al. 1994). Limitations of CT are the large volumes of contrast agents and timing problems.

Noninvasive 2D TOF MR venography is very efficient in the detection of portal and splenic vein stenosis. Examinations in two orthogonal imaging planes (e.g., coronal and transverse) are mandatory. Signal void due to saturation phenomena can be minimized if additional examinations in a plane perpendicular to the flow direction are performed. Under this condition the inflow effect reaches its maximum and the cross-section of the vessel can be estimated with high accuracy. In most cases stenoses are better seen on the individual slices than on projection angiograms, because the MIP algorithm may obscure lesions of minor degree.

In cases of portal or splenic vein stenosis due to pancreatic tumors (Fig. 18.4) or lymphadenopathy, MRI and single MRA slices are useful in clarifying the relationship of a mass to the abdominal vessels (Longmaid et al. 1990). This vascular information may be of major importance for surgical treatment planning.

In cases of splenoportal occlusion (Fig. 18.5) MRA shows not only the site of occlusion but also the course of collateral pathways that may have developed. The availability of projection angiograms is a clear advantage of MRA in comparison to US, which has difficulties in visualizing the path of collaterals. In contrast to US, MR projection angiograms can provide complete information on the amount and course of collaterals due to the larger field of view.

a 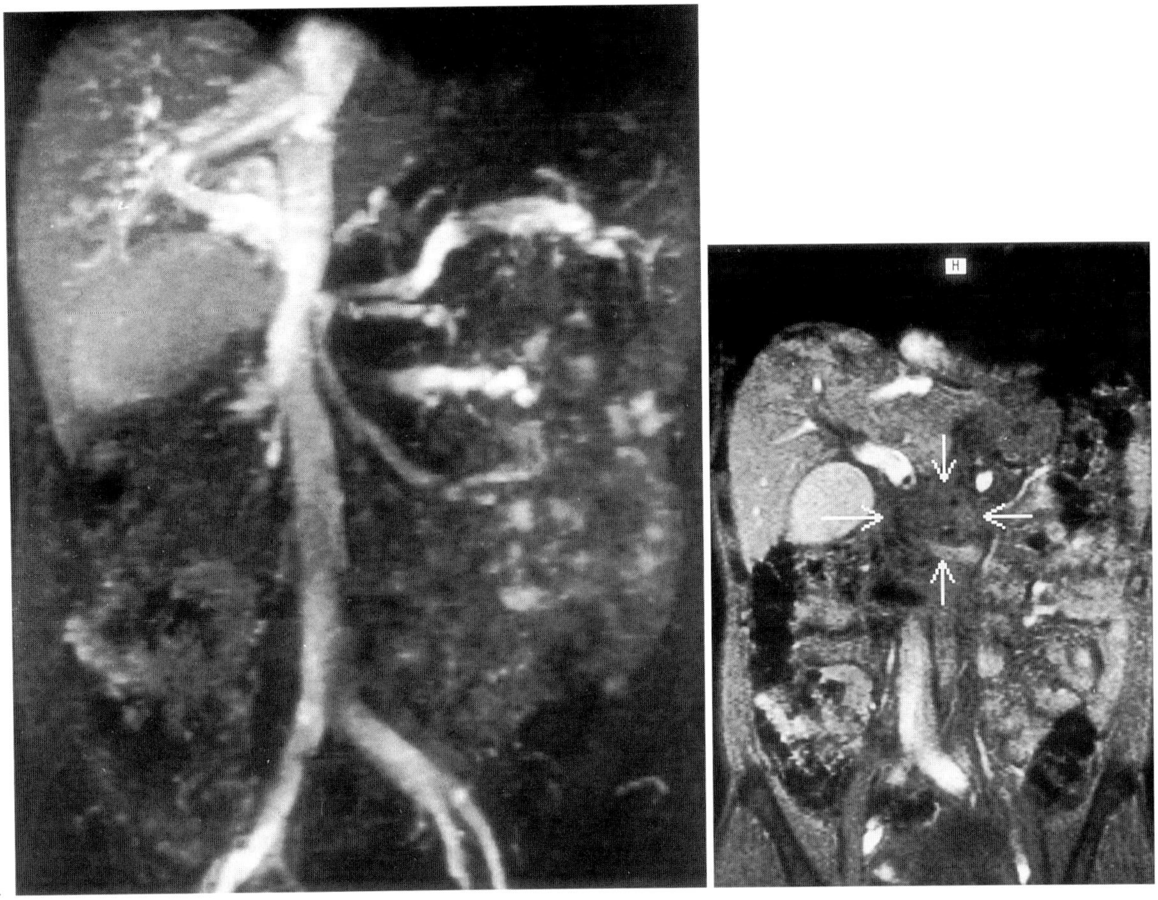 b

Fig. 18.4a,b. Stenosis of the splenic vein due to pancreatic carcinoma. **a** 2D TOF MIP image showing caudad displacement and tapering of the splenic vein near the confluence with the superior mesenteric vein. **b** On the individual slice the pancreatic tumor (*arrows*) compressing the splenic vein is visible

18.3.3 Portal and Mesenteric Vein Thrombosis

Portal vein thrombosis may be caused by trauma, abdominal malignancies, or inflammatory disorders. In US thrombosis usually appears as an echogenic structure within the vessel lumen. Color Doppler US may be superior to gray-scale US in cases of hypoechoic or anechoic thrombus (Ralls 1990; Tessler et al. 1991), but may not solve the problem of misinterpretation of insufficient acoustic window and an occluded vein (Tessler et al. 1991). In computed tomography (CT) thrombus is isodense or slightly hypodense in comparison to soft tissue and sometimes hyperdense when recently formed (Parvey et al. 1994). In MRI using SE sequences, intraluminal thrombus appears with isointense or hyperintense signal on T1-weighted and is typically hyperintense on T2-weighted images (Parvey et al. 1994).

Portal vein thrombosis may be detected safely by means of MRA. In acute cases the thrombus material may be displayed as a low signal area within the bright signal of flowing blood (Figs. 18.6, 18.7). Older thrombus material may show high signal on 2D TOF GE images, mimicking a patent lumen. In these cases additional black blood examinations may be helpful. Furthermore, flowing blood and clot material may be discriminated by means of the presaturation or bolus tracking technique (Edelman et al. 1989c).

Even the diagnosis of thrombosis of the SMV is possible with MRA. Scanning in a transaxial slice orientation is mandatory for this condition. Using the coronal slice orientation the SMV may be obscured in patients without any disease of the splenoportal system due to in-plane saturation. Thus, a considerable rate of false-positives may occur when using improper technique.

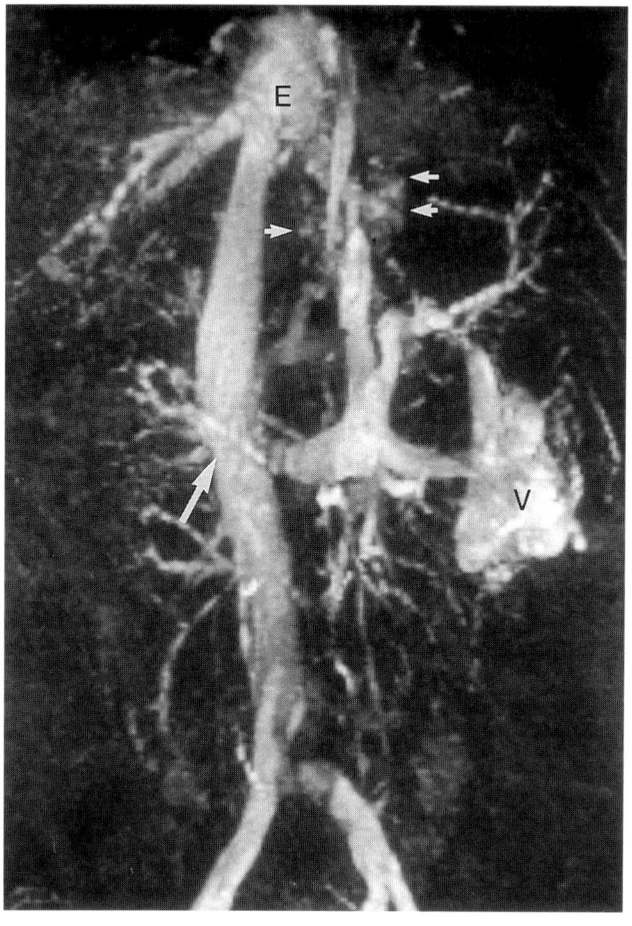

Fig. 18.5. Portal vein occlusion in a child suffering from myelodysplastic syndrome with splenomegaly. The main stem of the portal vein is occluded (*long arrow*) and collateralized by a few thin veins. *V*, varices in the splenic hilum; *short arrow*, gastric collaterals; *E*, esophageal varices

18.3.4 Portal Hypertension

In portal hypertension the major role of MRA is to evaluate the amount and course of collateral pathways. These can be divided into spontaneously developed portohepatic and portosystemic shunts. Portohepatic shunts may bypass an occluded or stenosed part of the portal venous system such as the splenic, superior mesenteric, or main portal vein and are characterized by hepatopetal flow. Portosystemic shunts can be divided into two groups, draining into either the superior (SVC) or the inferior vena cava (IVC). These subgroups have clinical implications for patients with portal hypertension. For example, esophageal and gastric varices, emptying into the SVC, are the main cause of life-threatening hemorrhage. Gastrorenal, splenoretroperitoneal, and mesenteric collaterals finally drain into the IVC. Recanalized paraumbilical veins may have connections to the SVC (via internal thoracic veins) or IVC.

In a surgically proven patient group, FINN et al. (1991a) found that MRA is an extremely accurate means of evaluating the portal venous system. The only method that may be superior to MRA is conventional angiography performed by transhepatic access. Indirect phlebography via selective arteriography of the SMA or the celiac trunk may miss collaterals (GEHL et al. 1990), particularly if improper technique, including an insufficient volume of contrast medium (OKUDA and TAKAYASU 1991), is used. Doppler US yields accurate information on the patency of the main portal vein and the flow direction within that vessel. The combination of real-time appearance and characteristic pulsed-Doppler waveforms allows better identification of the different vessels in the upper abdomen (WELTIN et al. 1985). However, limitations exist in the detection and display of the course of collaterals (FINN et al. 1991a; GEHL et al. 1990). Another problem is false-positives in the diagnosis of portal vein occlusions

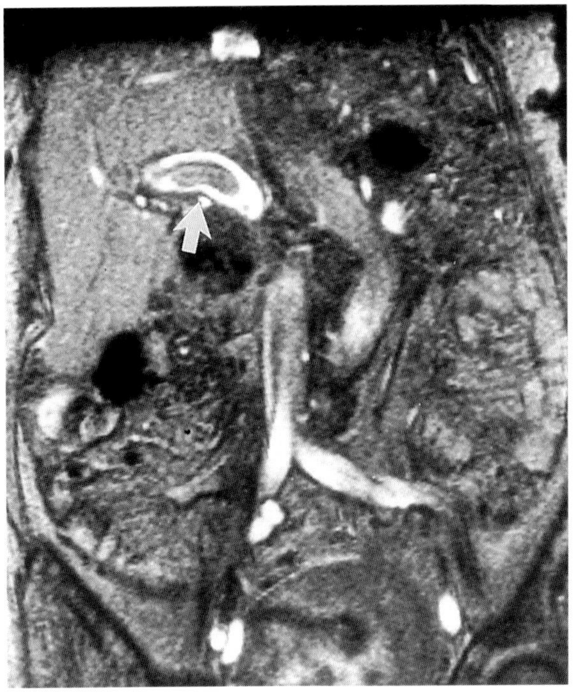

Fig. 18.6. Portal vein thrombosis. Thrombus is clearly demarcated as a low signal area (*arrow*) within the portal venous blood. Individual slice

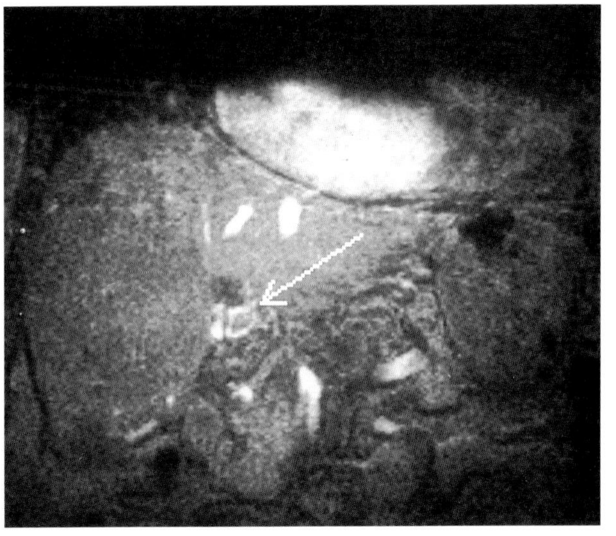

Fig. 18.7. Portal vein thrombosis in a patient with liver tumor. The portal vein runs nearly perpendicular to the slice in this coronal image. Thrombus (*arrow*) is surrounded by a thin rim of flowing portal venous blood

due to insufficient ultrasound penetration (Tessler et al. 1991).

Moreover, MRA has been shown to be superior to dynamic CT in the detection of collaterals (Gehl et al. 1990). CT angiography (CTA) with a helical scan technique is an interesting alternative method that has provided promising results (Rubin et al. 1993; Stehling et al. 1994). The disadvantages of this technique are problems with improper time settings and the need for relatively large amounts of contrast media. The value of intravenous CTA also may be limited in cases of extensive collateralization due to insufficient enhancement. CTA after intra-arterial contrast injection (CTAP) may give better results but is more invasive. Although superiority of CTAP to conventional angiography (which is performed in the same procedure) is reported (Savader et al. 1994), there will probably be no substantial benefit. Conventional angiography and CTA of the splenoportal vein tend to be contraindicated in patients with portal hypertension associated with both impaired hepatic and impaired renal function. In these cases application of iodinated contrast agents should be avoided.

Magnetic resonance angiography is able to show more esophageal varices than endoscopy (Johnson et al. 1991). This discrepancy can be explained by the fact that endoscopy only detects submucosal varices (Burcharth 1991) whereas in MRA both submucosal and periesophageal varices can be visualized. In terms of sensitivity in detecting collaterals, no significant difference can be observed between the transaxial and coronal imaging planes (Johnson et al. 1991).

Probably due to T2*-effects and pulsation-related inflow effect, ascites may show a bright signal on TOF GRE sequences. The resulting ghosting artifacts may lead to a considerable degradation of image quality. Nevertheless, diagnostic accuracy of 2D TOF MRA normally is not diminished substantially by ascites.

18.3.4.1 Prehepatic Block

Portal hypertension may occur as a sequel of stenosis or occlusion of the splenic or extrahepatic portal vein that is caused in most cases by thrombosis. These conditions are a frequent cause of portal hypertension in children (Fig. 18.8). Portohepatic and portosystemic shunts may occur spontaneously.

Cavernous transformation of the portal and splenic vein is characterized by a replacement of the occluded splenic and/or portal vein by a network of

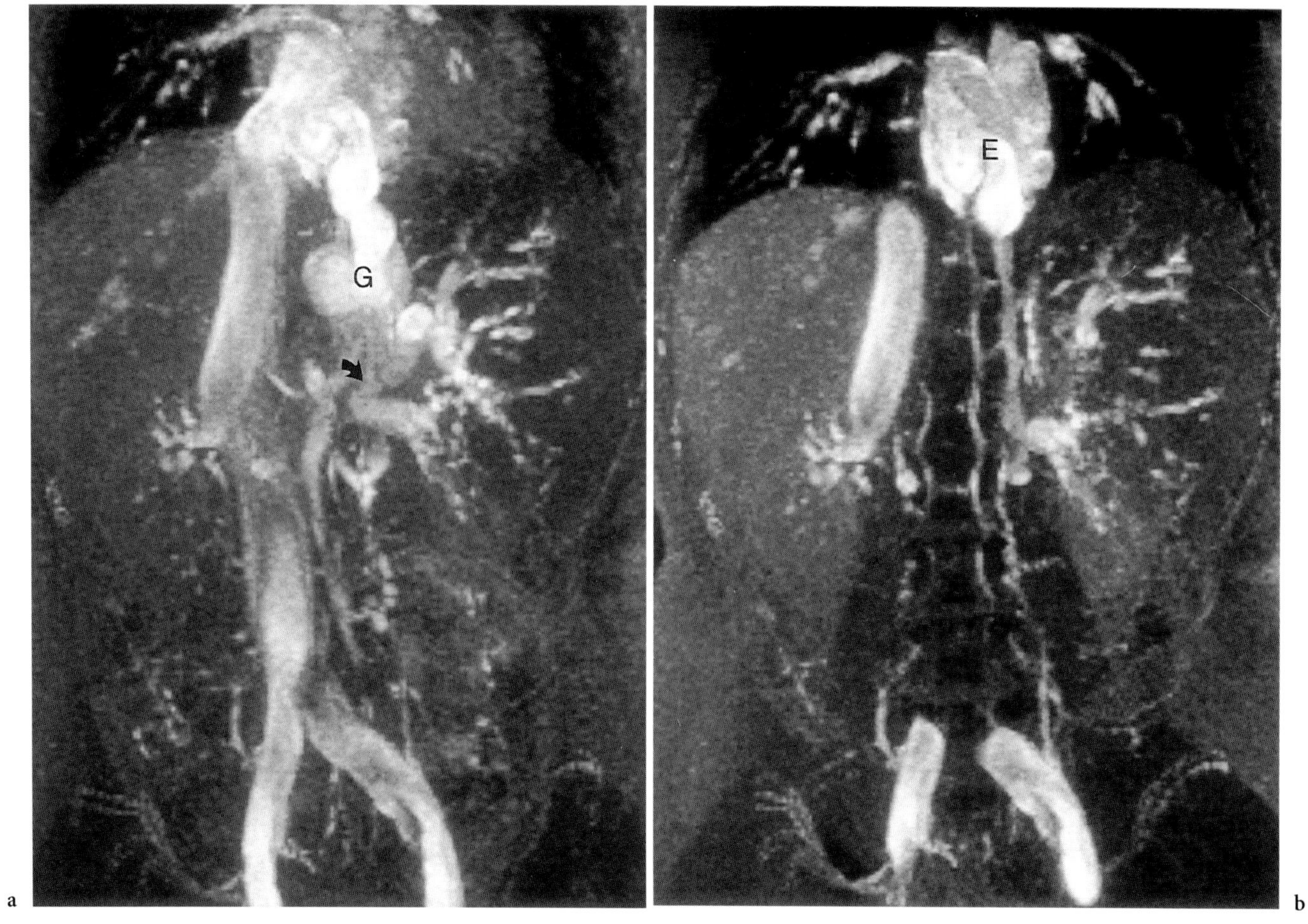

Fig. 18.8a,b. Prehepatic block in a child with portal vein occlusion after thrombosis. **a** Huge splenogastric (*G*) varices and small splenorenal shunt (*curved arrow*). Enormous esophageal varices (*E*) are better displayed on the targeted MIP angiogram (**b**)

collaterals running along the former course of these vessels (Fig. 18.9). Cavernous transformation may even develop in patients with incomplete obstruction (Okuda and Aoki 1991) or may only concern the intrahepatic branches while the main stem of the portal vein remains patent (Fig. 18.10). Two types are distinguished, the so-called primary, congenital form and the secondary form. The congenital form, if it exists at all, is very rare. In most cases the "primary" cavernous transformation is a sequel of clinically silent umbilical vein sepsis or portal vein thrombosis following intra- or retroperitoneal infections (Okuda and Aoki 1991).

Although duplex US and CT are reported to allow the diagnosis of this disorder (Weltin et al. 1985), in our experience both methods are of limited value in comparison to MRA. Both TOF and PC MRA are reported to depict this condition clearly and allow accurate visualization of the amount of venous collaterals (Nghiem et al. 1994a).

18.3.4.2 Presinusoid and Sinusoid Intrahepatic Block

Presinusoid block of the portal venous branches may be idiopathic. It can occur in myeloproliferative or granulomatous diseases and after intoxications. Reduction of the caliber of intrahepatic portal branches is regarded as a characteristic angiographic sign of presinusoid block (Deimer 1973). Arai et al. (1991) found small vessels running parallel to the second order branches of the intrahepatic portal vein in MRA examinations of patients with idiopathic portal hypertension (Banti's syndrome).

The most frequent cause of sinusoid block, often in combination with presinusoid and sometimes postsinusoid block, is liver cirrhosis. SE imaging is a rather efficacious method for detecting nodular alterations of liver parenchyma in cirrhosis. MRA provides additional information to T1- and T2-weighted

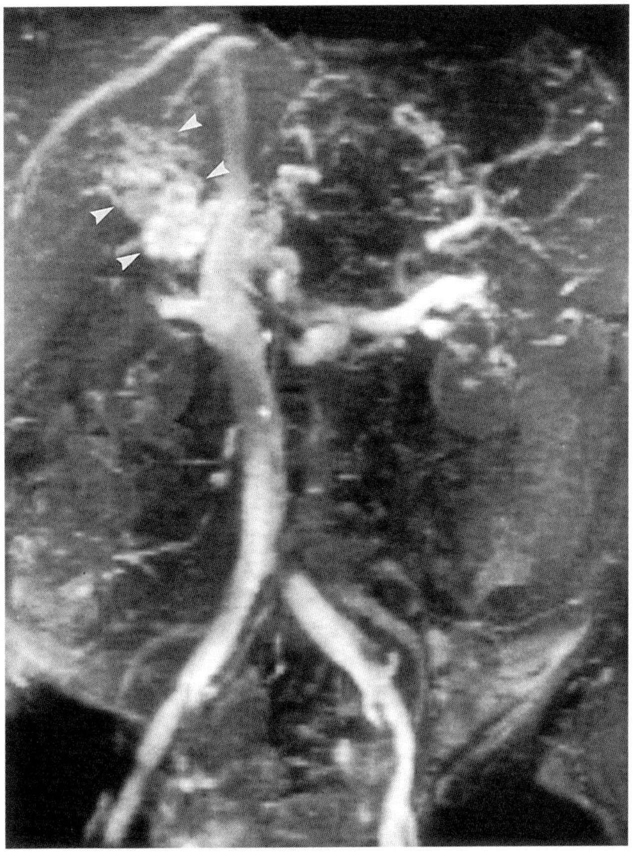

Fig. 18.9. Cavernous transformation of the portal vein. The main stem of the portal vein is occluded and replaced by a network of multiple small tortuous collaterals (*arrowheads*)

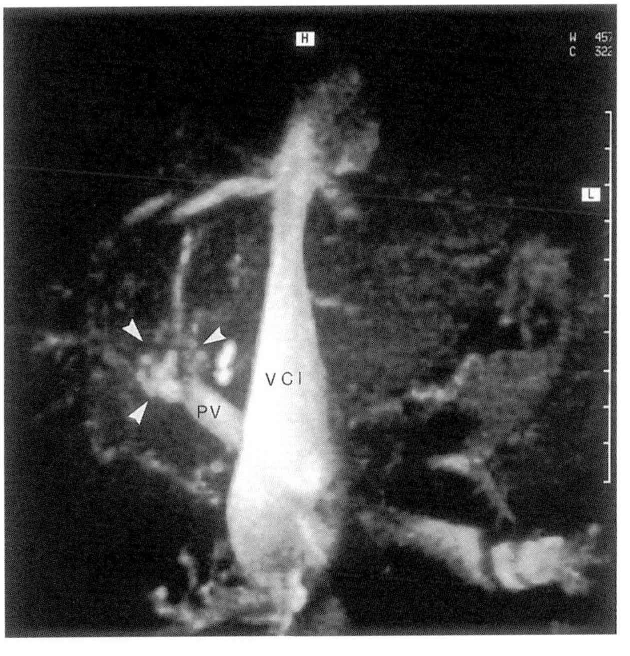

Fig. 18.10. Cavernous transformation of the intrahepatic branches of the portal vein. Young woman with episodes of hemorrhage due to espohageal varices. Liver biopsy and laboratory values were normal, and all large branches of the portal venous system were patent on indirect portography. This targeted projection image in the coronal plane shows that the main stem of the portal vein (*PV*) is patent, but nearly all of the intrahepatic branches of portal vein are replaced by varicose collaterals (*arrowheads*). 2D TOF examination with Helmholtz body coil. *VCI*, inferior vena cava

MR images. Frequently intrahepatic branches of the portal vein with narrowed calibers or a tortuous course can be seen (Fig. 18.11). Sometimes the left branch of the portal vein is found to be enlarged because of hypertrophy of the left hepatic lobe (Burcharth 1991). In cases of pure sinusoid block intrahepatic branches of the portal vein seem to be dilated (Deimer 1973).

18.3.4.3 Postsinusoid and Posthepatic Block

The main reasons for postsinusoid block are tumor obstruction, toxic veno-occlusive disease, and thrombosis of the hepatic veins. Sometimes the intrahepatic part of the IVC may be involved. The role of MRA in this vascular area is discussed in Chap. 17.

a

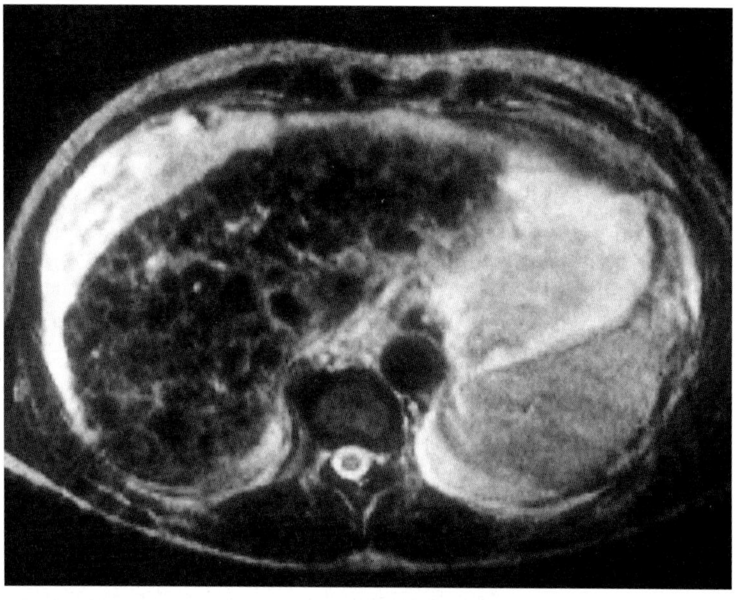

b

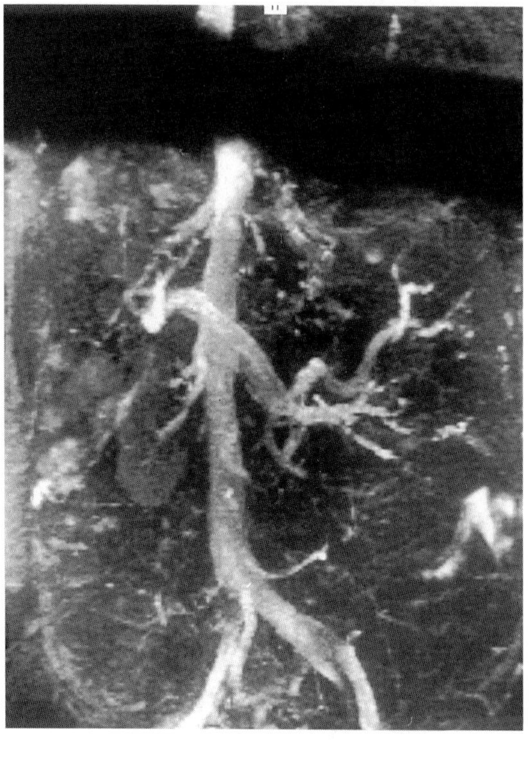

Fig. 18.11a,b. Liver cirrhosis. **a** T2-weighted SE image showing nodular degeneration of the liver parenchyma and ascites. **b** 2D TOF projection image displaying diminished caliber and slightly tortuous course of intrahepatic branches of the portal vein

18.3.5 Planning of Therapeutic Procedures

Knowledge of the condition of the splenoportal system is important for the planning of many interventional procedures. When a patient is considered to be a candidate for either portosystemic shunt placement or liver transplantation, knowledge of host portal anatomy and hemodynamic characteristics is of vital importance for the surgeon in planning the operative approach. Preexisting portal vein thrombosis, for example, influences the operative technique of liver transplantation (LANGREHR et al. 1994) and must be proved or ruled out preoperatively. In a recently published study by FINN et al. (1991a) the authors reported that MRA is an extremely accurate method for depicting the anatomy of the portal venous system and for demonstrating flow direction, thrombosis, varices, and the distribution of portosystemic collaterals.

The extent and course of spontaneous collateral pathways are important in the indication for shunt surgery. Esophageal and gastric varices (Fig. 18.12) are the main cause of life-threatening hemorrhage. Spontaneous splenorenal (Fig. 18.13) and gastrorenal shunts provoke hepatic encephalopathy (OKUDA and MATSUTANI 1991). Spontaneously developed splenorenal shunts and dilated paraumbilical veins (Fig. 18.14) may diminish the risk of hemorrhage. Thus, there may be no need for surgical intervention (RALLS 1990).

By means of a MR bolus tracking technique, flow reversal in the different parts of the splenoportal system can be proved (Fig. 18.15). In addition this method allows the determination of flow velocities. On this basis vascular flow maps can be established that can be used for planning surgical shunts. A reversed flow in the portal vein is considered a contraindication to shunt surgery (FINN et al. 1991a).

In many cases of portal hypertension, portal flow is diminished (D'ALIMONTE et al. 1993). In other patients it may be normal or increased. The increased, so-called hyperdynamic flow in the portal vein seems to be closely related to the probability of variceal

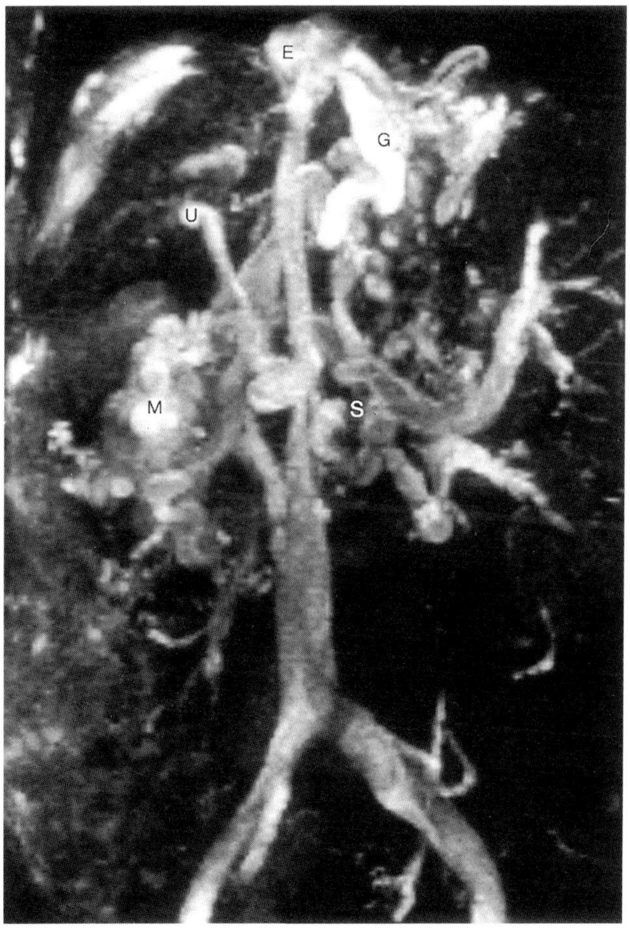

Fig. 18.12. Occluded portal vein after thrombosis due to cirrhosis. Besides mesenteric collaterals (*M*), huge gastric varices (*G*), esophageal varices (*E*), a widened paraumbilical vein (*U*), and a spontaneous splenorenal shunt (*S*) can be seen

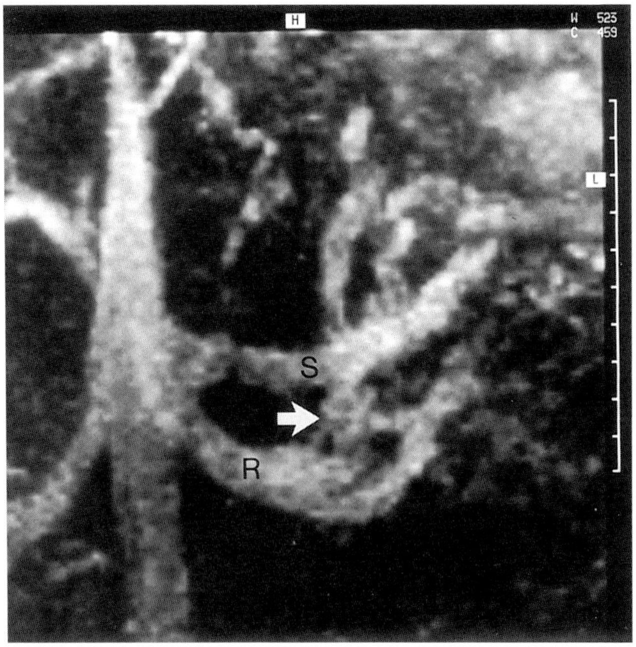

Fig. 18.13. Spontaneous splenorenal shunt (*arrow*) in a patient with liver cirrhosis. *S*, splenic vein; *R*, renal vein

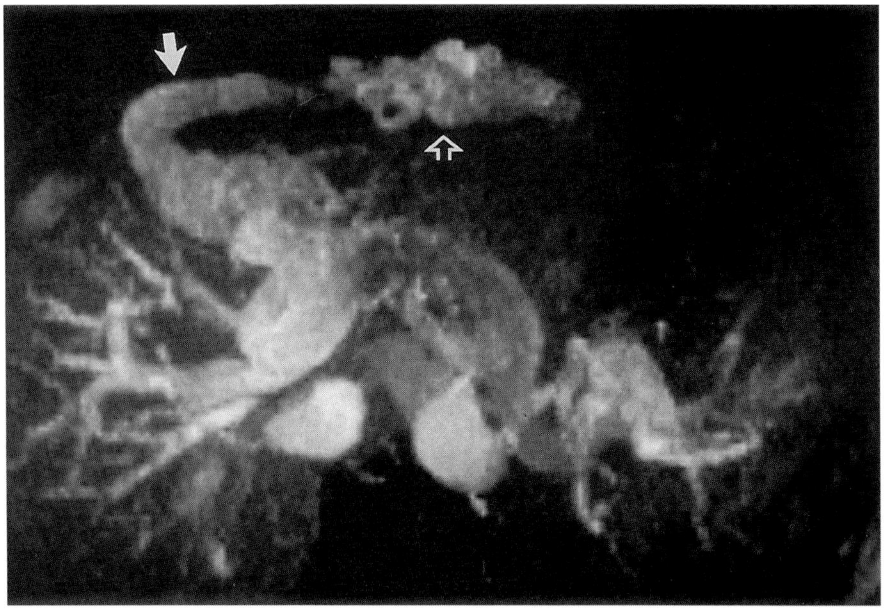

a

Fig. 18.14a–c. Paraumbilical portosystemic shunt. **a** projection image in the axial orientation showing the wide paraumbilical vein (*arrow*) and a network of vessels in the subcutis (*open arrow*). **b** Projection image in the coronal projection showing the paraumbilical vein (*arrow*). **c** The further course of this vessel (*arrow*) to the internal thoracic veins (*arrowheads*) can only be displayed on a targeted MIP image

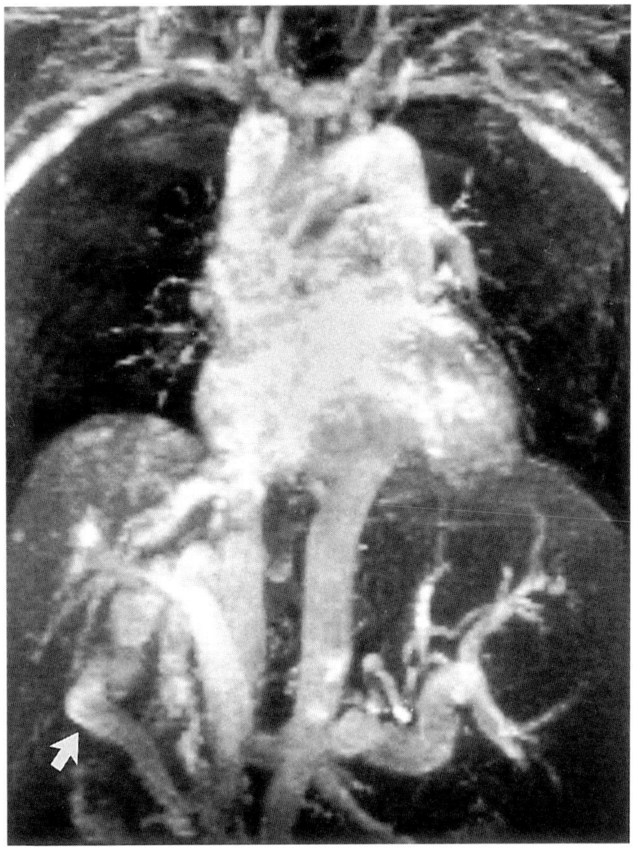

b

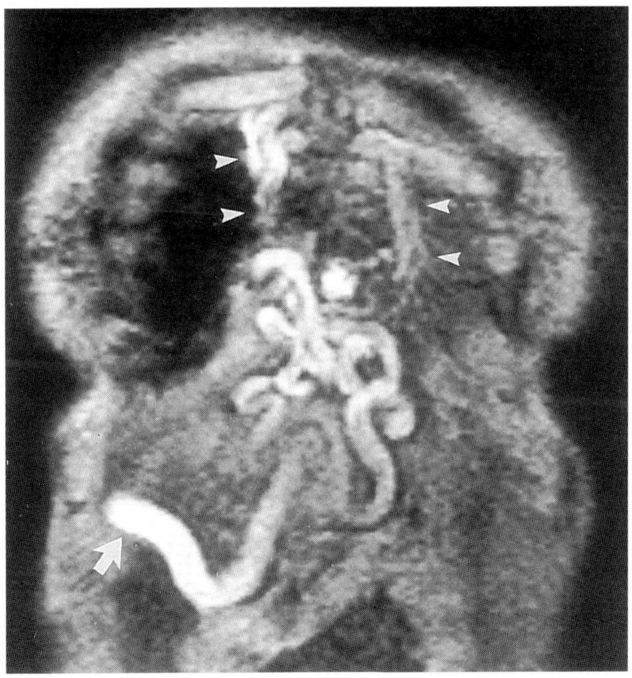

c

Fig. 18.14b,c

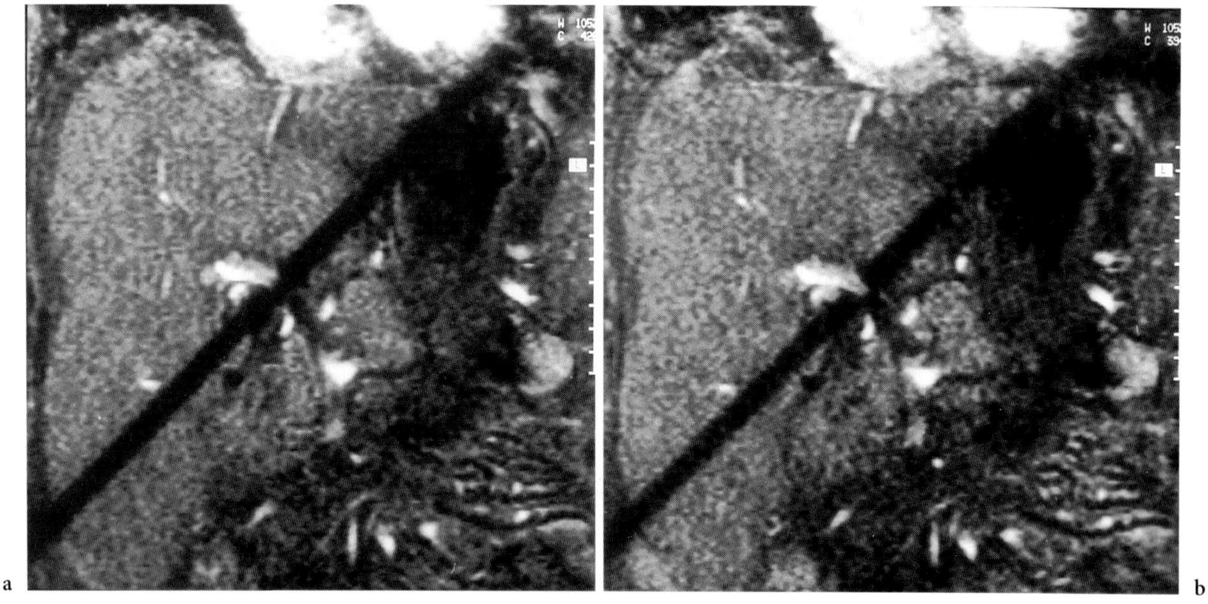

Fig. 18.15a,b. Bolus tracking series of a patient with liver cirrhosis and portal hypertension. Flow reversal in the portal vein is evident

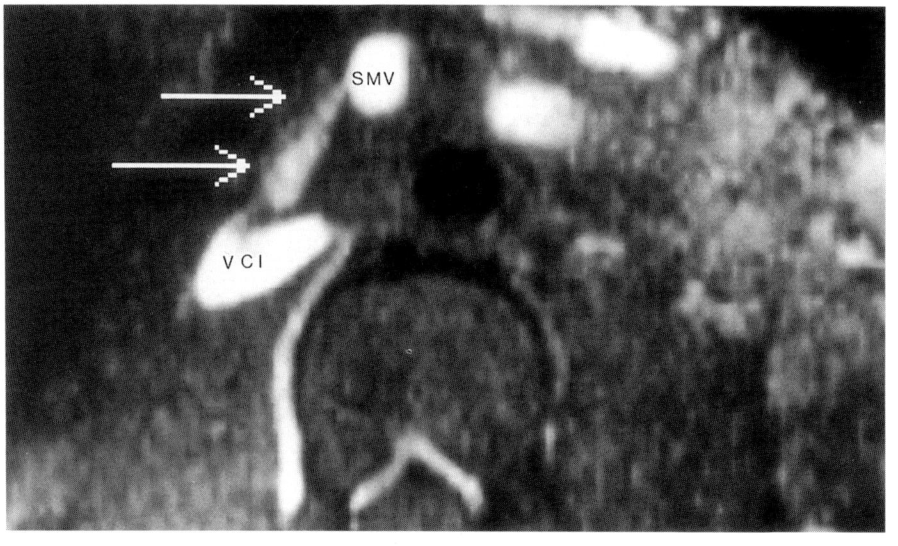

Fig. 18.16. Therapeutic mesocaval shunt (*arrow*). 2D TOF examination in the transverse orientation using a Helmholtz coil; projection image. Mesocaval shunt was performed between a peripheral branch of the superior mesenteric vein and the inferior vena cava. Transverse slice orientation is necessary to obtain a sufficient flow signal from the small vessels that are of interest. *VCI*, vena cava inferior; *SMV*, superior mesenteric vein

bleeding (Burkart et al. 1993). Thus, its verification, using MR flow quantification methods, may be an important prognostic factor.

Patency and diameter of the individual branches of the splenoportal system may determine the shunt location. For example, patency of the SMV is a sine qua non for mesentericocaval shunting. Some types of collaterals, such as widened paraumbilical veins, provide the opportunity to perform unusual, individually optimized portosystemic communications.

Magnetic resonance venography is useful for the planning of transjugular portosystemic shunts (TIPS). The shortest way to perform a stented communication between a hepatic vein and an intrahepatic branch of the portal vein can be estimated before beginning the intervention. Thus, the procedure can be performed within a shorter time and the number of needle punctures can be reduced (MUELLER et al. 1994).

While portacaval shunts can usually be evaluated by duplex US examination (FINN et al. 1987), this method is of limited value for the assessment of patency of the more commonly used splenorenal shunt, and especially the distal splenorenal (Warren) shunt (WARREN et al. 1982). In a study by FINN et al. (1991b) MR phlebography showed promising results in the follow-up of surgical shunt patency. In all shunt types size and patency of the anastomotic area could be estimated accurately (Fig. 18.16). Flow direction can be estimated by a presaturation or bolus tracking technique. Because of its noninvasive character MRA is a valuable tool for the detection of patency of iatrogenic shunts, particularly in the follow-up.

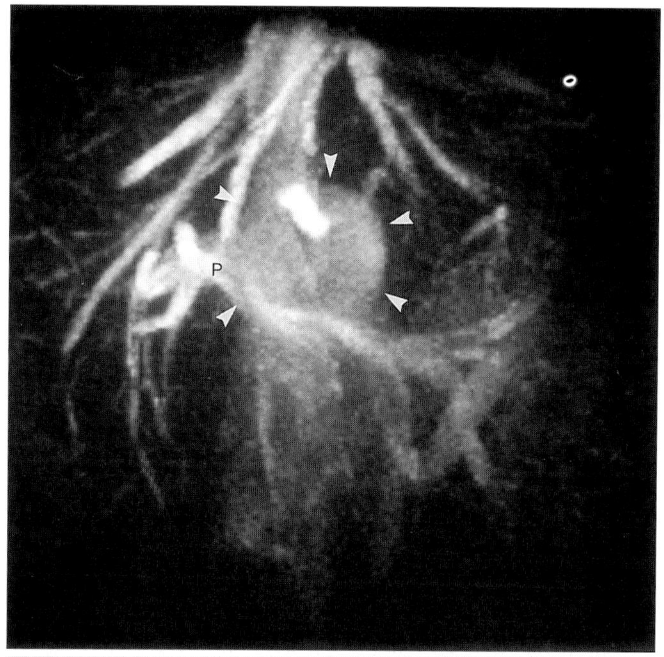

a

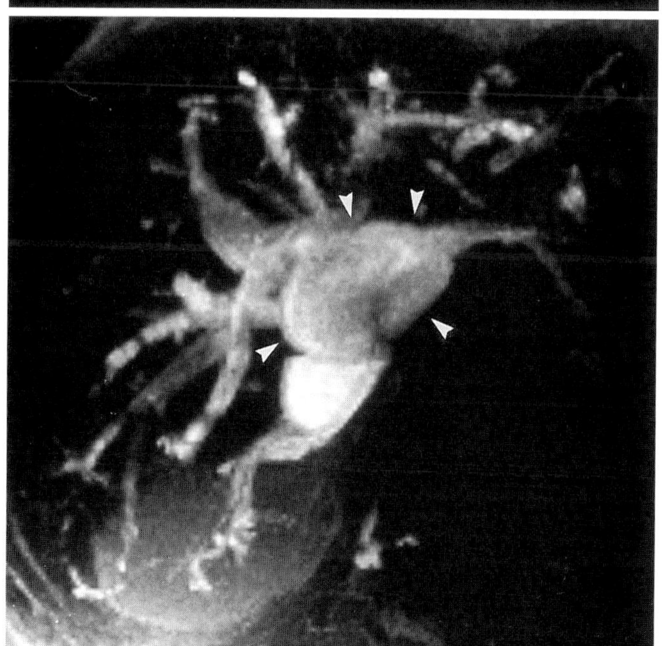

b

Fig. 18.17a,b. Aneurysm (*arrowheads*) of the portal vein in a patient suffering from liver cirrhosis. **a** Examination in the coronal slice orientation, projection image. **b** Transaxial examination, projection angiogram

18.3.6 Aneurysms

Portal vein aneurysms are rare. Most frequently they occur in patients with portal hypertension. Venous aneurysms sometimes can be accidentally detected with TOF MRA (Fig. 18.17). However, experience with arterial aneurysms suggests that they will be missed in many cases due to insufficient inflow effect under slow flow conditions. Black blood techniques and PC MRA may be superior to TOF MRA and show higher sensitivity.

18.3.7 Liver Tumors

As liver veins and portal veins are displayed simultaneously, MR venography may be used as a tool for treatment planning in hepatic tumor surgery. Tumors can be allocated to the individual liver segments defined by these venous structures (Bismuth 1982; Couinaud 1957). Smaller tumors, however, may not be displayed on the flow-weighted GRE images due to low soft tissue contrast. Thus, additional T2- or T1-weighted imaging may be necessary.

18.4 Conclusions

Two-dimensional TOF MR venography has proved to be a practicable method for imaging of the portal venous system. It can be performed within a short examination time and has been shown to be very effective in the visualization of the most important features of portal vein disease, particularly in portal hypertension. It is able to demonstrate extensive collateral pathways on large overview projection angiograms that can be used for planning endoscopic, radiological, or surgical interventions. Compared with operative findings, MRA has proved to be very accurate. Furthermore, it is completely noninvasive and does not need injection of contrast material. Due to all these factors MRA is superior to alternative imaging techniques except for transhepatic portography, a technique that is compromised by a mortality of about 2% (Lunderquist 1983). Thus, MRA of the splenoportal system can be considered the imaging method of choice in all cases in which US in unable to answer the clinical questions and transhepatic portography is considered to be too invasive.

The major drawback of the MRA methods currently in use is the limited display of small vascular structures. Further improvements are necessary to obtain techniques that yield higher signal-to-noise ratios and shorter data acquisition times. With such techniques it would be possible to increase in-plane resolution and diminish slice thickness preconditions for effective examination of small vessels. Inconsistencies of breathing, which could become a problem in this case, may presumably be overcome by sophisticated gating- or triggering methods.

References

Anderson CM, Saloner D, Tsuruda JS, Shapeero LG, Lee RE (1990) Artifacts in maximum-intensity-projection display of MR angiograms. AJR 154:623–629

Arai K, Matsui O, Kadoya J et al. (1991) MR imaging in idiopathic portal hypertension. J Comput Assist Tomogr 15:405–408

Arlart IP, Guhl L, Fauser L, Edelman RR, Kim D, Laub G (1991) Magnetic resonance angiography (MRA) of the abdominal veins. Radiologe 31:192–201

Berchtold R (1976) Portal hypertension. In: Allgöwer M (ed) General and special surgery. Springer, Berlin Heidelberg New York, pp 252–257

Bismuth H (1982) Surgical anatomy and anatomical surgery of the liver. World J Surg 6:3–9

Burcharth F (1991) Percutaneous transhepatic catheterization of the portal venous system. In: Okuda K, Benhamou JP (eds) Portal hypertension. Springer, Berlin Heidelberg New York, pp 127–137

Burkart DJ, Johnson CD, Morton MJ, Wolf RL, Ehman RL (1992) Volumetric flow rates in the portal venous system: measurement with cine phase-contrast MR imaging. AJR 16:1113–1118

Burkart DJ, Johnson CD, Ehman RL, Weaver AL, Ilstrup DM (1993) Evaluation of portal venous hypertension with cine phase-contrast MR flow measurements: high association of hyperdynamic portal flow with variceal hemorrhage. Radiology 188:643–648

Burns P, Taylor K, Blei AT (1993) Doppler flowmetry and portal hypertension. Gastroenterology 92:824–826

Couinaud C (1957) Le foie: études anatomiques et chirurgicales. Masson, Paris, pp 9–12

D'Alimonte PD, Cioni G, Cristani A, Ferrari A, Ventura E, Romagnoli R (1993) Duplex-doppler ultrasonography in the assessment of portal hypertension. Utility of the measurement of portal flow velocity. Eur J Radiol 17:126–129

Deimer E (1973) Intrahepatic obstruction in portal hypertension. Its radiological differentiation and clinical significance. Fortschr Roentgenstr 119:315–326

Dumoulin CL, Yucel EK, Vock P, Souza SP, Terroer F, Steinberg FL, Wegmueller H (1990) Two- and three-dimensional phase contrast MR angiography of the abdomen. J Comput Assist Tomogr 14:779–784

Edelman RR, Atkinson DJ, Silver MS, Loaiza FL, Warren SW (1988) FRODO Pulse sequences: a new means of eliminating motion, flow and wraparound artifacts. Radiology 166:231–236

Edelman RR, Mattle HP, Kleefield J, Silver MS (1989a) Quantification of blood flow with dynamic MR imaging and presaturation bolus tracking. Radiology 171:551–556

Edelman RR, Wentz KU, Mattle H, Zhao B, Liu C, Kim D, Laub G (1989b) Projection arteriography and venography: initial clinical results with MR. Radiology 172:351-357

Edelman RR, Zhao B, Liu C, Wentz KU, Mattle HP, Finn JP, McArdle C (1989c) MR angiography and dynamic flow evaluation of the portal venous system. AJR 153:755-760

Edelman RR, Mattle HP, Atkinson DJ, Hoogewoud HM (1990) MR angiography. AJR 154:937-946

Finn JP, Gibson RN, Dunn GD (1987) Duplex ultrasound in the evaluation of portocaval shunts. Clin Radiol 38:87-89

Finn JP, Edelman RR, Jenkins RL et al. (1991a) Liver transplantation: MR angiography with surgical validation. Radiology 179:265-269

Finn JP, Lewis WD, Edelman RR et al. (1991b) MR angiography in the evaluation of surgical portosystemic shunts. RSNA, Book of abstracts; p 289

Gehl HB, Bohndorf K, Klose KC, Günther RW (1990) Two-dimensional MR angiography in the evaluation of abdominal veins with gradient refocused sequences. J Comput Assist Tomogr 14:619-624

Guhl L, Arlart IP (1992) MR phlebography of the spleno-portal system: experience with 80 patients. SMRM, Book of abstracts, p 3125

Guhl L, Arlart IP, Hausmann R (1993) MR phlebography of the splenoportal system: increased image quality under use of a body-surface coil of Helmholtz type. Radiologe 33: 547

Johnson CD, Ehman RL, Rakela J, Ilstrup DM (1991) MR angiography in portal hypertension: detection of varices and imaging techniques. J Comput Assist Tomogr 15:578-584

Kraas E, Schreiber HW (1984) Surgery of the portal vein and portal hypertension. In: Demling L (ed) Clinical gastroenterology. Thieme, Stuttgart, pp 152-175

Langrehr JM, Keck H, Lohmann R, Blumhardt G, Bechstein UO, Neuhaus P (1994) Present state of orthotopic liver transplantation: indications, results, recommendations. Aktuel Probl Chir Orthop 29:41-46

Laub GA, Kaiser WA (1988) MR angiography with gradient motion refocusing. J Comput Assist Tomogr 12:377-382

Longmaid HE, Finn JP, Oldershaw J, Edelmann RR, Volpe J (1990) Presurgical staging of non-hepatic abdominal masses by MR angiography. SMRM, Book of abstracts, p 221

Lunderquist A (1983) In: Abrams HL (ed) Angiography. Little, Brown and Co, Boston, pp 1505

Mueller E, Laub G, Graumann R, Loeffler W (1988) RACE - real time aquisition and evaluation of pulsatile blood flow on a whole body MR unit. SMRM, Book of abstracts, p 729

Mueller E, Finelli D, Laub G (1989) Real time quantification of blood flow on a whole body MR unit using the RACE sequence. SMRM, Book of abstracts, p 894

Mueller MF, Siewert B, Kim D, Edelmann RR, Stokes KR, Finn JP (1994) The role of MR angiography before transjugular placement of a portosystemic stent shunt (TIPS). Fortschr Roentgenstr 160:312-318

Nabeshima M, Moriyasu F, Ono S, Nishikawa K, Kajimura K, Okuma M, Shimizu K (1993) Selective MR angiography of the liver. J Comput Assist Tomogr 17:730-734

Nghiem HV, Freeny PC, Winter TC, Mack LA, Yuan C (1994a) Phase-contrast MR angiography of the portal venous system: preoperative findings in liver transplant recipients. AJR 163:445-450

Nghiem HV, Mountford MC, Mack LA et al. (1994b) Preliver transplant evaluation of the portal venous system with phase contrast MR angiography. AJR 162 (suppl): 59

Okuda K, Aoki H (1991) Extrahepatic portal vein obstruction. In: Okuda K, Benhamou JP (eds) Portal hypertension. Springer, Berlin Heidelberg New York, pp 365-378

Okuda K, Matsutani S (1991) Portal-systemic collaterals: anatomy and clinical implications. In: Okuda K, Benhamou JP (eds) Portal hypertension. Springer, Berlin Heidelberg New York, pp 51-62

Okuda K, Takayasu K (1991) Angiography in the study of portal hypertension. In: Okuda K, Benhamou JP (eds) Portal hypertension. Springer, Berlin Heidelberg New York, pp 219-229

Okudira M (1991) Anatomy of the portal vein system and hepatic vasculature. In: Okuda K, Benhamou JP (eds) Portal hypertension. Springer, Berlin Heidelberg New York, pp 3-12

Otake S, Matsuo M, Kuroda Y (1990) Distinction of hepatic vein from portal vein by MR imaging. J Comput Assist Tomogr 14:201-204

Parvey HR, Raval B, Sandler CM (1994) Portal vein thrombosis: imaging findings. AJR 162:77-81

Ralls PW (1990) Color Doppler sonography of the hepatic artery and portal venous system. AJR 155:517-525

Reichardt W, Ingemanson S, Lunderquist A, Nobin A (1979) Selective mesenteric phlebography in patients with carcinoid tumors. Gastrointest Radiol 4:179-189

Rubin GD, Dake MD, Napel SA, McDonnel CH, Jeffrey RB (1993) Three-dimensional spiral CT angiography of the abdomen: initial clinical experience. Radiology 186:147-152

Savader BL, Fishman EK, Savader SJ, Cameron JL (1994) CT arterial portography vs pancreatic arteriography in the assessment of vascular involvement in pancreatic and periampullary tumors. J Comput Assist Tomogr 16:916-920

Spritzer CE, Pelc NJ, Lee JN, Evans AJ, Sostman HD, Riederer SJ (1990) Rapid MR imaging of blood flow with a phase-sensitive, limited flip angle, gradient-recalled pulse sequence: preliminary experience. Radiology 176:255-262

Stehling MK, Lawrence JA, Weintraub JL, Raptopoulos V (1994) CT angiography: expanded clinical applications. AJR 163:947-955

Tessler FN, Gehring BJ, Gomes AS, Perrella RR, Ragavendra N, Busuttil RW, Grant EG (1991) Diagnosis of portal vein thrombosis: value of color Doppler imaging. AJR 157:293-296

Tamada T, Moriyasu F, Ono S et al. (1989) Portal blood flow: measurement with MR imaging. Radiology 173:639-644

Warren WD, Millikan WJ, Henderson JM (1982) Ten years portal hypertensive surgery at Emory. Ann Surg 195:530-541

Weltin G, Taylor KJW, Carter AR, Taylor CR (1985) Duplex doppler: identification of cavernous transformation of the portal vein. AJR 144:999-1001

Williams DM, Cho KJ, Aisen AM, Eckhauser FE (1985) Portal hypertension evaluated by MR imaging. Radiology 157:703-706

19 Pelvic and Peripheral Veins

G.P. KRESTIN

CONTENTS

19.1 Introduction

Venous thrombosis in the lower extremity and pelvic veins is a serious complication that may occur especially following surgical interventions and traumatic events, as well as during pregnancy or after delivery. Pelvic and lower extremity veins are believed to be the origin of 85%–95% of pulmonary emboli. Other conditions like inborn or acquired anomalies or external compression by space-occupying lesions may also compromise blood flow in these vessels.

The traditional diagnostic workup for venous pathologies in the lower extremities and pelvis consists mainly of invasive venography with contrast material (FERRIS 1990; NAIDICH et al. 1988; REDMAN 1988). However, although rarely, complications such as systemic reactions to contrast material, tissue necrosis secondary to extravasation of contrast material, postvenographic thrombosis, and pulmonary embolism may be associated with venography Noninvasive methods like real-time ultrasonography (US), Doppler or duplex US, impedance plethysmography, and radionuclide studies have yielded varying results: impedance plethysmography and radionuclide studies have not been sufficiently accurate to warrant the abandonment of venography (CRONAN 1991; HAYT and BINKERT 1990); ultrasound methods are accurate but user dependent and do not allow for easy examination of the calf and pelvis (CRONAN 1991; YUCEL et al. 1991). Additionally, most of these methods do not provide information on the internal iliac veins and their branches or on the ovarian veins. Moreover, complete diagnosis of venous dysplasia often requires additional direct puncture for visualization of the dysplastic vessels and their drainage.

G.P. KRESTIN, MD, Department of Radiology, University Hospital Zürich, Rämistrasse 100, 8091 Zürich, Switzerland

19.2 Technical Considerations

Inflow magnetic resonance angiography (MRA) is the best method for demonstrating the lower extremity and pelvic veins. A spoiled gradient-echo sequence helps to minimize magnetization in the transverse plane of stationary tissues. The 2D technique is preferable in order to avoid saturation effects due to low flow velocities. For 2D time of flight (TOF) venography, flow compensation is routinely applied along the slice-select and readout directions. Higher order flow effects are less important than for MR arteriography since collaterals develop rapidly and abnormal vessels with normal flow patterns are more common than visible venous obstruction. Presaturation pulses above the examination plane (40–80 mm thick) are crucial in order to suppress arterial blood flow signal. Slice thickness should not exceed 3–4 mm in the lower extremity and 1.5–2 mm for assessment of the pelvic veins since otherwise, due to their oblique direction, edge artifacts may be disturbing. The best results can be achieved using the shortest possible TR and TE in combination with a relatively large flip angle (around 60°). An imaging matrix of 128 × 256 pixels in addition to a single data acquisition is sufficient for adequate visualization. Further increase in spatial resolution or in the number of excitations (NEX) leads to a substantial increase in the examination time without a significant improvement in image quality. In contrast to MRA of the upper abdomen, MRA of the lower extremities and pelvic veins does not require breath-holding (ANGLADE et al. 1989; GEHL et al. 1990a; SPRITZER et al. 1988, 1990; TOTTERMANN et al. 1990).

The phase-contrast (PC) technique has been applied to great advantage in imaging of upper abdominal veins like the portal venous system or renal veins. Superior stationary tissue signal suppression and less sensitivity to in-plane flow saturation make 3D PC MRA highly desirable, though at the cost of long acquisition times. With 2D PC MRA, signal-to-noise ratios may be limited (Tavares et al. 1989).

The obtained axial or coronal slices can be reconstructed in multiple projections using the maximum-intensity projection (MIP) algorithm. An overlapping of the acquired slices by 0.5–1 mm helps to improve the quality of the reconstructed MIP images. Such postprocessing techniques are particularly useful for demonstration purposes and for a better overview of the whole examined region. However, accurate diagnosis always requires evaluation of all the individual slices in order to rule out partial thrombosis and to avoid misinterpretations due to overlying vessels (Gehl et al. 1990b; Richter et al. 1993).

Administration of contrast material may improve the results of MRA, particularly in vessels with very slow flow, since the shortening of T1 relaxation achieved by the available paramagnetic gadolinium (Gd) chelates decreases the signal loss due to saturation effects. Flow-related dephasing artifacts are not, however, eliminated in contrast-enhanced TOF or PC MRA (Marchal et al. 1990). Therefore in imaging of vascular malformations a 3D fat-saturated gradient-echo sequence (SPGR or spoiled FLASH) after intravenous administration of 0.1 mmol/kg Gd-DOTA or Gd-DTPA allows excellent delineation of the venous structures (Krestin et al. 1992).

19.3 Applications of MRA

19.3.1 Venous Anatomy

The veins of the lower extremities may be classified into three categories: deep veins including the main trunks and muscle veins, superficial veins, and perforating veins.

The *deep veins* are enclosed by the deep fascia. In the calf, three main pairs of deep veins accompany the respective arteries (anterior tibial, posterior tibial, and peroneal veins). These three trunks merge in the upper third of the calf to form the popliteal vein, which is commonly single but may be paired in up to 20% of subjects. Multiple muscle veins (soleus, gastrocnemius) empty into the posterior tibial, peroneal, or popliteal veins. The main deep veins of the thigh, the superficial femoral vein and the profunda femoris vein, drain into the common femoral vein at the level of the groin. In the pelvis the internal iliac veins receive the venous drainage of the gluteal (superior and inferior), the internal pudendal, and the obturator veins and merge with the external iliac veins to form the common iliac veins. Testicular and ovarian veins drain at a higher level: on the right side directly into the inferior vena cava and on the left into the left renal vein. During pregnancy and in the postpartum period, ovarian veins are markedly enlarged (the diameter on the right may reach up to 2.5 cm). Variations in venous topography are common. Particularly duplication of deep venous trunks has to be considered in order to accurately asses thrombotic occlusion of only one branch.

The *superficial veins* consist mainly of the greater and lesser saphenous veins and multiple other smaller tributaries. The lesser (short) saphenous vein drains into the popliteal vein and the greater (long) saphenous vein empties into the femoral vein at the level of the groin. Above this level superficial veins play a clinical role only as collaterals following obstruction of the deep trunks.

The *perforating veins* of the lower extremity penetrate through the deep fascial planes and connect the deep veins with the superficial veins. They extend from the foot to the groin.

The deep veins of the lower extremity can be characterized as capacity vessels. They contain varying volumes of blood with only minimal changes in pressure. Venous blood return at rest depends on small local pressure gradients, on dynamic pressure transmitted from the arterial side, and on changes in intrathoracic pressure. Only with muscle contraction are dynamics of venous flow different because the veins of the calf are markedly compressed by the contracting muscles and refill with blood during muscle relaxation. This is the reason why accurate detection of the venous structures of the calf is not always possible by means of inflow MRA. Compression during the examination or administration of a contrast agent may improve the results. In contrast, the deep veins above the knee are well depicted. This applies to the superficial femoral and the deep femoral veins as well as the greater saphenous vein (Fig. 19.1).

In the pelvis the common iliac, the external iliac, and the internal iliac veins are adequately visualized in all patients (Gehl et al. 1990b; Richter et al. 1993;

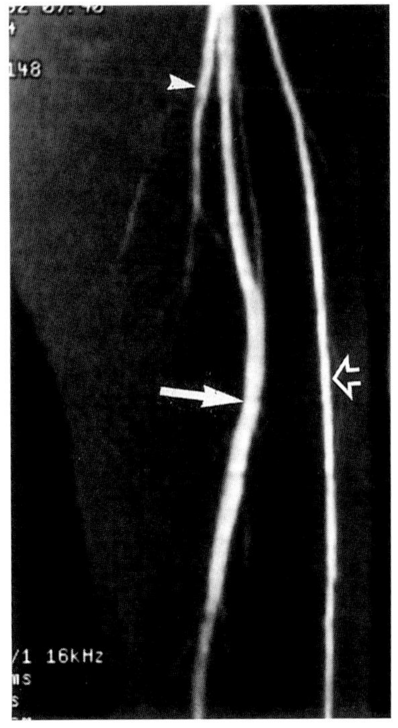

Fig. 19.1. Normal 2D TOF MRA of the veins of the right thigh showing the superficial femoral vein (*arrow*), the deep femoral vein (*arrowhead*), and the greater saphenous vein (*open arrow*)

SIEWERT et al. 1992). Even the branches of the internal iliac vein are detectable in 60% (pudendal veins) to 90% (gluteal veins) of cases (Fig. 19.2). The normal testicular or ovarian veins are usually depicted only in their proximal third. Visualization is better on the right, where drainage occurs directly into the inferior vena cava, than on the left, where the ovarian or testicular vein empties into the left renal vein. During pregnancy and shortly after delivery, depiction of patent ovarian veins is possible in all cases due to the markedly enlarged lumen (Fig. 19.3). These normal vascular structures can be visualized regardless of the position of the patient (prone, supine, or lateral decubitus) (RICHTER et al. 1993).

Some common artifacts of inflow MRA have to be taken into consideration in order to correctly evaluate MR venograms:

1. Compression of the left common iliac vein by the overlying artery occurs in one-third to one-half of cases, with signal loss in a short portion of the vessel without real morphological alteration (LANZER et al. 1991) (Fig. 19.15b). This phenomenon has to be distinguished from iliac "spurs" (see below).
2. Slow venous flow is a problem limiting the evaluation of pathological changes. The contrast between flowing blood and stationary tissue decreases at slow velocities due to saturation effects. Thin sections and a longer TR will mitigate those slow-flow effects (SPRITZER et al. 1990).
3. In 2D MRA the intraluminal signal is reduced to an intermediate intensity at the confluence of vessels where mixing of blood occurs. These flow-related artifacts are a potential source of a false-positive diagnosis of intraluminal thrombus. The problem may be reduced by the use of strict diagnostic cirteria. Bizarrely shaped intraluminal flow voids are common at the junction of large veins (GEHL et al. 1990a).
4. During respiration flow reversal may occur in veins of the pelvis and upper abdomen. Some of the individual sections may present a signal decrease in the vessel lumen. However, this kind of artifact affects only several images and the diagnosis will not be impaired in most cases (RICHTER et al. 1993).

19.3.2 Venous Anomalies

Simple *anatomical variants* such as duplication (Fig. 19.4), interruption, or hypoplasia are common

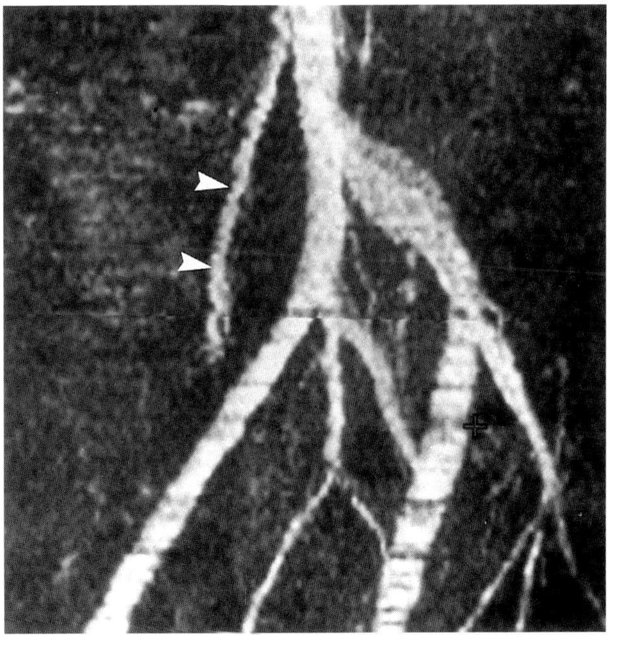

Fig. 19.2. Normal 2D TOF MRA of the pelvic veins in a 30° LAO projection showing the external and internal iliac veins, the branches of the internal iliac veins (superior and inferior gluteal posteriorly and the internal pudendal anteriorly), the common iliac veins and their confluence in the inferior vena cava, and the distal part of the superior mesenteric vein (*arrowheads*)

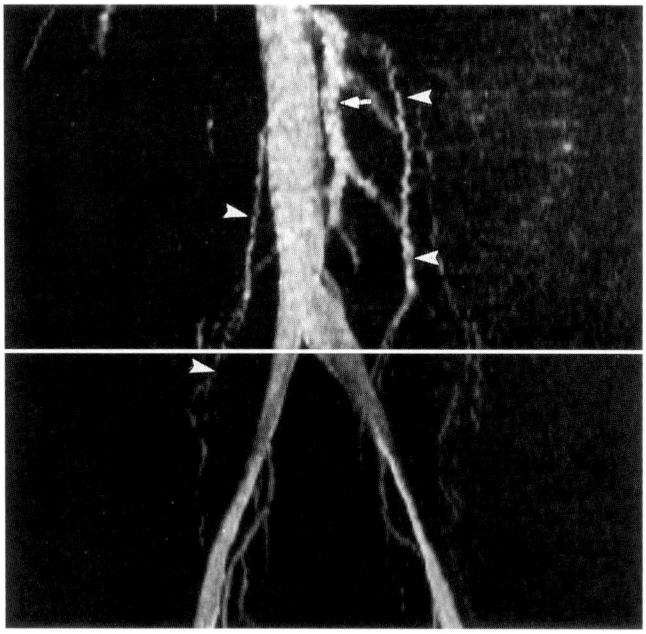

Fig. 19.3. Normal 2D TOF MRA of pelvic and lower abdominal veins (combination of two acquisition volumes) in a woman in the postpartum period showing the iliac veins (external, internal, and common), the inferior vena cava, the distal superior mesenteric vein (*arrow*), and the patent ovarian veins (*arrowheads*) draining into the inferior vena cava on the right and into the renal vein on the left. The renal veins are not visualized due to in-plane saturation effects

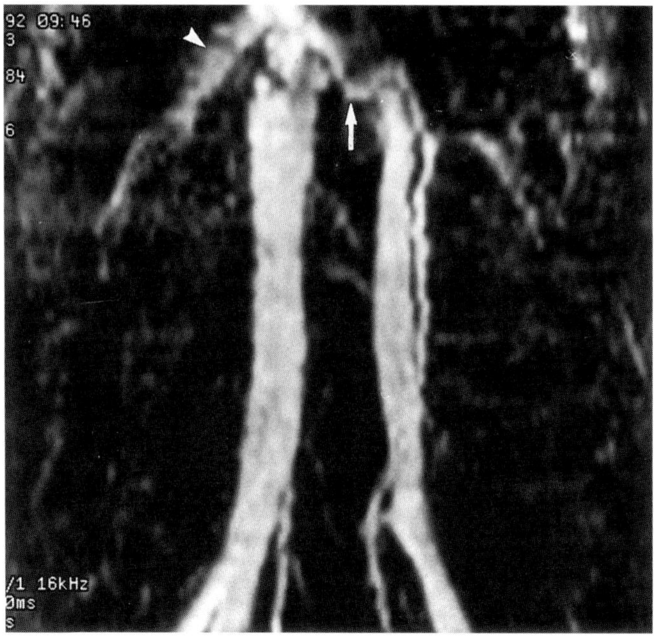

Fig. 19.4. Duplication of the inferior vena cava in the infrarenal segment. On the left blood from the pelvis drains into the left renal vein (*arrow*). The right renal vein is normal (*arrowhead*)

in any region of the venous trunks in the lower extremities as well as in the inferior vena cava but are relatively rare in the pelvis. Most of the normal variants have no practical relevance but anomalies of the inferior vena cava, such as an interruption with azygos/hemiazygos continuation, may have significant clinical implications (Fig. 19.5). Knowledge on duplication and transposition of the inferior vena cava as well as on the circumaortic and retroaortic left renal vein is important before performing surgery of the retroperitoneum. All these anomalies are clearly detectable on MRA and do not require further examinations (FINN et al. 1993; FRIEDLAND et al. 1992).

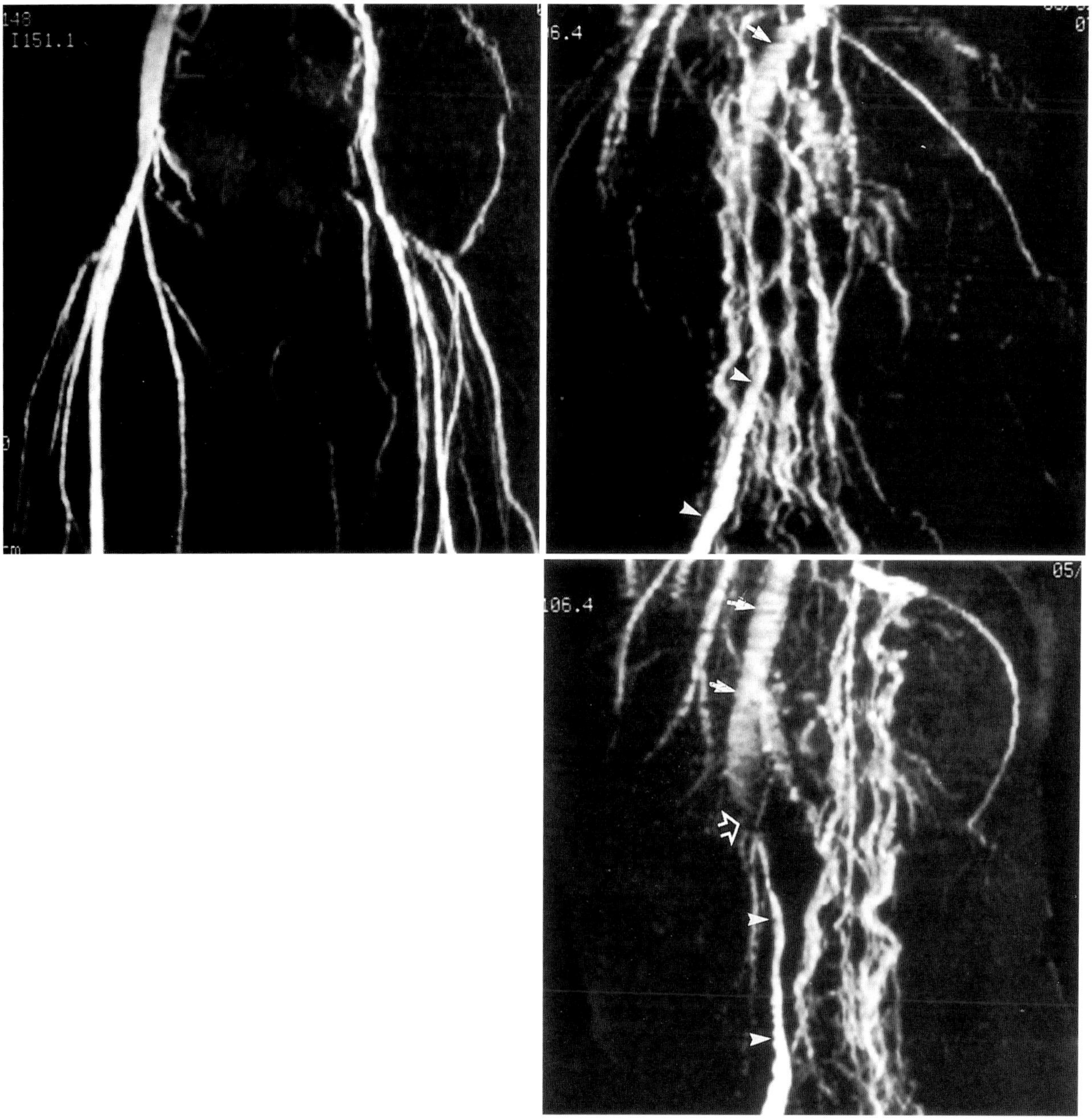

a b c

Fig. 19.5a–c. Congenital hypoplasia of the inferior vena cava. Surgical bypass of the intrahepatic segment. **a** Normal MR venogram of the thighs and the pelvis. **b** The right common iliac vein drains mainly into a hypoplastic inferior vena cava (*arrowheads*). The left iliac vein drains into paravertebral collaterals. There is flow in the surgical bypass (*arrows*). **c** The 30° LAO projection demonstrates a marked stenosis (*open arrow*) at the anastomosis between the hypoplastic inferior vena cava (*arrowheads*) and the bypass(*arrows*)

In about 20% of patients sail-shaped folds or "*spurs*" may be found in the left common iliac vein at the junction with the inferior vena cava. These soft tissue membranes protrude from either the medial or the lateral venous wall into the vascular lumen, which they narrow. Other spurs present as cords aligned sagittally, dividing the venous lumen. The spurs develop with age and have to be distinguished from congenital abnormalities of the iliocaval junction. The formation of the spur originates in a reaction of the venous wall to the local compression of the overlying right common iliac artery. At least partially, these spurs are the cause of the threefold higher incidence of thrombosis of the left lower extremity veins. Blood flow is altered at the level of the spur, which may cause signal loss in MRA (Fig. 19.6). Differentiation from simple compression by the overlying right common iliac artery is consequently difficult (Richter et al. 1993; Lanzer et al. 1991). Compression is the main cause of development of iliac spurs; therefore the distinction of the two entities is of limited significance. In patients with visible flow alteration at this level, a higher incidence of thrombosis has to be considered. The hemodynamic relevance of the spur can be quantified by flow measurements and comparison with the contralateral vein.

Congenital vascular malformations of the lower limbs include simple and cavernous hemangiomas, arteriovenous malformations with macro- or microfistulas, and venous angiomas. All these lesions originate in the abnormal development of the primitive vascular system. The clinical manifestation is extremely varied and extends from hardly visible, small strawberry-like birthmarks to clusters of massively enlarged vascular channels that may deform a whole extremity (Malan and Puglionisi 1965; Szilagyi et al. 1976).

The malformation can affect the venous system alone or can manifest itself clinically by enlarged venous structures fed by arteriovenous macro- or microfistulas (Malan and Puglionisi 1965; Szilagyi et al. 1976). Regardless of the type, vascular malformations present a difficult therapeutic challenge. Therefore, management of the disease requires detailed knowledge of the location and extent of the lesion, and precise delineation of its relationships to the surrounding anatomical and functional structures.

Real time US, color-coded Doppler US, contrast-enhanced computed tomography, conventional venography, and arteriography have all been advocated for the evaluation of these lesions, but each has significant limitations. Conventional venograms generally allow the detection of draining vessels. However, simple ascending phlebography of the deep venous system may be normal and visualization of purely venous superficial anomalies is possible in such cases only after direct puncture of the superficial venous channels. MR imaging is a valuable additional tool in the management of congenital vascular malformations of the lower limb (Cohen et al. 1986). It provides important complementary information concerning the location and extent of the lesion and may even replace conventional venography. The relationships to surrounding deep anatomical struc-

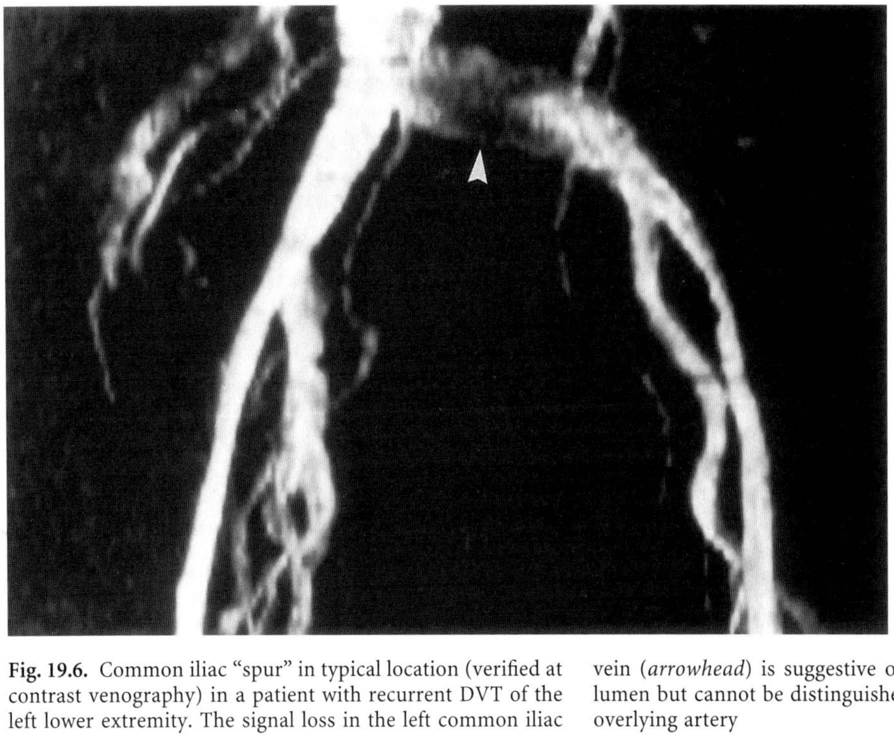

Fig. 19.6. Common iliac "spur" in typical location (verified at contrast venography) in a patient with recurrent DVT of the left lower extremity. The signal loss in the left common iliac vein (*arrowhead*) is suggestive of a narrowing of the vessel lumen but cannot be distinguished from compression by the overlying artery

tures, especially the joints, can be accurately demonstrated (Fig. 19.7a). The fat-saturated contrast-enhanced 3D gradient-echo sequence is particularly helpful in this regard (Fig. 19.7b). Additionally, this method allows for the assessment of most of the draining veins and their connections to the vascular anomaly. The results are similar to the findings on conventional venograms (KRESTIN et al. 1992). Plain MRA is merely of limited value for the evaluation of the disease (Fig. 19.7c). Large draining veins can be visualized on TOF MR venograms and tortuous venous anomalies may be assessed with PC MRA. On the other hand, arterial MRA cannot compete with conventional arteriography, since feeding vessels as well as arteriovenous shunts can be detected in only a limited number of cases (KRESTIN et al. 1992)

19.3.2.1 Acute and Chronic Venous Thrombosis and Postthrombotic Syndrome

Deep venous thrombosis (DVT) is the third most common cardiovascular disease after myocardial infarction and cerebrovascular stroke. The clinical relevance of this disorder results from the high risk of pulmonary embolism (FERRIS 1990). In patients with verified DVT, radionuclide lung scans may reveal pulmonary embolism in up to 50% of cases. More than 90% of pulmonary emboli originate from the lower extremities or the pelvic area and therefore early diagnosis of DVT and prevention of pulmonary embolism has to focus on this region. Clinical diagnosis of DVT is difficult since only half of the patients will have characteristic symptoms. On the other hand, in patients with clinically suspected DVT, thrombosis is present in only 50% of cases (HUISMAN et al. 1989).

The accurate diagnosis of acute DVT allows for optimal treatment, thereby reducing the risk of subsequent complications associated with DVT and with needless anticoagulation therapy. The "gold standard" contrast venography may demonstrate opacified patent vessels and thrombotic lesions in the deep veins of the calf, knee, thigh, and pelvis. Typical signs of thrombosis are the lack of opacification of single veins or groups of vessels, and the presence of constant filling defects and of augmented flow in collateral (commonly superficial) vessels (NAIDICH et al. 1988; REDMAN 1988). Although the indirect signs of nonopacification and flow diversion are highly suggestive of DVT, they can occasionally be caused by edema, cellulitis, hematoma, and muscle fiber rupture or may even be due to technical problems (puncture site). Additionally contrast venography is an invasive technique with the risk of adverse reactions to contrast material and a 2–6% rate of unsuccessful studies. Therefore over recent years many authors have called for less invasive and more reliable techniques (CRONAN 1991). Many of these methods, such as impedance plethysmography and blood pool or fibrinogen scanning, cannot compete with the results of venography: US with its different techniques (real-time compression US, duplex US and color-coded Doppler US) provides more promising results (CRONAN 1991; YUCEL et al. 1991). However, sensitivity for assessment of DVT below the knee is less than 70% and the examination is extremely operator dependent (YUCEL et al. 1991). Additionally, results are often limited in the pelvis and abdomen of obese patients due to overlying bowel structures.

The age of a thrombotic lesion can best be assessed using the radioactive fibrinogen uptake test (HAYT and BINKERT 1990). Morphological changes occur within 10 days to 2 weeks after acute onset of thrombosis. The clots may be completely or partially resorbed or the vessel may recanalize. Recanalization occurs within 6 months but may not be complete for up to 4 years and there are cases in which occlusion persists. Morphological assessment of the age of a thrombus is therefore difficult and only possible to a certain degree. Fresh clots are usually separated from the venous wall. They are surrounded by a thin layer of flowing blood. Within 48 h increasing adherence will occur, and the areas with flow will disappear. Finally, complete occlusion may result. On the other hand, shrinking and retraction of the thrombus may produce increased areas of normal blood flow (NAIDICH et al. 1988).

Permanent damage in the veins after healing of acute DVT may result in chronic venous insufficiency and postthrombotic syndrome. Ulcers may develop in up to 20% of patients within 5 years and in more than 50% within 10 years. A history of previous DVT can be obtained in only half of the cases since DVT may have been clinically silent. Residual anomalies may be detectable in patients with postthrombotic syndrome. Abnormal changes include narrowing or occlusion, multiple superficial or deep collateral veins, vessels with irregular margins, and venous dilatation.

A rare cause of impaired venous drainage with associated edema and stasis is the compression of a venous trunk. The levels of compression vary but the most common sites are the popliteal region (Baker's cyst, popliteal aneurysm), the groin (hernia, enlarged

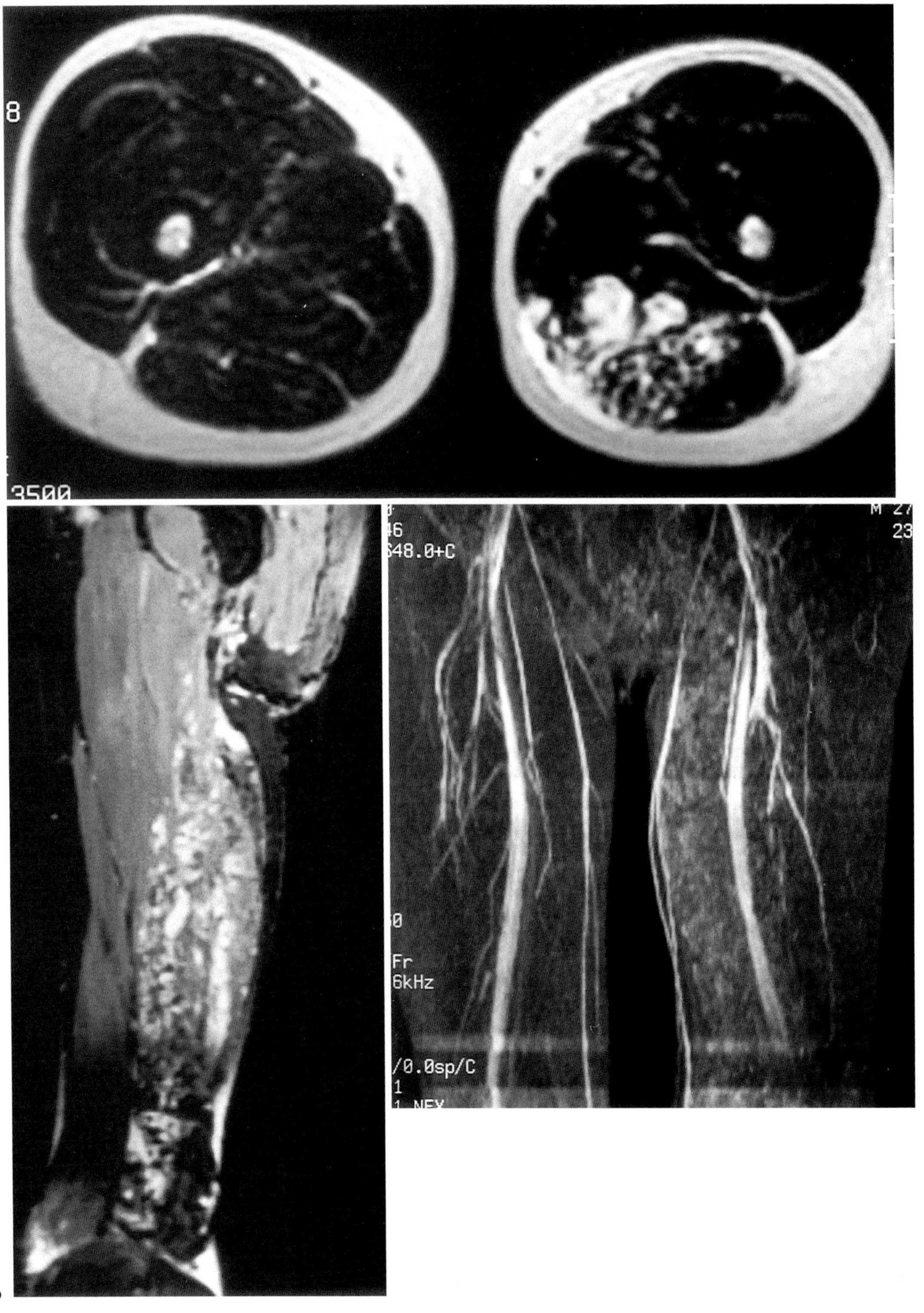

a b c

Fig. 19.7a–c. Congenital venous malformation of the thigh. **a** T2-weighted transverse image with clear visualization of pathological high-signal-intensity vascular structures in the semimembranosus muscle and perifascial edema. **b** Contrast-enhanced sagittal 3D gradient-echo sequence demonstrates the enhancing vessels in the muscles of the left thigh. **c** 2D TOF MR venogram of the thigh and groin showing normal deep and superficial veins. There is a diffuse signal increase within the soft tissues of the left thigh lymph nodes), and the iliac axis (enlarged uterus in pregnancy, tumors). While in most of the cases peripheral compression can be detected clinically, pelvic masses with consequent venous obstruction are visualized only by means of imaging methods.

Magnetic resonance angiography is a reliable, accurate noninvasive method for diagnosing DVT in the femoropopliteal and pelvic region (Figs. 19.8–19.12). MRA allows simultaneous visualization of the veins of both lower extremities and of pelvic vessels bilaterally. Signs of thrombosis in MRA are:

1. Complete lack of high-signal flowing blood in the region of a venous trunk
2. Irregularly shaped flow void
3. Eccentric filling defects with smooth margins in venous structures (on axial images)
4. Demonstration of collateral vessels in chronic thrombosis (GEHL et al. 1990a; RICHTER et al. 1993)

Postthrombotic syndromes and venous compressions manifest themselves by (Figs. 19.13, 19.14):

1. Inconsistent diameter of the vessels with smooth contours
2. Progressive decrease up to complete loss of signal intensity in the vessel lumen and signal increase distally
3. Pronounced collateral circulation (RICHTER et al. 1993; ERDMANN et al. 1990)

Several reports in the literature of the last 5 years have described the value of MRI and MRA in the detection of DVT. Even SE sequences provide a high

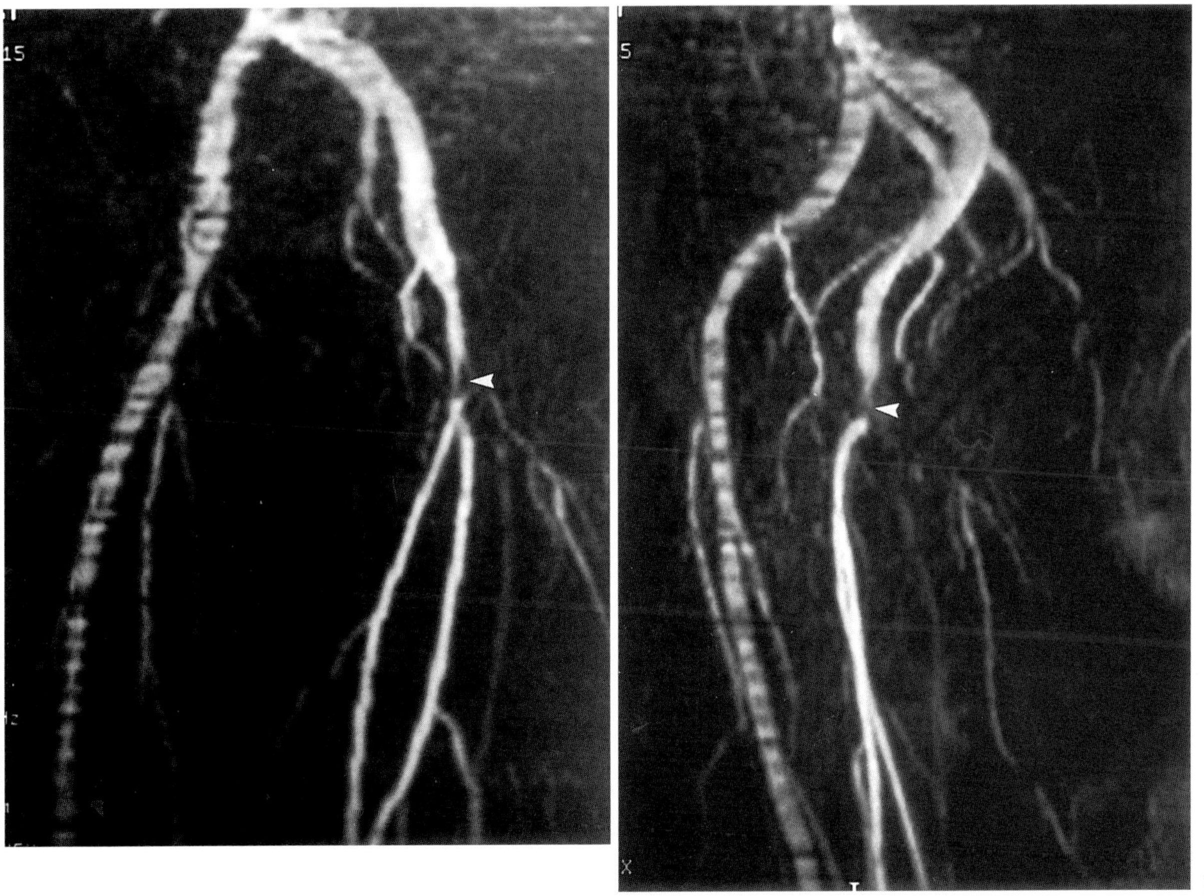

a b

Fig. 19.8a–c. Acute DVT of the left thigh. **a, b** 15° and 60° LAO projection venograms of the thigh and pelvis demonstrating occlusion of the deep femoral veins and collateral flow in two superficial vessels. There is a floating thrombus in the common femoral vein (*arrowhead*). **c** One of the multiple transverse gradient-echo images at the level of the thigh clearly demonstrating the lack of flow in the region of the superficial femoral vein on the left and some superficial collaterals (duplicated long saphenous and branches of the profunda femoris vein)

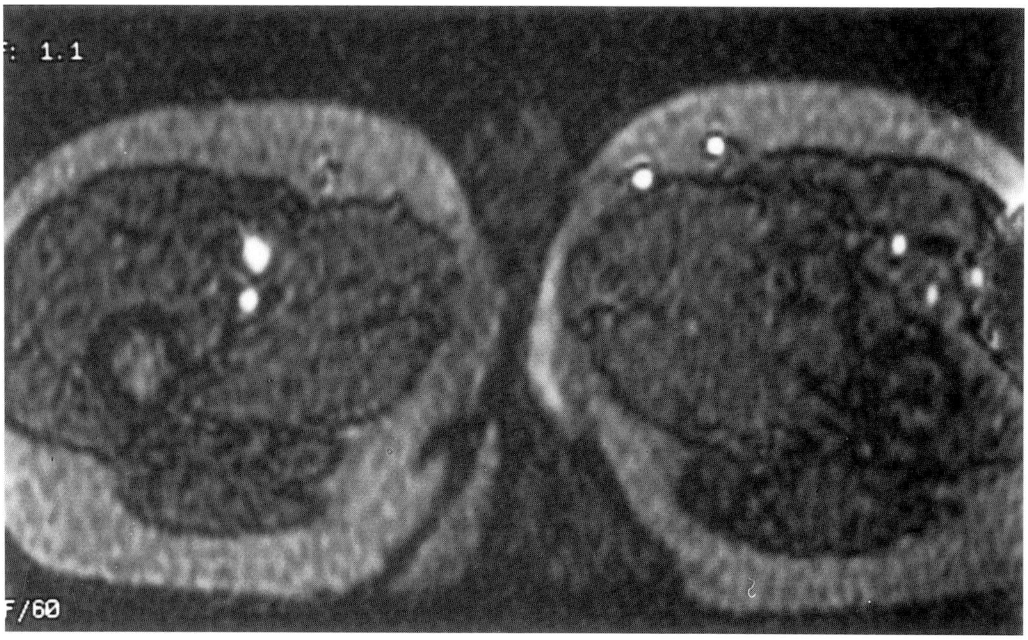

Fig. 19.8c

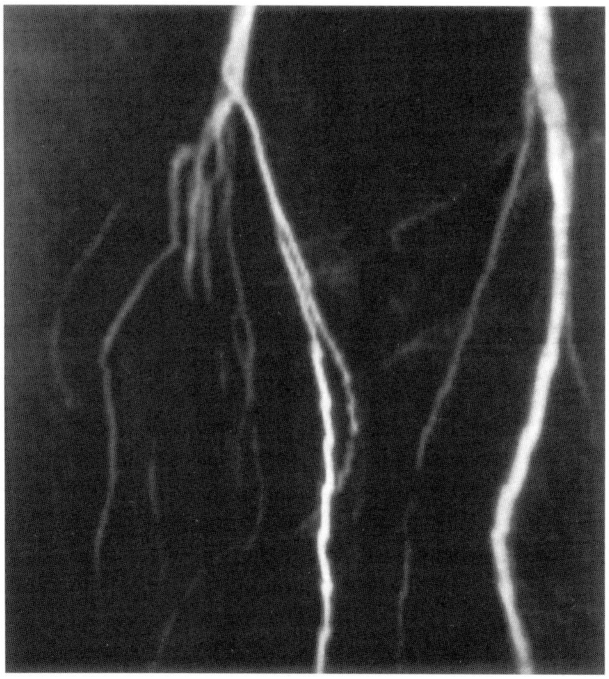

Fig. 19.9. Acute DVT of the right thigh with loss of flow in the superficial femoral vein and increased flow in the right saphenous vein as compared to the unaffected left thigh

sensitivity and specificity [90% and 100% respectively (ERDMANN et al. 1986, 1990; HRICAK et al. 1985)]. The results are similar with GRE sequences [sensitivity of 100% and specificity of 93% described by SPRITZER et al. (1990)]. In direct comparisons, GRE images, however, proved to be clearly superior to SE sequences (TOTTERMANN et al. 1990; EVANS et al. 1993; SOSTMAN et al. 1993).

Even using this imaging sequence, there are limitations in the evaluation of calf veins and some

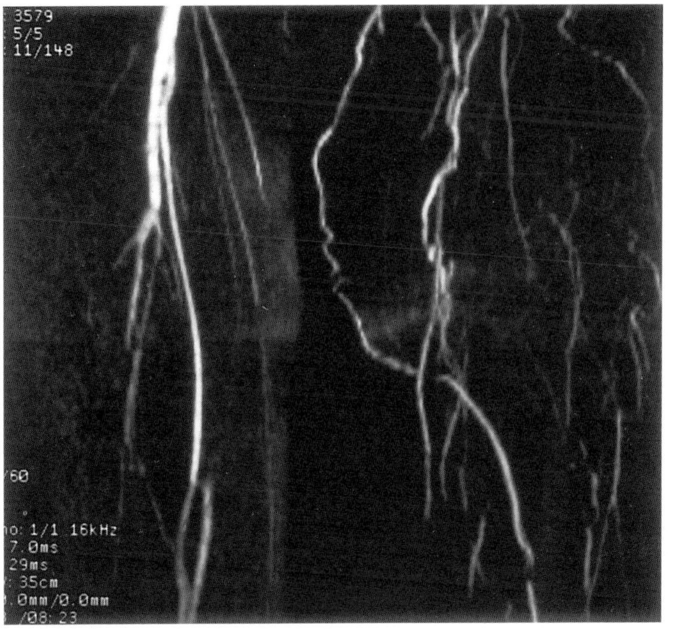

Fig. 19.10. Subacute DVT of the left thigh. There is no flow in the superficial femoral vein and in the greater saphenous vein but there are multiple other superficial collateral vessels on the left

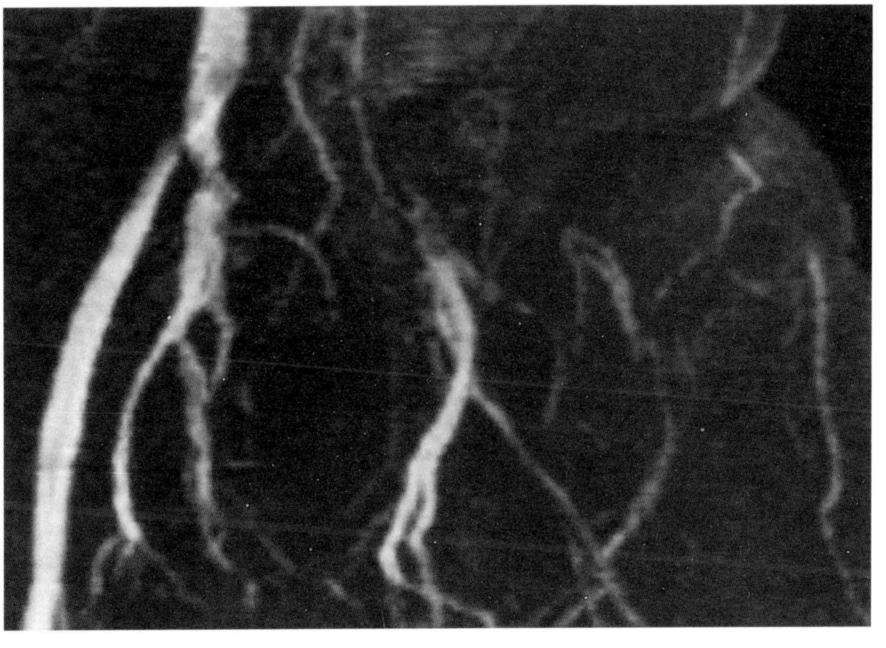

Fig. 19.11. Acute DVT of the pelvis. 15° LAO projection venogram showing complete occlusion of the left external and common iliac veins but a patent internal iliac vein on the left

authors do not consider MRA reliable enough below the knee (SIEWERT et al. 1992; LANZER et al. 1991). Therefore an improved imaging strategy is needed to produce flow images in the calf veins. In a recent prospective study a compression technique was proposed with an 87% sensitivity and a 97% specificity for the detection of thrombi in the calf, while the method reached almost a 100% accuracy in the thigh and in the pelvis (EVANS et al. 1993). Flow augmentation techniques (compression and decompression)

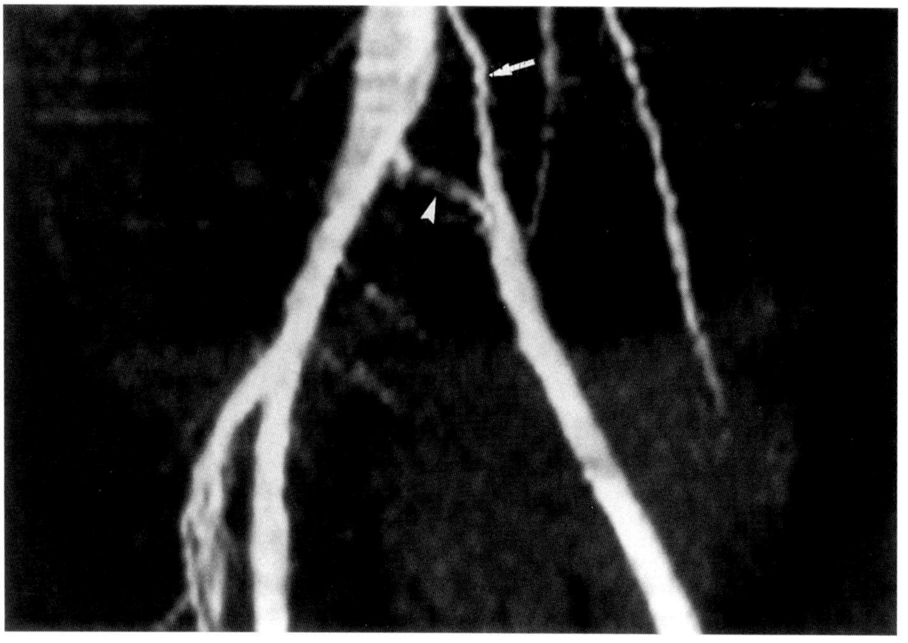

Fig. 19.12. Patient with pulmonary embolism. The 15° RAO projection venogram of the pelvis reveals complete occlusion of the left internal iliac vein, partial occlusion of the left common iliac vein (*arrowhead*), and some paravertebral collaterals on the left (*arrows*), but a patent right iliac axis and inferior vena cava

in addition to ultrafast imaging (EPI) have shown some promising preliminary results in this difficult region.

Magnetic resonance angiography of the lower extremity and pelvic veins may also provide additional unexpected findings. A more proximal extent of the thrombus as compared to conventional contrast venography was described in 52% of patients by ERDMANN et al. (1990). On the other hand, in 22% of cases, signs of unsuspected chronic disease and in 44% ancillary abnormalities were found (ERDMANN et al. 1990).

One of the most important alterations which may be diagnosed only with MRA is thrombosis of the internal iliac veins and their branches. Some authors describe such clinically unsuspected findings in 20%–25% of examined cases (GEHL et al. 1990b; RICHTER et al. 1993). Opacification of the inferior vena cava at contrast venography in acute thrombosis of pelvic veins may be limited: using MRA GEHL et al. (1990b) found undetected involvement of the inferior vena cava in 25% and clinically inapparent thrombosis of the opposite side in 7 of 32 evaluated patients

Evaluation of lower extremity and pelvic veins was performed in the majority of the published reports using axial slices. Interpretation limited to MIP reconstructions is not reliable in the diagnosis of DVT (RICHTER et al. 1993; SIEWERT et al. 1992). MRA can be used for monitoring anticoagulant and fibrinolytic therapy (FRANCIS et al. 1989) and the method has a role as the definitive examination when results of screening are unsatisfactory and even as a first-line examination in suspected pelvic thrombosis (ERDMANN et al. 1990).

Chronic disease, postthrombotic syndromes, and venous compressions can be reliably detected (Figs. 19.13, 19.14). All collateral vessels with sufficient flow are seen despite their drainage or connection to deep or superficial vessels. This represents a clear advantage compared to contrast venography, in which such collaterals may be poorly opacified or even completely missed. In many cases MRI provides information on the underlying cause of a venous compression by directly visualizing space-occupying lesions and their relationsship to surrounding structures (SIEWERT et al. 1992; ERDMANN et al. 1990).

19.3.2.2 Ovarian Vein Thrombosis

Septic puerperal ovarian vein thrombosis (SPOVT) is a potentially life-threatening condition in the

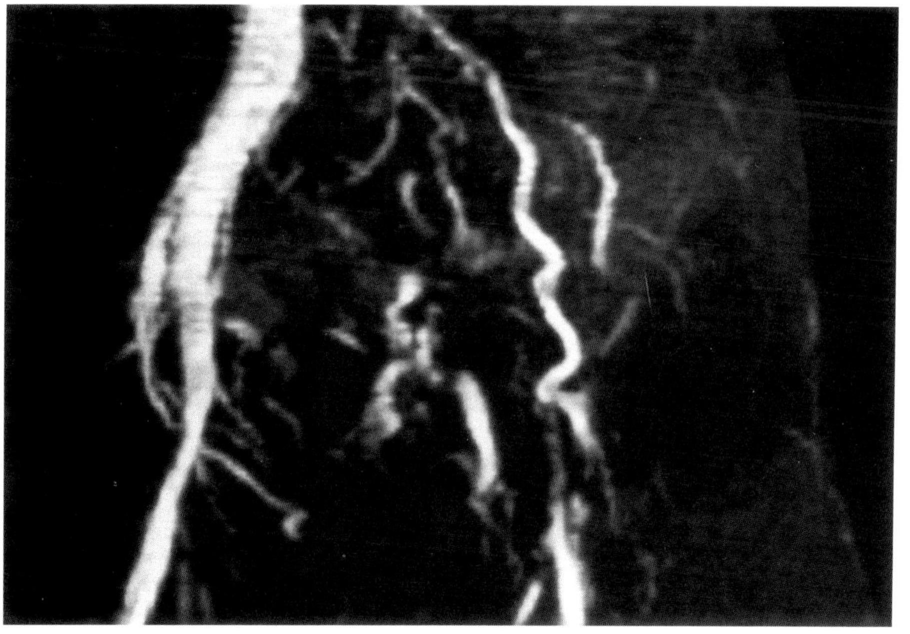

Fig. 19.13. Postthromobotic syndrome of the left iliac axis 6 months after acute DVT. There are some tortuous collaterals on the left and some enlarged branches of the internal iliac vein, but still complete occlusion of the external and common iliac veins on the left

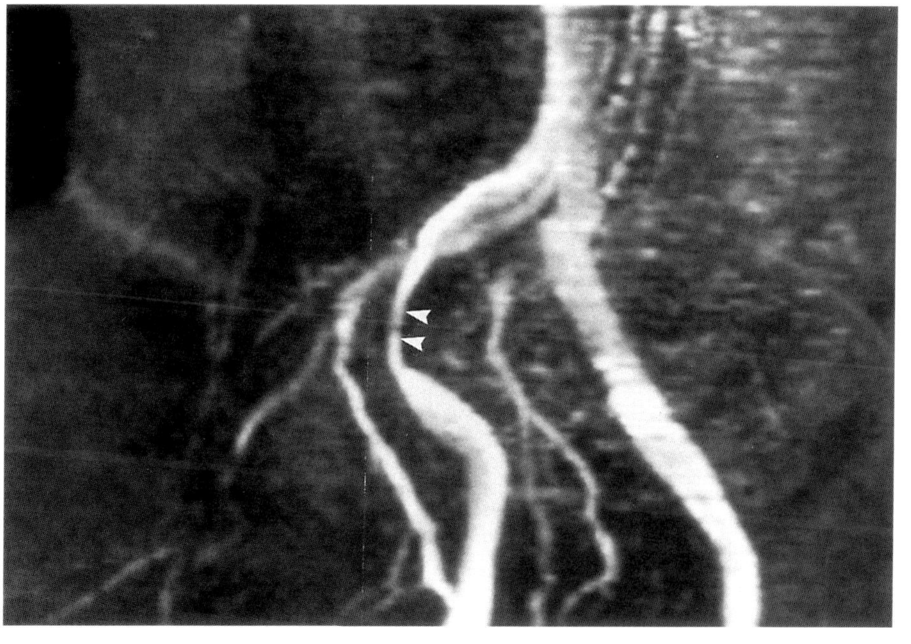

Fig. 19.14. Patient with pelvic lymphadenopathy and progressive edema of the right lower extremity. 30° RAO projection venogram shows compression of the right external iliac vein (*arrowheads*) but still preserved flow

postpartum period that may have fatal consequences without adequate treatement. Septic conditions in the postpartum period are characterized by fever above 38.5° C on at least two consecutive days during the first 2 weeks after delivery. Puerperal sepsis occurs in only 2%–3% of cases after vaginal delivery but in up to 95% following cesarean section. There are many underlying causes among which endometritis, infected hematoma, pelvic abscesses, and pyelonephritis are by far the most common.

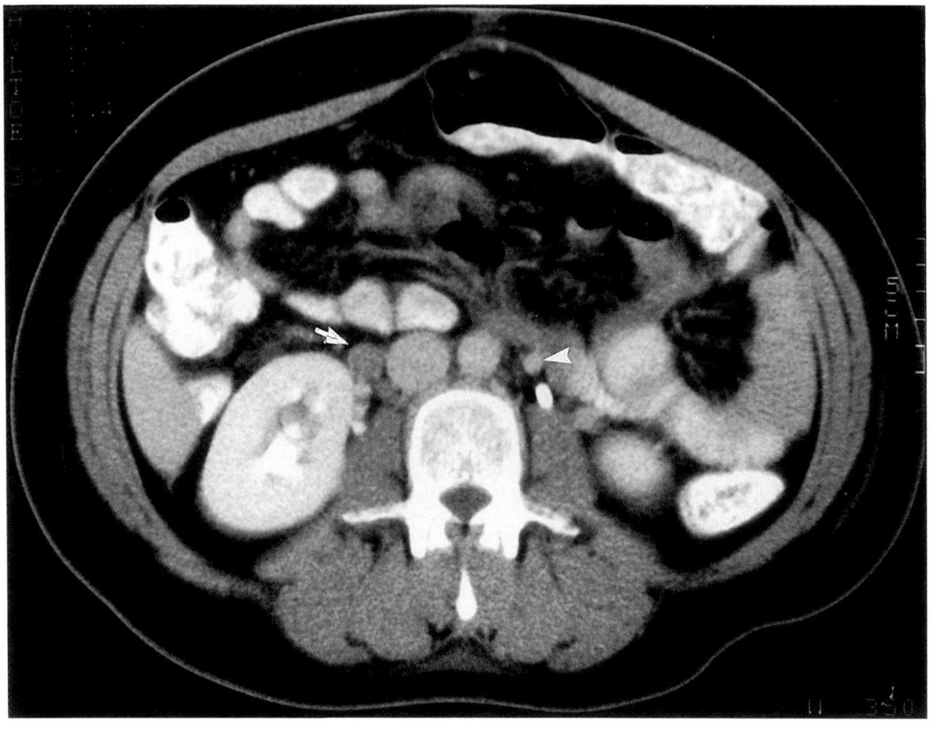

a

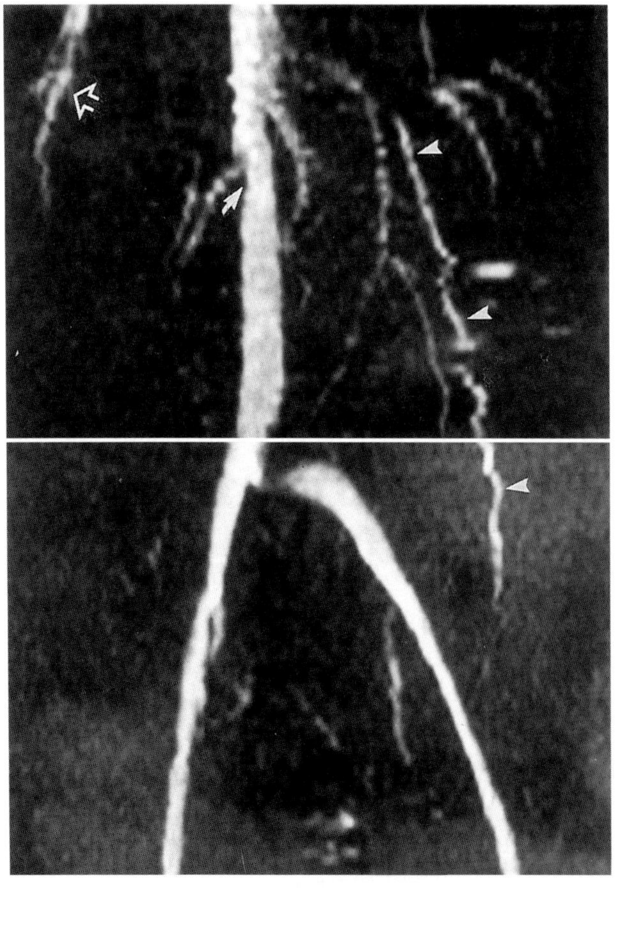

b

Fig. 19.15a,b. Young woman with postpartum fever. **a** CT shows an enlarged right ovarian vein (*arrow*) with slightly lower attenuation values as in the adjacent inferior vena cava. Normal left ovarian vein (*arrowhead*). **b** The projection MR venogram (combination of two acquisition volumes) shows complete occlusion of the right ovarian vein, a floating thrombus reaching the inferior vena cava below the level of the right renal vein (*arrow*), and a patent left ovarian vein (*arrowheads*). Note signal loss in the left common iliac vein due to compression by overlying right iliac artery. *Open arrow*: hepatic veins

However, 5%–20% may be due to septic thrombophlebitis of the ovarian veins, in which case additional anticoagulation (for approximately 3 weeks) is mandatory (COHEN et al. 1983; MINTZ et al. 1987).

This syndrome usually presents with fever and lower abdominal or flank pain, and occasionally with a palpable tender linear-shaped mass. The symptoms at the onset are consistent with endo- or parametritis, and the lack of response to antibiotic therapy often raises the suspicion of bacterial resistance or pelvic abscess. The diagnosis of SPOVT is therefore often made by exclusion (COHEN et al. 1983; MINTZ et al. 1987).

The pathogenesis of SPOVT can be explained by the Virchow triad, i.e., Blood flow stasis in the enlarged ovarian veins during pregnancy and after delivery, iatrogenic or bacterial injury of the vessel wall during delivery, and the state of hypercoagulabitlity during pregnancy, with some elevated clotting factors. SPOVT usually occurs (up to 90% of cases) on the right, since venous drainage here is more important and a common dextrotorsion of the enlarged uterus may lead to subsequent compression of the right ureter and the right ovarian vein at the level of the linea terminalis (COHEN et al. 1983; MINTZ et al. 1987).

The best imaging method for the diagnosis of SPOVT before the era of MRA was contrast-enhanced computed tomography (LEV-TOAFF et al. 1991). The characteristic finding here is a soft tissue tubular structure with an enhancing rim (inflamed vessel wall) and a relatively hypodense center (thrombus) with a diameter of 1–2.5 cm and a length of 5–15 cm, extending from the uterine cornua to the level of the renal veins. This sign, however, is not always detectable (Fig. 19.15a). Duplex-Doppler and color-coded Doppler US have limited value for the assessment of SPOVT due to the commonly overlying distended bowel structures in patients with puerperal sepsis.

Magnetic resonance angiography is an accurate method for evaluation of the postpartal ovarian veins (SALVADER et al. 1988). Due to the increased diameter during pregnancy and after delivery, these vessels are constantly depicted on 2D TOF MRA (Fig. 19.15b). In a prospective comparative study MRA proved superior to CT and duplex-Doppler US. Among 25 consecutive patients with septic fever in the postpartum period, SPOVT could be detected with MRA in eight cases, while US allowed the identification of only four cases and contrast-enhanced CT provided two false-negative and one false-positive diagnoses. MRA also showed additional unsuspected thrombosis of pelvic vessels (common and external ilac veins) in three patients.

19.4 Conclusion

Magnetic resonance angiography of the lower extremities and pelvic veins is a promising method for the assessment of venous disease. Using an adequate examination technique, the femoropopliteal region, the pelvic veins, and the ovarian veins in the postpartum period can be consistently visualized. MRA has already had an impact on the diagnosis of venous thrombosis in the pelvis since it allows noninvasive rapid simultaneous assessment of both external and common iliac trunks and also of the internal iliac veins and their branches, which are not visible on conventional examinations. Moreover, MRA should be considered the method of choice for rapid assessment or exclusion of septic thrombophlebitis of the ovarian veins in the postpartum period. MRA still has some limitations in the diagnosis of calf vein thrombosis, but further development of ultrafast imaging sequences and flow augmentation techniques should improve the results even in this area.

References

Anglade MC, Derhy S, Delvalle A, Roche A, Mathieu D (1989) Abdominal venous thrombosis: applications of gradient-echo magnetic resonance imaging. Diagn Intervent Radiol 1:61–67

Cohen MB, Pernoll ML, Gevirtz CM, Kerstein MD (1983) Septic pelvic thrombophlebitis: an update. Obstet Gynecol 62:83–89

Cohen JM, Weinreb JC, Redman HC (1986) Arteriovenous malformations of the extremities: MR imaging. Radiology 158:475–479

Cronan JJ (1991) Contemporary venous imaging. Cardiovasc Intervent Radiol 14:87–97

Erdmann WA, Weinreb JC, Cohen JM, Buja LM, Chaney C, Peshock RM (1986) Venous thrombosis: clinical and experimental MR imaging: Radiology 161:233–238

Erdmann WA, Jayson HT, Redman HC, Miller GL, Parkey RW, Peshock RW (1990) Deep venous thrombosis of extremities: role of MR imaging in the diagnosis. Radiology 174:425–431

Evans AJ, Sostman HD, Knelson MH, Spritzer CE, Newman GE, Paine SS, Beam CA (1993) Detection of deep venous thrombosis: prospective comparison of MR imaging with contrast venography. AJR 161:131–139

Ferris EJ (1990) Deep venous thrombosis and pulmonary embolism: correlative evaluation and therapeutic implications. AJR 159:1149–1152

List of Contributors

INGOLF P. ARLART, Professor Dr.
Ärztlicher Direktor
Radiologisches Institut
Katharinenhospital
Kriegsbergstraße 60
70174 Stuttgart
FRG

GEORG M. BONGARTZ, PD Dr.
Leitender Arzt für Röntgendiagnostik
der Universität Basel/Kantonsspital
Petersgraben 4
4031 Basel
Switzerland

H. BOSMANS, NMR Engineer
Dienst Radiologie
Katholieke Universiteit
Gasthuisberg
Herestraat 49
3000 Leuven
Belgium

CARLO CATALANO, MD
Department of Radiology
University of Rome "La Sapienza"
Policlinico Umberto I
Viale Regina Elena 324
00161 Rome
Italy

JÖRG F. DEBATIN, MD
MRI Center
Department of Medical Radiology
Zürich University Hospital
Rämistraße 100
8091 Zürich
Switzerland

ROBERT R. EDELMAN, MD
Department of Radiology
Beth Israel Hospital
Harvard Medical School
330 Brooklyn Avenue
Boston, MA 02215
USA

LOTHAR GUHL, Dr.
Radiologisches Institut
Katharinenhospital
Kriegsbergstraße 60
70174 Stuttgart
FRG

M. HAUSER, MD
MRI Center
Department of Medical Radiology
Zürich University Hospital
Rämistraße 100
8091 Zürich
Switzerland

RICHARD HAUSMANN, PhD
Siemens AG
Medizinische Technik
Henkestraße 127
91052 Erlangen
FRG

PD Dr. G.P. KRESTIN
Universitätsspital Zürich
Dept. Medizinische Radiologie
Rämistraße 100
8091 Zürich
Switzerland

Practical Approaches to Cancer Invasion and Metastases
A Compendium of Radiation Oncologists Responses to 40 Histories
Edited by A.R. Kagan with the assistance of R.J. Steckel

Radiation Therapy in Pediatric Oncology   Edited by J.R. Cassady

Radiation Therapy Physics   Edited by A.R. Smith

Late Sequelae in Oncology   Edited by J. Dunst and R. Sauer

Mediastinal Tumors. Update 1995   Edited by D.E. Wood and C.R. Thomas, Jr.

Thermoradiotherapy and Thermochemotherapy
Volume 1: Biology, Physiology, and Physics
Volume 2: Clinical Applications
Edited by M.H. Seegenschmiedt, P. Fessenden and C.C. Vernon
